Pheromones of Non-Lepidopteran Insects Associated with Agricultural Plants

Dedication

To Betty and Roy, Ros, Sally, Nicola and Robert.

Erratum

Page 287
Last complete sentence should read:
In the former, two components (*E2*-octenyl acetate and *E2,7*-octadienyl acetate) found in extracts from both sexes elicited strong GC-EAD responses from antennae of both sexes.

Pheromones of Non-Lepidopteran Insects Associated with Agricultural Plants

Edited by

Jim Hardie

Department of Biology
Imperial College at Silwood Park
Ascot, UK

and

Albert K. Minks

Research Institute for Plant Protection (IPO–DLO)
Wageningen
The Netherlands

CABI *Publishing*

CABI is a trading name of CAB International

CABI Head Office
Nosworthy Way
Wallingford
Oxfordshire OX10 8DE
UK

Tel: +44 (0)1491 832111
Fax: +44 (0)1491 833508
Email: cabi@cabi.org
Web site: www.cabi.org

CABI North American Office
875 Massachusetts Avenue
7th Floor
Cambridge, MA 02139
USA

Tel: +1 617 395 4056
Fax: +1 617 354 6875
Email: cabi-nao@cabi.org

A catalogue record for this book is available from the British Library, London, UK

Library of Congress Cataloging-in-Publication Data
Pheromones of non-lepidopteran insects associated with agricultural plants / edited by J. Hardie and A.K. Minks
p. cm.
Includes bibliographical references.
ISBN 0-85199-345-1 (alk. paper)
1. Insects—Physiology. 2. Agricultural pests—Physiology. 3. Pheromones.
I. Hardie, Jim. II. Minks, A.K., 1934- .
QL495.P485 1999
573.9'29157—dc21 99-17681
CIP

ISBN-13: 978-0-85199-345-4
ISBN-10: 0-85199-345-1

First published 1999
Transferred to print on demand 2006

Printed and bound in the UK by CPI Antony Rowe, Eastbourne

Contents

Contributors

Jeffrey R. Aldrich, *USDA–ARS, Insect Chemical Ecology Laboratory, Beltsville Agricultural Research Center West, Building 007, Beltsville, MD 20705, USA.*

Olle Anderbrant, *Department of Ecology, Ecology Building, Lund University, S-223 62 Lund, Sweden.*

Anne L. Averill, *Department of Entomology, University of Massachusetts, Amherst, MA 01003, USA.*

Robert J. Bartelt, *USDA–ARS, National Center for Agricultural Utilization Research, Bioactive Agents Research Unit, 1815 N. University St, Peoria, IL 61604, USA.*

Göran A. Birgersson, *Chemical Ecology, Department of Botany, Göteborg University, PO Box 461, S-405 30, Göteborg, Sweden.*

Ezra Dunkelblum, *Institute of Plant Protection, The Volcani Center, Agricultural Research Organization, PO Box 6, Bet Dagan 50250, Israel.*

Stephen P. Foster, *The Horticulture and Food Research Institute of New Zealand Ltd, Mt Albert Research Centre, Private Bag 92169, Auckland, New Zealand.*

Jim Hardie, *Aphid Biology Group, Department of Biology, Imperial College at Silwood Park, Ascot, Berks SL5 7PY, UK.*

Marion O. Harris, *The Horticulture and Food Research Institute of New Zealand Ltd, Mt Albert Research Centre, Private Bag 92169, Auckland, New Zealand.*

Ahmed Hassanali, *Chemical Ecology Department, International Centre of Insect Physiology and Ecology (ICIPE), PO Box 30772, Nairobi, Kenya.*

Manfred Kaib, *Department of Animal Physiology, University of Bayreuth, D–95440 Bayreuth, Germany.*

Yooichi Kainoh, *Institute of Agriculture and Forestry, University of Tsukuba, Tsukuba, Ibaraki 305–8572, Japan.*

Peter J. Landolt, *USDA–ARS, Yakima Agricultural Research Laboratory, 5230 Konnowac Pass Road, Wapato, WA 98951, USA.*

Walter S. Leal, *Laboratory of Chemical Prospecting, National Institute of Sericultural and Entomological Science, 1–2 Ohwashi, Tsukuba 305–8634, Japan.*

Heather L. McBrien, *Department of Entomology, University of California, Riverside,* CA 912521, USA.

Jocelyn G. Millar, *Department of Entomology, University of California, Riverside,* CA 912521, USA.

Albert K. Minks, *Research Institute for Plant Protection (IPO–DLO), PO Box 9060, 6700 GW Wageningen, The Netherlands.*

Tanya Pankiw, *Department of Entomology, University of California, Davis,* CA 95616, USA.

Jeff Pettis, *USDA–ARS, Honey Bee Research Laboratory, Building 476, Beltsville Agricultural Research Center East, Beltsville,* MD 20705, USA.

John A. Pickett, *Department of Biological and Ecological Chemistry, IACR-Rothamsted, Harpenden, Herts AL5 2JQ, UK.*

Rudy Plarre, *BAM–Federal Institute of Materials Research and Testing, Lab IV.II, Materials Protection against Biological Agents, Under den Eichen 87, D–12205, Berlin, Germany.*

Erika Plettner, *Department of Medicinal Chemistry, University of Utah, 30 S. 2000 E., Skaggs Hall, Salt Lake City,* UT 84112–5820, USA.

Eleanor M. Pow, *Department of Biological and Ecological Chemistry, IACR-Rothamsted, Harpenden, Herts AL5 2JQ, UK.*

Wilf Powell, *Department of Entomology and Nematology, IACR-Rothamsted, Harpenden, Herts AL5 2JQ, UK.*

Fredrik Schlyter, *Department of Chemical Ecology, Plant Protection Sciences, Swedish Agricultural University, PO Box 44, S–230 53 Alnarp, Sweden.*

Diane W.M. Smiley, *Department of Biological and Ecological Chemistry, IACR-Rothamsted, Harpenden, Herts AL5 2JQ, UK.*

Baldwyn Torto, *Chemical Ecology Department, International Centre of Insect Physiology and Ecology (ICIPE), PO Box 30772, Nairobi, Kenya.*

Desiree C. Vanderwel, *Department of Chemistry, University of Winnipeg, 515 Portage Avenue, Winnipeg, Manitoba, Canada R3B 2E9.*

Preface

Since the identification of the first insect pheromone from the silk moth, *Bombyx mori*, by Butenandt and co-workers in 1959, there has been a continuing interest in these very potent behaviour-modifying chemicals and in their potential role in integrated pest management. The major attention of pheromone researchers has always been directed to lepidopterous insects, as they provided the focus of the pioneering work and because they comprise the largest group of agricultural pests worldwide. In the most important books on insect pheromones that have appeared over the last 10 years, only moderate space has been reserved for pheromones of insects other than Lepidoptera.

In recent years, however, the research situation has changed and other insect orders are receiving increasing attention. We believe that it is now time to garner this profusion of fresh information on pheromones, to discuss the new state of the art, to evaluate the potential application for pest control and to discuss future developments. We are proud to inform you that this book contains contributions from authors who are the foremost pheromone experts in the various insect groups. We report on history and current progress in elucidating the biology and the chemistry, as well as on control strategies, initiated or envisaged. Chapters cover agriculture, horticulture and forestry as well as encompassing stored products. In addition, we emphasize that not only are pheromones of pest insects covered, but also pheromones of beneficials together with pheromones/kairomones of hosts and prey of beneficials, where this knowledge might facilitate biological control. The final chapter discusses the special functions of bee pheromones.

Inevitably, a multi-authored book will show some overlap between various chapters. We have attempted to reduce this to an acceptable level, but on several occasions a particular subject is discussed from different points of view and we believe this to be enlightening rather than disturbing. In these instances, we have inserted cross-references to other chapters.

We are convinced that this book will fill a real niche in our knowledge of insect pheromones and will find wide appeal. Its content includes research examples taken from work done all over the world. Its potential markets comprise academic and research institutes, learned society libraries and those companies involved in crop-protection research and product sales. Because of the attention to practical matters, we hope it will also find its way to extension services and consultancies in crop protection, biodiversity and ecosystem management.

We are very much indebted to a great number of colleagues who kindly helped us to review the manuscripts and we gratefully acknowledge: R.J. Bartelt (Peoria), E. Boller (Wädenswil), J.H. Borden (Burnaby), W. Budenberg (Nottingham), C.A.M. Campbell (East Malling), J. Chambers (York), M. Dicke (Wageningen), D.G. James (Yanco), B. Katsoyannos (Thessaloniki), Y. Le Conte (Avignon), M. Lenz (Canberra), K. Mori (Tokyo), J. Pasteels (Brussels), W. Powell (Harpenden), R.E. Rice (Parlier), Y. Roisin (Brussels), S.J. Seybold (Reno), S.J. Simpson (Oxford), L.E.M. Vet (Wageningen), R.A. Vickers (Indooroopilly), I.H. Williams (Harpenden), M.L. Winston (Burnaby) and T. Wyatt (Oxford).

And finally we thank T. Hardwick and A. West from CABI *Publishing* for their advice and smooth and pleasant cooperation.

Jim Hardie, Ascot, UK
Albert K. Minks, Wageningen, The Netherlands

About the Editors

Dr Jim Hardie is at present a Reader in Insect Physiology at Imperial College of Science, Technology and Medicine, London. His education was completed in the UK with a first degree from Brunel University and a PhD from the University of Birmingham. In 1989 he was awarded a DSc by the University of London. Jim has been with the Department of Biology at Imperial College's Silwood Park campus for over 20 years where his interests have been broadly engaged by aphid physiology, with particular interests in chemical ecology and photoperiodism. He is a Fellow of the Institute of Biology, the Royal Entomological Society and the Royal Microscopical Society and a Member of the International Society of Chemical Ecology as well as the Society for Experimental Biology. He sits on the Council of the Royal Entomological Society.

Dr Albert K. Minks received his PhD in Insect Endocrinology in 1967 (promoter Professor Jan de Wilde) from Wageningen Agricultural University, The Netherlands. From 1967 to 1976 he worked at the Laboratory for Insecticide Research. He was head of the Department of Entomology of the Research Institute for Plant Protection (IPO) from 1976 to 1988, and from 1988 to 1991 of the Department of Biological and Integrated Control at the same institute. In 1991, he resigned as department head and returned, as a senior scientist, to pheromone research. Albert began an active retirement in 1997, continuing his duties as Executive Editor of *Entomologia Experimentalis et Applicata*; as a member of the Editorial Board of the *European Journal of Entomology* and as Vice-President of IOBC/WPRS where he was also Convenor of the Publication Commission from 1987 to 1997.

I Pests

1 Fruit Flies

Peter J. Landolt[1] and Anne L. Averill[2]

[1] USDA–ARS, Yakima Agricultural Research Laboratory, 5230 Konnowac Pass Road, Wapato, WA 98951, USA; [2] Department of Entomology, University of Massachusetts, Amherst, MA 01003, USA

Introduction

The dipterous family Tephritidae includes both fruit-infesting as well as stem-, gall- and flower-infesting species. Tephritids associated with agriculture include many fruit-infesting species as crop pests and some stem- and flower-infesting species as potential biological control agents for weeds. There has been little research on pheromonal communication of tephritids used for biological control of weeds, but considerable work has been done on pheromones of pestiferous fruit-infesting tephritids (Fig. 1.1).

Fruit-infesting Tephritidae include some of the most damaging agricultural insect pests, with problem species in nearly every inhabited area of the world. These range from nearly monophagous species to polyphagous species that infest many different crop plants. The former include a number of species of *Rhagoletis* as well as the papaya fruit fly, *Toxotrypana curvicauda* (Gerstaecker). Many New World species of *Anastrepha* and Old World species of *Bactrocera* are important fruit pests with varying host breadth. Important polyphagous pests include the Mediterranean fruit fly, *Ceratitis capitata* (Wiedemann), the Oriental fruit fly, *Bactrocera dorsalis* (Hendel) and the melon fly, *Bactrocera cucurbitae* (Coquillet). Together, these and other fruit flies make many fruits unfit for human consumption and necessitate costly control and disinfestation measures to protect fruit production and satisfy quarantine requirements.

There is a long tradition of developing and using semiochemicals to protect agricultural crops from fruit flies (see review by Chambers, 1977), although there has been limited success with pheromones. Attractive materials, such as food baits, kerosene and essential oils, have been used for trapping fruit flies for many decades. Poison-bait sprays that incorporate attractive materials have been used for control and eradication since the turn of the century. Over the last 30 years, numerous advances in micro-analytical chemistry and behavioural experimentation, and a rapidly improving understanding of the behavioural ecology of fruit flies, have

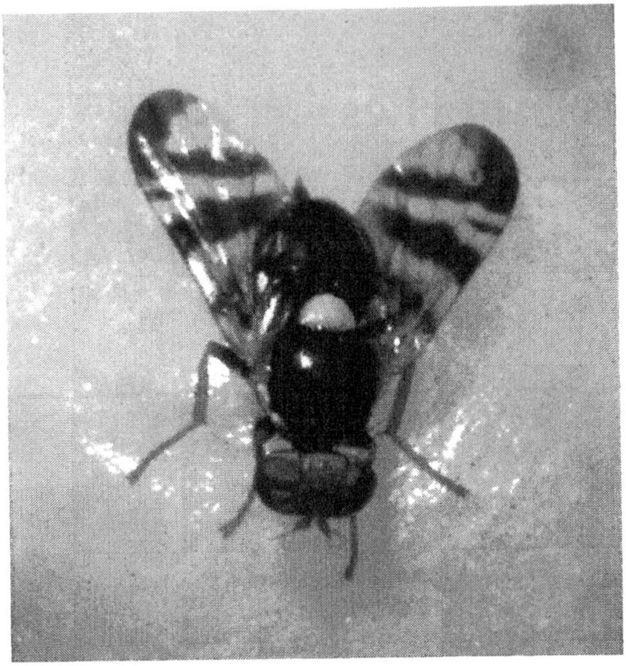

Fig. 1.1. Female of the European cherry fruit fly, *Rhagoletis cerasi.* (Courtesy of E.F. Boller.)

fuelled much interest in isolating and identifying behaviourally active chemicals (including pheromones) that are useful in the management and eradication of fruit flies.

This chapter is a summary and synthesis of our current knowledge of pheromones of fruit flies. We do not review chemicals for which there is no known communicative function, parapheromones, or semiochemicals emitted from foods and host plants. Pheromones of tephritids are of two categories: sex pheromones and host-marking (oviposition-deterring) pheromones. Nearly all research has been conducted on fruit-infesting species due to the need to develop methods of managing these agricultural pests. There are other reviews of note that include pheromones of fruit flies. A brief review of fruit fly pheromone research was written by Nation (1977). Chambers (1977) provided a history and summary of information on attractants used for detecting, monitoring and controlling pestiferous tephritids, and includes a summary of the information on fruit fly pheromones available at that time. Sivinski and Calkins (1986) reviewed pheromones and parapheromones (semiochemicals of unknown function and origin) of fruit flies and their use in control and eradication of fruit flies. The book *Fruit Flies: Their Biology, Natural Enemies and Control* (Robinson and Hooper, 1989) included chapters on fruit fly mating pheromones and host-marking pheromones. Fletcher and Kitching (1995) provided a comprehensive review of chemicals, including pheromones, found in fruit flies while Landolt and Heath (1996) summarized work conducted in Florida on sex attractant pheromones of tephritids.

Sex Pheromones

Sex pheromones include chemical signals that function over a close range and mediate courtship interactions but also include chemical signals that function over a longer distance and elicit orientation or attraction responses. Courtship pheromones may serve as species and sex recognition signals and may or may not elicit an overt behavioural response. Sex attractant pheromones generally serve to bring the sexes to the same location, at which other types of signalling (acoustic, visual, courtship pheromones) may occur. Most research on sex pheromones of fruit flies has focused on sex attractants because of their potential as lures useful for detection, monitoring or controlling pest species (Table 1.1).

Fruit fly sex pheromones are often produced and released by males in aggregations or leks and in concert with acoustic and visual displays. These pheromones may be attractive to both sexes, although presumably they are released by males to attract potential mates. A notable exception to this theme is the female pheromone of the olive fruit fly, *Bactrocera oleae* (Gmelin), that is attractive to males. Despite in-depth investigations into pheromone chemistry, there have been few successes in developing useful attractants based on these sex pheromones. Limitations and problems with the development of useful attractants based on tephritid sex pheromones are discussed below for individual species. This review of tephritid sex pheromones is organized by species and genus because the majority of studies have been conducted on a small number of pest species.

The Caribbean fruit fly, Anastrepha suspensa

Males of the Caribbean fruit fly, *Anastrepha suspensa* (Loew), form leks on foliage of host and non-host trees (Burk, 1983) from which they emit a combination of odours and sounds that are attractive to females and to other males (Webb *et al.*, 1983a). Although it has been hoped that duplication of the male scent may provide a lure useful in attracting and trapping female Caribbean fruit flies, a pheromone lure for this species is not at hand.

Table 1.1. Tephritid species where evidence of attractive sex pheromones has been reported.

Species	Reference
Anastrepha ludens (Loew)	Robacker *et al.* (1985), Robacker and Hart (1986)
Anastrepha suspensa (Loew)	Nation (1972, 1975, 1989), Perdomo *et al.* (1975)
Bactrocera dorsalis (Hendel)	Kobayashi *et al.* (1978)
Bactrocera oleae (Gmelin)	Haniotakis (1974, 1977)
Bactrocera tryoni (Froggatt)	Fletcher (1969)
Ceratitis capitata (Wiedemann)	Feron (1959), Landolt *et al.* (1992a)
Rhagoletis cerasi L.	Katsoyannos (1982)
Rhagoletis pomonella (Walsh)	Prokopy (1975b)
Toxotrypana curvicauda (Gerstaecker)	Landolt *et al.* (1985)

There is good evidence that male *A. suspensa* release a pheromone attractive to females and males. Female attraction to males, male odours and male extracts has been shown using a variety of types of laboratory assay. Nation (1972) demonstrated upwind movement of females in glass tubes in response to males with food placed near the upwind end of the tube. Although such movement could have been in response to odours, sounds or visual signals, subsequent demonstrations of response to extracts of males (Nation, 1975) support the hypothesis that female responses to males are due in part to a pheromonal signal from males. Nation (1975) demonstrated upwind movement of females from a plastic box as they entered traps baited with male extracts. Nation (1989) also demonstrated female aggregation in response to male extracts using a cage-top bioassay, in which treated versus control filter papers were placed on the top of screened cages containing flies, stimulating flies to accumulate under 'attractive' papers. Such aggregation assays may indicate attraction and/or arrest by flies in response to male pheromone in treatments. Sivinski and Heath (1988) used a similar assay to demonstrate female response to male odours trapped from air flowing over males. Webb *et al.* (1983a) showed that females released in a field cage could be captured in traps baited with males or male extracts, indicating attraction to male pheromone. Additionally, female Caribbean fruit flies released in the field were recaptured in traps baited with live males, indicating female attraction to males, although responses to male pheromone versus male acoustic signals were not differentiated (Perdomo *et al.*, 1975). Male Caribbean fruit flies may release attractive pheromones when puffing abdominal pouches that are associated with pleural exocrine glands, when they rub the abdominal tip on substrates and leave hindgut secretions (Sivinski *et al.*, 1994), and from the mouthparts and associated salivary glands (Nation, 1989, 1990). Puffing of the abdominal pouches occurs in a bimodal pattern during the day and is strongly affected by sugar feeding (Landolt and Sivinski, 1992). Epsky and Heath (1993a,b) verified a bimodal pattern of release of pheromone (two peaks of activity, in the early morning and again in the afternoon) and demonstrated that males release significantly more pheromone when provided with sugar. Sivinski *et al.* (1994) observed a similar bimodal pattern of male marking of leaf surfaces with pheromone (by rubbing the everted anal membrane on leaves) and demonstrated female visits to leaves marked by males.

Little research has been conducted on female Caribbean fruit fly responses to male pheromone. Sivinski and Heath (1988) compared virgin and mated female responses to male pheromone using a cage-top bioassay that may indicate short-range orientation and arrest, and found that mated females with access to oviposition sites (assumed to be ovipositing) became as responsive to male pheromone as unmated females. The male pheromone used was a headspace sample from calling males.

Eight compounds have been isolated and identified from extracts of males and from filtering of air passed over males: (Z)-3-nonenol (**1** in Fig. 1.2) and (Z,Z)-3,6-nonadienol (**2** in Fig. 1.2; Nation, 1983), anastrephin (*E*-hexahydro-*E*-4,7a-dimethyl-4-vinyl-2-[3H]-benzofuranone; **3** and **4** in Fig. 1.2) and epianastrephin (*E*-hexahydro-Z-4,7a-dimethyl-4-vinyl-2-[3H]-benzofuranone; **5** and **6** in Fig. 1.2) (Battiste *et al.*, 1983), suspensolide ([*E*,*E*]-4,8-dimethyl-3,8-decadien-10-olide) (Chuman *et al.*, 1988), β-bisabolene, ocimene, and *E*,*E*-α-farnesene (Rocca *et al.*,

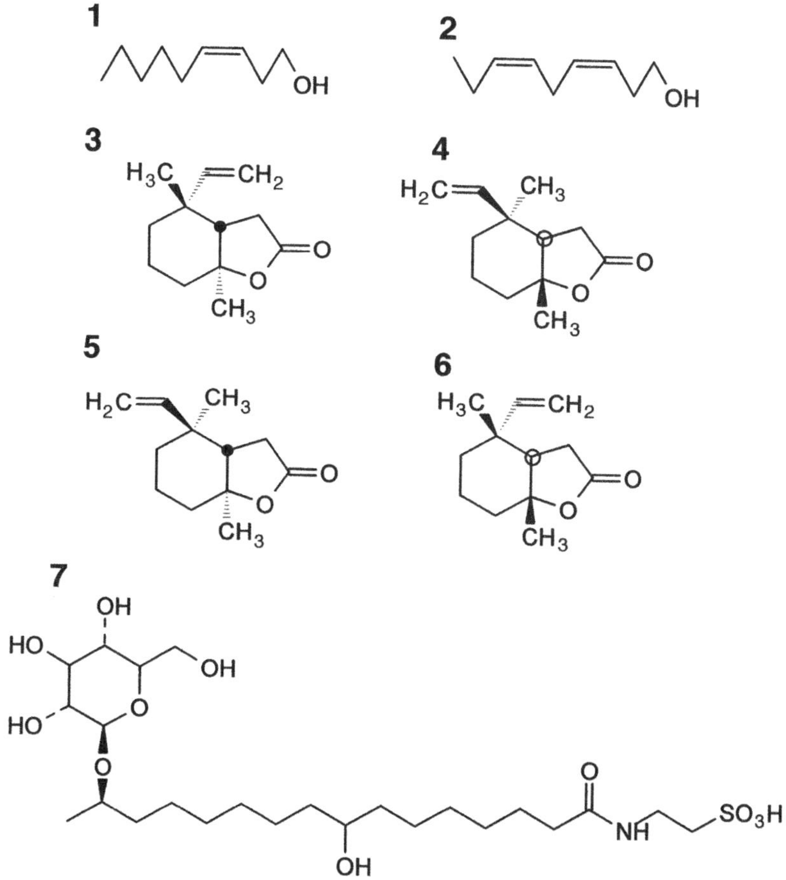

Fig. 1.2. Structures of fruit fly pheromones. (**1**) (*Z*)-3-nonenol; (**2**) (*Z,Z*)-3,6-nonadienol; (**3**) *S,S*-(−)-anastrephin; (**4**) (*R,R*)-(+)-anastrephin; (**5**) *S,S*-(−)-epianastrephin; (**6**) (*R,R*)-(+)-epianastrephin; (**7**) *N* [5(β-3-glucopyranosyl)oxy-8-hydroxypalmitol]-taurine.

1992). Four of these compounds – (Z)-nonenol, (Z,Z)-nonadienol, anastrephin and epianastrephin – are attractive to females in laboratory bioassays (Nation, 1983, 1989). None of these chemicals has yet been demonstrated to be attractive to Caribbean fruit flies in field trials. Progress in developing pheromone-based lures for A. *suspensa* is hampered in part by a lack of availability of pheromone chemicals in amounts suitable for field testing and methods of formulating blends comprised of compounds of differing volatilities. Also, it is not entirely clear what roles are played by male-produced pheromones in sex attraction and in the formation of aggregations in this species. Demonstrations of sex attraction are confounded by a mixture of chemical, acoustic and visual signals by males that make it difficult to determine what, if any, attraction responses in nature occur strictly to male odours and may then be expected to occur in response to synthetic odours presented in isolation.

The Mexican fruit fly, Anastrepha ludens

Males of the Mexican fruit fly, *Anastrepha ludens* (Loew), also form leks and exhibit chemical, acoustic and visual signalling (Robacker and Hart, 1985; Robacker *et al.*, 1991). Females of *A. ludens* are attracted to and arrested by extracts of males and oriented at distances up to 8 m to an extract of male abdomens (Robacker *et al.*, 1985; Robacker and Hart, 1986; Robacker and Moreno, 1988). In a field test, using an extract of male Mexican fruit flies as a source of pheromone, both male and female laboratory-reared, irradiated Mexican fruit flies were captured in traps baited with one or ten male-equivalent doses (Robacker and Wolfenbarger, 1988).

The responsiveness of female *A. ludens* to male pheromone (presented as a male abdominal extract) peaks in the afternoon (Robacker and Garcia, 1990), but is reduced following mating (Robacker *et al.*, 1985), following irradiation (Moreno *et al.*, 1991), and by the presence of an extract of fermented host fruit (Robacker and Garcia, 1990).

The blend of compounds produced or emitted by male Mexican fruit flies is similar in structure to those produced by male *A. suspensa*. Three chemicals isolated from extracts of male abdomens, ((Z)-3-nonenol, (Z,Z)-3,6-nonadienol and (*S*,*S*)-(−)-epianastrephin (**1**, **2** and **5** in Fig. 1.2; Esponda-Gaxiola, 1977; Battiste *et al.*, 1983; Nation, 1983; Stokes *et al.*, 1983) elicited short-range attraction or arrest in female Mexican fruit flies in a laboratory cage-top bioassay (Robacker and Hart, 1985). (Z,Z)-3,6-Nonadienol alone, combinations of either (Z)-3-nonenol or (Z,Z)-3,6-nonadienol with (S,S)-(−)-epianastrephin, and a six-component blend comprising (Z)-3-nonenol, (Z,Z)-3,6-nonadienol, (*S*,*S*)-(−)-epianastrephin (**5** in Fig. 1.2), (*R*,*R*)-(+)-epianastrephin (**6** in Fig. 1.2), (*S*,*S*)-(−)-anastrephin (**3** in Fig. 1.2) and (*R*,*R*)-(+)-anastrephin (**4** in Fig. 1.2) all elicited long-range attraction by female flies in a flight tunnel (Robacker, 1988). To date, these pheromone compounds have not been shown to be active as attractants to Mexican fruit flies in the field. Principal constraints to testing and development are similar to those for *A. suspensa*, including a lack of the amounts of the chemicals suitable for field testing as well as difficulties in formulation.

The papaya fruit fly, Toxotrypana curvicauda

The male of the papaya fruit fly, *T. curvicauda*, emits a sex pheromone that is attractive to females (Landolt *et al.*, 1985), as well as to other males (Landolt and Heath, 1990). The sex pheromone of the male is 2,6-methylvinyl pyrazine. This chemical was isolated and identified from headspace collections over calling males and elicits behavioural responses in females similar to their responses to calling males (Chuman *et al.*, 1987; Landolt and Heath, 1988). In laboratory assays, female attraction to sex pheromone coincides with reproductive maturation (beginning 6 days after adult emergence) and is not reduced following mating (Landolt and Heath, 1988).

The papaya fruit fly sex pheromone is a useful attractant when appropriately formulated and used with an appropriate visual target. Landolt *et al.* (1988) and Landolt and Heath (1990) demonstrated captures of males, unmated females and mated females on a fruit model trap (green sphere coated with adhesive) baited with 2,6-methylvinyl pyrazine loaded in glass capillary tubes. Effective lure release rates were 140–900 ng h^{-1}. Nearly all flies in papaya fields were captured on these traps when they were placed about 5 m apart in the field perimeter (Landolt *et al.*, 1991). A membrane formulation for controlled release of pheromone in combination with a green cylinder was also demonstrated to be an effective design for attracting and capturing papaya fruit flies (Heath *et al.*, 1996).

Male papaya fruit flies call from perches on immature papaya fruit where they defend small territories against other males and court females (Landolt and Hendrichs, 1983). Because female attraction to male pheromone was not reduced following mating (Landolt and Heath, 1988), host-seeking females may use male pheromone emitted from males on host fruit to locate fruit for oviposition. Female attraction to synthetic male pheromone is enhanced or synergized by the odour of immature papaya fruit (Landolt *et al.*, 1992b). A combination of host kairomones and sex pheromone may then constitute a more powerful attractant useful in trapping papaya fruit flies. However, host odours important to papaya fruit fly host-finding and sexual responses have not been identified.

The Mediterranean fruit fly, Ceratitis capitata

As in *Anastrepha* species, males of the Mediterranean fruit fly, C. *capitata* (medfly), form leks and emit odours and sounds that are attractive and stimulating to females (Prokopy and Hendrichs, 1979; Webb *et al.*, 1983b). Feron (1959) first demonstrated that females are attracted to a pheromone produced by males. Female medfly attraction to male pheromone was confirmed by Nakagawa *et al.* (1981), Landolt *et al.* (1992a), Jang *et al.* (1994) and Jang (1995). In a flight tunnel, females responded to male pheromone in air piped into the tunnel with zigzagging upwind flights within the odour plume (Landolt *et al.*, 1992a). Jang (1995) demonstrated unmated female preferential attraction to male pheromone (odour) over host fruit odour and mated female preferential attraction to host fruit odour over male pheromone.

Male medflies exhibit a pheromone-release behaviour pattern that includes the extrusion of the rectal lining, expansion of the pleural abdominal pouches, and wing fanning (Feron, 1959; McDonald, 1987). This behaviour occurs predominantly when males are in leks (Prokopy and Hendrichs, 1979). Males exhibit calling more frequently when in the company of other males and it is speculated that increased pheromone release by males in leks is the result of individual males competing with neighbouring males in the aggregation (McDonald, 1987). Males may call in response to the presence of another male in an attempt to attract the attention of any female present. Male medflies are also known to dab the abdominal tip to leaf surfaces while calling (Prokopy and Hendrichs, 1979), possibly applying pheromone to such a surface, as occurs in *Anastrepha suspensa* (Sivinski *et al.*, 1994). Pheromonal calling by males is also somewhat conditional on feeding, with a rapid decrease in the incidence of calling by males that are denied access to sugar (Landolt *et al.*, 1992a).

A large number of compounds isolated and identified from male extracts or male headspace have been considered as possible pheromones of the male

Mediterranean fruit fly (Jacobson *et al.*, 1973; Ohinata *et al.*, 1973, 1977; Baker *et al.*, 1985; Jang *et al.*, 1989; Heath *et al.*, 1991). However, behavioural activity indicating a pheromonal function has been demonstrated for only a small number of these compounds. Mature, unmated females oriented upwind in a wind tunnel to a three-component blend (Heath *et al.*, 1991) comprised of (*E*,*E*)-α-farnesene, geranyl acetate and ethyl-3-octenoate (Landolt *et al.*, 1992a) and to a five-component blend comprising (*E*,*E*)-α-farnesene, geranyl acetate, ethyl-3-octenoate, 1-pyrroline and ethyl acetate (Flath *et al.*, 1993; Jang *et al.*, 1994).

Field trials with compounds isolated and identified from male Mediterranean fruit flies have shown some promise in attracting flies to traps. Methyl (*E*)-6-nonenoate, isolated by Jacobson *et al.* (1973) from male *C. capitata*, attracted males to traps in field tests (Ohinata *et al.*, 1979). Baker *et al.* (1990) trapped males and females of released laboratory-reared flies with linalool, 2,3-dimethylpyrazine, 2,5-dimethylpyrazine, geranyl acetate and some combinations of these compounds that were found in anal secretions of males (Baker *et al.*, 1985). Heath *et al.* (1991) trapped feral flies in coffee fincas in Guatemala with a blend of (*E*,*E*)-α-farnesene, (*E*)-3-octenoate and geranyl acetate. Successful field trials with more complex blends of components that are indicated to be the male pheromone of the Mediterranean fruit fly have not been reported and no usable lure based on male sex pheromone has been developed.

The olive fruit fly, Bactrocera oleae

The olive fruit fly is unusual among tephritids in possessing a female-produced pheromone that is attractive to males. This species is also unusual in being one of the few tephritids for which a sex pheromone has been developed as a useful attractant.

Haniotakis (1974) was the first to demonstrate male olive fruit fly attraction to females, using a straight-tube olfactometer, and noted that female attractiveness to males was coincident with the age at which they mated. Subsequently, Haniotakis (1977) demonstrated male attraction to females and to males held within glass traps in the field.

The sex pheromone of the female olive fruit fly comprises four chemicals that are maximally attractive to males as a complete blend (Mazomenos and Haniotakis, 1981). The major component, (1,7)-dioxaspiro-[5,5]undecane (olean), was isolated from extracts of female rectal glands independently by Baker *et al.* (1980), Mazomenos and Pomonis (1983) and Garibaldi *et al.* (1983). Maximum attractiveness of this compound in laboratory experiments occurred at a dosage of 10 μg of a racemic mixture applied to a filter paper (Haniotakis and Pittara, 1994). It was determined by Haniotakis *et al.* (1986) that the *R*-(−)-enantiomer was active in attracting males, while the *S*-(+)-enantiomer was attractive to females in the laboratory, although not in the field. The three other components of the female pheromone, α-pinene, *n*-nonanol and ethyl dodecanoate, were identified by Mazomenos and Haniotakis (1985). Attractiveness of pheromone and pheromone components in field tests was indicated by formulating chemicals in rubber stoppers and baiting sticky-coated triangular traps (Baker *et al.*, 1980) or formulating chemicals in polyethylene vials and baiting traps comprising yellow rectangles coated with adhesive (Mazomenos and Haniotakis, 1985; Haniotakis *et al.*, 1986).

Olean has proved to be effective in the field as a lure for male olive fruit flies and as part of an integrated approach to controlling populations of olive fruit fly in olive groves. Haniotakis *et al.* (1983) showed that a combination of yellow sticky panels and a sex pheromone lure, comprising a racemic mixture of the enantiomers of olean, reduced pest populations and infestation of fruit. Subsequently, Haniotakis *et al.* (1991) developed a multi-faceted mass trapping approach to control *B. oleae*, comprising a food attractant (ammonium bicarbonate), sex attractant (racemic olean), phagostimulant (sugar), visual attractant (yellow panel) and pesticide. This approach reduced olive fruit fly populations in large olive groves and resulted in dramatic reductions in pesticides used to control olive fruit flies.

Other Bactrocera *species*

Although sex pheromones appear to exist in other species of *Bactrocera*, including agriculturally important species, no sex attractant pheromones have been isolated and identified, outside of the work summarized above on *B. oleae*.

Sex pheromones produced by males that are attractive to females are indicated in the oriental fruit fly and the melon fly (Kobayashi *et al.*, 1978; Poramarcom, 1988) and in the Queensland fruit fly, *Bactrocera tryoni* (Froggatt) (Fletcher, 1969; Fletcher and Giannakakis, 1973). Short-range orientation of females to live males and to rectal glands of males was demonstrated for all three species. Fitt (1981a) demonstrated short-range sex attraction and interspecific cross attraction among *Bactrocera opiliae* Drew & Hardy, *Bactrocera aquilonis* (May) and *Bactrocera jarvisi* (Tryon), using extracts of male glands as pheromone sources. Demonstration of long-range attraction to sex pheromone, either upwind-oriented flights in laboratory assays or responses in the field, are lacking for these species. However, methyl eugenol and cuelure are powerful attractants for the oriental fruit fly and melon fly, respectively, and there may be little incentive to research attractive sex pheromones for these species.

B. tryoni female responsiveness to male pheromone (male gland extract) coincides with female reproductive maturity and is maximized under dim light (Fletcher and Giannakakis, 1973). These authors also demonstrated a diel pattern to female responsiveness to male pheromone, with a peak in late afternoon and at dusk, and a steep decline in female response to male pheromone following mating.

Much work has focused of late on the chemistry of rectal gland secretions of species of *Bactrocera*, apparently because of the successful development of olean as a useful sex attractant for *B. oleae*. A summary of the compounds found in these secretions was provided by Metcalf (1990; see also Fletcher and Kitching, 1995), with much of the work revealing compounds similar to olean in structure (spiroacetals) and *N*-alkyl acylamides, such as *N*-3-methylbutylacetamide, found in the rectal glands of *B. dorsalis* (Ohinata *et al.*, 1982) among others (Bellas and Fletcher, 1979). However, there are no demonstrations of pheromonal functions for these compounds from rectal glands of other species of *Bactrocera*.

Rhagoletis species

Males of the apple maggot, *Rhagoletis pomonella* (Walsh), produce a pheromone that is attractive to unmated females (Prokopy, 1975b). In large field cages females were attracted to caged males and to cages from which males had recently been removed.

Males of the European cherry fruit fly, *Rhagoletis cerasi* L., also produce an odour that is attractive to unmated females. This was first demonstrated in a field cage assay, in which female flies released into a large cage were trapped on a sticky surface as they arrived at small cages of males, small cages that had previously contained males, or extracts of males (Katsoyannos, 1976, 1982). Female response to males was also demonstrated in a laboratory olfactometer assay in which female flies moved from a holding cage upwind into a plastic tube holding males. Female responsiveness to males in this assay was coincident with reproductive maturation. Katsoyannos (1982) also demonstrated the attraction of unmated females to males in the field, using male-baited traps. Raptapoulos *et al.* (1995) demonstrated attraction of female *R. cerasi* in a laboratory bioassay to male headspace samples. Several fatty acids isolated from male odour were found to be arrestants for females but none of the compounds isolated from male headspace were attractive when they were individually tested.

The sex pheromones of the apple maggot and European cherry fruit fly males have not been characterized or identified and glandular sources of the pheromone are not known.

Host-marking Pheromones

Host-marking pheromones (HMPs) have been documented in dozens of insect species within different families (Prokopy, 1981a). The family Tephritidae contains the greatest number of species known to deposit HMP (Boller, 1994, 1996). This includes *Anastrepha fraterculus* (Wiedemann), *A. suspensa*, *Ceratitis capitata* and 13 species within the genus *Rhagoletis*. Most recently, HMP has been reported in several tephritid species that utilize flower heads: *Chaetorellia australis* (Hering), *Paraceratitella eurycephala* Hardy, *Tephritis bardane* (Schrank) and *Orellia ruficauda* (Fabricius) (Table 1.2). In the case of the genus *Bactrocera*, studies have shown that while HMP deposition is absent, females avoid oviposition on exploited hosts in response to compounds in juice or cues produced by larvae (Cirio, 1971; Prokopy and Koyama, 1982; Fitt, 1984; Lo Scalzo *et al.*, 1994).

In all fruit fly species known to deposit HMP, the active compound(s) is released as a fluid trail on the host surface during the dragging of the ovipositor following oviposition. The host-marking behaviour itself is highly stereotyped, invariably occurring after each successful oviposition. However, observations of individual females show that there is considerable variation in the time and pattern involved in dragging, often ranging from a few seconds to a few minutes. Some of the factors contributing to variability in pheromone release include fly age, fly or host size, and previous marking on the host (see Averill and Prokopy, 1987b, 1988, 1989b; Papaj and Aluja, 1993). Different responses have also been reported

Table 1.2. Tephritid species where evidence of host-marking pheromones has been reported.

Species	Reference
Anastrepha fraterculus (Wiedemann)	Prokopy *et al.* (1982a)
Anastrepha suspensa (Loew)	Prokopy *et al.* (1977)
Ceratitis capitata (Wiedemann)	Prokopy *et al.* (1978)
Chaetorellia australis (Hering)	Pittara and Katsoyannos (1990)
Orellia ruficauda (Fabricius)	Lalonde and Roitberg (1992)
Paraceratitella eurycephala Hardy	Fitt (1981b)
Rhagoletis alternata (Fallèn)	Bauer (1986)
Rhagoletis basiola (Osten Sacken)	Averill and Prokopy (1981)
Rhagoletis berberis Curran	Mayes and Roitberg (1986)
Rhagoletis cerasi L.	Katsoyannos (1975)
Rhagoletis cingulata (Loew)	Prokopy *et al.* (1976)
Rhagoletis completa Cresson	Cirio (1972)
Rhagoletis cornivora Bush	Prokopy *et al.* (1976)
Rhagoletis fausta (Osten Sacken)	Prokopy (1975a)
Rhagoletis indifferens Curran	Mumtaz (1976), Prokopy *et al.* (1976)
Rhagoletis mendax Curran	Prokopy *et al.* (1976)
Rhagoletis pomonella (Walsh)	Prokopy (1972)
Rhagoletis tabellaria (Fitch)	Prokopy *et al.* (1976)
Rhagoletis zephyria Snow	Averill and Prokopy (1982)
Tephritis bardane (Schrank)	Straw (1989)

between wild and laboratory-reared C. *capitata* flies and when they originate from different geographic regions (Boller *et al.*, 1994).

The larvae of many tephritids known to deposit HMP feed enclosed within a plant structure of a host that is seasonally or spatially restricted, such as a fruit or flower bud. Further, larvae cannot move among hosts and must complete development in the host chosen by the female (Roitberg and Prokopy, 1987). Deposition of the pheromone mark is thought to benefit the marking female in two ways: (i) by preventing a fly searching locally in a patch from adding additional offspring to hosts that already contain eggs or larvae; and (ii) by providing information to conspecific foragers that a competitively superior offspring is already present (Roitberg and Mangel, 1988). Avoidance of occupied hosts may reduce competition among larvae (Averill and Prokopy, 1987b; but see also Lalonde and Roitberg, 1992) and may result in a tendency towards uniformity in dispersion of eggs among available hosts (Averill and Prokopy, 1989a). Long-range dispersal flights are increased in *R. pomonella* females following exposure to HMP-marked fruit, possibly enhancing female fly fitness when such flights lead to new unutilized oviposition sites (Roitberg *et al.*, 1984).

Of the tephritid HMPs examined to date, all are water soluble and have moderate residual activity (Averill and Prokopy, 1987a; Pittara and Katsoyannos, 1990). Some researchers have questioned the effectiveness of these pheromones as mediators of uniform egg dispersion and larval competition: pheromonal activity breaks down well before completion of larval development and, in climates with moderate to high rainfall, activity is rapidly lost (Averill and Prokopy, 1987a). From an applied standpoint, lack of persistence could be of concern if synthetic HMP is applied in a crop protection programme. However, as discussed below, applications of a synthetic formulation of *R. cerasi* HMP to cherry trees have shown suprisingly high persistence (Boller *et al.*, 1987; Boller and Hurter, 1998). From an evolutionary standpoint, a transient mark may make sense. Theoretical discussion and models developed by Roitberg and Mangel (1988) show that when benefit to an individual marking insect is measured in terms of larval survival, nearly all the benefit to a mother is obtained from even a short-lived mark. In sum, the main advantage of marking is that an insect can avoid wasting her eggs by ovipositing a second time in a host where she has just been.

Although females will generally avoid the use of hosts marked with pheromone, studies of individual females show that responses may be highly flexible owing to variability in the female or her environment (Papaj, 1993; Papaj and Messing, 1996; Randen and Roitberg, 1996). Recent consideration of a few species of tephritids when the hosts are difficult to penetrate, such as *C. capitata* on citrus, show that females prefer to exploit existing oviposition sites for egg-laying. Here, the interplay of the positive stimulus of an oviposition puncture and the negative stimulus of HMP renders decision making during host acceptance to be a far more dynamic process than previously thought (see Papaj, 1993).

Host-marking in several tephritids not only serves to communicate information regarding host utilization, but also plays a role in mating behaviour. Contact with the pheromone or components in the mark results in males remaining longer on fruits with HMP than on unmarked fruits for *C. capitata* (Prokopy and Hendrichs, 1979), *R. pomonella* (Prokopy and Bush, 1972), *R. cerasi* (Katsoyannos, 1975), *R. indifferens* Curran (Mumtaz and Aliniazee, 1983a) and *C. australis* (Pittara and Katsoyannos, 1990). The phenomenon of arrestment suggests that contact with HMP may cause males to remain in trees or parts of trees recently occupied by mating females. Interestingly, as determined by electrophysiological studies of *R. cerasi*, young males, which are reproductively active soon after emergence, are significantly more sensitive to HMP than older males or young females that are not reproductively mature (Städler *et al.*, 1994). Further, males of *A. fraterculus* were not arrested in the presence of HMP (Prokopy *et al.*, 1982a), perhaps because mating initiation is not confined to host trees in this species, as is the case in a number of other fruit fly species.

Identification of host-marking pheromones in Rhagoletis

The current knowledge of any host-marking pheromone system is greatest for *R. pomonella* and *R. cerasi*. The HMP system of *R. pomonella* was previously detailed in Prokopy (1981b). Regarding the site of production of HMP, the only work to date is that of Prokopy *et al.* (1982b) on *R. pomonella*, which suggests that the posterior half of the midgut is a principal site of production. The HMP is released together with gut contents during deposition of the pheromone trail, as well as during defecation. In studies of both *R. pomonella* and *R. cerasi*, faeces elicited both a strong behavioural and electrophysiological response (Prokopy *et al.*, 1982b; Boller and Hurter, 1985). Electrophysiological recordings using *R. pomonella* and *R. cerasi* have shown that on the ventral side of the tarsi of both males and females, there exist contact chemoreceptors that contain a receptor cell selectively sensitive to HMP (Bowdan, 1984; Crnjar and Prokopy, 1982; Städler *et al.*, 1987, 1994).

Collections of HMP deposited by *R. cerasi*, *R. indifferens*, *Rhagoletis mendax* Curran and *R. pomonella* have been made by solvent washes of surfaces of pheromone-marked ceresin wax domes or real fruit (see Averill and Prokopy, 1989b). Boller and Hurter (1985) reported excessive debris in extracts collected off wax domes and adopted a system wherein cages of *R. cerasi* females were lined with glass plates; following deposition of faeces for several weeks, the dry faeces were removed with a razor blade.

Although HMPs have been isolated in solvent extracts and partially purified in some species (Prokopy, 1981b; Mumtaz and Aliniazee, 1983b), the isolation, identification and synthesis of HMP components as well as the official registration of synthetic and biologically active derivatives have so far only been achieved in *R. cerasi* (Boller, 1994; Boller and Hurter, 1998). Faeces collection (described above), development of sensitive behavioural tests and reliance on electrophysiological bioassay for low quantities of test substance in the final steps of the identification process led to the identification of a complex molecule, *N*[5(β-3-glucopyranosyl)oxy-8-hydroxypalmitol]-taurine (**7** in Fig. 1.2) (Hurter *et al.*, 1987) and its synthesis (Ernst and Wagner, 1989). The molecule contains two chiral centres at the C-8 and C-15 positions and bioassays in the laboratory showed that two of the four stereoisomers were active and that a similar level of activity was observed using a mixture of these two active isomers (8*R*,15*R* and 8*S*,15*R*) (Boller and Aluja, 1992).

Application of host-marking pheromones

Prior to the identification of HMP from *R. cerasi*, initial field tests on trees using applications of crude washes of oviposition devices (wax domes) showed substantial reductions (ranging from 77 to 90% less) in cherry fruit fly infestation compared with the control (Katsoyannos and Boller, 1976, 1980; Boller, 1981). Following the identification of the HMP, semi-field trials showed that applications of a racemic mixture (comprising the 8*R*,15*R*, 8*R*,15*S*, 8*S*,18*R* and 8*S*,15*S* enantiomers) of *N* [5(β-3-glucopyranosyl)oxy-8-hydroxypalmitol]-taurine (Fig. 1.2, **7**) resulted in more rapid emigration of flies and lowered infestation rates of *R. cerasi* (Aluja and Boller, 1992a). In field trials (Aluja and Boller, 1992b), applications of the same mixture of enantiomers of the pheromone to trees reduced cherry fruit infestation by *c.* 90%. The largest reductions were achieved when entire trees were treated, but partial treatment of trees was also highly effective. In later field trials, Boller and Hurter (1998) detailed the conditions under which HMP can successfully be applied for *R. cerasi*. Because the flies aggregate in the tops of trees, thorough coverage of tree tops is the key to success, as is a low initial population of flies. They recommended two sprays timed 1 week apart, with the first application made just prior to the onset of oviposition.

As a result of their field studies, Aluja and Boller (1992b) concluded, as had Prokopy (1972) 20 years previously, that management of *R. cerasi*, as well as some other tephritids, may be accomplished with HMP if nine out of ten trees within orchards are treated. Females move along trees readily; thus, the untreated trees could serve as sites to capture deterred females with appropriate trapping systems (Boller, 1981). This approach has been effective in small field trials utilizing HMP application for *R. cerasi* control (Boller and Hurter, 1998). Build-up of HMP on fruit in trap trees could also be washed off with high-pressure water jets to expand the trees' effectiveness over the season (Katsoyannos, 1975). Such an approach also addresses a possible constraint of HMP application: loss in the level of female responsiveness to HMP may result from continuous exposure to marked plant material (Roitberg and Prokopy, 1987). To minimize this possibility in the application of *R. cerasi* HMP, Boller and Hurter (1998) recommend that synthetic HMP be applied to as little foliage as possible. This may allow recovery of the sensitivity of the HMP receptors during a fly's rest on untreated areas of the tree. Regarding the water solubility of HMP, field applications of synthetic pheromone, when compared with pheromone deposited by natural fly populations, showed a surprisingly high residual deterrent activity, even after high rainfall, and was sufficient to protect the fruits from infestation in field efficacy trials. This high level of residual activity is probably related to the higher concentration of synthetic HMP applied to fruits (Boller, 1994; Boller and Hurter, 1998).

Future Prospects

Despite the success in developing sex pheromone-based attractants as trap baits for the olive fruit fly and the papaya fruit fly, we lack good evidence of long-range attraction of females to pheromone produced by males in many species of fruit flies of economic interest. Long-range attraction is critical to the usefulness of any pheromone as an effective lure for use in monitoring or control. It is possible that some species of fruit flies do not use such pheromones as a way of locating potential mates at a distance. There are other key constraints to progress in developing sex pheromone lures for some species. First, the apparent complexity of the pheromone blends includes multiple chemicals that have complex structures with different isomers. Second, differences in volatilities of compounds complicates the formulation of chemicals to produce a lure for field trials. Progress with sex pheromones of fruit flies may require more rigorous behavioural studies to define adequately the behaviour mediated by pheromones as well as needing expensive efforts to supply multiple chemicals in precise formulation systems. With some species, combinations of male pheromones and odours from resources needed by females may comprise more powerful attractants and research on multiple odour sources may be required to develop the most useful lures or baits (Landolt, 1997). For example, attraction of the female papaya fruit fly is synergized by the combination of male sex pheromone and odours from immature papaya fruit (Landolt *et al.*, 1992b). However, Robacker and Garcia (1990) found a lower response of female Mexican fruit flies to a combination of an extract of males with a food attractant, compared with pheromone alone.

Field trials utilizing HMP (even as a crude extract) in tephritid management programmes are limited to *R. cerasi*. Here, the difficult synthesis of *R. cerasi* HMP is expensive; actual costs will only be established when commercial production is under way (Boller and Hurter, 1998). Further, although the problem is not insurmountable, bioassays for detection and measurement of HMP have shown high variability for both behavioural and electrophysiological assessments (Boller and Hurter, 1985; Städler *et al.*, 1987; Prokopy *et al.*, 1988; Boller and Aluja, 1992). Currently, there is a lack of forward movement toward the identification of additional HMPs. With other species, the main constraint appears to be in chemical identification, which has proved to be expensive and difficult. This is unfortunate, given the advantages of the approach as detailed by Boller (1981), one of which is that it is likely that HMPs could be applied with conventional spray equipment.

In the case of the Mediterranean fruit fly, while its HMP is still not identified, as pointed out by Papaj *et al.* (1992), HMP application in management systems may be complicated by the fly's exploitation of existing punctures or fruit wounds, perhaps necessitating a relatively high dose of HMP to achieve some level of fruit protection. Overcoming habituation would be of greater necessity here.

Finally, regarding flower-head infesting tephritids that may be considered potential agents of biological control, deposition of HMP by the fly may facilitate their successful distribution over large areas. Following the release of such flies and exhaustion of local resources, the presence of HMP should stimulate the movement of flies to unexploited weed patches (Pittara and Katsoyannos, 1990).

Acknowledgements

We thank R.J. Prokopy for use of his literature files and E.F. Boller for helpful comments and suggestions on an earlier draft of the manuscript.

References

Aluja, M. and Boller, E.F. (1992a) Host marking pheromone of *Rhagoletis cerasi*: foraging behavior in response to synthetic pheromonal isomers. *Journal of Chemical Ecology* 18, 1299–1311.

Aluja, M. and Boller, E.F. (1992b) Host marking pheromone of *Rhagoletis cerasi:* field deployment of synthetic pheromone as a novel cherry fruit fly management strategy. *Entomologia Experimentalis et Applicata* 65, 141–147.

Averill, A.L. and Prokopy, R.J. (1981) Oviposition-deterring fruit marking pheromone in *Rhagoletis basiola*. *Florida Entomologist* 64, 223–226.

Averill, A.L. and Prokopy, R.J. (1982) Oviposition-deterring fruit marking pheromone in *Rhagoletis zephyria*. *Journal of the Georgia Entomological Society* 17, 315–319.

Averill, A.L. and Prokopy, R.J. (1987a) Residual activity of the oviposition-deterring pheromone in *Rhagoletis pomonella* and female response to infested fruit. *Journal of Chemical Ecology* 13, 167–177.

Averill, A.L. and Prokopy, R.J. (1987b) Intraspecific competition in the tephritid fruit fly *Rhagoletis pomonella*. *Ecology* 68, 878–886.

Averill, A.L. and Prokopy, R.J. (1988) Factors influencing release of host-marking pheromone by *Rhagoletis pomonella* flies. *Journal of Chemical Ecology* 14, 95–111.

Averill, A.L. and Prokopy, R.J. (1989a) Distribution patterns of *Rhagoletis pomonella* (Diptera: Tephritidae) eggs in hawthorn. *Annals of the Entomological Society of America* 82, 38–44.

Averill, A.L. and Prokopy, R.J. (1989b) Host-marking pheromones. In: Robinson, A.S. and Hooper, G. (eds) *World Crop Pests*, Vol. 3A, *Fruit Flies, Their Biology, Natural Enemies and Control*. Elsevier, Amsterdam, pp. 207–219.

Baker, P.S., Howse, P.E., Ondarza, R.N. and Reyes, J. (1990) Field trials of synthetic sex pheromone components of the male Mediterranean fruit fly (Diptera: Tephritidae) in southern Mexico. *Journal of Economic Entomology* 83, 2235–2245.

Baker, R., Herbert, R., Howse, P.E. and Jones, O.T. (1980) Identification and synthesis of the major sex pheromone of the olive fly (*Dacus oleae*). *Journal of the Chemistry Society, Chemical Communications* 1, 52–53.

Baker, R., Herbert, R.H. and Grant, G.G. (1985) Isolation and identification of the sex pheromone of the Mediterranean fruit fly, *Ceratitis capitata* (Wied.). *Journal of the Chemistry Society, Chemical Communications* 6, 824–825.

Battiste, M.A., Strekowski, L., Vanderbilt, D.O., Visnick, M., King, R.W. and Nation, J.L. (1983) Anastrephin and epianastrephin, novel lactone components isolated from the sex pheromone blend of male Caribbean and Mexican fruit flies. *Tetrahedron Letters* 24, 2611–2614.

Bauer, G. (1986) Life-history strategy of *Rhagoletis alternata* (Diptera: Trypetidae), a fruit fly operating in a 'non-interactive' system. *Journal of Animal Ecology* 55, 785–794.

Bellas, T.E. and Fletcher, B.S. (1979) Identification of the major components in the secretion from the rectal pheromone glands of the Queensland fruit flies *Dacus tryoni* and *Dacus neonumeralis* (Diptera: Tephritidae). *Journal of Chemical Ecology* 5, 795–803.

Boller, E.F. (1981) Oviposition-deterring pheromone of the European cherry fruit fly: status of research and potential applications. In: Mitchell, E.R. (ed.) *Management of Insect Pests with Semiochemicals: Concepts and Practice*. Plenum Press, New York, pp. 457–462.

Boller, E.F. (1994) Host marking pheromones in fruit flies: state of the art. *IOBC/WPRS Bulletin* 17, 93–96.

Boller, E.F. (1996) Host marking pheromones in fruit flies. In: Aliniazee, M.T. and Long, L.E. (eds) *Biology and Control of the Cherry Fruit Flies: a World Wide Perspective*. Special Report 971, Oregon Agricultural Experimental Station, pp. 39–44.

Boller, E.F. and Aluja, M. (1992) Oviposition deterring pheromone in *Rhagoletis cerasi* L.: biological activity of synthetic isomers and HMP discrimination of two host races as measured by an improved laboratory bioassay. *Journal of Applied Entomology* 113, 113–119.

Boller, E.F. and Hurter, J. (1985) Oviposition deterring pheromone in *Rhagoletis cerasi*: behavioral laboratory test to measure pheromone activity. *Entomologia Experimentalis et Applicata* 39, 163–169.

Boller, E.F. and Hurter, J. (1998) The marking pheromone of the cherry fruit fly: a novel, non-toxic and ecologically safe technique to protect cherries against cherry fruit fly infestation. In: *Abstract Book, 2nd International Symposium on Insect Pheromones*, 30 March–3 April, 1998, Wageningen, pp. 99–101.

Boller, E.F., Schoni, R. and Bush, G.L. (1987) Oviposition deterring pheromone in *Rhagoletis cerasi*: biological activity of a pure single compound verified in semifield test. *Entomologia Experimentalis et Applicata* 45, 7–22.

Boller, E.F., Hippe, C. and Prokopy, R.J. (1994) Response of wild and laboratory reared *Ceratitis capitata* Wied. (Diptera: Tephritidae) flies from different geographic origins to a standard host marking pheromone solution. *Journal of Applied Entomology* 118, 84–91.

Bowdan, E. (1984) Electrophysiological responses of tarsal contact chemoreceptors of the apple maggot fly *Rhagoletis pomonella* to salt, sucrose and oviposition deterring pheromone. *Journal of Comparative Physiology* A154, 143–152.

Burk, T. (1983) Behavioral ecology of mating in the Caribbean fruit fly, *Anastrepha suspensa* (Loew) (Diptera: Tephritidae). *Florida Entomologist* 66, 330–344.

Chambers, D.L. (1977) Attractants for fruit fly survey and control. In: Shorey, H.H. and McKelvey, J.J. (eds) *Chemical Control of Insect Behavior.* Wiley-Interscience, New York, pp. 327–344.

Chuman, T., Landolt, P.J., Heath, R.R. and Tumlinson, J.H. (1987) Isolation, identification, and synthesis of male produced sex pheromone of the papaya fruit fly, *Toxotrypana curvicauda* Gerstaecker (Diptera: Tephritidae). *Journal of Chemical Ecology* 13, 1979–1992.

Chuman, T., Sivinski, J., Heath, R.R., Calkins, C.O., Tumlinson, J.H., Battiste, M.A., Wydra, R.L., Strekowski, L. and Nation, J.L. (1988) Suspensolide, a new macrolide component of male Caribbean fruit fly (*Anastrepha suspensa* [Loew]) volatiles. *Tetrahedron Letters* 50, 6561–6564.

Cirio, U. (1971) Reperi sul meccanismo stimolo-riposta nell'ovideposizione della *Dacus oleae* Gmelin (Diptera: Trypetidae). *Redia* 52, 577–600.

Cirio, U. (1972) Observazioni sul comportamento di ovideposizione della *Rhagoletis completa* (Diptera: Trypetidae) in laboratorio. *Proceedings of the Ninth Congress of the Italian Entomological Society*, 99–117.

Crnjar, R.M. and Prokopy, R.J. (1982) Morphological and electrophysiological mapping of tarsal chemoreceptors of oviposition-deterring pheromone in *Rhagoletis pomonella* flies. *Journal of Insect Physiology* 28, 393–400.

Epsky, N.D. and Heath, R.R. (1993a) Pheromone production by male *Anastrepha suspensa* (Diptera: Tephritidae) under natural light cycles in greenhouse studies. *Environmental Entomology* 22, 464–469.

Epsky, N.D. and Heath, R.R. (1993b) Food availability and pheromone production by males of *Anastrepha suspensa* (Diptera: Tephritidae). *Environmental Entomology* 22, 942–947.

Ernst, B. and Wagner, B. (1989) Synthesis of the oviposition-deterring pheromone (ODP) in *Rhagoletis cerasi*. *Helvetica Chimica Acta* 72, 165–171.

Esponda-Gaxiola, R.E. (1977) Contribucion al estudio quimico del atrayente sexual de la mosca Mexicana de la fruta, *Anastrepha ludens* (Loew). MSc thesis, Intituto Technologico y de Estudios Superiores Monterrey, Monterrey, N.L., Mexico.

Feron, M.M. (1959) Attraction chimique du mâle de *Ceratitis capitata* Wied. (Dipt., Trypetidae) pour la femelle. *Comptes Rendus de l'Académie des Sciences* 248, 2403–2404.

Fitt, G.P. (1981a) Inter- and intraspecific responses to sex pheromones in laboratory bioassays by females of three species of tephritid fruit flies from northern Australia. *Entomologia Experimentalis et Applicata* 30, 40–44.

Fitt, G.P. (1981b) Observations on the biology and behavior of *Paraceratitella eurycephala* (Diptera: Tephritidae) in Northern Australia. *Journal of the Australian Entomological Society* 20, 1–7.

Fitt, G.P. (1984) Oviposition behavior of two tephritid fruit flies, *Dacus tryoni* and *Dacus jarvisi*, as influenced by the presence of larvae in the host fruit. *Oecologia* 62, 37–46.

Flath, R.A., Jang, E.B., Light, D.M., Mon, R.T., Carvalho, L., Binder, R.G. and John, J.O. (1993) Volatile pheromonal emissions from the Mediterranean fruit fly: effects of fly age and time of day. *Journal of Agricultural and Food Chemistry* 41, 830–837.

Fletcher, B.S. (1969) The structure and function of the sex pheromone glands of the Queensland fruit fly *Dacus tryoni*. *Journal of Insect Physiology* 13, 1309–1322.

Fletcher, B.S. and Giannakakis, A. (1973) Factors limiting the response of females of the Queensland fruit fly, *Dacus tryoni*, to the sex pheromone of the male. *Journal of Insect Physiology* 19, 1147–1155.

Fletcher, B.S. and Kitching, W. (1995) Chemistry of fruit flies. *Chemical Reviews* 95, 789–828.

Garibaldi, P., Verotta, L. and Fanelli, R. (1983) Studies on the sex pheromone of *Dacus oleae*. Analysis of the substances contained in the rectal glands. *Experientia* 39, 502–505.

Haniotakis, G.E. (1974) Male olive fly attraction to virgin females in the field. *Annales Zoologie et Ecologie Animale* 9, 273–276.

Haniotakis, G.E. (1977) Sexual attraction in the olive fruit fly, *Dacus oleae* (Gmelin). *Environmental Entomology* 3, 83–86.

Haniotakis, G.E. and Pittara, I.S. (1994) Response of *Bactrocera* (*Dacus*) *oleae* males (Diptera: Tephritidae) to pheromones as affected by concentration, insect age, time of day, and previous exposure. *Environmental Entomology* 23, 726–731.

Haniotakis, G.E., Kozyrakis, M. and Hardakis, I. (1983) Applications of pheromones for the control of the olive fruit fly. In: *Proceedings of the International Conference on Plant Protection*, Vol. 4, Budapest, Hungary, July 4–9, 1983, pp. 164–171.

Haniotakis, G.E., Francke, W., Mori, K., Redlich, H. and Schurig, V. (1986) Sex-specific activity of (*R*)-(−)- and (*S*)-(−)-1,7-dioxaspiro[5,5]undecane, the major pheromone of *Dacus oleae*. *Journal of Chemical Ecology* 12, 1559–1568.

Haniotakis, G.E, Kozyrakis, M., Fitsakis, T. and Antonidaki, A. (1991) An effective mass trapping method for the control of *Dacus oleae* (Diptera: Tephritidae). *Journal of Economic Entomology* 84, 564–569.

Heath, R.R., Landolt, P.J., Tumlinson, J.H., Chambers, D.L., Murphy, R.E., Doolittle, R.E., Dueben, B.D., Sivinski, J. and Calkins, C.O. (1991) Analysis, synthesis, formulation, and field testing of three major components of male Mediterranean fruit fly pheromone. *Journal of Chemical Ecology* 17, 1925–1940.

Heath, R.R., Epsky, N.D., Jiminez, A., Dueben, B.D., Landolt, P.J., Meyer, W.L., Aluja, M., Rizzo, J., Camino, M., Jeronimo, J. and Baranowski, R.M. (1996) Improved pheromone-based trapping systems to monitor *Toxotrypana curvicauda* (Diptera: Tephritidae). *Florida Entomologist* 79, 37–48.

Hurter, J., Boller, E.F., Städler, E., Blattman, B., Buser, H.R., Bosshard, N.U., Damm, L., Kozlowski, M.W., Schoni, R., Raschdorf, F., Dahinden, R., Shlumf, E., Fritz, H., Richter, W.J. and Schreiber, J. (1987) Oviposition-deterring pheromone in *Rhagoletis cerasi* L.: purification and determination of the chemical constitution. *Experientia* 43, 157–164.

Jacobson, M., Ohinata, K., Chambers, D.L., Jones, W.A. and Fujimoto, M.S. (1973) Insect sex attractants. 13. Isolation, identification, and synthesis of sex pheromones of the male Mediterranean fruit fly. *Journal of Medical Chemistry* 16, 248–251.

Jang, E.B. (1995) Effects of mating and accessory gland injections of factory-mediated behavior in the female Mediterranean fruit fly, *Ceratitis capitata*. *Journal of Insect Physiology* 41, 705–710.

Jang, E.B., Light, D.M., Flath, R.A., Nagata, J.T. and Mon, T.R. (1989) Electroantennogram responses of the Mediterranean fruit fly, *Ceratitis capitata*, to identified constituents from calling males. *Entomologia Experimentalis et Applicata* 50, 7–19.

Jang, E.B., Light, D.M., Binder, R.G., Flath, R.A. and Carvalho, L.A. (1994) Attraction of female Mediterranean fruit flies to the five major components of male produced pheromone in a laboratory flight tunnel. *Journal of Chemical Ecology* 20, 9–20.

Katsoyannos, B.I. (1975) Oviposition-deterring, male-arresting, fruit-marking pheromone in *Rhagoletis cerasi*. *Environmental Entomology* 4, 801–807.

Katsoyannos, B.I. (1976) Female attraction to males in *Rhagoletis cerasi*. *Environmental Entomology* 5, 474–476.

Katsoyannos, B.I. (1982) Male sex pheromone of *Rhagoletis cerasi* L. (Diptera: Tephritidae): factors affecting release and response and its role in the mating behavior. *Zeitschrift für Angewandte Entomologie* 94, 187–198.

Katsoyannos, B.I. and Boller, E.F. (1976) First field application of oviposition-deterring marking pheromone of European cherry fruit fly, *Rhagoletis cerasi*. *Environmental Entomology* 5, 151–152.

Katsoyannos, B.I. and Boller, E.F. (1980) Second field application of oviposition-deterring marking pheromone of the European cherry fruit fly, *Rhagoletis cerasi* L. (Diptera: Tephritidae). *Zeitschrift für Angewandte Entomologie* 89, 278–281.

Kobayashi, R.M., Ohinata, K., Chambers, D.L. and Fujimoto, M.S. (1978) Sex pheromones of the oriental fruit fly and the melon fly: mating behavior, bioassay method, and attraction of females by live males and by suspected pheromone. *Environmental Entomology* 7, 107–112.

Lalonde, R.G. and Roitberg, B.D. (1992) Host selection behavior of a thistle-feeding fly: choices and consequences. *Oecologia* 90, 534–539.

Landolt, P.J. (1997) Sex attractant and aggregation pheromones of male phytophagous insects. *American Entomologist* 43, 12–22.

Landolt, P.J. and Heath, R.R. (1988) Effects of age, mating and time of day on behavioral responses of female papaya fruit fly, *Toxotrypana curvicauda* Gerstaecker (Diptera: Tephritidae) to synthetic sex pheromone. *Environmental Entomology* 7, 47–51.

Landolt, P.J. and Heath, R.R. (1990) Effects of pheromone release rate and time of day on catches of male and female papaya fruit flies (Diptera: Tephritidae) on fruit model traps baited with pheromone. *Journal of Economic Entomology* 85, 2040–2043.

Landolt, P.J. and Heath, R.R. (1996) Development of pheromone based trapping systems for monitoring and controlling tephritid fruit flies in Florida. In: Rosen, D., Bennett, F.D. and Capinera, J.L. (eds) *Pest Management in the Subtropics: Integrated Pest Management – A Florida Perspective*. Intercept Ltd, Andover, UK, pp. 197–207.

Landolt, P.J. and Hendrichs, J. (1983) Reproductive behavior of the papaya fruit fly, *Toxotrypana curvicauda* Gerstaecker (Diptera: Tephritidae). *Annals of the Entomological Society of America* 76, 413–417.

Landolt, P.J. and Sivinski, J. (1992) Effects of time of day, adult food, and host fruit on incidence of calling by male Caribbean fruit flies (Diptera: Tephritidae). *Environmental Entomology* 21, 382–387.

Landolt, P.J., Heath, R.R. and King, J.R. (1985) Behavioral responses of female papaya fruit flies, *Toxotrypana curvicauda* Gerstaecker (Diptera: Tephritidae) to male-produced sex pheromone. *Annals of the Entomological Society of America* 78, 751–755.

Landolt, P.J., Heath, R.R., Agee, H.R., Tumlinson, J.H. and Calkins, C.O. (1988) A sex pheromone-based trapping system for the papaya fruit fly, *Toxotrypana curvicauda* Gerstaecker (Diptera: Tephritidae). *Journal of Economic Entomology* 81, 1163–1169.

Landolt, P.J., Gonzalez, M., Chambers, D.L. and Heath, R.R. (1991) Comparison of field observations and trapping of papaya fruit fly in papaya plantings in Central America and Florida. *Florida Entomologist* 74, 408–414.

Landolt, P.J., Heath, R.R. and Chambers, D.L. (1992a) Oriented flight responses of female Mediterranean fruit flies to calling males, odor of calling males, and a synthetic pheromone blend. *Entomologia Experimentalis et Applicata* 65, 259–266.

Landolt, P.J., Reed, H.C. and Heath, R.R. (1992b) Attraction of female papaya fruit fly (Diptera: Tephritidae) to male pheromone and host fruit. *Environmental Entomology* 21, 1154–1159.

Lo Scalzo, R., Scarpati, M.L., Verzegnassi, B. and Vita, G. (1994) *Olea europaea* chemicals repellent to *Dacus oleae* females. *Journal of Chemical Ecology* 20, 1813–1823.

Mayes, C.F. and Roitberg, B.D. (1986) Host discrimination in *Rhagoletis berberis* (Diptera: Tephritidae). *Journal of the Entomological Society of British Columbia* 83, 39–42.

Mazomenos, B.E. and Haniotakis, G.E. (1981) A multicomponent female sex pheromone of *Dacus oleae* Gmelin: isolation and bioassay. *Journal of Chemical Ecology* 7, 437–444.

Mazomenos, B.E. and Haniotakis, G.E. (1985) Male olive fruit fly attraction to synthetic sex pheromone components in laboratory and field tests. *Journal of Chemical Ecology* 11, 397–405.

Mazomenos, B.E. and Pomonis, J.G. (1983) Male olive fruit fly pheromone: isolation, identification, and lab bioassays. In: *Proceedings CEC/IOBC Symposium*, Athens, Nov. 1982, pp. 96–103.

McDonald, P.T. (1987) Intraspecific stimulation of pheromone release by male Mediterranean fruit flies (Diptera: Tephritidae). *Annals of the Entomological Society of America* 80, 17–20.

Metcalf, R.L. (1990) Chemical ecology of Dacinae fruit flies (Diptera: Tephritidae). *Annals of the Entomological Society of America* 83, 1017–1030.

Moreno, D.S., Sanchez, M., Robacker, D.C. and Worley, J. (1991) Mating competitiveness of irradiated Mexican fruit fly (Diptera: Tephritidae). *Journal of Economic Entomology* 84, 1227–1234.

Mumtaz, M.M. (1976) Pheromonal communication in the western cherry fruit fly with special reference to oviposition. MSc thesis, Oregon State University, Corvallis, 58 pp.

Mumtaz, M.M. and Aliniazee, M.T. (1983a) The oviposition-deterring pheromone in the western cherry fruit fly, *Rhagoletis indifferens* Curran (Dipt., Tephritidae). 1. Biological properties. *Zeitschrift für Angewandte Entomologie* 96, 83–93.

Mumtaz, M.M. and Aliniazee, M.T. (1983b) The oviposition-deterring pheromone in the western cherry fruit fly, *Rhagoletis indifferens* Curran (Dipt., Tephritidae). 2. Chemical characterization and partial purification. *Zeitschrift für Angewandte Entomologie* 96, 93–99.

Nakagawa, S., Steiner, L.F. and Farias, G.J. (1981) Response of virgin female Mediterranean fruit flies to live mature normal males, sterile males, and trimedlure in plastic traps. *Journal of Economic Entomology* 74, 566–567.

Nation, J.L. (1972) Courtship behavior and evidence for a sex attractant in the male Caribbean fruit fly, *Anastrepha suspensa*. *Annals of the Entomological Society of America* 65, 1364–1367.

Nation, J.L. (1975) The sex pheromone blend of Caribbean fruit fly males: isolation, biological activity, and partial chemical characterization. *Environmental Entomology* 4, 27–30.

Nation, J.L. (1977) Pheromone research in tephritid fruit flies (Diptera: Tephritidae). *Proceedings of the International Society of Citriculture* 2, 481–485.

Nation, J.L. (1983) Sex pheromone of the Caribbean fruit fly: chemistry and field ecology. *Proceedings of the Fifth International Conference on Pesticide Chemistry* 2, 109–110.

Nation, J.L. (1989) The role of pheromones in the mating system of *Anastrepha* fruit flies. In: Robinson, A.S. and Hooper, G. (eds) *World Crop Pests*, Vol. 3A, *Fruit Flies, Their Biology, Natural Enemies, and Control*. Elsevier, Amsterdam, pp. 189–205.

Nation, J.L. (1990) Biology of pheromone release by male Caribbean fruit flies, *Anastrepha suspensa* (Diptera: Tephritidae). *Journal of Chemical Ecology* 16, 553–571.

Ohinata, K., Fujimoto, M.S., Chambers, D.L., Jacobson, M. and Kamakahi, D.C. (1973) Mediterranean fruit fly: bioassay techniques for investigating sex pheromones. *Journal of Economic Entomology* 66, 812–814.

Ohinata, K., Jacobson, M., Nakagawa, S., Fujimoto, M. and Higa, H. (1977) Mediterranean fruit fly: laboratory and field evaluations of synthetic sex pheromones. *Journal of Environmental Science and Health* A12, 67–78.

Ohinata, K., Jacobson, M., Nakagawa, S., Urago, T., Fujimoto, M. and Higa, H. (1979) Methyl (*E*)-6-nonenoate: a new Mediterranean fruit fly male attractant. *Journal of Economic Entomology* 72, 648–650.

Ohinata, K., Jacobson, M., Kobayashi, R.M., Chambers, D.L., Fujimoto, M.S. and Higa, H.H. (1982) Oriental fruit fly and melon fly: biological and chemical status of smoke produced by males. *Journal of Environmental Science and Health* A17, 29–38.

Papaj, D.R. (1993) Use and avoidance of occupied hosts as a dynamic process in tephritid flies. In: Bernays, E.A. (ed.) *Insect–Plant Interactions*, Vol. 5. CRC Press, Boca Raton, Florida, pp. 25–46

Papaj, D.R. and Aluja, M. (1993) Temporal dynamics of host-marking in the tropical tephritid fly, *Anastrepha ludens*. *Physiological Entomology* 18, 279–284.

Papaj, D.R. and Messing, R.H. (1996) Functional shifts in the use of parasitized hosts by a tephritid fly: the role of host quality. *Behavioral Ecology* 7, 235–242.

Papaj, D.R., Averill, A.L., Prokopy, R.J. and Wong, T.T.Y. (1992) Host-marking pheromone and use of previously established oviposition sites by the Mediterranean fruit fly (Diptera: Tephritidae). *Journal of Insect Behavior* 5, 583–598.

Perdomo, A.J., Baranowski, R.M. and Nation, J.L. (1975) Recapture of virgin female Caribbean fruit flies from traps baited with males. *Florida Entomologist* 58, 291–296.

Pittara, I.S. and Katsoyannos, B.I. (1990) Evidence for a host-marking pheromone in *Chaetorellia australis*. *Entomologia Experimentalis et Applicata* 54, 287–295.

Poramarcom, R. (1988) Sexual communication in the oriental fruit fly, *Dacus dorsalis* Hendel (Diptera: Tephritidae). PhD dissertation, University of Hawaii, 126 pp.

Prokopy, R.J. (1972) Evidence for a marking pheromone deterring repeated oviposition in apple maggot flies. *Environmental Entomologist* 1, 326–332.

Prokopy, R.J. (1975a) Oviposition-deterring fruit marking pheromone in *Rhagoletis fausta*. *Environmental Entomologist* 4, 298–300.

Prokopy, R.J. (1975b) Mating behavior in *Rhagoletis pomonella* (Diptera: Tephritidae). V. Virgin female attraction to male odor. *Canadian Entomologist* 107, 905–908.

Prokopy, R.J. (1981a) Epideictic pheromones influencing spacing patterns of phytophagous insects. In: Nordlund, D.A., Jones, R.L. and Lewis, W.L. (eds) *Semiochemicals: Their Role in Pest Control*. John Wiley & Sons, New York, pp. 181–214.

Prokopy, R.J. (1981b) Oviposition-deterring pheromone system of apple maggot flies. In: Mitchell, E.R. (ed.) *Management of Insect Pests with Semiochemicals: Concepts and Practice*. Plenum Press, New York, pp. 477–494.

Prokopy, R.J. and Bush, G.L. (1972) Mating behavior in *Rhagoletis pomonella* (Diptera: Tephritidae). III. Male aggregation in response to an arrestant. *Canadian Entomologist* 104, 975–983.

Prokopy, R.J. and Hendrichs, J. (1979) Mating behavior of *Ceratitis capitata* on a field caged host tree. *Annals of the Entomological Society of America* 72, 642–648.

Prokopy, R.J. and Koyama, J. (1982) Oviposition site partitioning in *Dacus cucurbitae*. *Entomologia Experimentalis et Applicata* 31, 428–432.

Prokopy, R.J., Reissig, W.H. and Moericke, V. (1976) Marking pheromones deterring repeated oviposition in *Rhagoletis* flies. *Entomologia Experimentalis et Applicata* 20, 170–178.

Prokopy, R.J., Greany, P.D. and Chambers, D.L. (1977) Oviposition-deterring pheromone in *Anastrepha suspensa*. *Environmental Entomology* 6, 463–465.

Prokopy, R.J., Zeigler, J.R. and Wong, T.T.Y. (1978) Deterrence of repeated oviposition by fruit-marking pheromone in *Ceratitis capitata* (Diptera: Tephritidae). *Journal of Chemical Ecology* 4, 55–63.

Prokopy, R.J., Malavasi, A. and Morgante, J.S. (1982a) Oviposition deterring pheromone in *Anastrepha fraterculus* flies. *Journal of Chemical Ecology* 8, 763–771.

Prokopy, R.J., Averill, A.L., Bardinelli, C.M., Bowdan, E.S., Cooley, S.S., Crnjar, R.M., Dundullis, E.A., Roitberg, B.A., Spatcher, P.J., Tumlinson, J.H. and Weeks, B.L. (1982b) Site of production of an oviposition-deterring pheromone component in *Rhagoletis pomonella* flies. *Journal of Insect Physiology* 28, 1–10.

Prokopy, R.J., Powers, P.J., Heath, R.R., Dueben, B.D. and Tumlinson, J.H. (1988) Comparative laboratory methods for assaying behavioral responses of *Rhagoletis pomonella* to host marking pheromone. *Journal of Applied Entomology* 106, 437–443.

Randen, E.J. van and Roitberg, B.D. (1996) The effect of egg load on superparasitism by the snowberry fly. *Entomologia Experimentalis et Applicata* 79, 241–245.

Raptapoulos, D., Gittaniotakis, G., Koutsaftikis, A., Kelly, D., and Mavraganis, V. (1995) Biological activity of chemicals identified from extracts and volatiles of male *Rhagoletis cerasi*. *Journal of Chemical Ecology* 21, 1287–1297.

Robacker, D.C. (1988) Behavioral responses of female Mexican fruit flies, *Anastrepha ludens*, to components of male-produced sex pheromone. *Journal of Chemical Ecology* 14, 1715–1726.

Robacker, D.C. and Garcia, J.A. (1990) Responses of laboratory-strain Mexican fruit flies, *Anastrepha ludens*, to combinations of fermenting fruit odor and male produced pheromone in laboratory bioassays. *Journal of Chemical Ecology* 16, 2027–2038.

Robacker, D.C. and Hart, W.G. (1985) (Z)-3-nonenol, (Z,Z)-3,6-nonadienol and (S,S)-(−)-epianastrephin: male produced pheromones of the Mexican fruit fly. *Entomologia Experimentalis et Applicata* 39, 103–108.

Robacker, D.C. and Hart, W.G. (1986) Behavioral responses of male and female Mexican fruit flies, *Anastrepha ludens*, to male-produced chemicals in laboratory experiments. *Journal of Chemical Ecology* 12, 39–47.

Robacker, D.C. and Moreno, D.S. (1988) Responses of female Mexican fruit flies at various distances from male-produced pheromone. *Southwestern Entomologist* 13, 95–100.

Robacker, D.C. and Wolfenbarger, D.A. (1988) Attraction of laboratory-reared, irradiated Mexican fruit flies to male-produced pheromone in the field. *Southwestern Entomologist* 13, 75–80.

Robacker, D.C., Ingle, S.J. and Hart, W.G. (1985) Mating frequency and response to male-produced pheromone by virgin and mated females of the Mexican fruit fly. *Southwestern Entomologist* 10, 215–221.

Robacker, D.C., Mangan, R.L., Moreno, D.S. and Moreno, A.M.T. (1991) Mating behavior and male mating success in wild *Anastrepha ludens* (Diptera: Tephritidae) on a field-caged host tree. *Journal of Insect Behavior* 4, 471–487.

Robinson, A.S. and Hooper, G. (eds) (1989) *World Crop Pests*, Vol. 3A, *Fruit Flies, Their Biology, Natural Enemies and Control*. Elsevier, New York, 372 pp.

Rocca, J.R., Nation, J.L., Strekowski, L. and Battiste, M.A. (1992) Comparison of volatiles emitted by male Caribbean and Mexican fruit flies. *Journal of Chemical Ecology* 18, 223–244.

Roitberg, B.D. and Mangel, M. (1988) On the evolutionary ecology of marking pheromones. *Evolutionary Ecology* 2, 289–315.

Roitberg, B.D. and Prokopy, R.J. (1987) Insects that mark host plants: an ecological, evolutionary perspective on host-marking chemicals. *BioScience* 37, 400–406.

Roitberg, B.D., Cairl, R.S. and Prokopy, R.J. (1984) Oviposition deterring pheromone influences distance dispersal in tephritid fruit flies. *Entomologia Experimentalis et Applicata* 35, 217–220.

Sivinski, J.M. and Calkins, C.O. (1986) Pheromones and parapheromones in the control of tephritids. *Florida Entomologist* 73, 123–128.

Sivinski, J. and Heath, R.R. (1988) Effects of oviposition on remating, response to pheromones, and longevity in the female Caribbean fruit fly, *Anastrepha suspensa* (Diptera: Tephritidae). *Annals of the Entomological Society of America* 81, 1021–1024.

Sivinski, J.M., Epsky, N. and Heath, R.R. (1994) Pheromone deposition on leaf territories by male Caribbean fruit flies, *Anastrepha suspensa* (Loew) (Diptera: Tephritidae). *Journal of Insect Behavior* 7, 43–51.

Städler, E., Shoni, R. and Kozlowski, M.W. (1987) Relative air humidity influences the function of the tarsal chemoreceptor cells of the cherry fruit fly (*Rhagoletis cerasi*). *Physiological Entomology* 12, 339–346.

Städler, E., Ernst, B., Hurter, J. and Boller, E.F. (1994) Tarsal contact chemoreceptor for the host marking pheromone of the cherry fruit fly, *Rhagoletis cerasi*: responses to natural and synthetic compounds. *Physiological Entomology* 19, 139–151.

Stokes, J.B., Uebel, E.C., Warthen, J.D. Jr, Jacobson, M., Flippen-Anderson, J.L., Gilardi, R., Spishakoff, L.M. and Wilzer, K.R. (1983) Isolation and identification of novel lactones from male Mexican fruit flies. *Journal of Agricultural and Food Chemistry* 31, 1162–1167.

Straw, N.A. (1989) Evidence for an oviposition-deterring pheromone in *Tephritis bardanae* (Schrank) (Diptera: Tephritidae). *Oecologia* 78,121–130.

Webb, J.C., Burk, T. and Sivinski, J. (1983a) Attraction of female Caribbean fruit flies, *Anastrepha suspensa* (Diptera: Tephritidae) to the presence of males and male-produced stimuli, in field cages. *Annals of the Entomological Society of America* 76, 996–998.

Webb, J.C., Calkins, C.O., Chambers, D.L., Schwienbacher, W. and Ross, K. (1983b) Acoustical aspects of behavior of Mediterranean fruit fly, *Ceratitis capitata*: analysis and identification of courtship sounds. *Entomologia Experimentalis et Applicata* 33, 1–8.

2 Gall Midges

Marion O. Harris and Stephen P. Foster

The Horticulture and Food Research Institute of New Zealand Ltd, Mt Albert Research Centre, Private Bag 92169, Auckland, New Zealand

Introduction

The family Cecidomyiidae (Diptera) is notable for both the pest status and the diverse biology of its species. Within the order Diptera, the family is classified in the suborder Nematocera, and within this suborder, in the infraorder Bibionomorpha and the superfamily Sciaroidea (Gagné, 1989). There are over 5000 described species of Cecidomyiidae, 'a small fraction of the number that must exist' (Gagné, 1989). These species show a tremendous diversity in feeding habits, including mycetophagy, zoophagy and phytophagy (Mamaev and Krivosheina, 1965). Among cecidomyiids that feed on plants are important agricultural pests, such as the Hessian fly, the brassica pod midge, the rice midge and the sorghum midge, all of which have been subjects of major research efforts over the last 100 years.

In the 1920s, a chance observation made by Cartwright (1922), of a Hessian fly female trapped in mud, led to the first experimental evidence that females of a cecidomyiid species could attract males over distances of several metres. He later placed newly emerged females in wire cages coated with a sticky material and, over a 4-day period, caught an average of 145 males per caged female. In a second test, Cartwright set up sticky screens at various distances downwind from caged virgin females and, over distances of 0.3–4 m, caught males in increasing numbers on screens closer to the females. Cartwright referred to this phenomenon as 'sexual attraction' but did not offer any explanation for the stimulus that triggered the attraction.

Today, in spite of the large number and the pest status of species in the family Cecidomyiidae, relatively little on their pheromones has been published. Although it has been proposed (Miller and Borden, 1984) that the Douglas-fir cone gall midge uses host-marking pheromones, the only pheromones that have been demonstrated conclusively are all sex pheromones. Of the ten species that have been shown to use sex pheromones, only one – the Hessian fly – has had an actual sex pheromone component identified (Foster *et al.*, 1991c). This chapter reviews aspects of the mating behaviour of plant-feeding cecidomyiids relevant to the use of sex pheromones, largely focusing on the species known to use such pheromones.

General Aspects of Reproductive Biology

Features in common with other phytophagous Diptera

Adult phytophagous cecidomyiids share a number of characteristics with other phytophagous Diptera. First, all known species exhibit sexual reproduction (Gagné, 1994). Second, they exhibit a range of specialization on host plants similar to other Diptera (Gagné, 1989, 1994), with most species being either oligophagous (e.g. Hessian fly, brassica pod midge, sorghum midge) or monophagous (e.g. apple leaf-curling midge, Douglas-fir cone gall midge), and a smaller number being polyphagous (e.g. *Prodiplosis longifila* Gagné, which feeds on buds of tomato, potato, green beans, cotton and other crops). Third, as is the case for species in two other dipteran families, the Tephritidae and the Agromyzidae, phytophagous cecidomyiid species develop on plants in simple or complex galls (Gagné, 1989). Fourth, their neonate larvae are tiny and cannot move from one plant to another (Gagné, 1989). This feature, in combination with a limited host range, leads to the fifth feature that cecidomyiids share with many other phytophagous Diptera: adult females are almost entirely responsible for finding suitable host plants (Åhman, 1985; Harris and Rose, 1989).

Distinctive aspects of cecidomyiid adult life

Size

In general, adult cecidomyiids are small, rarely attaining body or wing lengths of more than 2–3 mm or body weights of more than 1–2 mg (e.g. Bergh *et al.*, 1990). One probable consequence of this small size is that it restricts cecidomyiids to flying in clement weather conditions. Mated female Hessian flies have been shown to lose control over their flight direction in relatively light winds (speeds > 0.9 m s^{-1}; Withers and Harris, 1997).

Duration of adult life

Adult reproductive activities are completed in a relatively short period of time (Gagné, 1989). The normal life span of adults can be as short as 1–2 h (e.g. sorghum midge males: Harris, 1976), but is commonly 1–2 days, and rarely longer than 3–4 days. Virgin females live longer than mated females (e.g. Harris and Rose, 1991). Moreover, within a population, reproductive activities are often highly synchronized over a period of 1–2 weeks, coinciding with a developmental stage of the host plant, such as early vegetative growth (Gagné, 1994).

Behavioural repertoire

The behavioural repertoire of adults is limited by their short lifespan (Gagné, 1989, 1994). Essentially, males emerge, take flight, find and copulate with females, while females emerge with a full complement of mature eggs, attract a mate (generally mating only once) and search for and oviposit on host plants, usually at multiple sites. While most cecidomyiid adults have mouthparts that allow ingestion of liquids, Gagné (1994) has suggested that, except for drinking water, most phytophagous species probably do not feed as adults.

Unisexual progenies

Another distinctive feature of cecidomyiids is that individual females often produce offspring of only one sex (Barnes, 1931). In the Hessian fly, the cause of this phenomenon is the elimination of paternally derived chromosomes in the male during spermatogenesis (Gallun and Hatchett, 1969). When bisexual progeny do occur, individuals of the less well-represented sex are the result of non-disjunction of sex chromosomes during oogenesis (Stuart and Hatchett, 1991). It has been suggested that the function of unisexual progeny is to reduce the probability of inbreeding within a population (Stuart and Hatchett, 1991).

Antennal sensory organs

In addition to the sensilla commonly found on the antennae of other insects, adult cecidomyiids have curious, looped sense organs known as circumfila (Slifer and Sekhon, 1971; Lee and Lee, 1985; Solinas and Nuzzaci, 1987). On the sorghum midge (Slifer and Sekhon, 1971), each of the 24 nodes of the male flagellum is encircled by a circumfilum attached to the surface of the antennae by short stalks (Fig. 2.1). The external and internal structure of the loops of the circumfila are similar to those of thin-walled olfactory chemoreceptors. Female sorghum midges also have circumfila, but theirs lie closer to the surface of the antennae. In spite of the small size of the antennae, electroantennogram recordings can be obtained from male antennae stimulated with female sex pheromone (e.g. pea midge: Wall *et al.*, 1985; Hillbur and Löfqvist, 1996; Hessian fly: S.P. Foster, 1998, unpublished results; apple leafcurling midge: A. Zhang and W.L. Roelofs, Geneva, New York, 1998, personal communication).

Location and temporal patterning of adult eclosion

As a general guide, the majority of cecidomyiid species living in the temperate zone emerge as adults from pupation sites in the soil litter (Gagné, 1989), while the majority of cecidomyiids living in the tropics emerge from pupation sites on the host plant (Gagné, 1994). Peak emergence of adult cecidomyiids most commonly occurs at either dawn or dusk (Gagné, 1989) and is strongly influenced by temperature (Summers, 1975).

Both adult males and females typically exhibit unimodal emergence patterns, with males emerging slightly in advance of females (i.e. protandry: Gagné, 1989). In the sorghum midge (Summers, 1975), eclosion begins shortly after sunrise at 0700 h, with peak emergence of males and females occurring at 0845 h and 0930 h, respectively. No males or females emerge after 1430 h. The lower threshold temperature of eclosion for males (13–16°C) than that for females (18–20°C) appears to be responsible for the protandry in the sorghum midge (Summers, 1975).

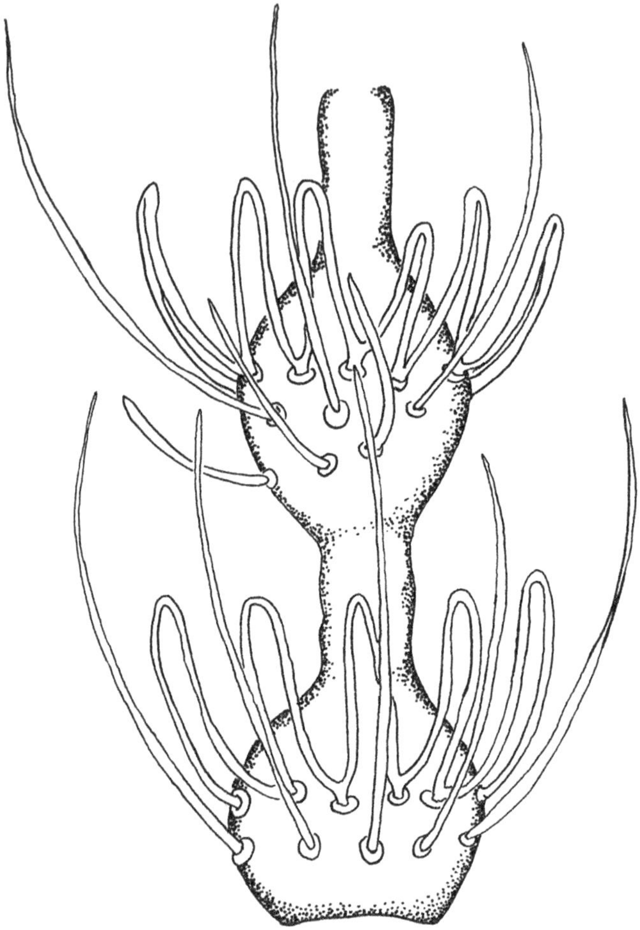

Fig. 2.1. Sense organs on an antennal flagellomere of the male sorghum midge. Each node is encircled by a looped circumfilum. (From Slifer and Sekhon, 1971.)

Not all cecidomyiids exhibit unimodal patterns of emergence. Male Hessian flies, for example, exhibit a bimodal pattern with most males emerging in the late afternoon, about 12 h before the peak of female eclosion (Bergh *et al.*, 1990). A smaller number of Hessian fly males emerge shortly after dawn, during the time of peak female eclosion.

Mating Behaviour

Phylogenetic aspects of mating behaviour

In Diptera, orientation to sex pheromones over long distances appears to be more common in the more primitive families (suborder Nematocera) than in more advanced families (Blomquist *et al.*, 1987). Thus, in the Sciaridae, a dipteran family closely related to the Cecidomyiidae (Wood and Borkent, 1989), males of some species have been shown to respond to volatile sex pheromones produced by conspecific females in Y-tube olfactometers, glass cylinders and wind tunnels (Kostelc *et al.*, 1980; Alberts *et al.*, 1981). In contrast, flies in the more advanced suborder Cyclorrhapha produce sex pheromones that typically do not elicit long-range orientation but rather release short-range courtship and copulatory behaviour (Shorey and Bartell, 1970; Carlson *et al.*, 1984) or increase mating strike activity (Mackley and Broce, 1981). One exception to this trend within the Cyclorrhapha is the olive fruit fly, in which males respond over distances of > 1 m to a female-produced sex pheromone (Haniotakis *et al.*, 1977; see Chapter 1).

The use of sex pheromones has been demonstrated for only ten species of Cecidomyiidae (Table 2.1) but presumably is much more extensive. The family Cecidomyiidae is usually divided into three subfamilies (Gagné, 1989, 1994): the Lestremiinae, the Porricondylinae and the Cecidomyiinae. All species in the Lestremiinae and Porricondylinae feed on fungus in decaying organic matter, the primitive feeding habit of the family Cecidomyiidae. There are no records of pheromone use in these two subfamilies, but it must be recognized that species in these subfamilies have not been the subject of studies on mating behaviour.

The Cecidomyiinae comprise the youngest subfamily of gall midges. Its 3850 described species (Gagné, 1989) comprise about 80% of all known cecidomyiid species (R.J. Gagné, Washington, DC, 1998, personal communication). Some of these are fungus feeders, but most are plant feeders or predators. The greatest proportion of species in the Cecidomyiinae can be divided into two monophyletic supertribes: the Lasiopteridi and the Cecidomyiidi. Behavioural evidence for the use of sex pheromones (Table 2.1) has been found in both of these supertribes. Species in the Lasiopteridi that use sex pheromones are in two different genera, *Dasineura* and *Mayetiola*, which are not closely related (R.J. Gagné, Washington, DC, 1998, personal communication). Genera in the supertribe Cecidomyiidi that use sex pheromones include representatives of two different tribes: *Contarinia*, *Thecodiplosis*, *Orseolia* and *Sitodiplosis*.

Evidence for sex pheromones

The use of a chemical attractant by cecidomyiid females was first suggested by H.F. Barnes, a British entomologist who made cecidomyiids his life's study. In a paper on the biology of the bat willow gall midge, Barnes (1932) observed that males were attracted to and excited by leaf tips that had previously been occupied by virgin females. Barnes proposed that mating in the bat willow gall midge was dependent on a 'chemotrophic factor'.

In the 1980s and 1990s, a number of studies were published which provided experimental evidence that cecidomyiid females emit a sex pheromone that attracts males. In these studies, chemical extracts of virgin females were made and tested in one or more of a variety of field and laboratory bioassays. These studies are summarized in Table 2.1.

Table 2.1. Cecidomyiid species that are known to use sex pheromones.

Midge species	Field trapping	Laboratory behavioural assays	Anatomical/ histological studies	Electro-antennograms
Hessian fly *Mayetiola destructor* (Say)	Cartwright (1922) Virgin females	MacKay and Hatchett (1984) Y-tube olfactometer Harris and Foster (1991) Wind tunnel	Solinas and Isidoro (1996)	
Brassica pod midge *Dasineura brassicae* Winnertz	Williams (1990) Virgin females	Williams and Martin (1986) Y-tube olfactometer	Isidoro *et al.* (1992)	
Blackcurrant leaf midge *Dasineura tetensi* (Rübsaamen)	Garthwaite and Wall (1986) Virgin females			
Apple leafcurling midge *Dasineura mali* Kieffer	Harris *et al.* (1996) Virgin females	Harris *et al.* (1996) Wind tunnel		
Douglas-fir cone gall midge *Contarinia oregonensis* Foote	Miller and Borden (1981) Virgin females	Miller and Borden (1984) Still-air cage assay		
Pea midge *Contarinia pisi* Winnertz.	Wall *et al.* (1985) Virgin females and extracts	Hillbur and Löfqvist (1996) Wind tunnel		Wall *et al.* (1985)
Sorghum midge *Contarinia sorghicola* (Coquillett)	Moura *et al.* (1988) Sharma and Vidyasagar (1992) Virgin females		Solinas and Isidoro (1991)	
Orange wheat blossom midge *Sitodiplosis mosellana* (Géhin)	Pivnick (1993) Virgin females and extracts	Pivnick (1993) Y-tube olfactometer		
Pine gall midge *Thecodiplosis japonensis* Uchida and Inouye		Lee and Lee (1985) Traps in cages		
Rice midge *Orseolia oryzae* (Wood-Mason)	Sain and Kalode (1985) Virgin females			

Male mating behaviour

For most of the ten species in Table 2.1, sex pheromone use has been demonstrated through tests that quantify the number of males that have arrived at a pheromone source (i.e. an endpoint of the behaviour: Harris and Foster, 1995). More detailed behavioural responses of males to sex pheromone have been documented for only three species. In one study, Miller and Borden (1984) applied an extract of virgin females of the Douglas-fir cone gall midge to a watch-glass and placed this watch-glass on the floor of a cage containing 20 males. This triggered an increase in male activity, particularly flight back and forth through the column of air above the female extract. Dishes treated with one female equivalent of pheromone extract elicited movement over the surface of the watch-glass and wing fanning from 27% and 18% of the males in the cage, respectively. Neither of these behavioural elements were observed on the solvent-treated control.

In two species, the Hessian fly and the pea midge, the flight responses of males to conspecific female sex pheromone have been observed in a wind-tunnel (Harris and Foster, 1991; Hillbur and Löfqvist, 1996). In response to female extract, the flight manoeuvres of male Hessian flies (Harris and Foster, 1991) were similar to those described for male moths (Baker, 1989), with upwind flight consisting of a combination of zigzags and relatively straight upwind progression. This flight was sometimes punctuated by casting (counterturning with no significant net progression upwind: Kennedy, 1986), with horizontal displacements of c. 10–35 cm, suggesting loss of the odour plume (Baker, 1989). Upon contacting the odour source, males arched the abdomen, in what appeared to be a copulatory attempt. These observations suggest that, during location of a source of sex pheromone, male Hessian flies may use behavioural mechanisms similar to those used by male moths (i.e. optomotor anemotaxis and self-steered counterturning: Kennedy, 1986; Baker, 1989; Baker and Vickers, 1997). Similar behavioural responses have been observed for the pea midge (Hillbur and Löfqvist, 1996).

Once a male locates a female, mating occurs quickly. The male Hessian fly, after landing near a female, orients parallel to the female while wing fanning and bends his abdomen sideways to make contact with the female's gonopore (McKay and Hatchett, 1984). Male and female Hessian flies remain *in copula* for about 10–20 s (McKay and Hatchett, 1984; Bergh *et al.*, 1992). Similar mating behaviour and short copulation times have been reported for other cecidomyiids (Barnes, 1935; Guppy, 1961; Miller and Borden, 1984). However, some cecidomyiid species do have longer copulation times – for example, the orange wheat blossom midge, which stays *in copula* for an average of 171 s (Pivnick and Labbé, 1992).

Given the opportunity, male cecidomyiids will mate many times (Gagné, 1989). The extent of this polygyny has been investigated in the Hessian fly (Bergh *et al.*, 1990, 1992). When put in a cage with 35 females (Fig. 2.2), small Hessian fly males (fresh weight 0.22 mg) successfully inseminated ten females, while larger males (weight 0.77 mg) inseminated all 35 females (Bergh *et al.*, 1990). When allowed to mate 30 females serially, males with fresh weights of 0.33–0.35 mg were capable of fertilizing from 2409 to 3445 eggs (Bergh *et al.*, 1992).

The distances that male cecidomyiids move away from eclosion sites probably varies from species to species. Flight activity of male sorghum midges is focused

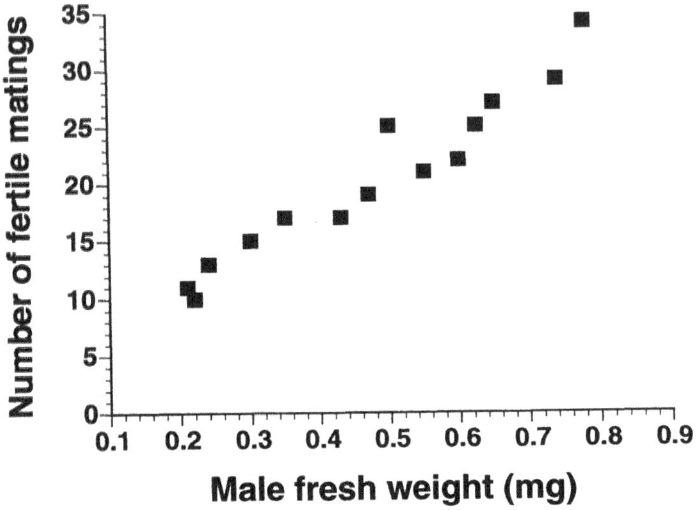

Fig. 2.2. Correlation of male Hessian fly fresh weight (mg) and the number of inseminations resulting in one or more fertilized eggs. (From Bergh *et al.*, 1990.)

around the sorghum head from which the male eclosed, with males only rarely following mated females from eclosion sites to egg-laying sites on other plants (Summers, 1975). However, Enoch (1891) found Hessian fly males 1.5 km away from the nearest barley or wheat field. In the case of the Douglas-fir cone gall midge (Miller and Borden, 1984), it has been suggested that a period of flight activity is required before males become responsive to sex pheromone – activity which could contribute to greater movement of males away from eclosion sites. Brassica pod midge males are caught in traps baited with host-plant odours (Murchie *et al.*, 1997) and, by exhibiting responses to host-plant odours as well as the female-produced sex pheromone, may move greater distances away from eclosion sites. Males of a congeneric species, the apple leafcurling midge, also respond to sex pheromones produced by virgin females but do not respond to host-plant odours (e.g. Harris *et al.*, 1996).

Another feature of the flight of cecidomyiid males is their restricted vertical movement. In several trapping studies of cecidomyiids that eclose from the soil litter (Wall *et al.*, 1985; Williams *et al.*, 1987; Pivnick, 1993), males were caught near ground level but not at greater heights. One apparent exception to limited vertical movement of males is *Contarinia pseudotsugae* Condr. (Condrashoff, 1962). After emerging from the soil litter near the base of a Douglas-fir tree, both males and females were observed flying upwards, following the trunk of the tree until they reached the foliage. Here the females stopped flying and sat on the foliage while males 'swarmed' around the tips of branches. Although it was not seen, the author concluded that mating took place up in the tree rather than on the ground near eclosion sites. Sex pheromone use has not been demonstrated for this species.

Behaviour of virgin females

From our own observations and a limited number of written accounts that describe female behaviour immediately after eclosion (Miller and Borden, 1984; Isidoro *et al.*, 1992; Pivnick and Labbé, 1992; M.O. Harris, 1998, unpublished results), it appears that female cecidomyiids mate near their eclosion sites. For example, in both the Hessian fly and the apple leafcurling midge, females typically walk short distances (1–5 cm), usually to a higher position on the surface soil litter or the plant, before stopping. Soon after stopping, females of these species, as well as many other species (e.g. Miller and Borden, 1984; Solinas and Isidoro, 1991; Isidoro *et al.*, 1992; Pivnick and Labbé, 1992), exhibit a characteristic posture referred to as 'calling'. This posture is characterized by the extension of the ovipositor (Fig. 2.3), a telescopic structure which, when extended to its full length, doubles the body length of the female (Solinas and Isidoro, 1991). Hessian fly females, on average, extend the ovipositor 9 min after eclosing from the pupal case (Bergh *et al.*, 1990).

Several lines of evidence indicate that calling plays a significant role in mating behaviour: (i) its frequency is positively correlated with the frequency of mating within caged groups of flies (McKay and Hatchett, 1984; Miller and Borden, 1984); (ii) calling behaviour is exhibited by virgin but not mated females (McKay and Hatchett, 1984); (iii) only females that call attract males (McKay and Hatchett, 1984); and (iv) the gland that has been identified as the site of production of sex pheromones in cecidomyiids (see discussion in next section; Solinas and Isidoro, 1991, 1996; Isidoro *et al.*, 1992) is only exposed when females call.

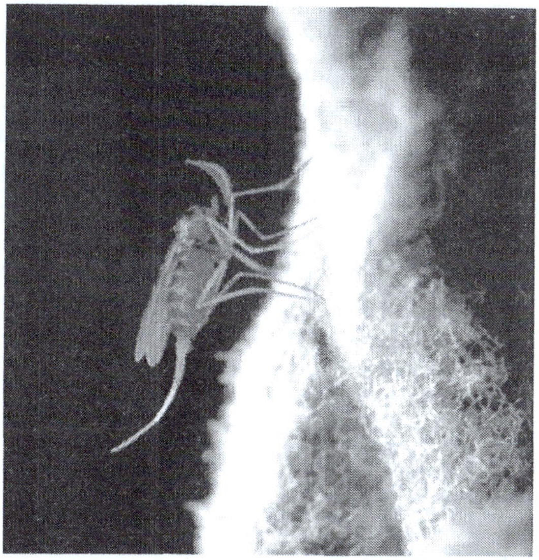

Fig. 2.3. Virgin apple leafcurling midge female exhibiting calling behaviour. Note the extension of the terminal abdominal segments. In a non-calling female, these segments are telescoped into the abdomen (courtesy of M. Heffer).

As mating usually ensues soon after the female ecloses in the field, the periodicity of calling behaviour in virgin females is best studied in the laboratory, where males can be excluded. Of the small number of species that have been studied (e.g. Douglas-fir cone gall midge, Hessian fly, orange wheat blossom midge, brassica pod midge), calling patterns range from erratic to highly regular. Orange wheat blossom midge females begin calling soon after eclosion in the late afternoon and exhibit peak calling during the scotophase (Pivnick and Labbé, 1992). On the day after eclosion, peak calling again occurs during scotophase, with only a small proportion of females calling during the first 8 h of the photophase. If mating does not occur on these first days of adult life, older females continue to call during scotophase for the remaining 3–4 days of their adult life span, but during the latter part of their lives are more likely to extend their calling into the first 8 h of the photophase.

Hessian fly females exhibit a more discrete pattern of calling (Bergh *et al.*, 1990). Females start calling soon after emergence (at 0500 h) and stop calling by about 1130 h (Fig. 2.4a). On the first day following eclosion, there is a strong inverse relationship between the time of female emergence and the duration of calling; thus females that emerge later call for shorter periods of time. Calling begins again the next day at about 0100 h, peaks between 0600 and 0800 h, then declines rapidly so that all females stop calling by 1230 h. On the third day and final day of adult life, half of the surviving virgin females call but the other half do not call and instead begin flying and ovipositing infertile eggs (Fig. 2.4b).

There appear to be substantial differences among species in the duration of individual bouts of calling. Individual females of Hessian fly (Bergh *et al.*, 1990) and orange wheat blossom midge (Pivnick and Labbé, 1992) call continuously for periods of up to 6–20 h, only retracting the ovipositor for short periods when disturbed. In other species, females exhibit a series of short bouts of calling within a particular period. For example, newly emerged female Douglas-fir cone midges call for periods of less than 1 min (Miller and Borden, 1984), with the duration of calling bouts increasing as the (virgin) female ages; females 30–36 h old call continuously for periods of up to 10 min. Our observations of the apple leafcurling midge indicate that it also calls for short periods, with the frequency of these short periods of calling being greatest during the morning hours (M.O. Harris, 1998, unpublished results).

Courtship behaviour between male and female cecidomyiids appears to be remarkably uncomplicated. The male attempts copulation (and generally succeeds) immediately after finding the virgin female, without any apparent movement by the female (Gagné, 1989, 1994). Only in the orange wheat blossom midge has slightly more complex behaviour been observed. Here, copulation attempts by the male are preceded by the female spreading her wings, which allows the male to climb on the female's back and initiate copulation (Pivnick and Labbé, 1992).

Female sex pheromone glands

The source of sex pheromone storage or release in a number of cecidomyiids has been localized to the terminal segments of the abdomen by dissection, solvent

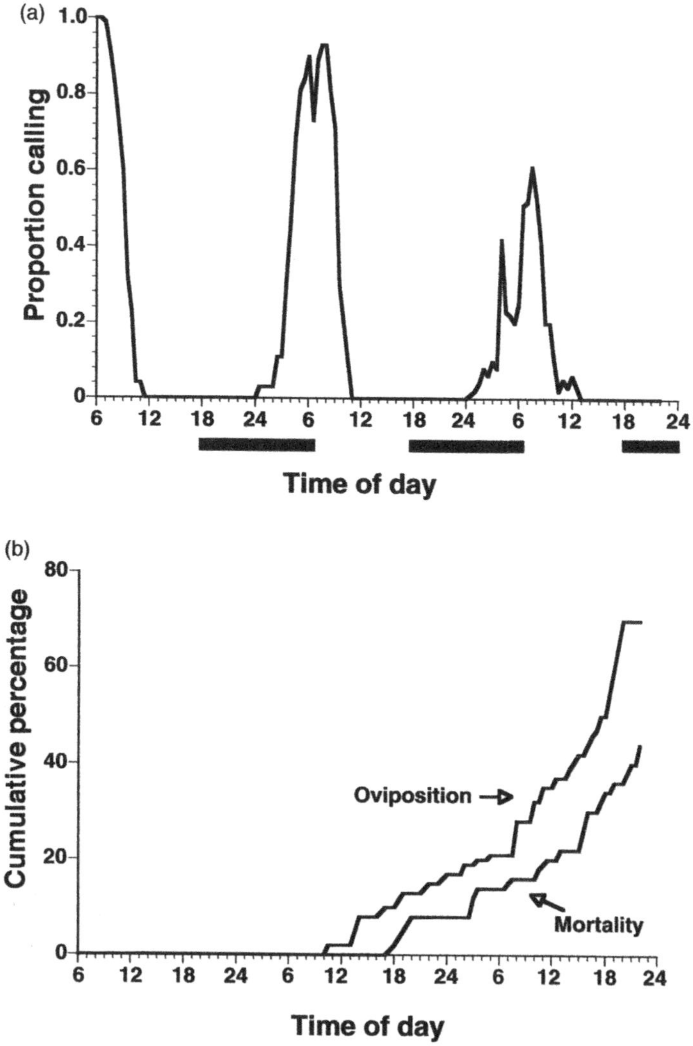

Fig. 2.4. (a) Daily patterns of calling behaviour by Hessian fly virgin females after emerging from pupal sites at 0500 h. Solid bars indicate scotophase. (b) Virgin females exhibit two to three daily periods of calling before they start laying infertile eggs and die. (From Bergh *et al.*, 1990.)

extraction and demonstration that these extracts elicit sexual responses from males (McKay and Hatchett, 1984; Miller and Borden, 1984; Williams and Martin, 1986).

In cecidomyiid species known to use sex pheromones (e.g. sorghum midge: Solinas and Isidoro, 1991; brassica pod midge: Isidoro *et al.*, 1992; Hessian fly: Solinas and Isidoro, 1996), secretory cells have been found in the epidermis of the intersegmental membrane that connects the eighth and ninth abdominal segments. These secretory cells have ultrastructural features typical of pheromone-producing cells, i.e. extensive smooth endoplasmic reticulum, numerous mitochondria with whorled cristae, and secretory vesicles. The secretory cells are hypertrophied only in virgin females; after mating occurs, secretory cells atrophy (Solinas and Isidoro, 1996).

The outer surface of secretory cells consists of grooves and minute protuberances, the latter covered with short microtrichia. Solinas and Isidoro (1996) have suggested that these surface features form a device for storing pheromone when the ovipositor is retracted and for releasing pheromone when the ovipositor is extended.

Behaviour of females after mating

As a general rule, female cecidomyiids appear to be monogamous (Gagné, 1989). In insects, monogamy in females generally results from one or both of two changes that can occur after the female mates: reduced attractiveness of mated females to males and an increased reluctance of females to accept copulation attempts by males (Ringo, 1996). In the Hessian fly, the reduced attractiveness of mated females is the result of the cessation of calling behaviour (McKay and Hatchett, 1984; Bergh *et al.*, 1992) as well as a reduction in sex pheromone production (Foster *et al.*, 1991a). In spite of their reduced attractiveness, mated Hessian fly females are sometimes approached by males (Bergh *et al.*, 1992) but generally avoid mating again by exhibiting various behaviours that are rarely exhibited by virgin females that are approached by males: wing fanning, walking or flying away from the male, and movements of the tip of the abdomen towards whatever substrate the female is sitting on. Mated female orange wheat blossom midges have also been observed to be unreceptive to copulatory attempts by males (Pivnick, 1993).

In the Hessian fly, the change from virgin to mated behaviour is induced by at least two factors (Bergh *et al.*, 1992): the mechanical act of mating, which has a short-term effect on female receptivity, and a male-produced substance that is transferred to the female during mating, which triggers the permanent transition to mated behaviour. The nature of the male-produced substance is unknown, but its effect appears to be unrelated to the presence of sperm in the female spermatheca (Bergh *et al.*, 1992).

The role of swarming in mating

Authors have sometimes referred to 'swarms' of adult midges at emergence, with some of them assuming that mating takes place in these swarms (e.g. apple leaf-curling midge: Todd, 1956; pinyon spindle midge: Houseweart and Brewer, 1972). In two cecidomyiid species, one a fungus feeder, *Anarete pritchardi* Kim. (Chiang *et al.*, 1978) and the other phytophagous species, *Janetiella coloradensis* Felt (Gillet and Brewer, 1983), swarming has been shown to be related to mating behaviour. In these species, males and females form mating pairs while flying within the swarm. In *A. pritchardi*, females appear to mate more than once in the swarm (Chiang *et al.*, 1978).

It is unclear whether mating during swarming and the use of female-produced sex pheromones are compatible mating strategies. The behaviour of virgin females of species that mate in swarms appears to differ from that of species that use sex pheromones. In species using sex pheromones, virgin females rarely fly and are generally inactive except for calling (Miller and Borden, 1984; Harris and Rose, 1991; Pivnick and Labbé, 1992). This can be contrasted with *A. pritchardi*, in which both males and females (presumably virgin) are attracted during flight to the contrasting edge of a highly reflective white surface against a dark background (i.e. a 'swarm marker'; Chiang *et al.*, 1978). The swarm grows in size as individuals arrive at the swarm marker and maintain their flight position relative to the marker. The frequency of mating pairs dropping out of the swarm increases with the size of the swarm.

The Sex Pheromone of the Hessian Fly

In 1991, the first identification of a sex pheromone component of a cecidomyiid species, the Hessian fly, was reported (Foster *et al.*, 1991c). The following section summarizes the identification of the sex pheromone component, as well as work on male flight responses to synthetic compounds and the biosynthesis and regulation of sex pheromone production in this species. Although a sex pheromone component has, to date, been reported only for the Hessian fly, studies on the pea midge are also under way. In the latter, two female-produced compounds have been isolated as potential sex pheromone components, and their chemical structure has been determined (Hillbur and Löfqvist, 1996; Y. Hillbur and J. Löfquist, Alnarp, 1999, personal communication).

Isolation and identification

Earlier work (McKay and Hatchett, 1984) had demonstrated the existence of a female-produced sex pheromone in the Hessian fly. A pentane extract of the terminal segments of the ovipositor of female Hessian flies was fractionated by preparative gas chromatography, and the area of activity isolated by bioassaying fractions using a Y-tube olfactometer. By hydrolysing and reacetylating the extract, and bioassaying the reaction products, the active compound(s) was determined to be an acetate ester. Mass spectrometric analysis of the area of activity revealed a more abundant compound with an M-60 fragment at m/z = 180, which corresponded to $(C_{13}H_{24})^+$, and a less abundant compound with an M-60 fragment at m/z = 182, which corresponded to $(C_{13}H_{26})^+$. Through various microreactions, the structures of the two compounds were determined as, respectively, a Δ10-tridecen-2-yl acetate and tridec-2-yl acetate (Foster *et al.*, 1991c). Independent syntheses of racemates of the two geometric isomers (Millar *et al.*, 1991) allowed the determination that the Δ10-tridecenyl acetate was a 10:90 mixture of the (*Z*)- and (*E*)-10-tridecen-2-yl acetate; only the latter elicited any significant behavioural response from male Hessian flies. Finally, stereoselective syntheses of the two enantiomers (Millar *et al.*, 1991) allowed determination that the natural pheromone component had exclusively the 2*S* configuration (Fig. 2.5).

Fig. 2.5. The known component of the sex pheromone of the Hessian fly, (2*S*)-(*E*)-10-tridecen-2-yl acetate. (From Foster *et al.*, 1991c.)

Behavioural responses to synthetic compounds

In a wind tunnel, male Hessian flies fly upwind to sources loaded with small amounts (2 ng) of the pheromone component, 2*S*-(*E*)-10-tridecen-2-yl acetate (Harris and Foster, 1991). The synthetic component elicits responses from a smaller proportion of the test population than the natural pheromone extract (Fig. 2.6), suggesting that there is at least one other component to the full sex pheromone blend. In tests of various other related compounds, including racemic (*Z*)-10-tridecen-2-yl acetate, racemic (*E*)-10-tridecen-2-ol, racemic tridecanyl acetate and 2*R*-(*E*)-10-tridecen-2-yl acetate, in binary mixtures with 2*S*-(*E*)-10-tridecen-2-yl acetate, no significant increase, over that from the single component alone, in behavioural responses of male Hessian flies occurs. Increasing the amounts of the (2*R*)-enantiomer results in a decrease in the proportion of males responding (Harris and Foster, 1991).

In field trapping trials in New Zealand wheat fields and pastures known to harbour Hessian flies, we have never caught any male Hessian flies in traps baited with the pheromone component, although this compound elicited large catches of

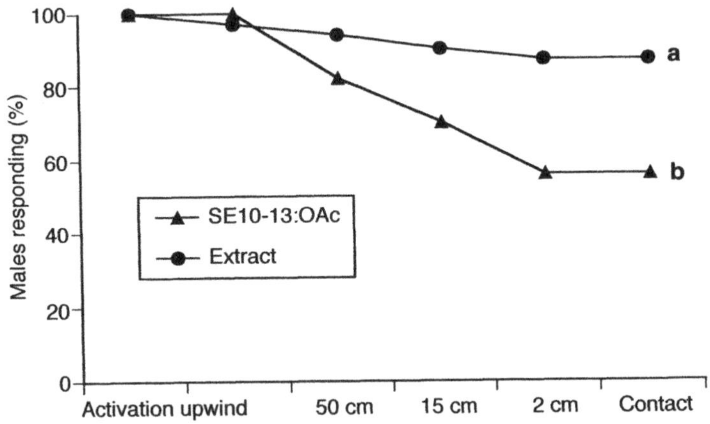

Fig. 2.6. Flight manoeuvres of male Hessian flies exposed to ten female equivalent sources of extract of female ovipositors (●) and 20 ng sources of (2*S*)-(*E*)-10-tridecen-2-yl acetate (SE10–13:OAc, ▲). Activation refers to the percentage of males taking flight. In the graph, the different letters accompanying the mean numbers of males contacting the two sources signify that these means were significantly different at $P < 0.01$. (From Harris and Foster, 1991.)

another, so far unidentified, cecidomyiid (S.P. Foster and M.O. Harris, 1998, unpublished results). Clearly, further study is needed to elucidate a likely additional component(s) of the sex pheromone of the Hessian fly.

Biosynthesis of sex pheromone

By following titres of the component, 2*S*-(*E*)-10-tridecen-2-yl acetate, in the pheromone gland of females, the regulation of sex pheromone biosynthesis in the Hessian fly was studied (Foster *et al.*, 1991a). In virgin females, pheromone was found to be biosynthesized in a cyclical fashion. Thus, early in the photophase and shortly after female eclosion, coinciding with the period of calling and mating, pheromone titre is high (*c.* 2 ng per female). It is lower (0.08 ng per female) some 8 h later in the photophase, a time when virgin females never call. Virgin Hessian fly females live for a maximum of 3 days and on each of these 3 days exhibit this diurnal pattern of pheromone biosynthesis. Pheromone titre declines over the 3-day life span, being highest for 1-day-old females (2 ng per female) and progressively lower in females that are 2 (1.5 ng per female) and 3 days old (1.1 ng per female).

Decapitation of virgin females results in a cessation of pheromone biosynthesis (Foster *et al.*, 1991a). Pheromone biosynthesis is partially restored when decapitated females are injected with either an extract of homogenized female head or with synthetic *Heliothis zea* pheromone biosynthesis activating neuropeptide (PBAN) (Foster *et al.*, 1991a). This suggests that a PBAN or a related peptide may, at least partially, control pheromone biosynthesis in the Hessian fly, as seen in many species of Lepidoptera (Raina, 1993; see chapters in Cardé and Minks, 1997). The identification of the unusual fatty acyl moiety, (*E*)-9-dodecenoate, in the pheromone gland suggested that it may be an intermediate in the biosynthesis of the pheromone component.

In further work on the effects of mating on pheromone biosynthesis, it was shown that mating induces permanent pheromonostasis in female Hessian flies (Foster *et al.*, 1991a). This effect is also induced by injecting male abdomen extract into virgin females, suggesting that the mating-induced pheromonostasis is triggered by the transfer of a chemical factor from the Hessian fly male to the female, as found for other Diptera (e.g. Chen *et al.*, 1988) and Lepidoptera (Kingan *et al.*, 1995).

In summary, pheromone biosynthesis and its regulation in the Hessian fly appears to be remarkably similar to that commonly observed in Lepidoptera and other insects (see chapters in Cardé and Minks, 1997).

Potential Uses for Sex Pheromones in the Management of Cecidomyiid Pests

Why cecidomyiids are difficult to control

Detection

Cecidomyiids are difficult to detect for several reasons. All life stages are extremely small. Adults have highly synchronized eclosion and flight activity, and may be present in a crop for only short periods because of their short life span. The larval stage typically feeds in sheltered places and often produces only subtle damage to the plant – damage which sometimes is only recognized after harvest (e.g. stunting of growth, smaller and fewer seeds: Barnes, 1946, and other volumes in the same series). Consequently, the presence of cecidomyiids in a crop is often detected only after they have become a major problem.

Sporadic outbreaks

Cecidomyiid pests are typically sporadic, rather than perennial, pests (Barnes, 1956), with outbreak populations building up quickly over 1–2 years, e.g. the Hessian fly in the southeastern USA in the 1980s (Buntin, 1990), the orange wheat blossom midge in Saskatchewan, Canada in 1983 (Hinks and Doane, 1988) and the apple leafcurling midge in New Zealand in the 1990s (Harris *et al.*, 1996). Often, there is little research or management devoted to these pests until a major outbreak occurs.

Aestivation, moisture and phenology

Cecidomyiids can survive for long periods when environmental conditions are unfavourable. For example, during dry hot periods, larvae enter a state of aestivation (Readshaw, 1966; Wellso, 1991). Furthermore, the life cycle of cecidomyiids is often strongly influenced by moisture. Moisture triggers the termination of larval aestivation in the soil (Readshaw, 1966) and appears to stimulate larvae to drop from the host plant before entering pupation sites in the soil (Todd, 1956). The strong effects of moisture on cecidomyiid development can reduce the accuracy of day-degree estimates of their life cycle development.

Diapause habits

The length of time cecidomyiids remain in diapause can be extremely variable. In some cecidomyiid species, diapausing larvae held for extended periods have produced adults as many as 13 years after larvae have fed on a plant (Barnes, 1956). This extended diapause complicates pest management for annual crops which are rotated each season to avoid pest problems (e.g. the brassica pod midge, a pest of oilseed rape in Europe: Williams *et al.*, 1987). An additional complication that diapause creates for pest management of cecidomyiids is that, in species that are regarded as multivoltine, a proportion of each generation, including the first generation, may enter diapause and therefore avoid control by mortality factors present or applied only for later generations (Barnes, 1956).

Evolution of virulent biotypes

While host-plant resistance has been very successful in controlling cecidomyiid pest populations (Barnes, 1956 and other volumes), in the Hessian fly the planting of a single resistant cultivar over large geographical areas has resulted in the evolution of populations that can survive on the formerly resistant cultivar (Sosa, 1981). The evolution of these populations, which are referred to as virulent biotypes, has necessitated the monitoring of Hessian fly populations and the development and deployment of cultivars with new resistance genes (Patterson *et al.*, 1992).

Sex pheromones and pest management

Sex pheromones are generally used in one of four ways in pest management programmes: (i) in traps for detecting the presence of the pest in a particular area and the timing of its reproductive activities; (ii) in traps for estimating population levels of a pest, so as to decide if preventive (e.g. host-plant resistance) or curative (e.g. pesticides) methods are necessary and/or economical; (iii) in an attract-and-annihilate method, in which the pest is attracted to a site where it is removed from the environment; and (iv) in a mating disruption method, in which a synthetic version of the pheromone is used to preclude the finding of mates (Foster and Harris, 1997). As relatively little work (especially compared with Lepidoptera) has been carried out on the sex pheromones of cecidomyiids it is not surprising that there has been relatively little work on their use in pest management systems. However, if the sex pheromones of the economically important species were to be identified, synthetic versions of the pheromones could find a prominent place in the management of these pests.

Sex pheromones of cecidomyiid pests would be extremely useful for detection of adult emergence and possibly also for estimating population levels. There are already several examples where traps baited with virgin cecidomyiid females have been used to catch large numbers of conspecific males in the field (Table 2.1). Relative to other methods for monitoring cecidomyiid pests (e.g. egg counts: Harris *et al.*, 1996; water or sticky traps for adults: Williams *et al.*, 1987; Williams, 1990; destructive sampling of plant material to find larvae or pupae: Barnes, 1956), monitoring by pheromone-baited traps is rapid and species-selective, and requires relatively little entomological expertise. Thus, traps baited with synthetic sex pheromones could be used to overcome the current difficulties of detecting cecidomyiid pests. For cecidomyiid pests that emerge as adults in the crop that their offspring will infest (e.g. species attacking perennial crops or crops that are grown in the same place each year), pheromone traps may also provide reliable estimates of population levels.

Compared with monitoring, the potential usefulness of cecidomyiid sex pheromones for direct control of populations by attract-and-annihilate or by mating disruption is less clear. Theoretically, both methods are feasible. Indeed, based on a laboratory study on the Douglas-fir cone gall midge, in which males were less likely to land and examine sources treated with ten versus two female equivalents of an extract of virgin females, Miller and Borden (1984) suggested that disruption of mate-finding behaviour by high concentrations of pheromone might be feasible. Both mating disruption and attract-and-annihilate methods appear to function more effectively when controlling small pest populations (Lanier, 1990; Cardé and Minks, 1995). They would, therefore, appear to have greater utility in preventing outbreaks (which may be economically unfeasible) rather than reducing population outbreaks once they are full-blown.

Another problem for the use of attract-and-annihilate or mating disruption methods against cecidomyiid pests could be the practice of crop rotation. For example, in the case of the brassica pod midge, the spring generation emerges as adults in fields planted with winter wheat, a crop that is commonly planted after oilseed rape has been harvested (Williams *et al.*, 1987). These adults mate in winter wheat fields before the mated females fly to neighbouring fields of flowering winter rape crops and lay eggs in pods (Sylvén, 1970; Williams *et al.*, 1987). Thus, for mating disruption to be effective in the spring, it would need to be carried out in the wheat fields rather than the rape fields. Such a practice may be difficult to implement if different farmers grow the two crops and are unwilling to control an insect which is not a pest of the crop they are currently growing. Moreover, given the extended diapause noted in many cecidomyiid pests, including the brassica pod midge (Williams *et al.*, 1987), the area covered by mating disruption might have to be much larger, covering all fields that had been planted to host plants in the previous 1–3 years. Perhaps, better prospects for direct control methods based on sex pheromones lie in the control of cecidomyiid pests that attack perennial crops or annual crops that are grown in the same place each year.

Conclusions

Female-produced sex pheromones perceived over a distance have been found in a limited number of phytophagous cecidomyiids but are probably widespread among species in the subfamily Cecidomyiinae, which contains about 80% of all known cecidomyiids. From the limited number of studies that have been carried out on the sex pheromone-based systems of cecidomyiids it appears that, in terms of behavioural responses and biosynthesis, they function much like the female-produced sex pheromones of Lepidoptera. The use of cecidomyiid pheromones in pest management has, so far, been limited by the chemical identification of attractive sex pheromone blends. To date, all effective trapping studies in the field have been conducted using either live virgin females or extracts of virgin females.

If they were to be identified and synthesized, the sex pheromones of cecidomyiids may prove to be useful in pest management, particularly for detecting economically threatening pest populations and for timing insecticidal sprays. Furthermore, sex pheromones could be used in combination with other semiochemicals that influence the behaviour of cecidomyiids. As many phytophagous cecidomyiids eclose as adults from pupation sites in the soil and must fly to find oviposition sites on plants, mated adult females presumably have well-developed host-finding behaviour. Furthermore, host-finding behaviour of mated cecidomyiid females is likely to include flight responses to host-plant chemicals, as many cecidomyiids are monophagous or specialize on a group of related plants (Gagné, 1989). Indeed, two cecidomyiid species – the brassica pod midge, which specializes on plants in the family *Brassicaceae* (Pettersson, 1976; Murchie *et al.*, 1997), and the apple leafcurling midge, which specializes on young apple foliage (Galanihe and Harris, 1997) – have been shown to exhibit upwind flight responses to host chemicals. When identified and synthesized, these host-plant chemicals could be used for monitoring cecidomyiid pests, either alone or in combination with sex pheromones. Host-plant chemicals could also prove more effective than sex pheromones for direct control of cecidomyiid populations by the attract-and-annihilate method, in that mated females, rather than adult males, would be targeted.

The chemical identification of cecidomyiid sex pheromones may also prove to be of benefit to the systematics of this diverse dipteran family. This has been the case in the Lepidoptera, where sex pheromones have been used to clarify the phylogenetic relationships among groups, as well as elucidating sibling species (Roelofs and Brown, 1982; Foster *et al.*, 1991b; Löfstedt and Kozlov, 1997). A precedent for this has already been set, with the demonstration that male *Dasineura* midges that developed as immatures on apple foliage will fly upwind to virgin females that developed on apple foliage but do not fly upwind to virgin females that developed on pear foliage (Harris *et al.*, 1996). This has led to the discovery of differences in male genitalia between midges that emerged from apple and pear, and the recognition that the apple leafcurling midge and the pear leafcurling midge are demonstrably distinct species (Gagné and Harris, 1998).

The sex pheromone chemistry and biology of the phytophagous Cecidomyiidae await further investigation.

Acknowledgements

The authors thank Drs Ray Gagné and Ingrid Williams for comments on the manuscript. This work was supported by a grant (Contract No. C06512) from the New Zealand Foundation for Research, Science and Technology.

References

Åhman, I. (1985) Oviposition behaviour of *Dasineura brassicae* on a high- versus a low-quality *Brassica* host. *Entomologia Experimentalis et Applicata* 39, 247–253.

Alberts, S.A., Kennedy, M.K. and Cardé, R.T. (1981) Pheromone-mediated anemotactic flight and mating behavior of the sciarid fly *Bradysia impatiens*. *Environmental Entomology* 10, 10–15.

Baker, T.C. (1989) Sex pheromone communication in the Lepidoptera: new research progress. *Experientia* 45, 248–262.

Baker, T.C. and Vickers, N.J. (1997) Pheromone-mediated flight in moths. In: Cardé, R.T. and Minks, A.K. (eds) *Insect Pheromone Research: New Directions*. Chapman and Hall, New York, pp. 248–264.

Barnes, H.F. (1931) The sex ratio at the time of emergence and the occurrence of unisexual families in the gall midges (Cecidomyiidae, Diptera). *Journal of Genetics* 24, 225–234.

Barnes, H.F. (1932) On the gall midges injurious to the cultivation of willows. I. The bat willow gall midge (*Rhabdophaga terminalis* H. Lw.). *Annals of Applied Biology* 19, 243–252.

Barnes, H.F. (1935) On the gall midges injurious to the cultivation of willows. II. The so-called 'shot hole' gall midges (*Rhabdophaga* spp.). *Annals of Applied Biology* 22, 86–105.

Barnes, H.F. (1946) *Gall Midges of Economic Importance*, Vol. 1, *Gall Midges of Root and Vegetable Crops*. Crosby Lockwood, London, 104 pp.

Barnes, H.F. (1956) *Gall Midges of Economic Importance*, Vol. VII, *Gall Midges of Cereal Crops*. Crosby Lockwood, London, 261 pp.

Bergh, J.C., Harris, M.O. and Rose, S. (1990) Temporal patterns of emergence and reproductive behavior of the Hessian fly (Diptera: Cecidomyiidae). *Annals of the Entomological Society of America* 83, 998–1004.

Bergh, J.C., Harris, M.O. and Rose, S. (1992) Factors inducing mated behavior in female Hessian flies (Diptera: Cecidomyiidae). *Annals of the Entomological Society of America* 85, 224–233.

Blomquist, W.C., Dillwith, J.W. and Adams, T.S. (1987) Biosynthesis and endocrine regulation of sex pheromone production in Diptera. In: Prestwich, G.D. and Blomquist, G.J. (eds) *Pheromone Biochemistry*. Academic Press, Orlando, Florida, pp. 217–250.

Buntin, G.D. (1990) Hessian fly (Diptera: Cecidomyiidae) management in winter wheat using systemic insecticides at planting. *Journal of Agricultural Entomology* 7, 321–331.

Cardé, R.T. and Minks, A.K. (1995) Control of moth pests by mating disruption: successes and constraints. *Annual Review of Entomology* 40, 559–585.

Cardé, R.T. and Minks, A.K. (1997) *Insect Pheromone Research: New Directions*. Chapman and Hall, New York, 684 pp.

Carlson, D.A., Nelson, D.R., Langley, P.A., Coates, T.W., Davis T. L. and Leegwater-van der Linden, M.E. (1984) Contact sex pheromone in the tsetse fly *Glossina pallidipes* (Austen): identification and synthesis. *Journal of Chemical Ecology* 10, 429–450.

Cartwright, W.B. (1922) Sexual attraction of the female Hessian fly (*Phytophaga destructor* Say). *Canadian Entomologist* 54, 154–155.

Chen, P.S., Stumm-Zollinger, E. , Aigaki, T. , Balmer, J., Bienz, M. and Bohlen, P. (1988) A male accessory gland peptide that regulates reproductive behavior of female *D. melanogaster*. *Cell* 54, 291–298.

Chiang, H.C., Mettler, B.J., Okubo, A. and Robbins, A.S. (1978) Coupling of midge individuals in a swarm, *Anarete pritchardi* (Diptera: Cecidomyiidae). *Annals of the Entomological Society of America* 71, 859–861.

Condrashoff, S.F. (1962) Bionomics of three closely-related species of *Contarinia* Rond. (Diptera: Cecidomyiidae) from galls on Douglas-fir needles. *Canadian Entomologist* 94, 376–394.

Enoch, F. (1891) The life history of the Hessian fly *Cecidomyia destructor* Say. *Transactions of the Entomological Society, London*, Part II, 329–366.

Foster, S.P. and Harris, M.O. (1997) Behavioral manipulation methods for insect pest-management. *Annual Review of Entomology* 42, 123–146.

Foster, S.P., Bergh, J.C., Rose, S. and Harris, M.O. (1991a) Aspects of pheromone biosynthesis in the Hessian fly, *Mayetiola destructor* (Say). *Journal of Insect Physiology* 12, 899–906.

Foster, S.P., Dugdale, J.S. and White, C.S. (1991b) Sex pheromones and the status of the greenheaded and brownheaded leafroller moths in New Zealand. *New Zealand Journal of Zoology* 17, 63–74.

Foster, S.P., Harris, M.O. and Millar, J.G. (1991c) Identification of the sex pheromone of the Hessian fly, *Mayetiola destructor* (Say). *Naturwissenschaften* 78, 130–131.

Gagné, R.J. (1989) *The Plant-feeding Gall Midges of North America*. Cornell University Press, Ithaca, New York, 356 pp.

Gagné, R.J. (1994) *The Gall Midges of the Neotropical Region*. Cornell University Press, Ithaca, New York, 352 pp.

Gagné, R.J. and Harris, M.O. (1998) The distinction between *Dasineura* spp. (Diptera: Cecidomyiidae) from apple and pear. *Proceedings of the Entomological Society of Washington* 100, 445–448.

Galanihe, L.D. and Harris, M.O. (1997) Plant volatiles mediate host-finding behavior of the apple leafcurling midge. *Journal of Chemical Ecology* 23, 2639–2655.

Gallun, R.L. and Hatchett, J.H. (1969) Genetic evidence of elimination of chromosomes in the Hessian fly. *Annals of the Entomological Society of America* 62, 1095–1101.

Garthwaite, D.G. and Wall, C. (1986) Further evidence for a female sex pheromone in the blackcurrant leaf midge *Dasineura tetensi*. *British Crop Protection Conference – Pests and Diseases*, Vol. 1. British Crop Protection Council, Surrey, UK, pp. 355–357.

Gillet, C. and Brewer, J.W. (1983) Biology of *Janetiella coloradensis* Felt, a needle gall midge on *Pinus edulis* Engelm. *Zeitschrift für Angewandte Entomologie* 95, 326–335.

Guppy, J.C. (1961) The life history, behaviour, and ecology of the clover seed midge, *Dasyneura leguminicola* (Lint.) (Diptera: Cecidomyiidae), in eastern Ontario. *Canadian Entomologist* 93, 59–73.

Haniotakis, G.E., Mazomenos, B.E. and Tumlinson, J.H. (1977) A sex attractant of the olive fruit fly, *Dacus oleae*, and its biological activity under laboratory and field conditions. *Entomologia Experimentalis et Applicata* 21, 81–87.

Harris, K.M. (1976) The sorghum midge. *Annals of Applied Biology* 84, 114–118.

Harris, M.O. and Foster, S.P. (1991) Wind tunnel studies of sex pheromone-mediated behavior of the Hessian fly (Diptera: Cecidomyiidae). *Journal of Chemical Ecology* 17, 2421–2435.

Harris, M.O. and Foster, S.P. (1995) Behavior and integration. In: Cardé, R.T. and Bell, W.J. (eds) *Chemical Ecology of Insects 2*. Chapman and Hall, New York, pp. 3–46.

Harris, M.O. and Rose, S. (1989) Temporal changes in the egglaying behavior of the Hessian fly. *Entomologia Experimentalis et Applicata* 53, 17–29.

Harris, M.O. and Rose, S. (1991) Factors influencing the onset of egglaying in a cecidomyiid fly. *Physiological Entomology* 16, 183–190.

Harris, M.O., Foster, S.P., Agee, K. and Dhana, S. (1996) Sex pheromone communication in the apple leafcurling midge (*Dasineura mali*). *Proceedings of the New Zealand Plant Protection Conference* 49, 52–58.

Hillbur, Y. and Löfqvist, J. (1996) Progress in the identification of the sex pheromone of the pea midge, *Contarinia pisi* (Diptera: Cecidomyiidae). *Abstract of the International Society of Chemical Ecology*, 13th Annual Meeting, Prague, Czech Republic, August 18–22, p. 190.

Hinks, C.F. and Doane, J.F. (1988) Observations on rearing and diapause termination of *Sitodiplosis mosellana* (Diptera: Cecidomyiidae) in the laboratory. *Journal of Economic Entomology* 81, 1816–1818.

Houseweart, M.W. and Brewer, J.W. (1972) Biology of a Pinyon spindle gall midge (Diptera: Cecidomyiidae). *Annals of the Entomological Society of America* 65, 331–336.

Isidoro, N., Williams, I.H., Solinas, M. and Martin, A. (1992) Mating behaviour and identification of the female sex pheromone gland in the brassica pod midge (*Dasineura brassicae* Winn.: Cecidomyiidae, Diptera). *Bolletino dell'Istituto de Entomologia 'Guido Grandi' della Università di Bologna* 47, 27–48.

Kennedy, J.S. (1986) Some current issues in orientation to odour sources. In: Payne, T.L., Birch, M.C. and Kennedy, C.E.J. (eds) *Mechanisms in Insect Olfaction*. Clarendon Press, Oxford, pp. 11–26.

Kingan, T.G., Bodnar, W.M., Raina, A.K., Shabanowitz, J. and Hunt, D.F. (1995) The loss of female sex pheromone after mating in the corn earworm moth *Helicoverpa zea*: identification of a male pheromonostatic peptide. *Proceedings of the National Academy of Sciences, USA* 92, 5082–5086.

Kostelc, J.G., Girard, J.E. and Hendry, L.B. (1980) Isolation and identification of a sex attractant of a mushroom-infesting sciarid fly. *Journal of Chemical Ecology* 6, 1–11.

Lanier, G.N. (1990) Principles of attraction-annihilation: mass trapping and other means. In: Ridgway, R.L., Silverstein, R.M. and Inscoe, M.N. (eds) *Behavior-modifying Chemicals for Pest Management: Applications of Pheromones and Other Attractants*. Marcel Dekker, New York, pp. 25–45.

Lee, C. and Lee, H.P. (1985) Studies on the sex pheromone and antennal ultrastructure of the pine gall midge (*Thecodiplosis japonensis* Uchida et Inouye). *Korean Journal of Entomology* 15, 31–40.

Löfstedt, C. and Kozlov, M. (1997) A phylogenetic analysis of pheromone communication in primitive moths. In: Cardé, R.T. and Minks, A.K. (eds) *Insect Pheromone Research: New Directions*. Chapman and Hall, New York, pp. 473–489.

Mackley, J.W. and Broce, A.B. (1981) Evidence of a female sex recognition pheromone in the screwworm fly. *Environmental Entomology* 10, 405–408.
Mamaev, B.M. and Krivosheina, N.P. (1965) *The Larvae of the Gall Midges (Diptera, Cecidomyiidae)*. A.A. Balkema, Rotterdam, 278 pp.
McKay, P.A. and Hatchett, J.H. (1984) Mating behavior and evidence of a female sex pheromone in the Hessian fly, *Mayetiola destructor* (Say) (Diptera: Cecidomyiidae). *Annals of the Entomological Society of America* 77, 616–620.
Millar, J.G., Foster, S.P. and Harris, M.O. (1991) Synthesis of the stereoisomers of the female sex pheromone of the Hessian fly, *Mayetiola destructor*. *Journal of Chemical Ecology* 17, 2437–2447.
Miller, G.E. and Borden, J.H. (1981) Evidence for a sex pheromone in the Douglas-fir cone gall midge. *Canadian Forest Service Research Notes* 1, 9–10.
Miller, G.E. and Borden, J.H. (1984) Reproductive behaviour of the Douglas-fir cone gall midge, *Contarinia oregonensis* (Diptera: Cecidomyiidae). *Canadian Entomologist* 116, 607–618.
Moura, J.I.L., Vilela, E.F. and Waquil, J.M. (1988) Sexual attraction in the field in the sorghum midge, *Contarinia sorghicola* (Coquillet), (Diptera: Cecidomyiidae). *Anais Society Entomology Brasil* 17, 125–132.
Murchie, A.K., Smart, L.E. and Williams, I.H. (1997) Responses of *Dasineura brassicae* and its parasitoids *Platygaster subuliformis* and *Omphale clypealis* to field traps baited with organic isothiocyanates. *Journal of Chemical Ecology* 23, 917–926.
Patterson, F.L., Foster, J.E., Ohm, H.W., Hatchett, J.H. and Taylor, P.L. (1992) Proposed system of nomenclature for biotypes of Hessian fly (Diptera: Cecidomyiidae) in North America. *Journal of Economic Entomology* 85, 307–311.
Pettersson, J. (1976) Ethology of *Dasyneura brassicae* Winn. (Dipt., Cecidomyiidae). I. Laboratory studies of olfactory reactions to the host-plant. *Symposia Biologica Hungarica* 16, 203–208.
Pivnick, K.A. (1993) Response of males to male sex pheromone in the orange wheat blossom midge, *Sitodiplosis mosellana* (Gehin) (Diptera: Cecidomyiidae). *Journal of Chemical Ecology* 19, 1677–1689.
Pivnick, K.A. and Labbé, E. (1992) Emergence and calling rhythms, and mating behaviour of the orange wheat blossom midge, *Sitodiplosis mosellana* (Gehin) (Diptera: Cecidomyiidae). *Canadian Entomologist* 124, 501–507.
Raina, A.K. (1993) Neuroendocrine control of sex pheromone biosynthesis in Lepidoptera. *Annual Review of Entomology* 38, 329–349.
Readshaw, J.L. (1966) The ecology of the swede midge, *Contarinia nasturtii* (Kieff.) (Diptera: Cecidomyiidae). I. Life history and influence of temperature and moisture on development. *Bulletin of Entomological Research* 56, 685–700.
Ringo, J. (1996) Sexual receptivity in insects. *Annual Review of Entomology* 41, 473–494.
Roelofs, W.L. and Brown, R.L. (1982) Pheromones and evolutionary relationships of Tortricidae. *Annual Review of Ecology and Systematics* 13, 395–422.
Sain, M. and Kalode, M.B. (1985) Traps to monitor gall midge population in rice. *Current Science* 54, 876–877.
Sharma, H.C. and Vidyasagar, P. (1992) Orientation of males of sorghum midge, *Contarinia sorghicola* to sex pheromones from virgin females in the field. *Entomologia Experimentalis et Applicata* 64, 23–29.
Shorey, H.H. and Bartell, R.J. (1970) Role of a volatile female sex pheromone in stimulating male courtship behavior in *Drosophila melanogaster*. *Animal Behavior* 18, 159–164.
Slifer, E.H. and Sekhon, S.S. (1971) Circumfila and other sense organs on the antenna of the sorghum midge (Diptera, Cecidomyiidae). *Journal of Morphology* 133, 281–302.

Solinas, M. and Isidoro, N. (1991) Identification of the female sex pheromone gland in the sorghum midge *Allocontarinia sorghicola* (Coq.) Solinas (Diptera: Cecidomyiidae). *Redia* 74, 441–466.

Solinas, M. and Isidoro, N. (1996) Functional anatomy of female sex pheromone gland of *Mayetiola destructor* Say (Diptera: Cecidomyiidae). *Entomologia, Università di Bari* 30, 43–54.

Solinas, M. and Nuzzaci, G. (1987) Antennal sensilla of *Mycodiplosis erysiphes* Ruebs. (Cecidomyiidae, Diptera). *Bolletino dell'Istituto de Entomologia 'Guido Grandi' della Università di Bologna*, 41, 173–194.

Sosa, O. (1981) Biotypes J and L of Hessian fly discovered in an Indiana wheat field. *Journal of Economic Entomology* 74, 180–182.

Stuart, J.J. and Hatchett, J.H. (1991) Genetics of sex determination in the Hessian fly. *Journal of Heredity* 82, 43–52.

Summers, C.G. (1975) Daily adult emergence in the sorghum midge, *Contarinia sorghicola*. *Environmental Entomology* 4, 495–498.

Sylvén, E. (1970) Field movement of radioactively labelled adults of *Dasyneura brassicae* Winn. (Dipt., Cecidomyiidae). *Entomologia Scandinavica* 1, 161–187.

Todd, D.H. (1956) A preliminary account of *Dasyneura mali* Kieffer (Cecidomyiidae: Dipt.), and an associated hymenopterous parasite in New Zealand. *New Zealand Journal of Science and Technology* A37, 462–464.

Wall, C., Pickett, J.A., Garthwaite, D.G. and Morris, N. (1985) A female sex-pheromone in the pea midge, *Contarinia pisi*. *Entomologia Experimentalis et Applicata* 39, 11–14.

Wellso, S.G. (1991). Aestivation and phenology of the Hessian fly (Diptera: Cecidomyiidae). *Environmental Entomology* 20, 795–801.

Williams, I.H. (1990) Pheromone monitoring of brassica pod midge (*Dasineura brassicae* Winn.). *SROP/WPRS Bulletin* 13, 40–45.

Williams, I.H. and Martin, A.P. (1986) Evidence for a female sex pheromone in the brassica pod midge *Dasineura brassicae*. *Physiological Entomology* 11, 353–356.

Williams, I.H., Martin, A.P. and Kelm, M. (1987) The phenology of the emergence of brassica pod midge (*Dasineura brassicae* Winn.) and its infestation of winter oil-seed rape (*Brassica napus* L.). *Journal of Agricultural Science, Cambridge* 108, 579–589.

Withers, T.M. and Harris, M.O. (1997) Influence of wind on Hessian fly (Diptera: Cecidomyiidae) flight and egglaying behavior. *Environmental Entomology* 26, 327–333.

Wood, D.M. and Borkent, A. (1989) Phylogeny and classification of the Nematocera. In: McAlpine, J.F. and Wood, D.M. (eds) *Manual of Nearctic Diptera*. Agriculture Canada, Hull, Quebec, pp. 1333–1370.

3 Scarab Beetles

WALTER S. LEAL

Laboratory of Chemical Prospecting, National Institute of Sericultural and Entomological Science, 1–2 Ohwashi, Tsukuba 305–8634, Japan

Introduction

The family Scarabaeidae, one of the eight families of the Scarabaeoidea, encompasses about 2000 genera and 25,000 species, usually placed in about 20 subfamilies. Four of these subfamilies, the monophyletic plant-feeding groups Rutelinae, Melolonthinae, Cetoniinae and Dynastinae, have the most notorious species because several of their members are agricultural pests throughout the world. Adult chafers are strictly phytophagous with different feeding habitats, although there are species that do not feed at all in the adult stage. Grubs feed on living roots and many are destructive to grasses, landscape plants and crops. Dynastinae, with conspicuous horns in males, usually attack stems or roots of plants, while Cetoniinae prefer the nectar, sap or juice of ripening fruits and vegetables (Sawada, 1991). Rutelinae and Melolonthinae mostly attack fresh leaves. Cetoniinae and Rutelinae are common flower visitors.

Scarabs are widespread and endemic species can be found in a wide range of environmental and economic niches, such as the European chafer, *Melolontha* sp. in Europe, *Cotinis nitida* (L.) and *Cyclocephala* spp. in the USA, the cupreous chafer, *Anomala cuprea* Hope in Japan and the grass grub, *Costelytra zealandica* (White) in New Zealand, to name just a few. Accidental introduction of exotic species has also led to a steady increase in the number of problems with scarab beetles around the world (Jackson, 1992), the most notable being the Japanese beetle *Popillia japonica* Newman in the USA and Azores, *Maladera matrida* Argaman in Israel, the oriental beetle *Exomala orientalis* (Waterhouse) in continental USA and Hawaii, and the Asiatic garden beetle *Maladera castanea* Argaman in the USA (Leal, 1998).

Although a few studies have been reported on the chemical ecology of dung beetles (Scarabaeinae), most of the emphasis of research programmes on chemical communication in scarabs has been focused on the subfamilies Cetoniinae, Melolonthinae, Dynastinae and Rutelinae because of their economic importance as agricultural and/or turf pests. In general, these research projects have been restricted to the development of attractants (pheromones or food-type compounds) for possible applications in management programmes. By contrast, research in our laboratory is aimed at increasing knowledge of the chemical ecology of scarab beetles in order to pave the way for the development of environmentally sound control strategies. We have taken a comprehensive approach, encompassing the characterization, synthesis and application of pheromones, and studies related to the biology, biosynthesis and physiology of pheromone production. In addition, we have been investigating the molecular mechanisms of early olfactory processing in order to explore new frontiers for pest control. This contribution summarizes some recent achievements in the identification, biology and application of scarab sex pheromones and the molecular basis of olfaction.

Chemical Communication in Scarab Beetles

Chemical communication involves the production and release of specific chemicals (semiochemicals) by the emitter, and the detection and olfactory processing of these signals leading to appropriate behavioural responses in the receiver. In scarab beetles, chemical attraction is the major means of sexual recruitment, with females normally being the emitters and males being the receivers; in this case these semiochemicals are referred to as sex pheromones.

Pheromone chemistry

Recent studies have led to the identification of sex pheromones of various species in the Rutelinae (Table 3.1) and Melolonthinae (Table 3.2). Initially, it was suggested that the sex pheromones of the former are fatty acid derivatives, whereas the semiochemicals of the latter are phenolic-, terpenoid- and amino acid-derived compounds (Leal, 1997). However, exceptions to this rule have recently been found in both groups. For example, the sex pheromone of the yellowish elongate chafer, *Heptophylla picea* Motschulsky, which has been identified as (*R*,*Z*)-7,15-hexadecadien-4-olide (Leal *et al.*, 1996c), is probably biosynthesized from stearic acid. On the other hand, the sex pheromone for *Phylloperttha diversa* Waterhouse (Rutelinae) is an intriguing alkaloid compound with medicinal properties (Leal *et al.*, 1997a).

Isolation of chemical communication channels

Utilizing pheromone blends that consist of just a few semiochemicals or even a single constituent, closely related scarab species have attained separate chemical communication channels and reproductive isolation. Some species that are geographically and/or temporally isolated even use the same sex pheromone system. *A. cuprea*, *A. octiescostata* Burmeister and *A. albopilosa sakishimana* Nomura produce a pheromone blend of buibuilactone and japonilure. Both *A. cuprea* and *A. octiescostata* occur on the mainland of the Japanese archipelago (Honshu), but the flight activity of the former occurs in summer, whereas the latter is active in early spring. *A. albopilosa sakishimana* is native to Miyako Island (Okinawa) in the south of Japan. Thus, for these species there was probably limited selection pressure

Table 3.1. Sex pheromones of rutelines.

Popillia japonica	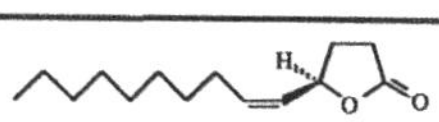	Tumlinson *et al.* (1977)
Anomala rufocuprea	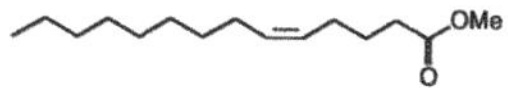	Tamaki *et al.* (1985)
A. cuprea	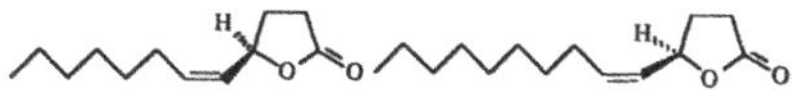	Leal (1991); Leal *et al.* (1993a)
A. schonfeldti		Leal *et al.* (1992b)
A. daimiana	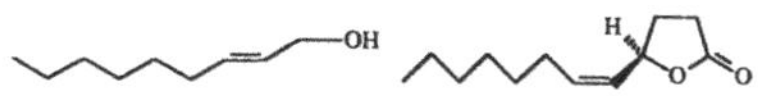	Leal *et al.* (1993b)
Exomala orientalis		Leal (1993), Leal *et al.* (1994c)
A. octiescostata	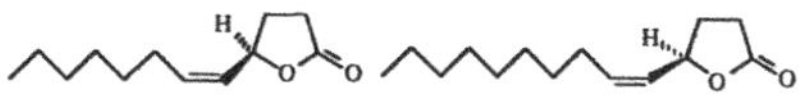	Leal *et al.* (1994a)
A. albopilosa sakishimana	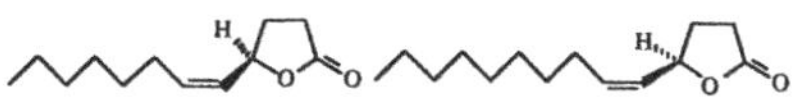	Leal *et al.* (1994b)
A. albopilosa albopilosa	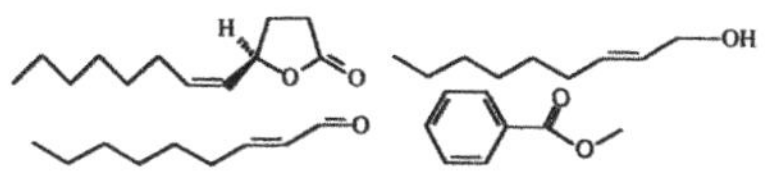	Leal *et al.* (1996b)
A. osakana	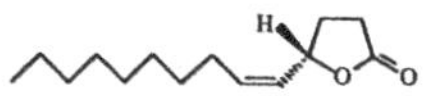	Leal (1996)
Phyllopertha diversa	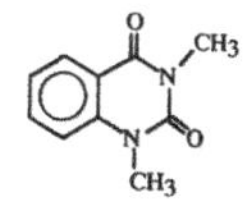	Leal *et al.* (1997a)

Table 3.2. Sex pheromones of melolonthines.

Costelytra zealandica	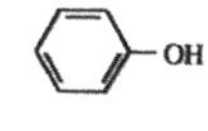	Henzell and Lowe (1970)
Holotrichia parallela	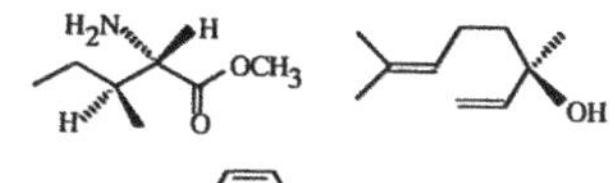	Leal *et al.* (1992a, 1993c)
H. consanguinea	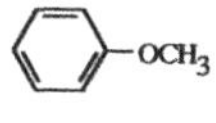	Leal *et al.* (1996a)
Heptophylla picea	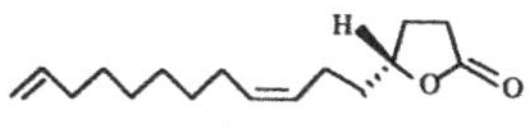	Leal *et al.* (1996c)
Maladera matrida	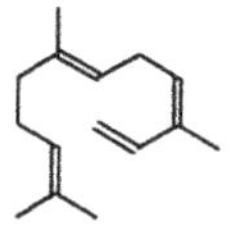	Yarden *et al.* (1996)
Phyllophaga anxia	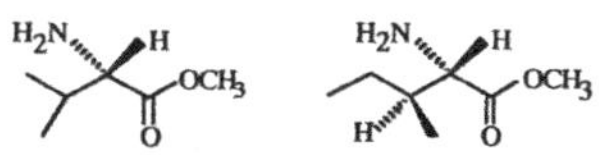	Zhang *et al.* (1997)
P. elemans	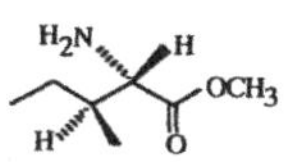	Leal *et al.* (unpublished data)

for the evolution of sex pheromone systems with different constituents (Leal, 1997). On the other hand, *A. albopilosa albopilosa* Hope, which shares the same habitat and has a common flight season with *A. cuprea*, has a more complex pheromone blend. In addition to the major constituent, buibuilactone, it utilizes 2-(E)-nonenol, 2-(E)-nonenal and methyl benzoate (Leal *et al.*, 1996b) (Fig. 3.1). Although the chemical nature of the pheromones provides species-specific chemical signals for these two species, temporal differences in mating activities also play a part in their reproductive isolation. While the diel rhythm of pheromone release in *A. albopilosa albopilosa* shows a peak at the beginning of the photophase (Leal *et al.*, 1996b), pheromone release in *A. cuprea* reaches a peak late in the photophase (W.S. Leal, 1998, unpublished results). Species specificity of the chemical communication channels of the northern and southern masked chafers, *Cyclocephala borealis* Arrow and *C. lurida* Bland, is also attained by temporal rather than chemical differences in the pheromone systems (Potter, 1980).

Agonist–antagonist activities of enantiomeric pheromones

In order to achieve effective chemical communication in an environment rich in potential chemical cues, insects must overcome the problem of recognizing the signal buried in background noise. Moths have overcome this problem by utilizing blends of similar compounds, where the exact ratio and the specific signature of the whole blend are critical factors for eliciting sexual behaviour. By contrast, a number of scarab beetles utilize single constituents as their sex pheromones.

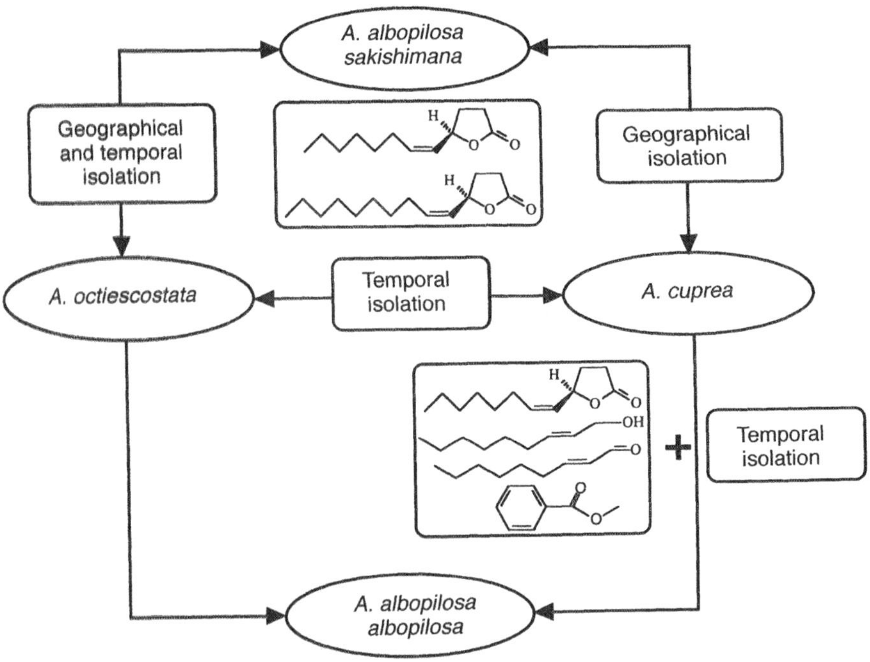

Fig. 3.1. Mechanism of separation of the chemical communication channels among *Anomala* species native to Japan.

However, in some of these cases the pheromone molecule has a chiral centre and chirality plays a pivotal role in determining the pheromone specificity. The identification of the sex pheromone of the Japanese beetle, (*R*)-japonilure, led to the first recognition of the importance of chirality in chemical communication in scarab beetles. Male response to the synthetic pheromone was dramatically reduced by the presence of the (*S*)-stereoisomer (Tumlinson *et al.*, 1977), but the ecological significance of this inhibition remained speculative. The recent identification of the sex pheromone for the Osaka beetle, *Anomala osakana* Sawada, revealed an interesting mechanism of mutual agonist–antagonist activities of enantiomeric pheromones. The Osaka beetle produces and responds only to (*S*)-japonilure, the activity of which is completely inhibited by the presence of the Japanese beetle sex pheromone, (*R*)-japonilure (Leal, 1996). Chiral gas chromatography with an electroantennographic detector (Leal *et al.*, 1995) demonstrated that both the Japanese and the Osaka beetles had olfactory receptor neurones tuned to each enantiomer of japonilure. The two semiochemicals are recognized by the same species; the enantiomer produced by conspecific females is detected as a sex pheromone (agonist), whereas the other species' sex pheromone is recognized as a behavioural antagonist (Fig. 3.2).

The occurrence of specific olfactory receptor neurones in the antennae of scarab beetles, leading to the ability to detect the stereoisomer not produced by females of the same species, is more an exception than a rule. Recently, it has been demonstrated that the antennae of *A. cuprea* and *A. octiescostata* do not possess olfactory receptor neurones that can detect the non-natural stereoisomer of the

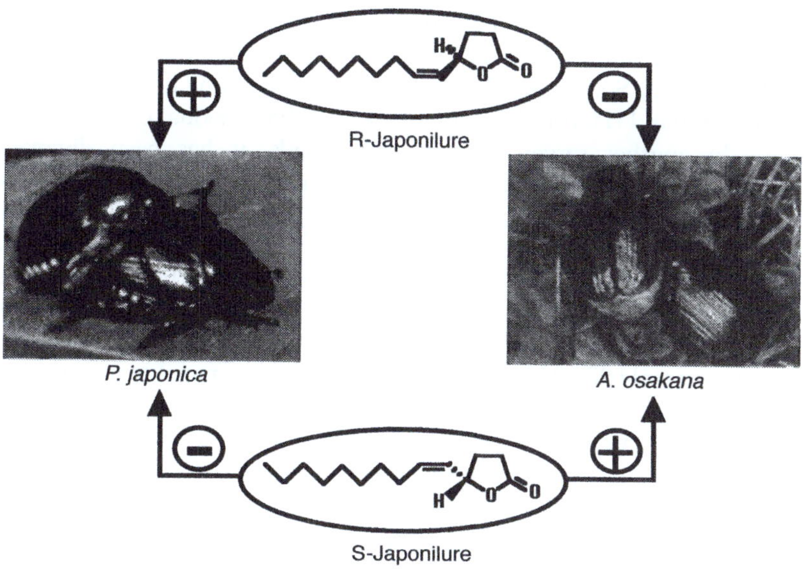

Fig. 3.2. Schematic view of a mechanism of mutual agonist–antagonist activities of enantiomeric pheromones.

pheromone (W.S. Leal, 1998, unpublished results). In addition, field experiments have suggested that both *Holotrichia parallela* (Motschulsky) and *Heptophylla picea* are also 'blind' to the non-natural stereoisomer of their pheromone. These findings shed more light on the evolutionary significance of the mutual behavioural agonist–antagonist activities of enantiomeric pheromones in the Japanese and the Osaka beetles.

Other attractants

The search for plant kairomones that could have a direct role in scarab population control and could be used to survey some species was carried out a few decades before the first insect pheromone was identified. These programmes have been successful in developing attractants for the Japanese beetle (reviewed in Leal, 1998), which are now utilized in combination with japonilure. Generally, plant kairomones are less attractive to scarab beetles than sex pheromones, but there are a few exceptions. Methyl anthranilate, for example, has been demonstrated to be a 2.4 times more potent attractant for males of *Anomala rufocuprea* Motschulsky than the female-produced methyl (Z)-5-tetradecenoate, whereas female catches with the kairomone were 65 times higher (Imai *et al.*, 1997). The proportion of females in pheromone-baited traps (4.7%) increased dramatically in the traps loaded with kairomone (57.5%). More importantly, a combination of the kairomone with the synthetic sex pheromone showed a synergistic effect. Total capture of beetles in traps baited with sex pheromone plus kairomone was 5 times higher than the synthetic sex pheromone alone and 1.8 times more than the kairomone only. Interestingly, the attractiveness of this semiochemical showed tremendous variation among populations of the soybean beetle in eight geographical regions of the Japanese archipelago. Methyl anthranilate was not active at Memuro (Hokkaido, North), showed only low attractiveness at Kushira (Kagoshima, South) and had medium to very strong activity in various parts of the mainland (Honshu) (T. Imai, Tokyo, 1998, personal communication).

In another survey of floral compounds, it has been discovered that 2-phenylethanol is a potent attractant to the long-legged chafer, *Hoplia communis* Waterhouse (Imai *et al.*, 1998). Adults of this species, which normally feed on whitish flowers, reach high populations in May and become a nuisance; they are highly attracted to any whitish surfaces, especially the T-shirts of golf players. During the peak of the flight season, traps became full of beetles in 1 h, and > 20,000 beetles were captured in a single trap per day. The total catch over 6 days, approximately 90,000 beetles, was close to the estimated maximum population per 1000 m^2 (Imai *et al.*, 1998).

By and large, the search for attractants is ecologically oriented, but sometimes the lead for new potential lures is quite unexpected; for example, *A. octiescostata* is attracted to a cigarette brand, Mild Seven® (Leal, 1998). It was initially observed that this species is attracted to cigarette stubs; later experiments showed that beetles are also attracted to hexane extracts and to the smoke generated by the cigarette of the same brand. It is not known whether the attraction is due to chemicals derived from the tobacco leaves or from flavour compounds used in cigarette preparation (W.S. Leal, 1998, unpublished results).

Normally, lures are developed after the isolation of the active compounds from natural sources, but sometimes the sources themselves may be used for attracting beetles. In Indonesia, the observation that adults of *Leucopholis rorida* Fabricius are attracted by the smell of crushed fruits of red pepper, *Capsicum annuum* L., provided a method for the collection of beetles to be utilized for feeding chickens and ducks (Jongeleen and Mahrub, 1979).

Application of Pheromones

Some of the scarab pheromones are now commercially available for monitoring their populations in golf courses and agricultural areas. Although there is no direct evidence that application of pheromones has an effect on suppressing scarab populations, some pheromone systems have been used in Japan for pest control. It is difficult to obtain accurate data on the effect of a treatment with pheromones (or other semiochemicals) on scarab populations. In lepidopteran insects, the effect of the treatment can be estimated by the number of eggs laid on treated versus control areas. This is technically very difficult (if not impossible) with scarab beetles, since females lay eggs in the soil. We are left with the choices of estimating the effect of these treatments by the 'decrease in the number of beetles captured' in an area treated and/or 'decrease in larval populations' following the treatment(s). One has to be careful when interpreting these data because these estimates may be misleading. First, lower numbers of beetles captured may not necessarily be translated into a decrease in the next adult population. Secondly, larval populations of scarab beetles do fluctuate a great deal over the years even in areas with no treatment at all. These drawbacks may be retarding progress in the application of pheromones and other semiochemicals for control of scarab populations. It is highly desirable that these programmes be conducted for a reasonable length of time so that accurate conclusions can be drawn.

The identification of pheromones for various species created a stimulus for the application of pheromones in monitoring and for tentative programmes of mass trapping. Initially, farmers and administrators of golf courses were reluctant to test the feasibility of semiochemicals, because they would be able to attack only one or two species out of a number of scarab species. The identification of pheromones and kairomones for the most important species and the fact that environmentally sound strategies are badly needed, stimulated the use of pheromones in pest monitoring and management. Fortunately, some research groups motivated by this change have started large-scale projects in order to find ways and means of using pheromones and other semiochemicals for the control of scarab beetles.

The research group at Japan Tobacco launched a project aimed at the development of a mass trapping programme for the control of the soybean beetle, *A. rufocuprea* (T. Imai, Tokyo, 1998, personal communication). This is an economically important species in Japan, for which both the pheromone (Tamaki *et al.*, 1985) and a kairomone (Imai *et al.*, 1997) systems have been developed. It was reasoned that a combination of the two types of lure could have an effect on the soybean beetle population, given the fact that both male and female beetles are captured (Imai *et al.*, 1997). Capture of males only is unlikely to have a direct effect on future beetle populations. Preliminary experiments with marked insects by Imai and collaborators demonstrated that beetles are recaptured in pheromone-baited traps within 150 m of their release point, in agreement with the figures obtained by Nakano and Tamaki (1986). Utilizing the Yamamura method (Yamamura *et al.*, 1992), they estimated the populations of soybean beetles in a few agricultural areas in Japan. In Hatano-city, Kanagawa Prefecture, for example, the population was estimated to be 316 beetles per hectare in mid August (T. Imai, Tokyo, 1998, personal communication). Within this level of population, the optimal density of traps baited with 3 mg of the pheromone, methyl Z-(5)-tetradecenoate (Tamaki *et al.*, 1985) and 2 g of the kairomone, methyl anthranilate (Imai *et al.*, 1997), was estimated to be one trap per hectare.

Experiments have been carried out for 1–2 years in different areas, where the populations of soybean beetle in the year(s) before the treatments were estimated on the basis of the larval population in the soil. Treatments were conducted in summer and the larval populations were measured in the following autumn. Results of the experiments conducted on 14 ha in Maki-machi, Niigata Prefecture, were highly promising. Catches of the soybean beetle decreased significantly inside the treated area after the population started reaching a peak (6 August) (Fig. 3.3). For example, a survey on 13 August showed that traps on the border of the treated area (0 m) captured on average 1386 beetles, whereas the average figures were significantly lower for traps inside the area: 842, 768 and 630 beetles per trap at 100, 200 and 300 m, respectively (Fig. 3.3).

In addition, surveys of the larval population showed a dramatic decrease in the numbers of grubs in the soil following the treatment season. Provided that the same data can be obtained in the following season(s), it may be correct to claim that semiochemical treatment has an effect on the population of the soybean beetle. The effect of the treatment may be enhanced by increasing the amount of the pheromone per trap. This is feasible, given the low cost of the pheromone system and the fact that catches in traps baited with 10 mg of the pheromone were significantly higher than in traps with 3 mg, and there was no saturation even with 30 mg of the pheromone per trap (Nakano *et al.*, 1986).

Another promising strategy is the use of beetles to disseminate entomopathogenic fungi into the larval breeding habitats. An inoculation chamber has been designed, which fits between the top of a standard Trece Catch Cam™ Japanese beetle trap and its holding canister (Klein and Lacey, 1999). Beetles, which are attracted to the trap, fall through a hole in the inoculation chamber and land on a mesh screen. The trap has been tested with *Metarhizium anisopliae* (Metschnikoff) as the pathogen, formulated with bran flour. Beetles emerging from the device in the field were captured and returned to the laboratory, where the presence of conidia and mortality of adult beetles from the fungus were confirmed. Mortality after incubation for 7 days was as high as 82%, whereas that of a control group was only 6.7% during the same period. In addition, contaminated beetles passed the fungus to untreated beetles in the same feeding containers. Seventeen days after they were placed together, 90% of the beetles in the container with the contaminated adults were dead, while the mortality in the control group was 30% in the same period. More importantly, a field trial in Missouri indicated that this autodissemination technique may have an impact on future adult populations.

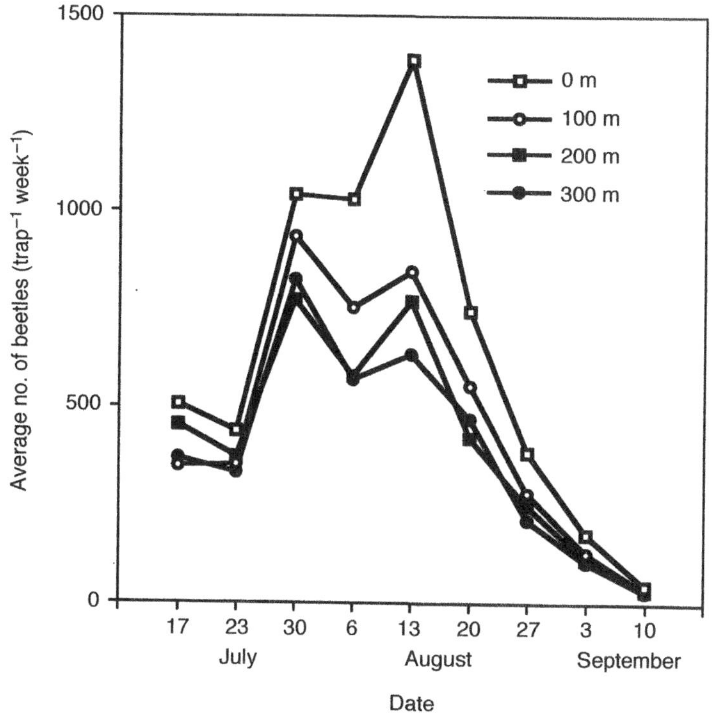

Fig. 3.3. Catches of the soybean beetle, *A. rufocuprea,* in pheromone- and kairomone-baited traps in Maki-machi (Niigata Prefecture, Japan) in a 14 ha field treated with one trap per hectare in the summer of 1997 (T. Imai, Tokyo, 1998, personal communication). Surveys at 1-week intervals showed that traps on the border of the treated area (0 m) captured significantly more beetles than traps inside the treated area (100, 200 and 300 m away from the border) during the peak of flight activity. Catches in the central area were almost half of the captures on the border at the peak of the season (13 August).

Surveys of the larval population after the treatment showed a slightly higher number of larvae of *P. japonica* per sample in the control than in the treated area (M.G. Klein, Ohio, 1998, personal communication).

Pheromone Biology

Sex pheromone glands of rutelines have been identified by extracting various parts of the body, analysing the extracts by gas chromatography–mass spectrometry (GC–MS) and by observations of gland morphology. Female-specific epithelial cells line the inner surfaces of the anal plate and two apical sternites, and the volatile compounds produced in these cells are probably released through pores that appear in the cuticular layer adjacent to these glandular cells only in females (Tada and Leal, 1997). On the other hand, scarab beetles in the genera *Holotrichia* and *Phyllophaga* display a typical calling behaviour by extruding the abdominal tip, exposing a ball-shaped sac (Leal *et al.*, 1992a, 1993c, 1996a). Sex pheromones of a few melolonthines have been identified from extracts of these abdominal tissues, but the fine structures of the gland remained unknown. Analysis of extracts from the abdominal sac, as well as from the anal plate and the two apical sternites of *H. parallela*, showed that the pheromone was present only in the abdominal sac (Tada and Leal, 1997). Conversely, the pheromone of *H. picea* was detected in the anal plate plus two apical sternites, no trace being detected in the abdominal tips of females (Tada and Leal, 1997).

Hoyt and Osborne (1971) demonstrated that in another melolonthine species, *C. zealandica*, sex pheromone is produced by symbiotic bacteria in colleterial glands, that lie beneath the seventh sternite, ventral in position to the vagina. Recently, Jackson and co-workers (T. Jackson, Lincoln, 1998, personal communication) isolated and identified the phenol-producing bacteria in this species and observed the occurrence of similar bacteria-filled glands in other species of melolonthines, suggesting that this symbiosis may be a common feature in the pheromone biology of this subfamily.

We have re-investigated the eversible pheromone-producing organ of *H. parallela*. Through careful dissection of the organ and GC–MS analysis of tissue extracts, it was determined that the pheromones were produced in the posterior part of the organ, whereas paired accessory glands were not involved in pheromone production. Observations under the light microscope revealed a layer of cuticular epithelium composed of columnar cells responsible for the pheromone production (Kim and Leal, 1998).

It is intriguing that three species in the same subfamily have remarkably different pheromone biology. Although their pheromone glands are so distinct, *Heptophylla picea* and *Holotrichia parallela* produce pheromone *de novo*. Neither (*R*,*Z*)-7,15-hexadecadien-4-olide nor L-isoleucine methyl ester has been isolated from any other natural sources. On the other hand, the production of ubiquitous phenol by symbiotic bacteria in the colleterial glands of *C. zealandica* has been confirmed by independent research groups in New Zealand (Hoyt and Osborne, 1971; T. Jackson, Lincoln, 1998, personal communication). Although phenol is clearly an attractant for *C. zealandica* (Henzell, 1970), the chemistry of the sex pheromone for this species should be revisited before we can make any speculation regarding the difference in the biological features.

Pheromone Perception

The detection of chemical signals in insects takes place in the antennae where they are 'translated' into electrophysiological signals. Chemical communication is achieved after further olfactory processing, which ultimately leads to appropriate behavioural responses in the receiver. The early olfactory processing takes place in olfactory sensilla, which in scarab beetles are olfactory plates (sensilla placodea).

It is widely believed that the slow potentials are triggered by second messengers generated by interaction of odour molecules with odour receptors present in the membranes of olfactory neurones. Later, the slow potential is transformed into nervous activity (spikes), which is transmitted through the axon of olfactory receptor neurones to the antennal lobe. As a critical step just before such interactions take place, the odorants, largely hydrophobic in nature, have to pass the aqueous barrier of the sensillar lymph surrounding the neuronal cells. A class of small proteins (< 20 kDa), discovered almost simultaneously in insects (Vogt and Riddiford, 1981) and vertebrates (Pelosi *et al.*, 1982), is considered to participate in transport, protection and/or inactivation of chemical signals from the environment in the early stage of olfactory processing. These proteins are present in the sensillar lymph of insects (and nasal mucus of vertebrates) in enormous concentrations (*c.* 10 mM in insects) and provide a milieu that may facilitate the entrance of hydrophobic molecules into the aqueous environment. It is postulated that at least in Lepidoptera there are two classes of such proteins: one involved in the recognition of pheromones, pheromone-binding proteins (PBPs); and another one participating in the recognition of general odorants, odorant-binding proteins (OBPs) (Vogt and Riddiford, 1981; Vogt *et al.*, 1990; Steinbrecht *et al.*, 1992; Pelosi, 1994; Pelosi and Maida, 1995). The sensory system must be fast enough to follow the dynamics of odour fluctuation during insect flight. It has been suggested that pheromone-degrading enzymes play the major role in pheromone inactivation (Vogt and Riddiford, 1981; Vogt *et al.*, 1985).

Pheromone-binding proteins

In our attempt to disclose the molecular basis of olfaction in scarab beetles, we isolated antenna-specific proteins from a number of scarab species. In all cases there was one major antenna-specific protein, which was found not only in male, but also in female antennae (H. Wojtasek and W.S. Leal, 1998, unpublished results). The lack of sexual difference in the profile of antenna-specific proteins is consistent with the observation that female beetles can also detect, and even respond behaviourally, to their own sex pheromone (Leal *et al.*, 1994c, 1996a). Thus, one would expect that both males and females of scarab beetles would have the same machinery for the detection of sex pheromones.

The N-terminal sequences of the antennae-specific proteins from various scarab species were highly conserved within the family and showed moderate homology to putative OBPs from *Drosophila melanogaster* Macquart (McKenna *et al.*, 1994) (47%), *Lygus lineolaris* Palisot (Dickens *et al.*, 1995) (45–50%) and the ABPX protein from *Bombyx mori* (Moore) (Krieger *et al.*, 1996) (30–35%). It is worth noting that species that utilize unrelated pheromone structures have putative PBPs with low homologies (Table 3.3). When two proteins were detected, the comparison was done with the one expressed more strongly. The percentage amino acid identity (AA) was based on the 20 bases of the N-terminal sequences of each protein in a pair-wise comparison to the antenna-specific protein of A. *octiescostata*, which utilizes buibuilactone and japonilure as sex pheromone constituents. Species that utilize unrelated pheromone structures, such as the aromatic alkaloid (*P. diversa*) (Leal *et al.*, 1997a) and the γ-lactone (*P. japonica*) (Tumlinson *et al.*, 1977), have putative PBPs with very low homologies (Table 3.3) (H. Wojtasek, J.-F. Picimbon and W.S. Leal, 1998, unpublished results). Such differences could not be

Table 3.3. Relationship between pheromone structures and homology of putative pheromone binding proteins of some scarab species (AA = amino acid identity).

	AA(%)	
Anomala octiescostata	–	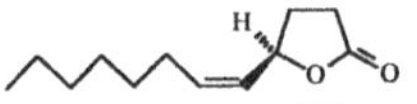
Popillia japonica	100	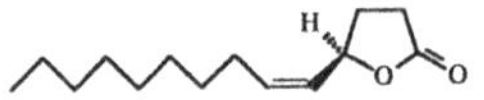
Anomala osakana	100	
Anomala schonfeldti	100	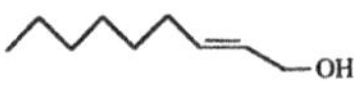
Exomala orientalis	93	
Heptophylla picea	84	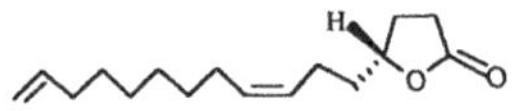
Holotrichia parallela	17	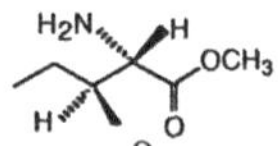
Phyllopertha diversa	8	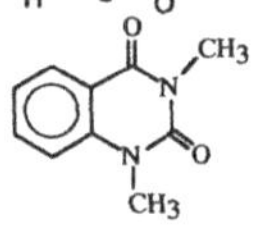

discovered in moths, the model for most of the previous PBP studies, because lepidopteran pheromones have similar structures containing a common long-chain hydrophobic moiety, except for disparlure. By contrast, sex pheromones of scarab beetles have both structures with a hydrophobic long chain, as in japonilure, and also those that lack this feature, like the amino acid derivative and the alkaloid compound in *Holotrichia parallela* and *P. diversa*, respectively.

The relationship between the pheromone structures and the N-terminal amino acid sequences of the antenna-specific proteins suggests that PBPs may possess odour-specificities for certain types of chemical structures. However, our data do not support the hypothesis that PBPs play the role of 'finely tuned filters', as previously suggested (Du and Prestwich, 1995; Breer, 1997). The *A. osakana* and *P. japonica* model provides a unique system to test this hypothesis as stereochemical discrimination is the ultimate refinement one would expect in terms of specificity.

In order to test the ligand specificity of the pheromone-binding proteins and address the question of selectivity of PBPs, we prepared a set of pheromone ligands in tritiated form. Both enantiomers of japonilure [Z-5-(decenyl)oxacyclopentan-2-one] and (*R*)-buibuilactone [(*R*,Z)-(−)-5-(octenyl)oxacyclopentan-2-one] have been prepared in enantiomeric pure forms by stereoselective synthesis of the alkyne precursors followed by catalytic reduction (S. Kuwahara and W.S. Leal, 1998, unpublished results) and used in binding experiments.

Due to the presence of highly active pheromone-degrading enzymes in the antennal extracts (see below), binding experiments with crude extracts were unrewarding (Leal *et al.*, 1997b). This difficulty was overcome by the purification of the putative pheromone-binding protein. In the isolated form, the PBP from *A. osakana* bound both enantiomeric forms of japonilure (Wojtasek *et al.*, 1998).

Interestingly, in both *A. osakana* and *P. japonica* we could detect only one PBP in the antennal extracts; the proteins from the two species showed 96% similarity (Wojtasek *et al.*, 1998). Because of the limited sensitivity of Coomassie blue, one cannot rule out the possibility of the presence of other antenna-specific proteins expressed at low levels. However, electrophysiological experiments suggest that if two PBPs were involved in the signal transduction of the enantiomers of japonilure they would be expressed at nearly the same level. Single sensillum recording in the antennae of the Japanese and Osaka beetles showed that enantiospecific receptor neurones respond equally to (*R*)- and (*S*)-japonilure (Wojtasek *et al.*, 1998). These findings and the observation that a single PBP from *A. osakana* bound both enantiomeric forms of japonilure (apparently with the same affinity) suggested that in the antennae of these species the same PBP may recognize both the pheromone and the 'stop signal' compound.

In a model borrowed from the study of bacterial chemotaxis, it has been proposed that OBPs do not only solubilize specific pheromones, but could also trigger the olfactory receptors, when bound to odour molecules (Pelosi, 1994). Prestwich and Du (1997) further proposed that interaction from both the bound ligand and ligated protein would be required for receptor activation. Under the hypothesis that the olfactory receptors interact with a PBP–pheromone complex, it is conceivable that the same PBP from scarab beetles could encode different signals upon binding different enantiomers. This could be achieved if each ligand evoked a different conformational change of the PBP upon binding, thus enabling recognition of the PBP–odorant complex by two different olfactory receptors (Fig. 3.4).

Pheromone-degrading enzymes

We observed that antennal protein extracts from several species could degrade buibuilactone and japonilure, even from beetles that do not use this group of compounds as their pheromones. Activity-staining of native polyacrylamide gels showed that species possessing high levels of antenna-specific esterases possessed a significant preference for the (*R*)-stereoisomer. However, this specificity seemed one-sided, since species that utilize the (*S*)-enantiomer did not degrade it more quickly (Wojtasek and Leal, 1996, 1997).

The esterase involved in the degradation of the lactones was resistant to classical esterase inhibitors, which was consistent with results obtained previously for antennal esterases in lepidopteran species. However, one of the tested compounds, 3-octylthio-1,1,1-trifluoropropanone (OTFP), was active at higher concentrations and completely inhibited the action of the enzyme (H. Wojtasek and W.S. Leal, 1998, unpublished results).

Interestingly, the recently identified alkaloid pheromone from *P. diversa*, 1,3-dimethyl-2,4-(1*H*,3*H*)-quinazolinedione, was readily degraded by antennal enzymes of this species (Wojtasek and Leal, 1998). On the other hand, beetles that utilize lactones as their pheromones possess little or no ability to metabolize the alkaloid. Because of the unusual structure of the alkaloid, it is not surprising that only species that utilize it in chemical communication have developed the ability to degrade it. Lactones and esters, on the other hand, are common odorous

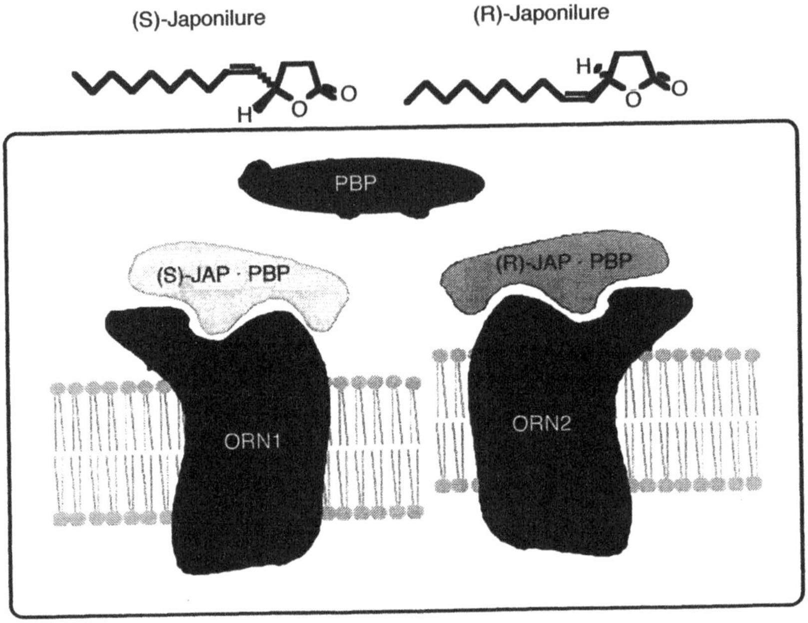

Fig. 3.4. Schematic view of an olfactory model of enantiomeric discrimination in some species of scarab beetles. One single pheromone-binding protein (PBP) would bind both enantiomeric forms of japonilure; different conformations of the ligand–PBP complex would activate different olfactory receptor neurones (ORN) co-localized in olfactory plates.

compounds, and thus the presence of appropriate enzymes in a large number of species is expected (Wojtasek and Leal, 1997). These data suggest that antennal pheromone-degrading enzymes in scarab beetles have evolved to participate in the deactivation of pheromones or other odorants.

Conclusions

Our knowledge of the chemical communication of scarab beetles has increased dramatically over the last decade. Utilization of the increasingly sophisticated analytical instrumentation and newly established techniques in chemical ecology will allow us to characterize the pheromones of many other species in the near future. This will not only pave the way for the utilization of semiochemicals in integrated pest management programmes, but also will give a better understanding of the evolution of chemical communication in this group. On the other hand, our studies on the molecular basis of olfaction in scarab beetles have shed new light on the early olfactory processing of pheromones in insect antennae. The fact that species which utilize pheromones with no structural similarities also have antenna-specific proteins (putative PBPs) with low homologies suggests a direct relationship between pheromone structure and binding proteins and gives support to the hypothesis that PBPs act as pre-filters. The change in conformation upon binding may play a significant part in the recognition of the PBP–pheromone complex by the olfactory receptors. Hence, we suggest that different enantiomers of japonilure evoke different conformational changes in the pheromone–PBP complex, thus enabling interaction with different olfactory receptors. The discovery of pheromone-degrading enzymes specific to a scarab species that utilizes an unusual type of pheromone suggests that antennal enzymes are involved in the deactivation of chemical signals. A better understanding of the perireceptor events and pheromonal signal transduction in insects will open the door for the development of new environmentally sound strategies for insect pest control.

Acknowledgements

Financial support from the Programme for Promotion of Basic Research Activities for Innovative Biosciences (BRAIN) and from a special coordination fund for promoting science and technology by the Science and Technology Agency of Japan is acknowledged with much gratitude. I am indebted to Mr T. Imai (Japan Tobacco) for sharing unpublished data, in particular those displayed in Fig. 3.3. Also, I would like to thank my past and present collaborators and members of my group who made a great contribution to this work.

References

Breer, H. (1997) Molecular mechanism of pheromone reception in insect antennae. In: Cardé, R.T. and Minks, A.K. (eds) *Insect Pheromone Research: New Directions*. Chapman and Hall, New York, pp. 115–130.

Dickens, J.C., Callahan, F.E., Wergin, W.P. and Erbe, E.F. (1995) Olfaction in a hemimetabolous insect: antennal-specific protein in adult *Lygus lineolaris* (Heteroptera: Miridae). *Journal of Insect Physiology* 41, 857–867.

Du, G. and Prestwich, G.D. (1995) Protein structure encodes the ligand binding specificity in pheromone binding proteins. *Biochemistry* 34, 8726–8732.

Henzell, R.F. (1970) Phenol, an attractant for the male grass grub beetle (*Costelytra zealandica*) (White) (Scarabaeidae: Coleoptera). *New Zealand Journal of Agricultural Research* 13, 294–296.

Henzell, R.F. and Lowe, M.D. (1970) Sex attractant of the grass grub beetle. *Science* 168, 1005–1006.

Hoyt, C.P. and Osborne, G.O. (1971) Production of an insect sex attractant by symbiotic bacteria. *Nature* 230, 472–473.

Imai, T., Tsuchiya, S., Maekawa, M., Fujimori, T. and Leal, W.S. (1997) Methyl anthranilate, a novel attractant for the soybean beetle, *Anomala rufocuprea* Motschulsky (Coleoptera: Scarabaeidae). *Applied Entomology and Zoology* 32, 45–48.

Imai, T., Maekawa, M., Tsuchiya, S. and Fujimori, T. (1998) Field attraction of *Hoplia communis* to 2-phenylethanol, a major volatile component from host flowers, *Rosa* spp. *Journal of Chemical Ecology* 24, 1491–1497.

Jackson, T.A. (1992) Scarabs – pests of the past or the future? In: Jackson, T.A. and Glare, T.R. (eds) *Use of Pathogens in Scarab Pest Management*. Intercept, Andover, UK, pp. 1–10.

Jongeleen, F.J.J. and Mahrub, E. (1979) The biology of two species of noxious Scarabaeidae from Indonesia. *Zeitschrift für Angewandte Entomologie* 87, 247–254.

Kim, J.-Y. and Leal, W.S. (1998) The eversible pheromone-producing organ in the Melolonthine beetle, *Holotrichia parallela* (Coleoptera: Scarabaeidae: Melolonthinae). In: *Proceedings, Annual Meeting of Japan Society of Applied Entomology*, Nagoya, 31 March–2 April, p. 127.

Klein, M.G. and Lacey, L.A. (1999). An attractant trap for autodissemination of entomopathogenic fungi into populations of the Japanese beetle, *Popillia japonica* (Coleoptera: Scarabaeidae). *Biocontrol Science and Technology* 9, 151–158.

Krieger, J., von Nickisch-Roseneck, E., Mameli, M. and Breer, H. (1996) Binding proteins from the antennae of *Bombyx mori*. *Insect Biochemistry and Molecular Biology* 26, 297–307.

Leal, W.S. (1991) (*R*,*Z*)-5-(−)-(Oct-1-enyl)oxacyclopentan-2-one, the sex pheromone of the scarab beetle *Anomala cuprea*. *Naturwissenschaften* 78, 521–523.

Leal, W.S. (1993) (*Z*)- and (*E*)-Tetradec-7-en-2-one, a new type of sex pheromone from the oriental beetle. *Naturwissenschaften* 80, 86–87.

Leal, W.S. (1996) Chemical communication in scarab beetles: reciprocal behavioral agonist–antagonist activities of chiral pheromones. *Proceedings of the National Academy of Sciences USA* 93, 12112–12115.

Leal, W.S. (1997) Evolution of sex pheromone communication in plant-feeding scarab beetles. In: Cardé, R.T. and Minks, A.K. (eds) *Insect Pheromone Research: New Directions*. Chapman and Hall, New York, pp. 503–513.

Leal, W.S. (1998) Chemical ecology of phytophagous scarab beetles. *Annual Review of Entomology* 43, 39–61.

Leal, W.S., Matsuyama, S., Kuwahara, Y., Wakamura, S. and Hasegawa, M. (1992a) An amino acid derivative as the sex pheromone of a scarab beetle. *Naturwissenschaften* 79, 184–185.

Leal, W.S., Hasegawa, M. and Sawada, M. (1992b) Identification of *Anomala schonfeldti* sex pheromone by high-resolution GC-behavior bioassay. *Naturwissenschaften* 79, 518–519.

Leal, W.S., Sawada, M. and Hasegawa, M. (1993a) The scarab beetle *Anomala cuprea* utilizes the sex pheromone of *Popillia japonica* as a minor component. *Journal of Chemical Ecology* 19, 1303–1313.

Leal, W.S., Sawada, M. and Hasegawa, M. (1993b) The scarab beetle *Anomala daimiana* utilizes a blend of two other *Anomala* spp. sex pheromones. *Naturwissenschaften* 80,181–183.

Leal, W.S., Sawada, M., Matsuyama, S., Kuwahara, Y. and Hasegawa, M. (1993c) Unusual periodicity of sex pheromone production in the large black chafer *Holotrichia parallela*. *Journal of Chemical Ecology* 19, 1381–1391.

Leal, W.S., Hasegawa, M., Sawada, M., Ono, M. and Ueda, Y. (1994a) Identification of *Anomala octiescostata* (Coleoptera: Scarabaeidae) sex pheromone. *Journal of Chemical Ecology* 20, 1643–1655.

Leal, W.S., Kawamura, F. and Ono, M. (1994b) The scarab beetle *Anomala albopilosa sakishimana* utilizes the same sex pheromone blend as a closely related and geographically isolated species, *Anomala cuprea*. *Journal of Chemical Ecology* 20, 1667–1676.

Leal, W.S., Hasegawa, M., Sawada, M. and Ono, M. (1994c) Sex pheromone of oriental beetle, *Exomala orientalis*: identification and field evaluation. *Journal of Chemical Ecology* 20, 1705–1718.

Leal, W.S., Shi, X., Liang, D., Schal, C. and Meinwald, J. (1995) Application of chiral gas chromatography with electroantennographic detection to the determination of the stereochemistry of a cockroach sex pheromone. *Proceedings of the National Academy of Sciences USA* 92, 1033–1037.

Leal, W.S., Yadava, C.P.S. and Vijayvergia, J.N. (1996a) Aggregation of the scarab beetle *Holotrichia consanguinea* in response to female-released pheromone suggests a secondary function hypothesis for semiochemical. *Journal of Chemical Ecology* 22, 1557–1566.

Leal, W.S., Hasegawa, M., Sawada, M., Ono, M. and Tada, S. (1996b) Scarab beetle *Anomala albopilosa albopilosa* utilizes a more complex sex pheromone system than a similar species A. *cuprea*. *Journal of Chemical Ecology* 22, 2001–2010.

Leal, W.S., Kuwahara, S., Ono, M. and Kubota, S. (1996c) (*R*,*Z*)-7,15-Hexadecadien-4-olide, sex pheromone of the yellowish elongate chafer, *Heptophylla picea*. *Bioorganic and Medicinal Chemistry* 4, 315–321.

Leal, W.S., Zarbin, P.H.Z., Wojtasek, H., Kuwahara, S., Hasegawa, M. and Ueda, Y. (1997a) Medicinal alkaloid as a sex pheromone. *Nature* 385, 213.

Leal, W.S., Wojtasek, H. and Miyazawa, M. (1997b) Putative pheromone binding proteins of scarab beetles do not bind pheromones. *Chemical Senses* 40, 730.

McKenna, M.P., Hekmat-Scaefe, D.S., Gaines, P. and Carlson, J.R. (1994) Putative *Drosophila* pheromone-binding proteins expressed in a subregion of the olfactory system. *Journal of Biological Chemistry* 269, 16340–16347.

Nakano, Y. and Tamaki, Y. (1986) Use of sex pheromone trap for the survey of occurrence and biology of the soybean beetle, *Anomala rufocuprea* Motschulsky (Coleoptera: Scarabaeidae). *Japanese Journal of Applied Entomology and Zoology* 30, 260–267 (in Japanese).

Nakano, Y., Tamaki, Y. and Sugie, H. (1986) Field evaluation of synthetic sex-attractant pheromone of the soybean beetle, *Anomala rufocuprea* Motschulsky (Coleoptera: Scarabaeidae). *Japanese Journal of Applied Entomology and Zoology* 30, 254–259 (in Japanese).

Pelosi, P. (1994) Odorant-binding proteins. *Critical Reviews in Biochemistry and Molecular Biology* 29, 199–228.

Pelosi, P. and Maida, R. (1995) Odorant-binding proteins in insects. *Comparative Biochemistry and Physiology* 111B, 503–514.

Pelosi, P., Baldaccini, N.E. and Pisanelli, A.M. (1982) Identification of a specific olfactory receptor for 2-isobutyl-3-methoxypyrazine. *Biochemical Journal* 201, 245–248.

Potter, D.A. (1980) Flight activity and sex attraction of Northern and Southern masked chafers in Kentucky turfgrass. *Annals of the Entomological Society of America* 73, 414–417.

Prestwich, G.D. and Du, G. (1997) Pheromone-binding proteins, pheromone recognition, and signal transduction in moth olfaction. In: Cardé, R.T. and Minks, A.K. (eds) *Insect Pheromone Research: New Directions*. Chapman and Hall, New York, pp. 131–143.

Sawada, H. (1991) *Morphological and Phylogenetical Study on the Larvae of* Pleuristica lamelicornia *in Japan*. Tokyo University of Agriculture Press, Tokyo, 289 pp.

Steinbrecht, R., Ozaki, A.M. and Zeigelberger, G. (1992) Immunocytochemical localization of pheromone-binding protein in moth antennae. *Cell and Tissue Research* 270, 287–302.

Tada, S. and Leal, W.S. (1997) Localization and morphology of the sex pheromone glands in scarab beetles (Coleoptera: Rutelinae, Melolonthinae). *Journal of Chemical Ecology* 23, 903–915.

Tamaki, Y., Sugie, H. and Noguchi, H. (1985) Methyl (Z)-5-tetradecenoate: sex attractant pheromone of the soybean beetle, *Anomala rufocuprea* Motschulsky (Coleoptera: Scarabaeidae). *Applied Entomology and Zoology* 20, 359–361.

Tumlinson, J.H., Klein, M.G., Doolittle, R.E., Ladd, R.E. and Proveaux, A.T. (1977) Identification of the female Japanese beetle sex pheromone: inhibition of male response by an enantiomer. *Science* 197, 789–792.

Vogt, R.G. and Riddiford, L.M. (1981) Pheromone binding and inactivation by moth antennae. *Nature* 293, 161–163.

Vogt, R.G., Riddiford, L.M. and Prestwich, G.D. (1985) Kinetic properties of a sex pheromone-degrading enzyme: the sensillar esterase of *Antheraea polyphemus*. *Proceedings of the National Academy of Sciences USA* 82, 8827–8831.

Vogt, R.G., Prestwich, G.D. and Lerner, M.R. (1990) Odorant-binding-protein subfamilies associate with distinct classes of olfactory receptor neurons in insects. *Journal of Neurobiology* 22, 74–84.

Wojtasek, H. and Leal, W.S. (1996) Can the pheromone binding proteins of scarab beetles distinguish enantiomers? In: *Proceedings of the Annual Meeting of the International Society of Chemical Ecology*, Prague, 18–22 August, p. 16.

Wojtasek, H. and Leal, W.S. (1997) Pheromone degradation in scarab beetles. *Chemical Senses* 40, 824.

Wojtasek, H. and Leal, W.S. (1998) Pheromone degradation in the pale-brown chafer, *Phyllopertha diversa* (Coleoptera: Scarabaeidae): identification of an insect olfactory cytochrome P450. In: *Proceedings of the Third International Symposium on Molecular Insect Science*, Snowbird, Utah, 5–10 June, p. 143.

Wojtasek, H., Hansson, B.S. and Leal, W.S. (1998) Attracted or repelled? – A matter of two neurons, one pheromone binding protein, and a chiral center. *Biochemical and Biophysical Research Communications* 250, 217–222.

Yamamura, K., Wakamura, S. and Kozai, S. (1992) A method for population estimation from a single release experiment. *Applied Entomology and Zoology* 27, 9–17.

Yarden, G., Shani, A. and Leal, W.S. (1996) (Z, E)-α-Farnesene – an electroantennogram-active component of *Maladera matrida* volatiles. *Bioorganic and Medicinal Chemistry* 4, 297–303.

Zhang, A., Robbins, P.S., Leal, W.S., Linn, C.E. Jr, Villani, M.G. and Roelofs, W.L. (1997) Essential amino acid methyl esters: major sex pheromone components of the cranberry white grub, *Phyllophaga anxia* (Coleoptera: Scarabaeidae). *Journal of Chemical Ecology* 23, 231–245.

4 Sap Beetles

ROBERT J. BARTELT

USDA–ARS, National Center for Agricultural Utilization Research, Bioactive Agents Research Unit, 1815 N. University St, Peoria, IL 61604, USA

Introduction

The sap beetle genus *Carpophilus* (Coleoptera: Nitidulidae) is an important and interesting group. These insects are small in size (typically less than 5 mm in length) and cryptic in habit, but they are frequently abundant, feeding mostly on fruits or other plant materials that are ripening, decomposing or fermenting. They occupy a wide range of habitats throughout the world and infest a variety of agricultural products, both before and after harvest. One important species, the driedfruit beetle, is shown in Fig. 4.1. Reference works describing these beetles include Hinton (1945) and Connell (1956, 1991). Williams *et al.* (1983) compiled a useful bibliography.

The beetles have been significant and persistent pests of dried fruit such as figs and dates in both the New and Old Worlds. Damage includes both direct feeding injury by adults and larvae and transmission of fruit-degrading microorganisms (Hinton, 1945; Lindgren and Vincent, 1953). Dried fruit crops usually require fumigation in warehouses after harvest to kill sap beetles and other insects. *Carpophilus* beetles are also pests of peaches, nectarines and apricots, particularly in Australia. The insects penetrate the skins of ripening fruit and cause rapid breakdown (James *et al.*, 1995).

The beetles have become pests of corn (maize) in the USA. Losses of sweet corn are due to feeding damage and contamination of canned products with insect fragments. In field corn, direct feeding damage can usually be tolerated, but the beetles are vectors of certain microorganisms that have caused great concern, such as *Aspergillus flavus* Link: Fr. (Lussenhop and Wicklow, 1990). This fungus produces aflatoxin, a chemical that poses a serious threat to human and animal health when contaminated corn is consumed (reviews by Dowd, 1991, 1995; Wicklow, 1991).

The agricultural impact of these beetles has increased as the use of broad-spectrum insecticides for other pests has been reduced. For example, sap beetles were hardly evident in Australian stone fruits until growers began to control the

 Pheromones of Non-Lepidopteran Insects Associated with Agricultural Plants (eds J. Hardie and A.K. Minks)

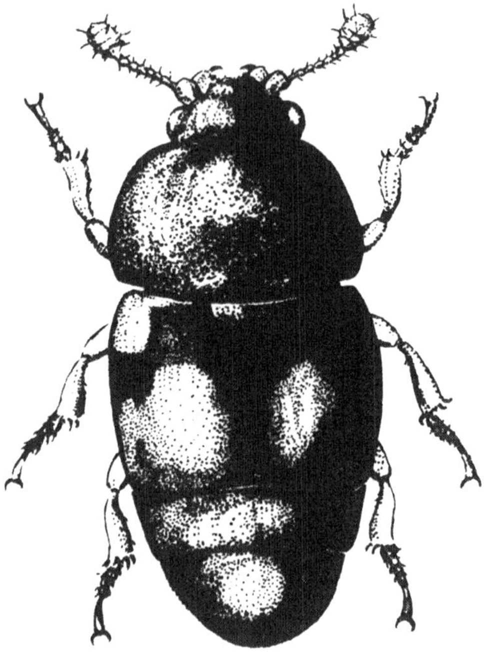

Fig. 4.1. The driedfruit beetle, *Carpophilus hemipterus* (L.). (Drawn by Pongsak Thiemmedh and reprinted from Williams *et al.*, 1983.)

oriental fruit moth (*Cydia molesta* Busck) by mating disruption with pheromones instead of with insecticides. Now that the beetles are released from incidental insecticide pressure, their numbers and economic importance have increased greatly (James *et al.*, 1995).

Similar shifts could develop elsewhere. A pheromone-based mating disruption method is being developed for the carob moth, currently the major pest of dates in California (Baker *et al.*, 1991), and corn has been genetically engineered to produce a bacterial protein (from *Bacillus thuringiensis* Berliner) that kills moth larvae. Both strategies can be expected to reduce insecticide use and will allow sap beetles to proliferate in these crops. Finally, the future elimination of methyl bromide as a warehouse fumigant (Taylor, 1994) will make dried fruit crops more susceptible to beetle damage.

There is a need for new, selective and environmentally acceptable tools for reducing the economic impact of these beetles. One promising approach is the use of pheromones, and during the last decade a great deal has been learned about chemical communication in these insects.

Pheromones have been studied in nine *Carpophilus* species. Four of these belong to the nitidulid complex attacking fruit in southern California but also occurring widely throughout the world; these are *C. hemipterus* (L.), *C. mutilatus* Erichson, *C. freemani* Dobson and *C. obsoletus* Erichson. *C. lugubris* Murray is a particularly common sap beetle in corn in the Midwestern USA; C. *antiquus* Melsheimer and C. *brachypterus* Say also occur in this region but are less abundant. C. *dimidiatus* (F.) is a major member of the sap beetle complex affecting corn in the southeastern USA (where C. *freemani* and C. *mutilatus* are abundant as well). Finally, C. *davidsoni* Dobson is an Australian species that is at present a major pest of peaches, apricots and other stone fruits.

Pheromone Identification

Wind-tunnel bioassays for aggregation pheromones

The existence of a male-produced pheromone was initially demonstrated for six of the above species in a wind tunnel. The first to be studied was C. *hemipterus*. Flying beetles of both sexes were clearly more attracted to males feeding on artificial diet than to females on diet or to diet alone, and bioassays with collected volatiles produced similar results (Bartelt *et al.*, 1990a). Bioassays with collected volatiles were subsequently used to demonstrate pheromones in male C. *lugubris* (Bartelt *et al.*, 1991), C. *freemani* (Bartelt *et al.*, 1990b), C. *obsoletus* (Petroski *et al.*, 1994a), C. *mutilatus* (Bartelt *et al.*, 1993a) and C. *davidsoni* (Bartelt and James, 1994) as well. The final three species, C. *antiquus*, C. *brachypterus* and C. *dimidiatus*, were not reared in sufficient numbers to allow wind-tunnel studies.

The wind tunnel was 1.4 m long and 0.6 m × 0.6 m in cross-section. The walls and top were of transparent plastic, but the floor had a rougher surface (plywood) so that beetles falling on their backs could right themselves. The air flow rate was 0.3 m s^{-1}, the temperature was about 27°C, and light was provided with four 40 W fluorescent bulbs mounted 0.1 m above the top. The wind tunnel was stocked with up to 1000 beetles of mixed sexes and 0–1 week old. The insects began to fly spontaneously after several hours without food, and as many as 100 would be in the air at any instant. Two odour sources to be compared (usually pieces of filter paper to which test chemicals or control solvents had been applied) were hung side by side in the upwind end of the wind tunnel, and the numbers of beetles alighting on these during a set period of time (typically 2–5 min) were recorded (Bartelt *et al.*, 1990a). Beetles entering an active odour plume would approach the source with a deliberate, upwind, casting flight and, after landing, would often remain on the source until the test was over. There was a pause of 1–5 min between consecutive tests, and as many as 30–50 tests could be conducted with one group of beetles during a bioassay session of several hours. A favourable characteristic of this system was that the ratio of landing counts for two treatments remained remarkably stable when the test was replicated, both during a bioassay session and from day to day (e.g. Bartelt *et al.*, 1992b). Fresh bioassay beetles were used for each session. While the assay procedure did not permit detailed observation of individual behaviour, it did allow rapid evaluation of samples, and the approach of a hundred or more beetles to an odour source within minutes was not unlike the responses sometimes observed around pheromone traps in the field (R.J. Bartelt, 1998, unpublished results).

Chromatographic recognition and isolation of pheromones

All volatile collections and derived chromatographic fractions were analysed by gas chromatography (GC). The key objective for each species was to associate the biological activity of male-derived samples with one or more GC peaks. Two approaches were used to recognize the pheromone peaks. The first was to compare male-derived and respective female-derived samples by GC, regarding any sex-specific compounds as possible pheromone components. The second approach was bioassay-guided chromatographic purification, which sought to remove inert compounds from a sample until only the pure, biologically active pheromone component(s) remained. Both approaches were used simultaneously whenever possible, but only the male/female comparisons could be used for C. *antiquus*, C. *brachypterus* and C. *dimidiatus* because of limited beetle numbers.

Chromatographic pheromone purification usually involved two steps (Bartelt *et al.*, 1990a). The first was column chromatography or high-performance liquid chromatography (HPLC) on silica, which separated the volatile collections into fractions by polarity. The second step was HPLC on a silica column coated with silver nitrate, which separated compounds by numbers of double bonds and substitution patterns.

Bioassay-guided purification of the first *Carpophilus* pheromone was complicated by a synergistic relationship with food odours. It was difficult to demonstrate the activity of the C. *hemipterus* pheromone once it became chromatographically separated from the food scent that was also present in the volatile collections (Bartelt *et al.*, 1990a). Consequently, food odours (or a synthetic component of these, such as propyl acetate) were routinely added to each bioassay treatment, and the food odour itself became the control when pheromone activity was being evaluated. A variety of esters, alcohols and other compounds were found to synergize the pheromone of C. *hemipterus* (Dowd and Bartelt, 1991). Fortunately, it was possible to obtain useful bioassay results for the other species without a food odour being present, but synergism was still observed whenever tests with food odour were made (Bartelt *et al.*, 1990b; Petroski *et al.*, 1994a).

For every species, the least polar (hexane) fraction from the silica column contained at least one male-specific compound (by GC), suggesting that each had a hydrocarbon pheromone. The pheromonal nature of these fractions was corroborated in the wind tunnel whenever the tests could be performed (Bartelt *et al.*, 1990a,b, 1991, 1993a; Bartelt and James, 1994; Petroski *et al.*, 1994a). The HPLC column was useful for separating complex blends of male-specific hydrocarbons from saturated impurities and also for separating blend components from each other (Bartelt *et al.*, 1990b, 1992b; Bartelt and James, 1994; Bartelt and Weisleder, 1996).

Details of Carpophilus pheromones

Twenty-two male-specific, conjugated triene and tetraene hydrocarbons (**1–22**) were eventually identified from the nine *Carpophilus* species (Table 4.1). These include relatively abundant components with clear activity in at least one species

Table 4.1. Structures and relative abundances of male-specific compounds identified from *Carpophilus* species.[a]

Structure		Species								
		hem	bra	obs	lug	dav	fre	mut	ant	dim
Components with demonstrated activity for at least one species										
1	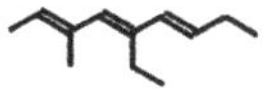	3	–	–	–	**100**	**100**	–	–	–
2	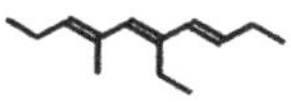	–	–	–	–	**7**	2	**10**	–	–
3	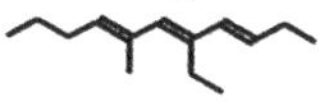	–	–	–	–	1	< 1	**100**	–	–
4	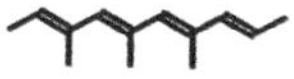	**100**	**100**	–	–	< 1	–	–	–	–
5		**8**	**11**	–	–	< 1	–	–	–	–
6	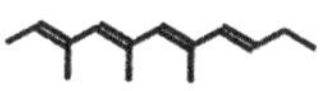	**13**	**6**	**100**	**10**	**9**	< 1	–	–	–
7	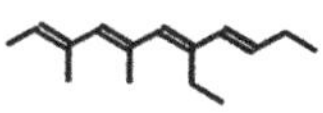	**2**	**3**	–	**100**	**31**	**4**	–	–	–
8	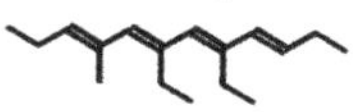	–	–	–	–	< 1	–	–	**100**	**100**
Minor components of uncertain biological importance										
9		–	–	–	–	< 1	–	–	–	–
10	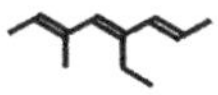	–	–	–	–	< 1	< 1	–	–	–
11		–	–	–	–	< 1	1	–	–	–
12	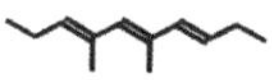	–	–	–	–	< 1	–	–	–	–
13	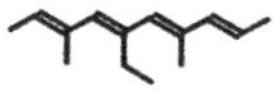	3	–	–	–	–	–	–	–	–
14	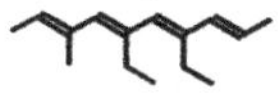	< 1	–	–	–	–	–	–	–	–
15	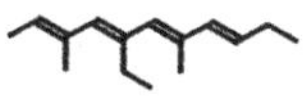	< 1	–	–	–	–	–	–	–	–
16	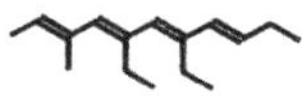	–	–	–	–	< 1	–	–	–	–
17	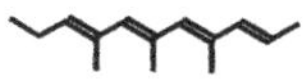	6[b]	**4**	–	–	–	–	–	–	–
18	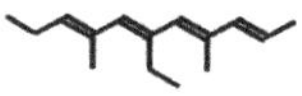	1	–	–	–	–	–	–	–	–
19	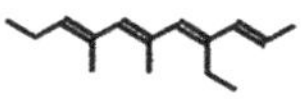	1	–	–	–	–	–	–	–	–
20	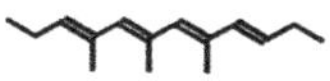	1	–	–	–	< 1	–	–	–	–
21	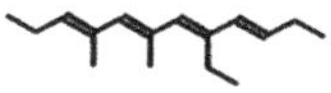	–	–	–	–	< 1	< 1	–	–	–
22	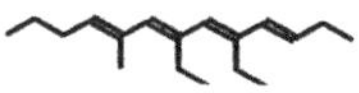	–	–	–	–	–	–	–	–	7

[a]For each species, components normally used in field baits shown in bold type. hem, *C. hemipterus*; bra, *C. brachypterus*; obs, *C. obsoletus*; lug, *C. lugubris*; dav, *C. davidsoni*; fre, *C. freemani*; mut, *C. mutilatus*; ant, *C. antiquus*; dim, *C. dimidiatus*.
[b]Used in field when *C. hemipterus* and *C. brachypterus* pheromones were compared.

and also minor components with uncertain biological importance. All compounds have their alkyl branches on alternate carbons, and all double bonds have the E configuration. Each of these hydrocarbons was previously unknown.

The major component and one minor component in C. *hemipterus* were identified as tetraenes **4** and **6**, respectively (Bartelt *et al.*, 1990a). Bartelt *et al.* (1992b) identified nine additional tetraenes and one triene. Of these hydrocarbons, only tetraenes **4–7** were active in the wind-tunnel bioassay (Bartelt *et al.*, 1992b); a blend of these four was active in the field in California (Bartelt *et al.*, 1992a). Williams *et al.* (1995) learned that C. *brachypterus* males produce five of the tetraenes also present in C. *hemipterus* (**4–7** and **17**), but in slightly different proportions. They also confirmed the attraction of C. *brachypterus* to a synthetic blend of these in a field study in Ohio. The only male-specific compound encountered with C. *obsoletus* was tetraene **6**; the synthetic pheromone was active in wind-tunnel tests and in the field in California (Petroski *et al.*, 1994a). The major component of the C. *lugubris* pheromone was identified as tetraene **7** by Bartelt *et al.* (1991). They also noted a response of the species to tetraene **6**, and this compound was later found to be a minor component (see Bartelt *et al.*, 1993b). Activity of the synthetic pheromone was verified in the field in Illinois (Bartelt *et al.*, 1991, 1993b).

Fifteen trienes and tetraenes were identified from C. *davidsoni*, at levels spanning nearly five orders of magnitude (Bartelt and James, 1994; Bartelt and Weisleder, 1996). A synthetic mixture of the four most abundant hydrocarbons (**1**, **2**, **6** and **7**) was active in the field in New South Wales in Australia (Bartelt and James, 1994). Activity of the minor components has not been assessed. The hydrocarbons from C. *freemani* include eight compounds that also occur in C. *davidsoni* (Bartelt *et al.*, 1990b). The two most abundant components, triene **1** and tetraene **7**, were synergistic in the wind tunnel and a synthetic mixture of these was active in Californian field tests (Bartelt *et al.*, 1990b). The pheromone of C. *mutilatus* consists of two trienes (**2** and **3**) previously encountered in C. *freemani* and C. *davidsoni*, and field activity of these was demonstrated in California (Bartelt *et al.*, 1993a). Males also emit tetradecanal (Bartelt *et al.*, 1993a; Bartelt and Weisleder, 1996), which was not active in the wind tunnel.

Only the 17-carbon tetraene **8** has been identified from C. *antiquus*; it was attractive in field tests in Illinois (Bartelt *et al.*, 1993b). The same compound was isolated from C. *dimidiatus*, along with a minor tetraene (**22**). Tetraene **8** was active in the field in South Carolina, but **22** was not tested (Bartelt *et al.*, 1995b).

In all of the above field tests, both sexes of beetles responded to the pheromones about equally. Activity of *Carpophilus* pheromones versus appropriate controls was also documented in other field studies in Ohio (Lin *et al.*, 1992; Williams *et al.*, 1993), California (Bartelt *et al.*, 1994a,b, 1995a), Israel (Blumberg *et al.*, 1993) and Australia (James *et al.*, 1993, 1994).

Chemical identification

Identification of the pheromone components always relied on mass spectrometry (MS) and sometimes on proton nuclear magnetic resonance (NMR) (Bartelt *et al.*,

1990a,b, 1993a,b) and ultraviolet (UV) spectroscopy (Bartelt *et al.*, 1990a) and on MS of saturated derivatives (prepared by catalytic hydrogenation) (Bartelt *et al.*, 1990a,b, 1991, 1992b). The procedure was to use the available (sometimes limited) spectral and chromatographic data to propose likely structures and then to synthesize these alternatives until one was found to match the natural compound by all criteria.

For example, MS (before and after hydrogenation), UV and NMR data suggested that the major pheromone component for *C. hemipterus* was an (8*E*)-3,5,7-trimethyl-2,4,6,8-decatetraene. The configurations at the tri-substituted double bonds could not be determined from the spectra, however. The (2*E*,4*E*,6*E*,8*E*) isomer (compound **4** in Table 4.1) was chosen for initial synthesis because it was expected to be the most stable, and the synthetic and natural compounds matched in all respects, completing the identification (Bartelt *et al.*, 1990a).

For the initial species, every effort was made to maximize the amount of spectral data for these unusual and somewhat complicated pheromone structures. However, as the library of model compounds grew and certain structural generalizations became evident (such as the ubiquitous *E* configuration at double bonds and a branching pattern related to biosynthesis), it became easier to 'guess' structures and synthesize the proper compounds just from the GC retention times and mass spectra.

Small amounts of *Z* isomers were found in natural samples (Bartelt *et al.*, 1990a, 1992b), but these were believed to be artefacts because the 'freshest' samples always had smaller amounts and because pure synthetic all-*E* tetraenes isomerized slightly during storage. The *Z* isomers were generally not active in the bioassays (Bartelt *et al.*, 1990a, 1992b), but large amounts of the synthetic (2*E*,4*E*,6*Z*) isomer of triene **1**, which was never found in natural samples, did partially inhibit the response of *C. freemani* to its major pheromone component (all-*E* **1**) in the wind tunnel (Petroski and Weisleder, 1997).

Synthesis and spectra

A general method for synthesizing the 22 compounds in Table 4.1 is illustrated in Fig. 4.2 (Bartelt *et al.*, 1990c, 1993a,b). The starting material was always an aldehyde. One or more Wittig–Horner condensations each extended the chain by two carbons, introduced an *E* double bond and a methyl or ethyl branch, and provided a functional group (an ester) at which subsequent elaboration would be directed. This ester was reduced to an alcohol with lithium aluminum hydride and the resulting alcohol was oxidized to an aldehyde with manganese dioxide to prepare for the next chain-extending cycle. The final reaction was a Wittig condensation that created the di-substituted double bond and attached a terminal alkyl group of proper size. An alternative procedure for the final Wittig reaction with greater *E* selectivity has also been reported (Petroski, 1997).

Each synthetic all-*E* hydrocarbon has been characterized with respect to GC retention (on a non-polar DB-1 capillary column), mass spectrum and proton NMR spectrum. The chromatographic and spectral data are located in the same

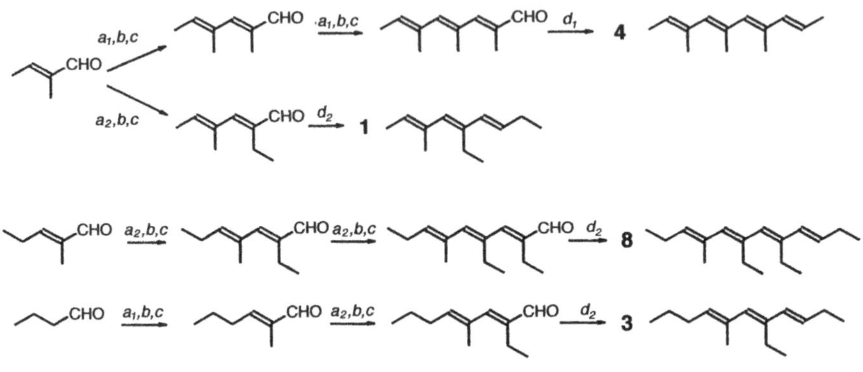

Fig. 4.2. Synthetic scheme for selected triene and tetraene hydrocarbons. $\mathbf{a_1}$ and $\mathbf{a_2}$ = Wittig–Horner condensations with triethyl 2-phosphonopropionate and triethyl 2-phosphonobutyrate, respectively; ***b*** = reduction of ester with lithium aluminum hydride; ***c*** = partial oxidation of alcohol with manganese dioxide to aldehyde; $\boldsymbol{d_1}$ and $\boldsymbol{d_2}$ = Wittig condensations with ethyltriphenylphosphonium bromide and propyltriphenylphosphonium bromide, respectively. Compound numbers are as in Table 4.1.

references as the identifications and syntheses. Interpretation of mass spectra of these compounds has been discussed (Bartelt *et al.*, 1992b). Proton NMR spectra are diagnostic for the compounds and were invaluable in confirming the structures of the synthetic hydrocarbons. Nuclear Overhauser NMR experiments allowed the configurations at the tri-substituted double bonds to be firmly established (Bartelt *et al.*, 1990a, 1991, 1992b). Carbon-13 spectra were obtained for compounds **1** (Petroski *et al.*, 1994b), **3**, **6** and **7** (Bartelt and Weisleder, 1996), and **4** and **5** (Bartelt *et al.*, 1990c).

Pheromone Biology

Pheromone emission from *Carpophilus* males usually began 1–4 days after placement of adults on diet medium (Bartelt *et al.*, 1993b; Bartelt and James, 1994; Nardi *et al.*, 1996). Separated males produced pheromone at far higher rates than males in groups (Bartelt *et al.*, 1993b, 1995b; Bartelt and James, 1994; Petroski *et al.*, 1994a). For example, mean emission from single *C. dimidiatus* males in flasks with artificial diet was 0.6 µg per day, but the amount was only 0.001 µg per male per day from groups of 50 males, based on GC analyses of volatile collections using a quantitative internal standard. Typical pheromone production rates for isolated *Carpophilus* males on artificial diet were in the range of 0.1–3 µg per day (Bartelt and James, 1994; Petroski *et al.*, 1994a; Bartelt *et al.*, 1995b; R.J. Bartelt, 1998, unpublished results).

In *C. davidsoni*, pheromone composition as well as pheromone amount varied with beetle numbers. The ratio of tetraenes to trienes in the pheromone from groups of males was only about 20% as high as in individual males (Bartelt and

James, 1994). In the field, pheromone with a smaller proportion of tetraene was less attractive (Bartelt and James, 1994). Thus, pheromone blend may contribute to a density regulating mechanism under natural conditions because the pheromone from crowded males is qualitatively less attractive. Finally, mating appears to influence pheromone production; emission from individual *C. freemani* males decreased by about 90% within 1 day of introducing females (Nardi *et al.*, 1996).

A model for pheromone function under natural conditions is that a male beetle would begin to release its pheromone within days of arriving at a site suitable for feeding and colonization. His pheromone, together with volatiles from the host site, would be highly attractive to other beetles flying in that area. Beetles would alight at the location of the male, but they would also alight at other, unoccupied food sources in the immediate area (James *et al.*, 1996b; D.G. James, 1998, unpublished results). While the original male would cease to emit pheromone when joined by females, the new males at the adjacent feeding sites would begin to produce it, mediating a spread of the infestation through the host resource.

Site of pheromone production

The site of pheromone production in male *C. freemani* was found to be the posterior portion of the abdomen (Dowd and Bartelt, 1993). Microscopic examination revealed a male-specific tissue within the abdomen that was composed of very large round cells (100 μm diameter). These cells were connected directly to the tracheal system (Dowd and Bartelt, 1993). Electron microscopy revealed cell structures and organelles typical for secretory cells and demonstrated that the secretory 'end apparatus' of each cell was connected by a fine duct to a tracheal branch (Nardi *et al.*, 1996). These features are shown diagrammatically in Fig. 4.3. The secretory cells were about 1000 times larger in volume in beetles emitting pheromone than in those that were not (Nardi *et al.*, 1996).

Pheromone biosynthesis and composition

The pheromone hydrocarbon structures suggested a polyketide biosynthesis involving just acetate, propionate and butyrate acyl units (Bartelt *et al.*, 1990b, 1991, 1992b). This suggestion was demonstrated experimentally with 12-carbon triene **1** (Table 4.1) from *C. freemani* (Petroski *et al.*, 1994b) and later with the tetraenes and other trienes of *C. davidsoni* and *C. mutilatus* (Bartelt and Weisleder, 1996) (see Fig. 4.4 for example). The biosynthetic experiments exploited the tolerance of these beetles to large amounts of organic acids in their diets. Acids labelled with deuterium or carbon-13 became incorporated into the pheromones of the beetles, and locations of the heavy isotope atoms were then determined by NMR and MS. The biochemical reactions were proposed to be as in fatty acid biosynthesis, except that the double bonds are not reduced (Fig. 4.4; Petroski *et al.*, 1994b). The carboxyl carbon of the final acyl unit is removed during biosynthesis (Bartelt and Weisleder, 1996), but the mechanism for this is not known.

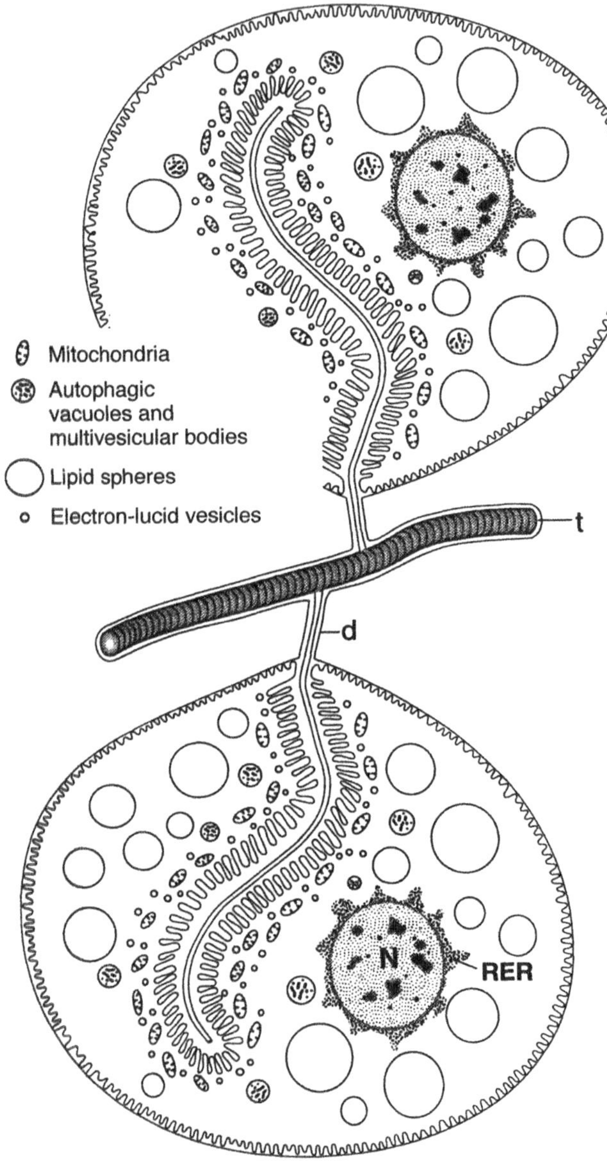

Fig. 4.3. Diagrammatic view of the large pheromone-secreting cells from *C. freemani* males, based on study by electron microscopy. Cell diameter is about 100 μm. **N** = nucleus; **RER** = rough endoplasmic reticulum; **t** = tracheal branch; **d** = ductule, which is the outlet from the secretory 'end apparatus' that transits each cell. (Drawn by J.B. Nardi and reprinted with permission from Nardi *et al.*, 1996.)

Another biosynthetic issue is the origin of complex hydrocarbon blends, such as the 11 different tetraenes of C. *hemipterus* (Table 4.1). While there could be separate biosynthetic systems for each of the C. *hemipterus* tetraenes, a more

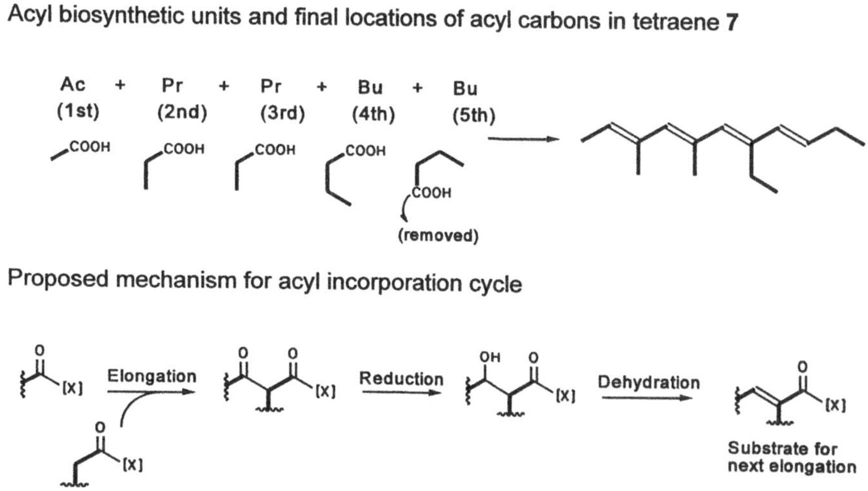

Fig. 4.4. Biosynthesis of tetraene **7** (see Table 4.1) from acyl precursors. **Ac** = acetate; **Bu** = butyrate; **Pr** = propionate; **[X]** is an unknown acyl carrier such as coenzyme A.

economical explanation, which fits the data, is that a single biosynthetic system exists with imperfect selectivity for acyl units (Table 4.2; Bartelt *et al.*, 1992b). If the most abundant tetraene, **4**, represents the normal product, then tetraenes **5**, **6**, **13** and **17** represent instances of one acyl substitution (or biosynthetic 'mistake'). There are six possible ways in which two of these substitutions could exist in one compound, and these are represented by **7**, **14**, **15**, **18**, **19** and **20**. Occurrence of two substitutions would be rarer than just one, and the observed abundances reflected this expectation. Tetraenes with three or four substitutions would be even rarer, and these were not detected. Substitutions were never observed for the second acyl unit, which was unfailingly propionate. Related arguments can be made for the patterns of hydrocarbons in *C. freemani* and *C. davidsoni* (Bartelt *et al.*, 1990b; Bartelt and Weisleder, 1996).

The functional pheromone of *C. hemipterus* appears to be a subset of the male-specific hydrocarbons (**4**–**7** in Table 4.2; Bartelt *et al.*, 1992b). The compounds of *C. hemipterus* that were active had structural features in common; as drawn in Table 4.2, the left-hand portions of the active compounds were identical. Petroski and Vaz (1995) used computer-based molecular modelling to correlate molecular shape with biological activity.

If some of the male-specific hydrocarbons are merely biosynthetic artefacts, then lack of behavioural activity in some cases would not be surprising. However, the inactive structures would represent a reservoir of male-specific compounds that could become pheromone components over evolutionary time once the beetles began to sense them. Natural selection operating both on the frequency of particular acyl substitutions during biosynthesis and on the antennal sensitivity to various structures could eventually shift pheromone composition. Such shifts must have occurred during speciation in the genus. It is striking that the major

Table 4.2. Origins of male-specific tetraenes of *C. hemipterus* from simple acyl biosynthetic units, relative abundances, and bioassay activities in wind tunnel (Bartelt *et al.*, 1992b).

Acyl unit[a]							Relative	Bioassay
1st	2nd	3rd	4th	5th		Structure[b]	abundance	activity
Main pattern of acyl units (tetraene **4**)								
Ac	Pr	Pr	Pr	Pr	**4**	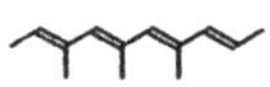	100	Yes
One acyl substitution, relative to tetraene **4**								
Pr*	Pr	Pr	Pr	Pr	**17**	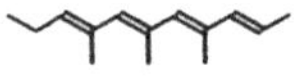	6	No
Ac	Pr	Bu*	Pr	Pr	**13**	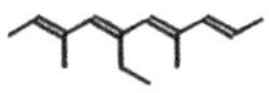	3	No
Ac	Pr	Pr	Bu*	Pr	**5**	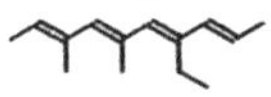	8	Yes
Ac	Pr	Pr	Pr	Bu*	**6**	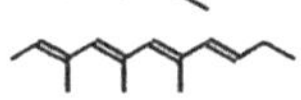	13	Yes
Two acyl substitutions, relative to tetraene **4**								
Pr*	Pr	Bu*	Pr	Pr	**18**	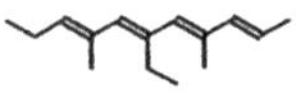	1	No
Pr*	Pr	Pr	Bu*	Pr	**19**	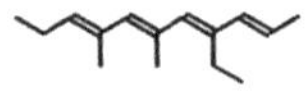	1	No
Pr*	Pr	Pr	Pr	Bu*	**20**	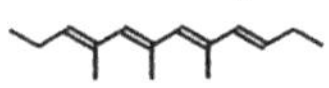	1	No
Ac	Pr	Bu*	Bu*	Pr	**14**	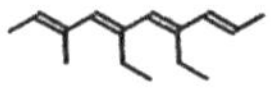	0.2	No
Ac	Pr	Bu*	Pr	Bu*	**15**	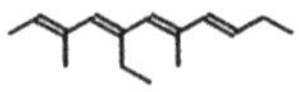	0.5	No
Ac	Pr	Pr	Bu*	Bu*	**7**	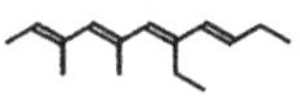	2	Yes

[a]Ac, acetate; Pr, propionate; Bu, butyrate. Deviations from the main incorporation pattern indicated by *.
[b]Structure numbers as in Table 4.1.

components of all species in Table 4.1 occur in other species as minor male-specific compounds. The genus could be very useful for studying the evolution of pheromone systems.

Factors affecting field activity of pheromones

A number of factors affected the performance of synthetic pheromones in the field, such as trap type, trap placement, pheromone formulation and the presence of food-related odours. The traps used in most studies were of a wind-directed, funnel design (Fig. 4.5) and were constructed from PVC plumbing pipe, screen wire and other common materials (Dowd *et al.*, 1992). The modular design allows any trap section to be modified, if necessary, without having to replace the entire trap. Other successful trap designs included Mason jars with plastic funnels attached at the mouth (known as 'NIT' traps: Williams *et al.*, 1993), Japanese beetle traps

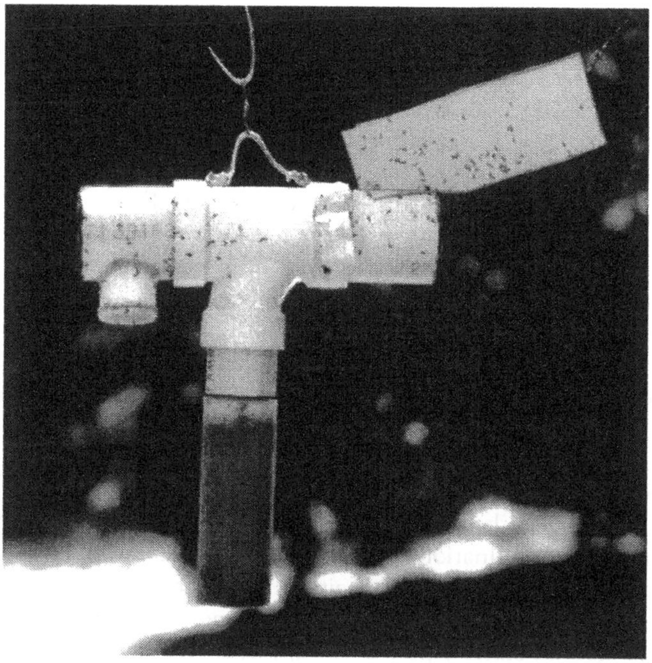

Fig. 4.5. Wind-directed funnel trap ('Dowd' trap) in operation in a date garden in California. The fin orients the trap so that the opening (just below the fin) is accessible to beetles approaching the trap from downwind. The pheromone and food baits are held in the trap compartment at the left; screens prevent the insects from contacting the baits. Attracted beetles eventually enter the bottle attached at the bottom of the trap. This trap contains about 100,000 *C. mutilatus*, captured over a period of 3 days. (Photograph by Richard S. Vetter.)

(Williams *et al.*, 1993) and various bucket, funnel and other traps of commercial or laboratory design (James *et al.*, 1996a).

Effects of trap height varied among species. For *C. hemipterus* and *C. mutilatus*, wind-directed 'Dowd' traps 3 m above the ground captured 3–4 times more beetles than those 0.3 m above the ground, but *Carpophilus humeralis* (Fabricius), responding to the pheromones of *C. hemipterus* and *C. mutilatus*, was caught 16 times more frequently at 0.3 m than at 3 m (Bartelt *et al.*, 1994a). In Ohio, *C. lugubris* was trapped more effectively in 'NIT' traps at 1 m above the ground than at ground level (Williams *et al.*, 1993).

Rubber septa were used as pheromone dispensers. A practical problem was that components of multi-component pheromones sometimes varied greatly in volatility. Therefore, the amounts of the less volatile components were increased so that the septa would release blends similar to the natural proportions listed in Table 4.1 (e.g. Bartelt and James, 1994; Williams *et al.*, 1995). The amount of pheromone per septum was usually 500 μg but was varied in some experiments. The synthetic pheromones used in field tests were typically 70–80% pure. The major impurities were Z isomers of the pheromone components (Bartelt *et al.*, 1990c), which were not inhibitory at such low levels.

Effects of pheromone dose on trap catch have been studied. High pheromone doses never became repellent. In California, all doses of C. *hemipterus* pheromone from 15 to 15,000 μg per septum were significantly active in the field; and for C. *mutilatus*, all doses from 50 to 15,000 μg were active (Bartelt *et al.*, 1994a). Attractiveness increased with pheromone dose throughout the range for both species. Similar trends were noted for these species in Australia when 500 μg and 5000 μg doses were compared (James *et al.*, 1994).

Septa were normally used in the field for 2 weeks and then replaced. Activity decreased during this time, inferred from trapping experiments in which septa of different ages were present simultaneously. In an all-season study in southern California, trap catches of C. *freemani*, which has the most volatile of the pheromones, were only 9–13% as high during the second week as they were during the first week (Bartelt *et al.*, 1994b). Catches decreased more slowly when less volatile pheromones were used. Catches of C. *hemipterus* declined to 54–66% during the second week, and catches of C. *mutilatus* to 36–39% (Bartelt *et al.*, 1994b). Other studies provided similar values (Bartelt *et al.*, 1992a, 1994a).

Synergism between the pheromones and food odours was remarkable. In one study, the combination of C. *hemipterus* pheromone and food volatiles caught over 75 times more beetles than either the pheromone or food alone (Bartelt *et al.*, 1994b). Typical examples of food–pheromone synergism are shown in Fig. 4.6. The numbers of beetles captured in traps with pheromone plus food could be dramatic. For example, over 100,000 C. *mutilatus* per trap were caught in just 3 days in a planting of dates in California (Bartelt *et al.*, 1994a).

Many kinds of food-type materials were used successfully in the field as pheromone synergists. These included fermenting fig juice (Bartelt *et al.*, 1990b, 1992a), rotting grapefruit (Blumberg *et al.*, 1993), blends of synthetic compounds typical of yeast fermentations (Lin *et al.*, 1992), empirical blends of synthetic compounds (Bartelt *et al.*, 1992a) and fermenting bread dough (see Williams *et al.*, 1993). The bread dough has been the 'standard' source of food odours in research studies because it is easy to use, effective and fairly reproducible. Except where indicated above, it was used in all of the cited field studies.

Cross-attraction

Cross-attraction to the synthetic *Carpophilus* pheromones was observed in field studies (Table 4.3). Much of this was probably due to pheromones having components in common. The pheromones of C. *hemipterus* and C. *brachypterus* differ only slightly in proportions of components, and both species responded about equally to both synthetic mixtures (Williams *et al.*, 1995). Another example is that the pheromones of C. *hemipterus* and C. *lugubris* contain minor amounts of the only known pheromone component of C. *obsoletus*, and C. *obsoletus* was readily attracted to these synthetic pheromones as well as to its own. Nevertheless, all the species showed a general preference for their own blends. Even with nearly identical pheromones, cross-attraction would still not occur in nature if the species have temporal or geographical separation. For example, C. *brachypterus* flies earlier in spring than C. *hemipterus* in Illinois (R.J. Bartelt, 1998, unpublished results), and

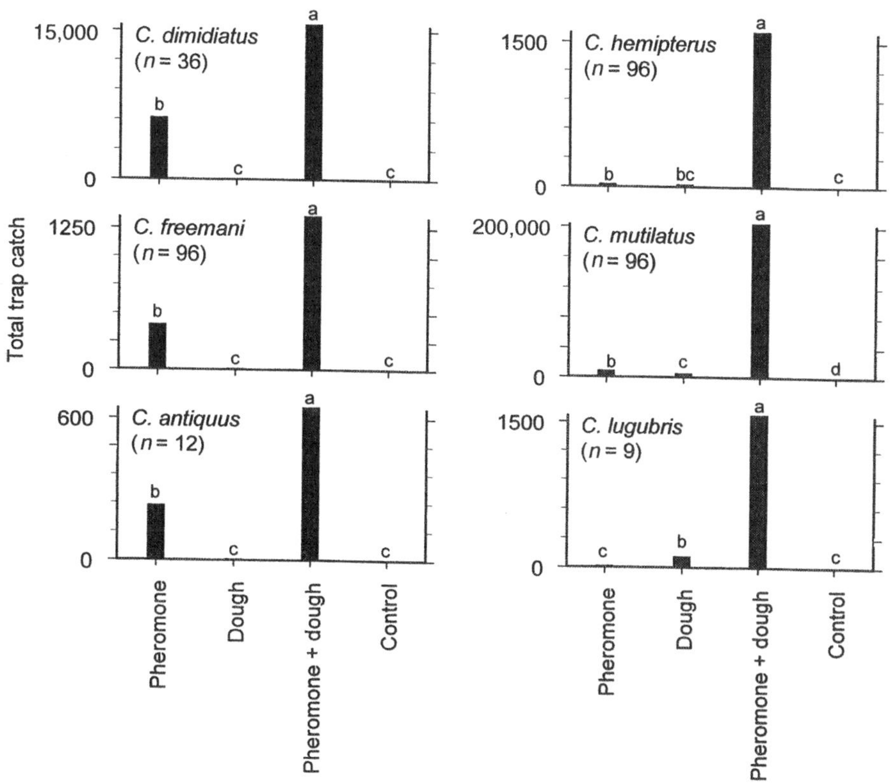

Fig. 4.6. Synergistic responses of six *Carpophilus* species to their pheromones plus fermenting bread dough in the field. In each panel, trap catch totals accompanied by the same letter were not significantly different. *C. dimidiatus*: Bartelt *et al.* (1995b); *C. hemipterus, C. freemani* and *C. mutilatus*: Bartelt *et al.* (1994b); *C. antiquus*: Bartelt *et al.* (1993b); *C. lugubris*: Bartelt *et al.* (1991).

C. *antiquus* has a more northern range than C. *dimidiatus* (Bartelt *et al.*, 1993b, 1995b).

Some of the observed cross-attraction may have been an artefact of unnaturally high pheromone releases from septa; for example, the response threshold of C. *mutilatus* to the pheromone of C. *hemipterus* was at least ten times higher than to its own (Bartelt *et al.*, 1994a). Furthermore, the ratios of components emitted from septa were not fully stable over time (Bartelt and James, 1994); therefore, the preference for a species' own blend may not have been as clear as possible, which would make cross-attraction appear relatively strong.

One case of cross-attraction seems to have a kairomonal basis. C. *antiquus* responds very clearly to the C. *lugubris* pheromone, yet these species do not have any components in common. In sampled corn ears, C. *antiquus* was seldom found except when C. *lugubris* was also present (Bartelt *et al.*, 1993b). Thus, C. *antiquus* appears to use the C. *lugubris* pheromone as a cue for locating a host site.

Table 4.3. Cross attraction in the field to pheromones of nine *Carpophilus* species[a].

Responding species	hem/ bra	obs	lug	dav	fre	mut	ant/ dim
	Pheromone tested						
Species for which pheromones are known							
C. hemipterus	+++	+	+	++	±	–	–
C. brachypterus	+++	+	±	nt	–	–	–
C. obsoletus	++	+++	++	nt	–	–	nt
C. lugubris	+	++	+++	nt	+	–	–
C. davidsoni	+	+	+	+++	+	+	nt
C. freemani	±	±	+	nt	+++	+	–
C. mutilatus	+	+	+	+	+	+++	–
C. antiquus	+	+	++	nt	+	–	+++
C. dimidiatus	±	±	±	nt	±	–	+++
Species for which pheromones are not known							
C. humeralis	+	+	+	nt	±	±	+
C. corticinis	–	+	++	nt	+	–	–
C. marginatus	+	+	+	nt	+	–	–
C. marginellus	+	+	++	++	++	–	±
C. sayi	+	+	++	nt	+	–	–
C. gaveni	±	nt	nt	+	nt	+	nt
Colopterus sp.	+	+	+	nt	+	–	–
Stelidota geminata	–	nt	–	nt	–	nt	nt
Haptoncus luteolus	–	–	–	nt	–	–	–
Glischrochilus quadrisignatus	–	nt	–	nt	–	nt	nt

[a]Relative degree of attraction: +++ > ++ > + > ± > –, where +++ = response to species' own pheromone plus dough, and –, response to dough control; nt, not tested. hem, *C. hemipterus*; bra, *C. brachypterus*; obs, *C. obsoletus*; lug, *C. lugubris*; dav, *C. davidsoni*; fre, *C. freemani*; mut, *C. mutilatus*; ant, *C. antiquus*; dim, *C. dimidiatus*.

Responses of other species may provide clues about unstudied pheromones. *C. corticinis* Erichson, *C. marginellus* Motschulsky and *C. sayi* Parsons were particularly attracted to the pheromone of *C. lugubris* and to other pheromones containing tetraene **7** (Bartelt *et al.*, 1995b; James *et al.*, 1995; Williams *et al.*, 1995). *C. humeralis* was generally attracted to pheromones that contained primarily tetraenes (Blumberg *et al.*, 1993; Bartelt *et al.*, 1994b), while *C. gaveni* Dobson responded best to pheromones with trienes (James *et al.*, 1995). Efforts to demonstrate pheromones in the laboratory with *C. marginellus* and *C. humeralis* have been unsuccessful.

Sap beetles of other genera were also captured during trapping studies. *Colopterus* sp. – probably *Colopterus truncatus* (Randall) – which belongs to the same subfamily as *Carpophilus*, Carpophilinae, clearly responded to several of the pheromones (Williams *et al.*, 1995). However, *Stelidota geminata* (Say), *Haptoncus luteolus* (Erichson) and *Glischrochilus quadrisignatus* (Say), which belong to other subfamilies, were captured only because of the food co-attractants and did not respond to any of the *Carpophilus* pheromones (Williams *et al.*, 1993; Bartelt *et al.*,

1994b). Other instances of cross attraction were reported by Bartelt *et al.* (1992a) and James *et al.* (1993, 1994).

Practical Uses

The pheromones provide a rapid, simple and sensitive monitoring method for the beetle populations. Due to their small size and cryptic nature, the beetles are difficult to sample in other ways. Flight periods of *Carpophilus* species have been monitored with pheromone traps during long-term studies in the USA (Bartelt *et al.*, 1994b, 1995b) and in Australia (James *et al.*, 1994, 1995). Trap catches, together with additional data for temperature and rainfall, can be used to predict the likelihood of economic damage to Australian stone fruits by *C. davidsoni* (James *et al.*, 1997). Abreu (1997) demonstrated in Brazil that *C. dimidiatus* and *C. obsoletus* could be detected in cocoa warehouses with pheromone traps. Monitoring traps have also been used to survey beetles in different habitats for infestation by natural enemies such as nematodes (Dowd *et al.*, 1995); these efforts resulted in the discovery of a new nematode species (Poinar and Dowd, 1997).

The pheromones may be useful for direct beetle control as well as for monitoring. The potential for mass trapping as a control technique was suggested by the massive captures of *C. mutilatus* in Californian date gardens (Bartelt *et al.*, 1994a,b) and by the fact that both sexes are captured (rather than just males as with moths). In Australia, mass trapping has been tried for the protection of stone fruit crops. Traps were able to provide significant protection for ripening apricots when placed in a cordon around the planting (at best, 5% infestation in protected area versus 85–90% infestation in perimeter and control areas: James *et al.*, 1996b). However, traps placed directly in the trees to be protected caused heightened infestations because many beetles responding to the traps colonized fruit in the immediate area instead of being captured (James *et al.*, 1996b). Optimal trap placement will be critical to the success of mass-trapping efforts, and additional research is needed.

The existence of cross-attraction suggested that it might be possible to combine several pheromones into one lure. Managing the sap beetle complex in any crop could be simpler if multi-species lures were effective. The idea was explored by field testing all two-way and three-way combinations of the pheromones for *C. hemipterus*, *C. mutilatus* and *C. freemani*. These combinations were as effective as the separate pheromones for all species except for a modest decrease in the response of *C. freemani* (Bartelt *et al.*, 1995a). Three-way pheromone combinations were also used successfully in Australia for *C. davidsoni*, *C. mutilatus* and *C. hemipterus* (James *et al.*, 1996a).

Pheromones can be used in conjunction with various microorganisms to manipulate or manage beetle populations. An autoinoculating device has been invented that causes feral beetles to come into contact with whatever micro-organism or material the experimenter intends (Vega *et al.*, 1995). For example, *C. lugubris* in field corn can be induced to pick up a formulation of the harmless bacterium, *Bacillus subtilis* Ehrenberg (Cohn). The beetles then disperse into the crop and colonize ears, typically those that have been previously damaged, introducing *B. subtilis* as they go. These sites would normally be subject to infection by *Aspergillus flavus*, which produces aflatoxin, but *A. flavus* does not become established when *B. subtilis* is present. This approach would capitalize on the natural behaviour of a potential pest to protect the crop from a dangerous fungal metabolite and would require very little input from the grower.

Direct control of the beetles could be exerted by inoculating attracted beetles with insecticidal nematodes (Dowd *et al.*, 1995) or a fungus such as *Beauveria bassiana* (Balsamo) Vuillemin (Vega *et al.*, 1995). The advantage of such inoculation rather than immediate killing with an insecticide is the potential for horizontal transmission of the pathogen to other beetles in the immediate area, magnifying the effect of the treatment.

Finally, the pheromones may be of use in enhancing a beneficial activity of the beetles. Nitidulids are primary pollinators of custard apple (*Annona*) flowers, but poor pollination has often been a problem (Gazit *et al.*, 1982; Nagel *et al.*, 1989). The pheromones could be placed in *Annona* trees to increase beetle density and enhance pollination.

Conclusions

Study of the *Carpophilus* pheromones has revealed a series of new compounds with potent biological activity and good pest management potential. The research has provided initial information in the areas of pheromone biology and biosynthesis. *Carpophilus* would be a good model group for further, in-depth exploration of a male-produced pheromone system because the beetles are easily reared and manipulated, because many species are available for comparative studies, and because fascinating relationships between food and pheromones are experimentally accessible. The pheromones have shown good promise to be useful pest management tools, but a number of practical needs must still be addressed. Key among these is the availability of long-lasting, commercial pheromone formulations. These, together with corresponding commercial traps and synthetic 'host' volatile formulations, would allow the development of solid new management methodology for these beetles to continue.

Acknowledgements

The author is grateful to colleagues at NCAUR, at other locations in the USA, in Israel, in Brazil, and especially in Australia, for their contributions to the *Carpophilus* pheromone project.

References

de Abreu, J.M. (1997) Attraction of *Carpophilus dimidiatus* and *C. obsoletus* (Coleoptera: Nitidulidae) to aggregation pheromones in stored cocoa. *Agrotrópica* 9, 41–48.

Baker, T.C., Francke, W., Millar, J.G., Löfstedt, C., Hansson, B., Du, J.-W., Phelan, P.L., Vetter, R.S., Youngman, R. and Todd, J.L. (1991) Identification and bioassay of sex pheromone components of carob moth, *Ectomyelois ceratoniae* (Zeller). *Journal of Chemical Ecology* 17, 1973–1988.

Bartelt, R.J. and James, D.G. (1994) Aggregation pheromone of Australian sap beetle, *Carpophilus davidsoni* (Coleoptera: Nitidulidae). *Journal of Chemical Ecology* 20, 3207–3219.

Bartelt, R.J. and Weisleder, D. (1996) Polyketide origin of pheromones of *Carpophilus davidsoni* and *C. mutilatus* (Coleoptera: Nitidulidae). *Bioorganic and Medicinal Chemistry* 4, 429–438.

Bartelt, R.J., Dowd, P.F., Plattner, R.D. and Weisleder, D. (1990a) Aggregation pheromone of driedfruit beetle, *Carpophilus hemipterus*: windtunnel bioassay and identification of two novel tetraene hydrocarbons. *Journal of Chemical Ecology* 16, 1015–1039.

Bartelt, R.J., Dowd, P.F., Shorey, H.H. and Weisleder, D. (1990b) Aggregation pheromone of *Carpophilus freemani* (Coleoptera: Nitidulidae): a blend of conjugated triene and tetraene hydrocarbons. *Chemoecology* 1, 105–113.

Bartelt, R.J., Weisleder, D. and Plattner, R.D. (1990c) Synthesis of nitidulid beetle pheromones: alkyl-branched tetraene hydrocarbons. *Journal of Agricultural and Food Chemistry* 38, 2192–2196.

Bartelt, R.J., Dowd, P.F. and Plattner, R.D. (1991) Aggregation pheromone of *Carpophilus lugubris*: new pest management tools for the nitidulid beetles. In: Hedin, P.A. (ed.) *Naturally Occurring Pest Bioregulators*. American Chemical Society Symposium Series 449, American Chemical Society, Washington, DC, pp. 27–40.

Bartelt, R.J., Dowd, P.F., Vetter, R.S., Shorey, H.H. and Baker, T.C. (1992a) Responses of *Carpophilus hemipterus* (Coleoptera: Nitidulidae) and other sap beetles to the pheromone of *C. hemipterus* and host-related coattractants in California field tests. *Environmental Entomology* 21, 1143–1153.

Bartelt, R.J., Weisleder, D., Dowd, P.F. and Plattner, R.D. (1992b) Male-specific tetraene and triene hydrocarbons of *Carpophilus hemipterus*: structure and pheromonal activity. *Journal of Chemical Ecology* 18, 379–402.

Bartelt, R.J., Carlson, D.G, Vetter, R.S. and Baker, T.C. (1993a) Male-produced aggregation pheromone of *Carpophilus mutilatus* (Coleoptera: Nitidulidae). *Journal of Chemical Ecology* 19, 107–118.

Bartelt, R.J., Seaton, K.L. and Dowd, P.F. (1993b) Aggregation pheromone of *Carpophilus antiquus* (Coleoptera: Nitidulidae) and kairomonal use of *C. lugubris* pheromone by *C. antiquus*. *Journal of Chemical Ecology* 19, 2203–2216.

Bartelt, R.J., Vetter, R.S., Carlson, D.G. and Baker, T.C. (1994a) Influence of pheromone dose, trap height, and septum age on effectiveness of pheromones for *Carpophilus mutilatus* and *C. hemipterus* (Coleoptera: Nitidulidae) in a California date garden. *Journal of Economic Entomology* 87, 667–675.

Bartelt, R.J., Vetter, R.S., Carlson, D.G. and Baker, T.C. (1994b) Responses to aggregation pheromones for five *Carpophilus* species (Coleoptera: Nitidulidae) in a California date garden. *Environmental Entomology* 23, 1534–1543.

Bartelt, R.J., Vetter, R.S., Carlson, D.G., Petroski, R.J. and Baker, T.C. (1995a) Pheromone combination lures for *Carpophilus* (Coleoptera: Nitidulidae) species. *Journal of Economic Entomology* 88, 864–869.

Bartelt, R.J., Weaver, D.K. and Arbogast, R. (1995b) Aggregation pheromone of *Carpophilus dimidiatus* (F.) (Coleoptera: Nitidulidae) and responses to *Carpophilus* pheromones in South Carolina. *Journal of Chemical Ecology* 21, 1763–1779.

Blumberg, D., Goldenberg, S., Bartelt, R.J. and Williams, R.N. (1993) Responses to synthetic aggregation pheromones, host-related volatiles, and their combinations by *Carpophilus* spp. (Coleoptera: Nitidulidae) in laboratory and field tests. *Environmental Entomology* 22, 837–842.

Connell, W.A. (1956) *Nitidulidae of Delaware*. Bulletin Number 318 (Technical), University of Delaware, Agricultural Experiment Station, Newark, 67 pp.

Connell, W.A. (1991) Sap beetles (Nitidulidae, Coleoptera). In: Gorham, J.R. (ed.) *Insect and Mite Pests in Food*, Vol. 1. Agricultural Handbook Number 65, United States Department of Agriculture and United States Department of Health and Human Services, United States Government Printing Office, Washington, DC, pp. 151–174.

Dowd, P.F. (1991) Nitidulids as vectors of mycotoxin-producing fungi. In: Shotwell, O.L. and Hurburgh, C.R., Jr (eds) *Aflatoxin in Corn: New Perspectives*. Research Bulletin 599, Iowa Agriculture and Home Economics Experiment Station, Iowa State University, Ames, pp. 335–342.

Dowd, P.F. (1995) Sap beetles and mycotoxins in maize. *Food Additives and Contaminants* 12, 497–508.

Dowd, P.F. and Bartelt, R.J. (1991) Host-derived volatiles as attractants and pheromone synergists for driedfruit beetle, *Carpophilus hemipterus*. *Journal of Chemical Ecology* 17, 285–308.

Dowd, P.F. and Bartelt, R.J. (1993) Aggregation pheromone glands of *Carpophilus freemani* (Coleoptera: Nitidulidae) and gland distribution among other sap beetles. *Annals of the Entomological Society of America* 86, 464–469.

Dowd, P.F., Bartelt, R.J. and Wicklow, D.T. (1992) Novel insect trap useful in capturing sap beetles (Coleoptera: Nitidulidae) and other flying insects. *Journal of Economic Entomology* 85, 772–778.

Dowd, P.F., Moore, D.E., Vega, F.E., McGuire, M.R., Bartelt, R.J., Nelson, T.C. and Miller, D.A. (1995) Occurrence of a mermithid nematode parasite of *Carpophilus lugubris* (Coleoptera: Nitidulidae) in central Illinois. *Environmental Entomology* 24, 1245–1251.

Gazit, S., Galon, I. and Podoler, H. (1982) The role of nitidulid beetles in natural pollination of annona in Israel. *Journal of the American Horticultural Society* 107, 849–852.

Hinton, H.E. (1945) *A Monograph of the Beetles Associated with Stored Products*. Jarrold and Sons, Norwich, 443 pp.

James, D.G., Bartelt, R.J., Faulder, R.J. and Taylor, A. (1993) Attraction of Australian *Carpophilus* spp. (Coleoptera: Nitidulidae) to synthetic pheromones and fermenting bread dough. *Journal of the Australian Entomological Society* 32, 339–345.

James, D.G., Bartelt, R.J. and Faulder, R.J. (1994) Attraction of *Carpophilus* spp. (Coleoptera: Nitidulidae) to synthetic aggregation pheromones and host-related coattractants in Australian stone fruit orchards: beetle phenology and pheromone dose studies. *Journal of Chemical Ecology* 20, 2805–2819.

James, D.G., Faulder, R.J. and Bartelt, R.J. (1995) Fauna and seasonal abundance of *Carpophilus* spp. (Coleoptera: Nitidulidae) in four stone fruit growing regions of southeastern Australia as determined by pheromone-trapping. *Journal of the Australian Entomological Society* 34, 327–333.

James, D.G., Bartelt, R.J. and Moore, C.J. (1996a) Trap design effect on capture of *Carpophilus* spp. (Coleoptera: Nitidulidae) using synthetic aggregation pheromones and a coattractant. *Journal of Economic Entomology* 89, 648–653.

James, D.G., Bartelt, R.J. and Moore, C.J. (1996b) Mass-trapping of *Carpophilus* spp. (Coleoptera: Nitidulidae) in stone fruit orchards using synthetic aggregation pheromones and a coattractant: development of a strategy for population suppression. *Journal of Chemical Ecology* 22, 1541–1556.

James, D.G., Faulder, R.J., Vogele, B., Bartelt, R.J. and Moore, C.J. (1997) Phenology of *Carpophilus* spp. (Coleoptera: Nitidulidae) in stone fruit orchards as determined by pheromone trapping: implications for prediction of crop damage. *Australian Journal of Entomology* 36, 165–173.

Lin, H., Phelan, P.L. and Bartelt, R.J. (1992) Synergism between synthetic food odors and the aggregation pheromone for attracting *Carpophilus lugubris* in the field (Coleoptera: Nitidulidae). *Environmental Entomology* 21, 156–159.

Lindgren, D.L. and Vincent, L.E. (1953) Nitidulid beetles infesting California dates. *Hilgardia* 22, 97–118.

Lussenhop, J. and Wicklow, D.T. (1990) Nitidulid beetles (Nitidulidae: Coleoptera) as vectors of *Aspergillus flavus* in pre-harvest maize. *Transactions of the Mycological Society of Japan* 31, 63–74.

Nagel, J., Peña, J.E. and Habeck, D. (1989) Insect pollination of atemoya in Florida. *Florida Entomologist* 72, 207–211.

Nardi, J.B., Dowd, P.F. and Bartelt, R.J. (1996) Fine structure of cells specialized for secretion of aggregation pheromone in a nitidulid beetle *Carpophilus freemani* (Coleoptera: Nitidulidae). *Tissue and Cell* 28, 43–52.

Petroski, R.J. (1997) Improved preparation of *Carpophilus freemani* aggregation pheromone. *Synthetic Communications* 27, 3279–3289.

Petroski, R.J. and Vaz, R. (1995) Insect aggregation pheromone response synergized by 'host-type' volatiles. Molecular modeling evidence for close proximity binding of pheromone and coattractant in *Carpophilus hemipterus* (L.) (Coleoptera: Nitidulidae). In: Reynolds, C.H., Holloway, M.K. and Cox, H.K. (eds) *Computer-Aided Molecular Design*. American Chemical Society Symposium Series 589, American Chemical Society, Washington, DC, pp. 197–210.

Petroski, R.J. and Weisleder, D. (1997) Inhibition of *Carpophilus freemani* aggregation pheromone response by a Z-double-bond pheromone analog. *Journal of Agricultural and Food Chemistry* 45, 943–945.

Petroski, R.J., Bartelt, R.J. and Vetter, R.S. (1994a) Male-produced aggregation pheromone of *Carpophilus obsoletus* (Coleoptera: Nitidulidae). *Journal of Chemical Ecology* 20, 1483–1493.

Petroski, R.J., Bartelt, R.J. and Weisleder, D. (1994b) Biosynthesis of (2*E*,4*E*,6*E*)-5-ethyl-3-methyl-2,4,6-nonatriene: the aggregation pheromone of *Carpophilus freemani* (Coleoptera: Nitidulidae). *Insect Biochemistry and Molecular Biology* 24, 69–78.

Poinar, G.O. and Dowd P.F. (1997) *Psammomermis nitidulensis* n. sp. (Nematoda: Mermithidae) from sap beetles (Coleoptera: Nitidulidae) with biological observations and a key to the species of *Psammomermis*. *Fundamental and Applied Nematology* 20, 207–211.

Taylor, R.W.D. (1994) Methyl bromide – is there any future for this noteworthy fumigant. *Journal of Stored Products Research* 30, 253–260.

Vega, F.E., Dowd, P.F. and Bartelt, R.J. (1995) Dissemination of microbial agents using an autoinoculating device and several insect species as vectors. *Biological Control* 5, 545–552.

Wicklow, D.T. (1991) Epidemiology of *Aspergillus flavus* in corn. In: Shotwell, O.L. and Hurburgh, C.R., Jr (eds) *Aflatoxin in Corn: New Perspectives*. Research Bulletin 599, Iowa Agriculture and Home Economics Experiment Station, Iowa State University, Ames, pp. 315–328.

Williams, R.N., Fickle, D.S., Kehat, M., Blumberg, D. and Klein, M.G. (1983) *Bibliography of the Genus* Carpophilus *Stephens (Coleoptera: Nitidulidae)*. Research Circular 278, Ohio State University, Ohio Agricultural Research and Development Center, Wooster, 95 pp.

Williams, R.N., Fickle, D.S., Bartelt, R.J. and Dowd, P.F. (1993) Responses by adult Nitidulidae (Coleoptera) to synthetic aggregation pheromones, a coattractant, and effects of trap design and placement. *European Journal of Entomology* 90, 287–294.

Williams, R.N., Ellis, M.S. and Bartelt, R.J. (1995) Efficacy of *Carpophilus* aggregation pheromones on nine species in northeastern Ohio and identification of the pheromone of *C. brachypterus*. *Entomologia Experimentalis et Applicata* 77, 141–147.

5 Weevils

ROBERT J. BARTELT

USA–ARS, National Center for Agricultural Utilization Research, Bioactive Agents Research Unit, 1815 N. University St, Peoria, IL 61604, USA

Introduction

The weevils (Curculionidae) constitute the largest family of beetles, numbering over 40,000 species worldwide (Arnett, 1993), and they utilize a vast range of hosts and habitats. A well-developed 'snout' is a characteristic of many members of the family (Fig. 5.1). Some weevil species are serious agricultural pests, and there is a constant need for more effective and environmentally safer ways to minimize their impact. As with other insect groups, this need has spurred investigation of pheromones as potential pest management tools. The current state of knowledge for weevil pheromones varies broadly among species. For some there is only an initial hint that a pheromone exists, but for others there have been detailed behavioural and physiological studies, chemical identifications, pheromone syntheses and commercial development. The pheromone of the boll weevil, *Anthonomus grandis* Boheman, has been the most extensively studied and used so far.

The majority of reported weevil pheromones are produced by males and usually attract both sexes; these aggregation pheromones appear to function over considerable distances; e.g. up to 10 m for the boll weevil (Cross and Mitchell, 1966). However, pheromones have also been reported from females. These include long-range sex pheromones that attract primarily (or exclusively) males, short-range or contact pheromones (operating, for example, within a body length) that allow recognition by conspecific males, and oviposition-deterring pheromones that females use to mark the hosts on which they have laid eggs. Finally, there are examples of weevils using the pheromones of other beetle species as kairomones.

Behaviourally, weevils are relatively difficult to study. They respond to pheromones with short, hopping flights and walking instead of with the sustained, upwind, hovering flights typical of moths, and laboratory bioassays for weevil pheromones are often slow or subtle. Many weevil pheromones consist of blends of components that act synergistically, and the accidental loss of just one key component during chromatographic manipulation could make all of the remaining

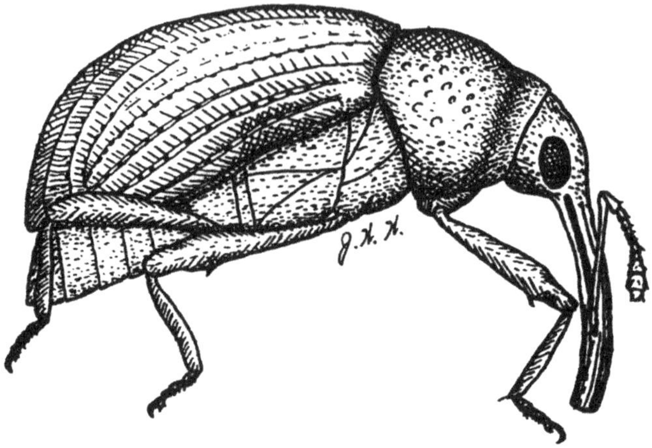

Fig. 5.1. Lateral view of typical weevil (*Anthonomus eugenii* Cano). (Drawn by J.W. Wilson and reprinted from Goff and Wilson, 1937.)

components behaviourally undetectable. Furthermore, activity of the pheromones can be difficult to observe unless appropriate host volatiles are present also. Using laboratory bioassays to guide chromatographic pheromone purification was successful for the boll weevil (Tumlinson *et al.*, 1969) and sweetpotato weevil (Heath *et al.*, 1986) but the approach can be fraught with difficulties.

An alternative research approach that avoids many of the problems with behavioural assays employs gas chromatography with electroantennographic detection (GC–EAD) and has recently yielded rapid progress (e.g. Perez *et al.*, 1997). With GC–EAD it has been possible to recognize exactly which compounds in complex volatile samples have sensory activity. This information, combined with GC comparisons of male- and female-derived volatiles, has allowed quick recognition of sex-specific pheromone components. Subsequent analyses using GC–mass spectrometry, chiral-phase GC columns and appropriate synthetic standards for comparison have led to complete pheromone identifications, including stereochemistry at asymmetric centres, without ever having to isolate the natural compounds in pure form. This approach has also reduced the amount of time and numbers of insects required for pheromone identification. Of course, any pheromone identification ultimately requires behavioural verification with synthetic compounds, preferably under natural field conditions, but such tests become easier when the putative, 'complete' pheromone blend is well defined at the outset.

This chapter surveys 34 weevil species that have been reported to produce or respond to pheromones. These species belong to 11 subfamilies (following the organization of Arnett, 1993). Species infesting stored products and forest trees are also included for comparison. The species and subfamilies are listed in Table 5.1, along with information about hosts, types of pheromones reported, and whether chemical identification has been completed. The literature references are given in the narrative sections dealing with the taxa. These sections are ordered alphabetically by genus.

The cited references primarily cover the pheromones from the initial detection up through chemical identification and verification that the synthetic pheromone has bioassay activity. Additional references are also included for novel findings in pheromone biology and for interactions involving pheromones and host volatiles. Chemical synthetic strategies are not discussed, but many of the cited papers report elegant syntheses of weevil pheromones in addition to their identifications. Reviews are cited to cover the extensive practical development that has attended the pheromones of the boll weevil and sweetpotato weevil.

Chemistry is presented where it is known. The identified pheromone compounds fall into two categories: ten-carbon compounds with terpenoid branching and compounds of various sizes, apparently of fatty-acid (polyketide) origin. These are shown in Figs 5.2 and 5.3, respectively. These pheromone structural categories are consistent within subfamilies, and this pattern is reflected in the organization of the figures. The biosynthetic research on the (terpenoid) pheromone of the boll weevil is reviewed and possible biosynthesis of pheromones from fatty-acid precursors is discussed by Vanderwel and Oehlschlager (1987).

Weevil Pheromones

The boll weevil, Anthonomus grandis *Boheman*

The first curculionid pheromone to be studied was that of the boll weevil, a devastating pest of New World cotton that attacks the flower buds (squares). The initial hint that a pheromone existed in males was that female weevils could be seen moving toward males in the field from as far away as 10 m, but males exhibited no such long-range response (Keller *et al.*, 1964; Cross and Mitchell, 1966). This observation triggered a period of intense research on weevil behaviour, behavioural bioassays and pheromone extraction methods (reviewed by Hardee and Mitchell, 1997) that culminated in the chemical identification and synthesis of the boll weevil pheromone (Tumlinson *et al.*, 1969). The most effective laboratory bioassay was of the two-choice type and involved an arena 30 cm in diameter into which 25 weevils were placed; volatiles from two sources were drawn into the arena from opposite sides, and the number of insects drawn to each source was recorded after 1 or 16 h (Hardee *et al.*, 1967). The pheromone consists of two major terpene alcohols (**1**, named grandisol, and **2**, Fig. 5.2) and two minor terpene aldehydes (**3** and **4**, Fig. 5.2). A positive optical rotation in the *cis* chiral alcohol **1** indicated that one enantiomer was present in excess, but the absolute configuration was unknown. The mixture of all four compounds, including racemic **1**, was active in laboratory bioassays and greatly exceeded the activity of any one component (Tumlinson *et al.*, 1969). The synthetic pheromone was active in the field as well and attracted both sexes (Hardee *et al.*, 1972).

The absolute configuration of **1** was established as (1*R*,2*S*) by synthesis of the enantiomer that had the same optical rotation as the natural compound (Hobbs and Magnus, 1976; Mori, 1978). Both enantiomers of **1** were initially reported to be active in bioassays (Mori *et al.*, 1978), but a re-examination with very pure

Table 5.1. Summary of weevil species (Curculionidae) for which pheromones have been reported.

Species	Subfamily	Principal damage	Pheromone[a]	Chem. ID?
Anthonomus grandis Boheman (boll weevil)	Curculioninae	Cotton buds, bolls	Male Female (SR) Ovip. Det.	Yes Yes? No
Anthonomus eugenii Cano (pepper weevil)	Curculioninae	Pepper pods	Male	Yes
Ceutorhynchus assimilis Paykull (cabbage seed weevil)	Ceutorhynchinae	Cabbage seed pods	Female (LR) Ovip. Det.	No No
Ceutorhynchus floralis Paykull	Ceutorhynchinae	Shepherd's purse seed pods	Ovip. Det.	No
Conotrachelus nenuphar (Herbst) (plum curculio)	Cryptorhynchinae	Unripe fruit of apple, plum	Male Ovip. Det.?	Yes
Cosmopolites sordidus (Germar) (banana weevil)	Rhynchophorinae	Banana plant corms	Male	Yes
Curculio caryae (Horn) (pecan weevil)	Curculioninae	Unripe nuts of pecan	Male	Yes
Cylas formicarius (Fabricius) (sweetpotato weevil)	Apioninae (= Apionidae)	Sweet potato vines, roots, tubers	Female (LR)	Yes
Cylas brunneus (Fabricius) (a sweetpotato weevil)	Apioninae (= Apionidae)	Sweet potato vines, roots, tubers	Female (LR)	No
Cylas puncticollis (Boheman) (a sweetpotato weevil)	Apioninae (= Apionidae)	Sweet potato vines, roots, tubers	Female (LR)	No
Dynamis borassi (Fabricius) (a palm weevil)	Rhynchophorinae	Inflorescence and crown of palms	Male Kairomone	Yes Yes
Diaprepes abbreviatus (L.) (West Indian sugarcane rootstock borer)	Otiorrhynchinae	Citrus foliage	Male? Female (LR)?	No No
Hylobius abietis (L.) (pine weevil)	Hylobiinae	Pine seedling foliage	Male? Female (SR)	No No
Hylobius pales (Herbst) (pales weevil)	Hylobiinae	Pine seedling foliage	Kairomone	Partial
Hypera meles (Fabricius) (cloverhead weevil)	Hyperinae	Clover seed head	Unknown	Partial?
Hypera postica (Gyllenhal) (alfalfa weevil)	Hyperinae	Lucerne (alfalfa) foliage, buds	Unknown	Partial?
Metamasius hemipterus (L.) (West Indian sugarcane borer)	Rhynchophorinae	Sugarcane, banana, palm stems	Male	Yes
Pissodes approximatus Hopkins (northern pine weevil)	Pissodinae	Pine shoots	Male	Yes
Pissodes strobi (Peck) (white pine weevil)	Pissodinae	Pine terminal shoots	Male	Yes
Pissodes nemorensis Germar (deodar weevil)	Pissodinae	Pine boles	Male	Yes
Pissodes terminalis Hopping (lodgepole terminal weevil)	Pissodinae	Pine terminal shoots	Ovip. Det.?	No
Pissodes schwarzi Hopkins (Yosemite bark weevil)	Pissodinae	Pine boles, root collars	Male	No
Rhabdoscelus obscurus (Boisduval) (New Guinea sugarcane weevil)	Rhynchophorinae	Sugarcane	Male	No
Rhynchophorus bilineatus (Montrouzier) (a palm weevil)	Rhynchophorinae	Palm stems	Male	Yes
Rhynchophorus cruentatus (Fabricius) (palmetto weevil)	Rhynchophorinae	Palm stems	Male	Yes
Rhynchophorus ferrugineus (Olivier) (an Asian palm weevil)	Rhynchophorinae	Palm stems	Male	Yes
Rhynchophorus palmarum (L.) (American palm weevil)	Rhynchophorinae	Palm stems	Male	Yes

Table 5.1 (*Continued*).

Species	Subfamily	Principal damage	Pheromone[a]	Chem. ID?
Rhynchophorus phoenicis (F.) (African palm weevil)	Rhynchophorinae	Palm stems	Male	Yes
Rhynchophorus vulneratus (Panzer) (an Asian palm weevil)	Rhynchophorinae	Palm stems	Male	Yes
Sitona lineatus L. (pea and bean weevil)	Brachyderinae	Legume roots	Male	Yes
Sitophilus granarius (L.) (granary weevil)	Rhynchophorinae	Stored grains	Male	Yes
Sitophilus oryzae (L.) (rice weevil)	Rhynchophorinae	Stored rice	Male	Yes
Sitophilus zeamais Motschulsky (maize weevil)	Rhynchophorinae	Stored grains	Male	Yes
Smicronyx fulvus LeConte (red sunflower seed weevil)	Erirrhininae	Sunflower immature seeds	Male	No

[a]Pheromone types reported: male-produced pheromones function, or appear to function, over long range. Female-produced pheromones can be long-range (LR) or short-range (SR). 'Ovip. Det.' indicates oviposition-deterring pheromone deposited by female. 'Kairomone' indicates attraction to pheromone of another species. Uncertainty indicated by '?'.

enantiomers established that the unnatural (1*S*,2*R*) isomer was behaviourally inert (Dickens and Mori, 1989).

Host compounds such as (*E*)-2-hexenol, (*Z*)-3-hexenol, 1-hexanol and various terpenes enhanced the activity of the pheromone in bioassays (Hedin *et al.*, 1979; Dickens, 1989; Dickens *et al.*, 1990). The pheromone has been developed as a tool for boll weevil detection, population monitoring and population suppression (reviewed by Ridgway *et al.*, 1990; Hardee and Mitchell, 1997).

Female boll weevils may also possess a pheromone, but it acts over a shorter distance than that from males (Cross and Mitchell, 1966). Alcohols **1** and **2** (Fig. 5.2) were detected in the faecal material (frass) from females, but only at about 1% of the level occurring in male frass. In laboratory bioassays these alcohols, plus certain cotton terpenes also found in frass, elicited responses primarily from males (McKibben *et al.*, 1977; Hedin *et al.* 1979). Interpretation of these data as a female-produced pheromone assumes that the thousands of female weevils used in the studies were sexed with virtually perfect accuracy, so that the presence of small amounts of the male pheromone alcohols was not merely a technical artefact. It is not known if there are any active female-derived compounds that are not present in frass.

Finally, there is evidence for an oviposition-deterring pheromone (host-marking pheromone) from female boll weevils (Stansly and Cate, 1984, and references therein). It appears to be associated with the frass plug that the female uses to seal the oviposition wound in a host bud. Other females avoid laying eggs in buds that are so marked. This effect is especially evident in weevil populations that subsist on a native host (*Hampea nutricia* Fryxell) which has particularly small buds, but it is less obvious on cotton, whose larger buds can support more than one weevil larva. Stansly and Cate (1984) felt that the habits of the species may have changed after it adopted cotton as a host. The chemistry of the deterrent is unknown.

CURCULIONINAE

Anthonomus grandis

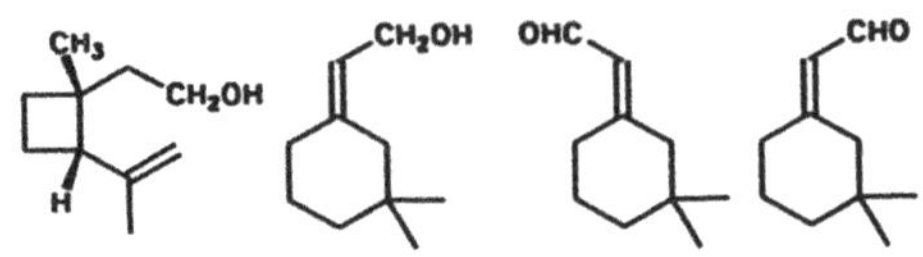

(1*R*,2*S*)-**1** **2** **3** **4**

Anthonomus eugenii

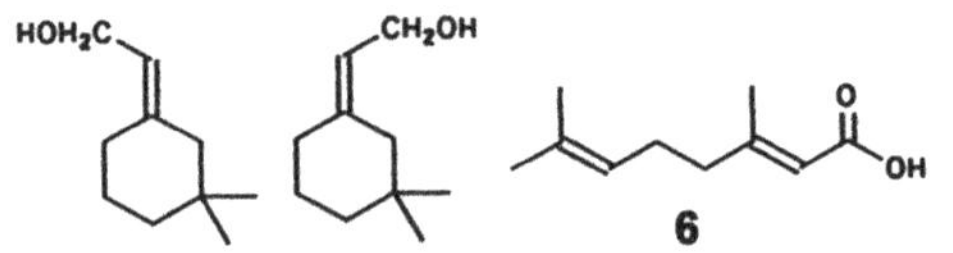

5 **2** **6**

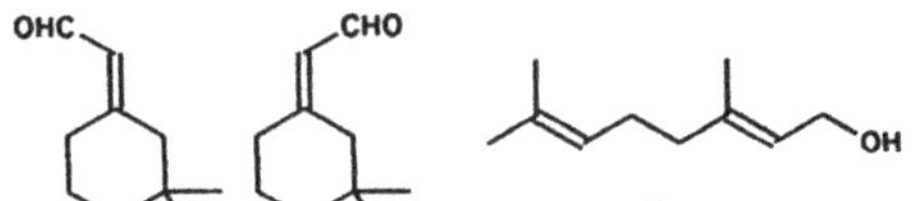

3 **4** **7**

Curculio caryae

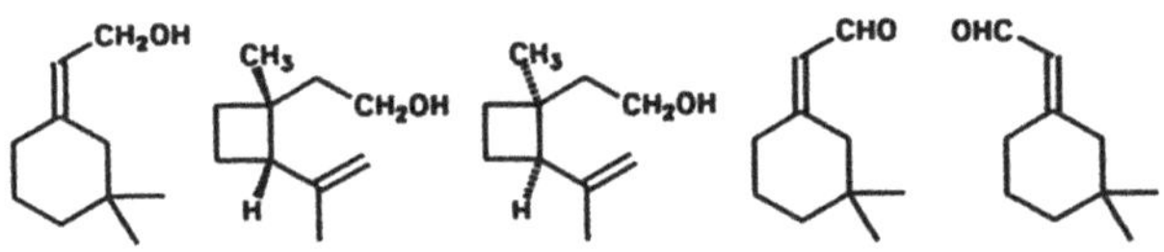

2 (1*R*,2*S*)-**1** (1*S*,2*R*)-**1** **3** **4**

CRYPTORHYNCHINAE

Conotrachelus nenuphar

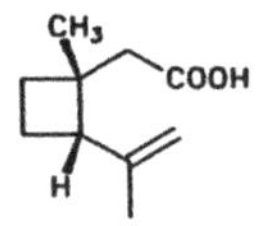

(1*R*,2*S*)-**8**

PISSODINAE

Pissodes approximatus

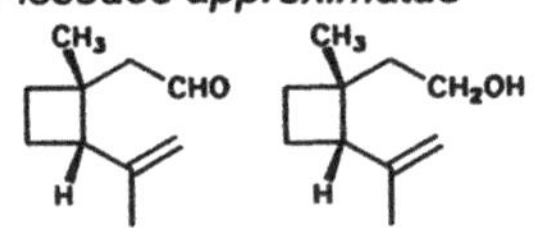

9* **1***

PISSODINAE (continued)

P. nemorensis

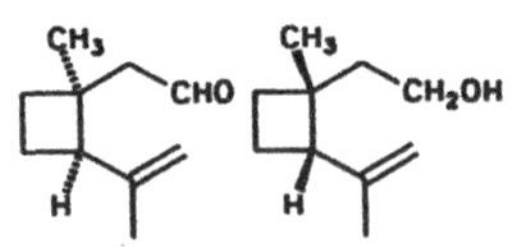

(1*S*,2*R*)-**9** (1*R*,2*S*)-**1**

P. strobi

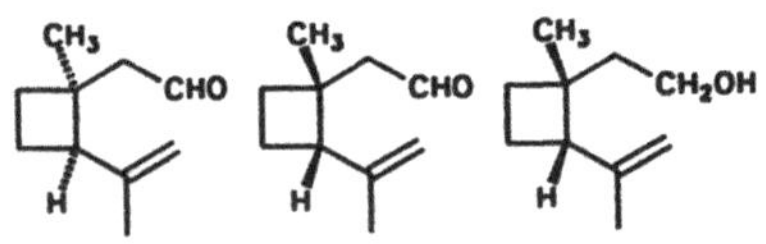

(1*S*,2*R*)-**9** (1*R*,2*S*)-**9** (1*R*,2*S*)-**1**

Fig. 5.2. Weevil pheromones having terpenoid structures. For each species, the first compound is the most abundant. * = absolute stereochemistry unknown.

The pepper weevil, Anthonomus eugenii *Cano*

The pepper weevil is a major pest of peppers (*Capsicum* spp.) in the southern USA, Mexico and Central America. As with *A. grandis*, males produce a pheromone that attracts both sexes (Coudriet and Kishaba, 1988; Patrock *et al.*, 1992). Eller *et al.* (1994) identified six male-specific oxygenated terpenoids from volatile collections of feeding weevils (Fig. 5.2), including three boll weevil pheromone components (**2**, **3** and **4**), an additional cyclohexylidene alcohol (**5**), geranic acid (**6**) and geraniol (**7**). Overall attraction to sticky traps baited with synthetic pheromone was clear in field experiments, but there was some inconsistency among trials. Additional compounds from *A. eugenii* or its host may still be required for optimal response.

Ceutorhynchus *species*

The cabbage seed weevil, *C. assimilis* Paykull, is an important pest of oilseed rape and other brassicas in Europe. Larvae feed in developing seed pods. There is evidence for a female-produced long-range attractant (Evans and Bergeron, 1994). The weevils overwinter as unmated adults, and in spring the weevils disperse into a flowering brassica crop where feeding, mating and oviposition occur. Unmated overwintered females were attractive to overwintered males and, to a lesser extent, to overwintered females in both laboratory and field tests. However, newly emerged females or males of any age were not attractive. The two-choice laboratory bioassays were conducted within a cage that held 30 weevils and four small containers; two of these had ten weevils of one sex and the other two were empty. Numbers of weevils moving to within 1 cm of the containers were recorded after 10 min. The field bioassays employed transparent sticky traps; some of these were baited with live weevils that were confined in small cloth bags with host plant. The chemistry of the pheromone is not yet known.

The species also has an oviposition-deterring pheromone. Each ovipositing female 'brushes' the surface of the seed pod with the end of its abdomen after it has deposited the egg. With this activity a pheromone is laid down that deters further oviposition on that pod for about 2 h (Kozlowski *et al.*, 1983; Ferguson and Williams, 1991). This chemical signal would allow a female weevil exploring a seed head to recognize and avoid those pods in which she has already laid eggs. The pheromone is apparently produced by a gland associated with the seventh urotergite and can be collected either by extracting this tissue directly or by inducing a female weevil to mark the outside of a narrow glass tube and then rinsing it with solvent. Chemical identification is not complete, but fairly high polarity is suggested (Mudd *et al.*, 1997). *C. floralis* Paykull, a related, weed-infesting species, also has an oviposition-deterring pheromone (Kozlowski, 1989).

The plum curculio, Conotrachelus nenuphar *(Herbst)*

The plum curculio is a major pest of stone and pome fruits in eastern North America. A male-specific compound was detected by GC comparisons of volatile collections from feeding males and females. This was identified as a terpene acid (grandisoic acid, **8** in Fig. 5.2) (Eller and Bartelt, 1996). Compound **8** is the acid analogue of grandisol (**1**) from the boll weevil. The chirality of grandisoic acid was established as (1*R*,2*S*) by reduction of **8** to its alcohol, followed by comparison with natural **1** from boll weevils on a chiral GC column. Preliminary field tests of the racemic acid indicated significant but subtle attraction of both sexes to the compound. Butkewich *et al.* (1987) tested for an oviposition-deterring pheromone from females, but results were inconclusive.

The banana weevil, Cosmopolites sordidus *(Germar)*

The banana weevil is a large, almost flightless species and is among the most serious pests of bananas and plantains in all major banana-growing areas of the world. Larvae burrow in the rhizome, reducing yield, weakening the plant and making it susceptible to disease. Laboratory behavioural studies (using a two-choice, static air, pitfall bioassay) and electroantennogram (EAG) studies indicated the existence of a male-produced pheromone to which both sexes respond (Budenberg *et al.*, 1993). Beauhaire *et al.* (1995) detected six male-specific compounds with EAG activity in volatile collections. They identified and synthesized the most abundant of these, a novel bicyclic ketal (sordidin, **26** in Fig. 5.3), for which Mori *et al.* (1996) determined the absolute stereochemistry. Ndiege *et al.* (1996) and Jayaraman *et al.* (1997) developed a larger-scale synthesis of racemic sordidin that made field testing possible. The latter paper also established that three of the six EAG-active compounds from the weevil are diastereoisomers of sordidin (**27–29**, Fig. 5.3) (two remain unidentified). Racemic mixtures of the four diastereoisomers, in combination with banana stem tissue, were attractive in the field (Ndiege *et al.*, 1996; Jayaraman *et al.*, 1997).

The pecan weevil, Curculio caryae *(Horn)*

The pecan weevil is found in temperate North America. Larvae develop in the maturing nuts and cause serious economic damage. Pheromone research began on this species in the early 1970s. Hedin *et al.* (1997) summarized the older literature and provided new details about the pheromone. There are five male-specific compounds which are attractive to females and somewhat attractive to males. Four of these, (1*R*,2*S*)-**1** and **2–4**, are also in the boll weevil pheromone (Fig. 5.2). The fifth is (1*S*,2*R*)-**1**, the enantiomer of grandisol that is not found in the boll weevil. Ratios of components determine whether a blend will be more attractive to the pecan weevil or to the boll weevil (Hedin *et al.*, 1997).

Cylas *species*

The sweetpotato weevil, *C. formicarius* (F.), is an important pest of sweet potatoes in the tropical and warmer temperate regions of both the New and Old

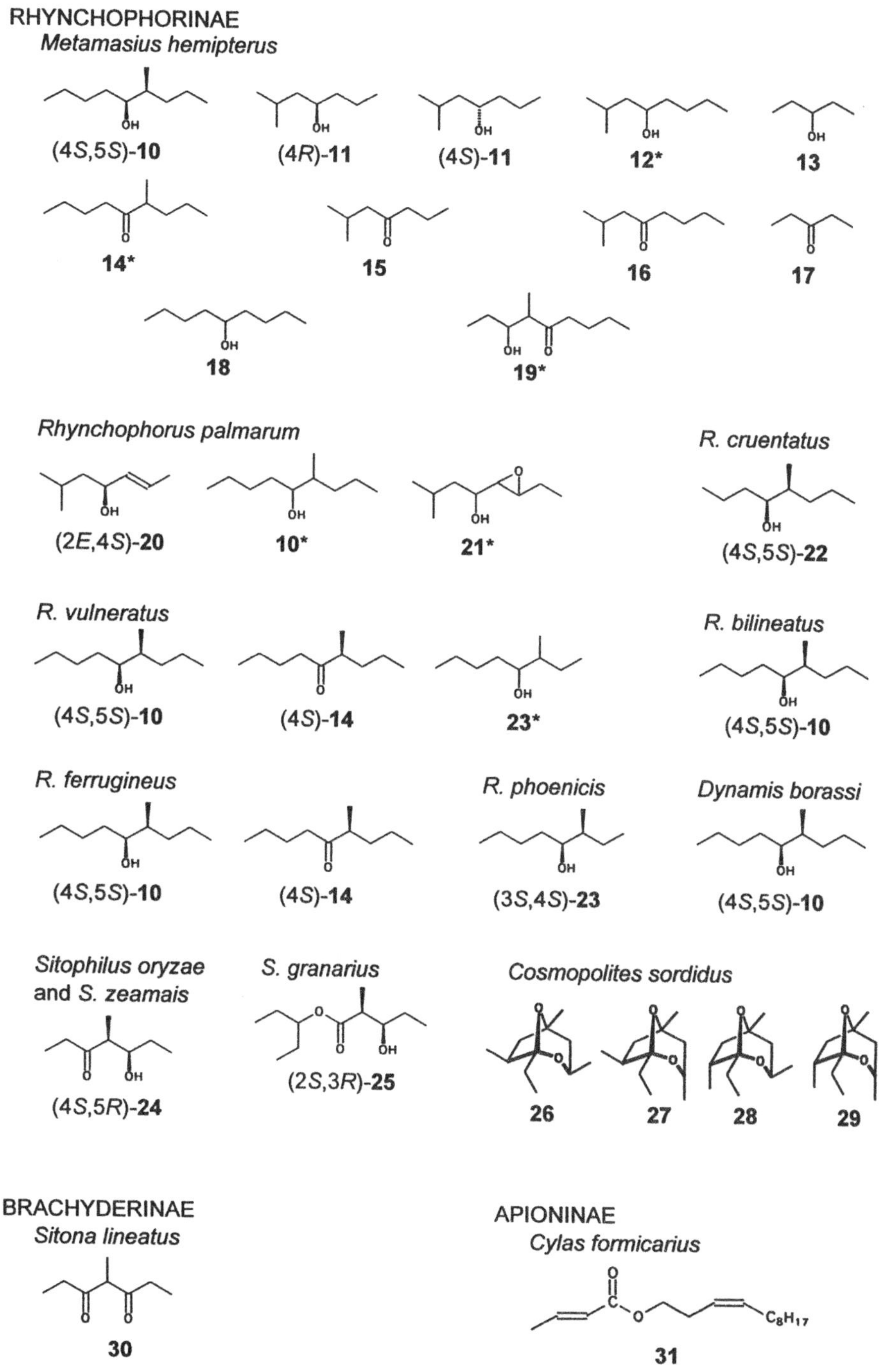

Fig. 5.3. Weevil pheromones of probable polyketide or fatty-acid origin. For each species, the first compound is the most abundant. * = stereochemistry at asymmetric centre(s) unknown.

Worlds. As noted by Arnett (1993), some authors elevate the subfamily to which *Cylas* belongs (Apioninae) to family status (Apionidae). Larvae bore into the vines, roots and tubers. Even slight feeding induces the production of terpenes by the tuber, imparting a bitter taste. Coffelt *et al.* (1978) demonstrated the existence of a female-produced long-range pheromone to which only males respond. Their laboratory bioassay involved the application of a sample to a glass rod which was placed within 1 cm of an isolated test insect. A positive response was antennal elevation and locomotion. This behaviour is the first step in a sequence that normally leads to mate location. Heath *et al.* (1986) used the bioassay to guide chromatographic purification of the pheromone from volatile collections. The pheromone was identified as the ester, (Z)-3-dodecenyl (E)-2-butenoate (**31**, Fig. 5.3). Synthetic **31** was remarkably active in the field; individual traps baited with as little as 10 μg on a rubber septum frequently caught over 1000 weevils per day (Jansson *et al.*, 1989) and even 10 ng was active. Flights of males toward the traps were short (< 1 m) and lasted only a few seconds. Some individuals landed on the traps, but most landed nearby and walked on to them.

The pheromone has been developed as monitoring tool, and there has been considerable effort to optimize trapping parameters such as trap design, trap location, pheromone dose, pheromone dispenser and bait longevity (Proshold *et al.*, 1986; Jansson *et al.*, 1989, 1992, 1993; Mason *et al.*, 1990). A review of the *C. formicarius* pheromone in pest management (Jansson *et al.*, 1991) presents much of this practical information in detail; it also explains how the pheromone could be useful for both population monitoring purposes and for mass trapping. Mass trapping for control has shown promise in Okinawa (Yasuda, 1995).

Smit *et al.* (1994) found that two African sweetpotato weevils, *Cylas brunneus* (F.) and *C. puncticollis* (Boheman), also have female-produced long-range pheromones. In both species, males but not females were clearly attracted to conspecific females in traps with sweet potato; traps with sweet potato only, or with sweet potato plus males, were ignored. The species were not cross-attractive, nor did they respond to synthetic pheromone for *C. formicarius*. Volatiles collected from females of both species elicited responses from males in the laboratory bioassay of Coffelt *et al.* (1978). No chemistry has been reported.

The sugarcane rootstalk borer, Diaprepes abbreviatus *(L.)*

The sugarcane rootstalk borer is native to the West Indies and now occurs in Florida also. This large weevil (2 cm) is a foliage feeder as an adult but a root feeder in the larval stage. Citrus and sugarcane are important hosts, and adults have been observed to aggregate in citrus foliage. Schroeder (1981), Beavers *et al.* (1982), Jones and Schroeder (1984) and Harari and Landolt (1997) presented evidence for both male-produced and female-produced pheromones in this species. The chemicals may be associated with frass, but none have been identified. Study of pheromones in this species has been complicated by strong responses to food plant odours.

The palm weevil, Dynamis borassi *(F.)*

This South American palm weevil is a serious pest of coconut palms. It damages inflorescences and crown tissue and vectors the nematode that causes the red ring disease. Using the GC–EAD technique, Giblin-Davis *et al.* (1997) detected a male-produced aggregation pheromone to which the antennae of both sexes react, and they identified it as (4*S*,5*S*)-4-methyl-5-nonanol [(4*S*,5*S*)-**10**; Fig. 5.3]. Racemic synthetic **10** was not attractive by itself in the field, but it clearly synergized the attractiveness of sugarcane stem tissue for both sexes. Females (but not males) also responded to the pheromones of certain sympatric palm weevils of the genus *Rhynchophorus* (**20** and **22** in Fig. 5.3) when these chemicals were combined with plant tissue; this suggests that ecological interactions among these species may be important in overcoming the defences of the host palms (Giblin-Davis *et al.*, 1997). Field traps were 4 l white plastic containers with entry slots cut in the sides and hung in palm trees; pheromone dispensers and/or plant tissue were placed inside, along with insecticide.

Hylobius *species*

The pales weevil, *H. pales* (Herbst), is a North American species that breeds in pine stumps and logs but feeds in the adult stage on pine foliage (see also Chapter 6). The species causes serious seedling mortality when new trees are planted in infested areas. Phillips (1990) reported that *H. pales* was strongly attracted to pine bolts from which pheromones of certain scolytid beetles were being released. Thus, *H. pales* may use scolytid pheromones as kairomones during host selection.

The pine weevil, *H. abietis* (L.), is a widespread Palearctic species that causes serious damage in the same way as *H. pales*. Selander (1978) presented evidence for a volatile pheromone from males to which both sexes respond and for a female-produced short-range pheromone that elicits mating behaviour in males. Pheromone emission from males appeared to depend on their physiological state. Tilles *et al.* (1986) found no clear evidence for a long-range pheromone in this species and disputed earlier claims in the literature to the contrary. Tilles *et al.* (1988) corroborated existence of the short-range pheromone from females that stimulates males to mate. Compounds from females were identified (Kalo and Nederström, 1986, and references therein), but biological activity of these was not established.

Hypera *species*

The clover head weevil, *H. meles* (F.), and the alfalfa weevil, *H. postica* (Gyllenhal), are important pests of the respective legume hay crops in North America. Larvae feed on stems, buds and leaves. Hedin *et al.* (1988) isolated boll weevil pheromone components **2**, **3** and **4** (Fig. 5.2) from extracts of mixed sex *H. meles* adults and **3** and **4** from *H. postica*. A variety of other, related terpenes were also reported from the extracts, but no behavioural tests were done.

The West Indian sugarcane weevil, Metamasius hemipterus *(L.)*

The West Indian sugarcane weevil is found from Florida through the West Indies and into Uruguay and northern Argentina and is a pest on sugarcane, banana, pineapple and various palm species. It is suspected of being a vector of red ring disease, caused by the nematode *Rhadinaphelenchus cocophilus* Cobb (Ramirez-Lucas *et al.*, 1996a). Rochat *et al.* (1993) compared volatile collections from males and females and detected five male-specific compounds. The major one was identified as 4-methyl-5-nonanol (**10**, Fig. 5.3). Three of the four minor ones were also identified (Fig. 5.3): 2-methyl-4-heptanol (**11**), 2-methyl-4-octanol (**12**) and 5-nonanol (**18**). The four synthetic racemic compounds caused strong EAG responses (Ramirez-Lucas *et al.*, 1996a), but a fifth, identified in their study as 3-hydroxy-4-methyl-5-nonanone (**19**), did not.

Perez *et al.* (1997) also investigated the species, but they studied the subspecies, *M. h. sericeus* (Olivier). Technical differences were that they used the GC–EAD technique to aid in recognizing active compounds in the volatile collections, and chiral GC to study stereochemistry. They reported all of the male-specific compounds found earlier (Ramirez-Lucas *et al.*, 1996a) except for **18** and **19**, but in addition they found 3-pentanol (**13**), 4-methyl-5-nonanone (**14**), 2-methyl-4-heptanone (**15**), 2-methyl-4-octanone (**16**) and 3-pentanone (**17**) (Fig. 5.3). It is not known if these reported differences are due to the techniques used or to the geographical origin of the weevils studied (Giblin-Davis *et al.*, 1996a). The absolute stereochemistry of **10** was established as (4*S*,5*S*) (Mori *et al.*, 1993a), but both the (4*R*) and (4*S*) isomers of **11** were present, in a 6:4 ratio (Perez *et al.*, 1997).

Field experiments indicated that a combination of racemic major alcohol **10** and minor alcohols **11** or **12** accounted for most of the activity (Ramirez-Lucas *et al.*, 1996b; Perez *et al.*, 1997). However, ketones **14–17** generally suppressed trap catch (Perez *et al.*, 1997). Ethyl acetate, ethyl propionate and ethyl butyrate, EAG-active volatiles from fermenting sugarcane, enhanced the effect of the pheromone, but not to the extent of sugarcane itself (Perez *et al.*, 1997). Combinations of pheromone, sugarcane and ethyl acetate were more attractive in the field than any binary combinations or single compounds. Various field trap styles were effective, but all had plastic containers which the responding weevils had to enter (Giblin-Davis *et al.*, 1996b).

Pissodes *species*

These nearctic weevils are all pests of conifers, feeding beneath the bark as larvae. Fontaine and Foltz (1982) demonstrated the existence of a male-produced pheromone in *P. nemorensis* Germar (the deodar weevil) in field tests with live weevils (see also Chapter 6). Volatile collections and gut extracts from feeding male *P. strobi* (Peck) (white pine weevil), *P. approximatus* Hopkins (northern pine weevil) and *P. nemorensis* contained grandisol (**1**), which is also part of the boll weevil pheromone, and the corresponding aldehyde (grandisal, **9** in Fig. 5.2) (Booth *et al.*, 1983; Phillips *et al.*, 1984). The compounds were absent from females. A mixture of the two synthetic racemic chemicals was only slightly active by itself, but it clearly enhanced the responses of the weevils to pine bolts (Booth *et al.*, 1983; Phillips *et al.*, 1984). Semiochemical interactions among these species were investigated (Phillips and Lanier, 1986). Hibbard and Webster (1993) determined the chirality of **1** and **9** for *P. nemorensis* and *P. strobi*. Both species produce nearly 100% (1*R*,2*S*)-**1**; *P. nemorensis* emits nearly 100% (1*S*,2*R*)-**9**, but *P. strobi* emits about 60% (1*S*,2*R*)-**9** and 40% (1*R*,2*S*)-**9**. Maclauchlan *et al.* (1993) presented evidence for a male-produced pheromone in a fourth *Pissodes* species, *P. schwarzi* Hopkins (the Yosemite bark weevil). Females of a fifth species, *P. terminalis* Hopping (the lodgepole terminal weevil) may produce an oviposition-deterring pheromone (Maclauchlan and Borden, 1995).

The New Guinea sugarcane weevil, Rhabdoscelus obscurus *(Boisduval)*

The New Guinea sugarcane weevil is a major pest of sugarcane in Hawaii. The species was reported by Chang and Curtis (1972) to have a male-produced pheromone that attracts primarily females, based on laboratory bioassays and preliminary field tests. Pheromone activity was associated with male abdomen, hindgut and faecal material. No chemistry has been reported.

Rhynchophorus *species*

The species of this genus are large (2–3 cm) and are very damaging to coconut palm (*Cocos nucifera* L.), oil palm (*Elaeis guineensis* Jacq.) and a number of other palm species in both the New and Old World tropics. Adults are attracted to damaged, stressed and sometimes healthy palms, where eggs are laid. Larvae burrow into the stem or crown tissue. Male-produced aggregation pheromones have been intensively studied during this decade and are now known for six *Rhynchophorus* species (reviewed by Giblin-Davis *et al.*, 1996a).

The first *Rhynchophorus* pheromone was demonstrated in the neotropical species *R. palmarum* (L.) (the American palm weevil) by Rochat *et al.* (1991a), who worked with living insects. In the laboratory (two-choice pitfall bioassays) and in the field (with large plastic containers as traps), both sexes were clearly more attracted to males plus palm stem tissue than to males alone or to palm tissue alone. Attraction to females was not demonstrated. The pattern of synergism between pheromone and host volatiles was subsequently found to occur throughout the genus.

Rochat *et al.* (1991b) compared male- and female-derived volatiles and found two male-specific compounds that were active by GC–EAD. The major one of these was identified as (2*E*)-6-methyl-2-hepten-4-ol (rhynchophorol, **20** in Fig. 5.3). The racemic synthetic compound was active in the laboratory bioassay. After synthesizing both enantiomers, Oehlschlager *et al.* (1992) compared the natural and synthetic compounds as diastereoisomeric, (*S*)-*O*-acetyl lactate derivatives by GC, and they learned that the weevils produce only the (2*E*,4*S*) enantiomer. In field tests, (2*E*,4*S*)-**20** was active, but the unnatural (2*E*,4*R*) isomer was not. Significantly, the (2*E*,4*R*) isomer did not inhibit response, suggesting that the racemic alcohol would be as useful as the more expensive pure enantiomer for pest management purposes.

Additional minor male-specific compounds were also identified. Rochat *et al.* (1993) determined one of these as 2,3-epoxy-6-methyl-4-heptanol (**21**, Fig. 5.3), and Hallett *et al.* (1993) found another to be 4-methyl-5-nonanol (**10**, Fig. 5.3).

Oehlschlager *et al.* (1993) demonstrated that racemic **20** and palm stem tissue were highly synergistic in field trials, and they optimized trap and bait parameters for a large-scale practical test. Oehlschlager *et al.* (1995a) described a 17-month mass trapping trial in Costa Rica during which the weevil population density in a 30 ha commercial plantation was substantially reduced.

The male *R. palmarum* are stimulated to release pheromone when they are exposed to ethyl acetate vapours, a volatile released from palm tissue (Jaffé *et al.*, 1993; Sánchez *et al.*, 1996). The pheromone originates in glands in the prothorax and is released in a droplet of liquid from the mouth. Capillary action then carries the pheromone through a covering of hairs to the surface of the rostrum ('snout'), where the chemicals evaporate (Sánchez *et al.*, 1996).

Hallett *et al.* (1993) investigated *R. ferrugineus* (Olivier) and *R. vulneratus* (Panzer), two very closely related sympatric Asian palm weevils. Two EAD-active male-specific compounds were detected in volatile collections from each species. The most abundant one in both species was 4-methyl-5-nonanol (**10**, ferrugineol). The less abundant compound in both species was 4-methyl-5-nonanone (**14**, ferrugineone). About the same time, *R. vulneratus* was also studied by Rochat *et al.* (1993), who found minor amounts of 3-methyl-4-octanol (**23**, Fig. 5.3) in addition to **10**. Mori *et al.* (1993a) established that natural **10** from *R. vulneratus* has the (4*S*,5*S*) configuration by GC comparisons of natural and synthetic isomers on a chiral column. Perez *et al.* (1996) confirmed this conclusion and also learned that *R. ferrugineus* makes exactly the same stereoisomer, (4*S*,5*S*)-**10**. Furthermore, they learned that the minor ketone (**14**) in both species has the (4*S*) configuration. Perez *et al.* (1996) demonstrated that both species respond well in the field to (4*S*,5*S*)-**10** as well as to racemic **10**, but not to (4*R*,5*R*)-**10**. Ketone **14** did not affect trap catch in any way. No definitive differences between chemical communication systems in these species have been found. The taxa are distinguished from each other primarily on the basis of colour, but they are otherwise so similar in form, habits and distribution that they may represent just one species (Perez *et al.*, 1996).

Volatiles from *R. phoenicis* (F.) (the African palm weevil) were examined by Rochat *et al.* (1993) and Gries *et al.* (1993). One male-specific compound was detected and was identified as 3-methyl-4-octanol (phoenicol, **23**). Mori *et al.* (1993b) and Perez *et al.* (1994) established the stereochemistry as (3*S*,4*S*). The (3*S*,4*S*) isomer was most active in the field, but the presence of the (3*R*,4*R*) isomer was not detrimental (Perez *et al.*, 1994; Rochat *et al.*, 1995). Gries *et al.* (1994) found that the ethyl propionate emitted from palm tissue was an important synergist of the *R. phoenicis* pheromone.

Weissling *et al.* (1993) presented evidence for a male-produced pheromone in the palmetto weevil, *R. cruentatus* (F.). This is the only *Rhynchophorus* species occurring in North America, and is a significant pest of stressed ornamental palms. In volatile collections, Weissling *et al.* (1994) identified the single male-specific compound that was strongly active by GC–EAD as 5-methyl-4-octanol (cruentol, **22** in Fig. 5.3). The synthetic compound was active in the field, but only as a synergist of palm-tissue volatiles. Giblin-Davis *et al.* (1994) evaluated a series of fermentation volatiles for synergistic activity with **22**; ethyl acetate, several other ethyl esters and ethanol were effective but did not fully account for the activity of fermenting plant tissue. Natural **22** has the (4*S*,5*S*) configuration (Perez *et al.*, 1994).

A palm weevil found in New Guinea, *R. bilineatus* (Montrouzier), was studied by Oehlschlager *et al.* (1995b). A male-produced pheromone was detected from male-produced volatiles by GC–EAD and identified as (4*S*,5*S*)-**10**, the same as the major component in *R. ferrugineus*, *R. vulneratus* and *Metamasius hemipterus*. It was active in the field when synergized by sugarcane pieces.

The pea and bean weevil, Sitona lineatus *L.*

The pea and bean weevil occurs in western Europe, the Middle East and western North America. Feeding adults create characteristic semicircular notches in the margins of host plant leaves (peas, beans and vetches), but more serious damage is caused by the larvae which live in the soil and feed on the root nodules. A male-produced pheromone to which both sexes respond was demonstrated in the species. Identified as 4-methyl-3,5-heptanedione (**30** in Fig. 5.3), it was synthesized and shown to be active in the field (Blight *et al.*, 1984; Blight and Wadhams, 1987). Smart *et al.* (1994) discussed using the pheromone to attract the weevils to a trap crop, where they would be killed, while at the same time protecting the main crop with an antifeedant such as azadirachtin.

Sitophilus *species*

Male-produced aggregation pheromones, to which both sexes respond, were demonstrated for three widely distributed species of stored-product weevils: *S. oryzae* (L.), *S. zeamais* Motschylsky and *S. granarius* (L.) (Phillips and Burkholder, 1981; Faustini *et al.*, 1982; Walgenbach *et al.*, 1983; see also Chapter 7). The pheromone for both *S. oryzae* and *S. zeamais* is (4*S*,5*R*)-5-hydroxy-4-methyl-3-heptanone [sitophilure, (4*S*,5*R*)-**24** in Fig. 5.3] (Schmuff *et al.*, 1984; Phillips *et al.*, 1985; Walgenbach *et al.*, 1987b). The pheromone for *S. granarius* is (2*S*,3*R*)-1-ethylpropyl-2-methyl-3-hydroxypentanoate [sitophilate, (2*S*,3*R*)-**25** in Fig. 5.3] (Phillips *et al.*,1987, 1989). Using the GC–EAD technique, Chambers *et al.* (1996) detected an additional active compound in *S. granarius* that has not yet been identified. Walgenbach *et al.* (1987a) and Phillips *et al.* (1993) found that the pheromone and grain-derived volatiles were synergistic for *S. zeamais* and *S. oryzae*. Effects of age, crowding, starvation and pheromone dose on *S. zeamais* responses were studied by Walgenbach and Burkholder (1986).

The red sunflower seed beetle, Smicronyx fulvus *LeConte*

The red sunflower seed weevil is a serious pest of sunflowers in the northern Great Plains of North America. The larvae feed on the developing seeds. Roseland *et al.*

(1990) indicated the existence of a male-produced pheromone to which females respond. They observed that male weevils are the first to arrive on sunflowers of the appropriate developmental stage, and females typically follow within 1–45 min. Furthermore, traps containing live males on sunflowers were over ten times more attractive to females than sunflowers alone or females on sunflowers. Females were not attractive. The pheromone could be extracted from males with hexane, but the chemistry has not been studied.

Conclusions

The study of pheromones in this huge and important family is still in its early stages, but there are now well-developed examples to guide further research. There is a rich vein of interesting biology and chemistry waiting to be tapped.

References

Arnett, R.H. Jr (1993) *American Insects: a Handbook of the Insects of America North of Mexico.* Sandhill Crane Press, Gainesville, Florida, 850 pp.

Beauhaire, J., Ducrot, P.H., Malosse, C., Rochat, D., Ndiege, I.O. and Otieno, D.O. (1995) Identification and synthesis of sordidin, a male pheromone emitted by *Cosmopolites sordidus*. *Tetrahedron Letters* 36, 1043–1046.

Beavers, J.B., McGovern, T.P. and Adler, V.E. (1982) *Diaprepes abbreviatus*: laboratory and field behavioral and attractancy studies. *Environmental Entomology* 11, 436–439.

Blight, M.M. and Wadhams, L.J. (1987) Male-produced aggregation pheromone in pea and bean weevil, *Sitona lineatus* (L.). *Journal of Chemical Ecology* 13, 733–739.

Blight, M.M., Pickett, J.A., Smith, M.C. and Wadhams, L.J. (1984) An aggregation pheromone of *Sitona lineatus*: identification and initial field studies. *Naturwissenschaften* 71, 480.

Booth, D.C., Phillips, T.W., Claesson, A., Silverstein, R.M., Lanier, G.N. and West, J.R. (1983) Aggregation pheromone components of two species of *Pissodes* weevils (Coleoptera: Curculionidae): isolation, identification, and field activity. *Journal of Chemical Ecology* 9, 1–12.

Budenberg, W.J., Ndiege, I.O. and Karago, F.W. (1993) Evidence for volatile male-produced pheromone in banana weevil *Cosmopolites sordidus*. *Journal of Chemical Ecology* 19, 1905–1916.

Butkewich, S.L., Prokopy, R.J. and Green, T.A. (1987) Discrimination of occupied host fruit by plum curculio females (Coleoptera: Curculionidae). *Journal of Chemical Ecology* 13, 1833–1841.

Chambers, J., Van Wyk, C.B., White, P.R., Gerrard, C.M. and Mori, K. (1996) Grain weevil, *Sitophilus granarius* (L.): antennal and behavioral responses to male-produced volatiles. *Journal of Chemical Ecology* 22, 1639–1654.

Chang, V.C.S. and Curtis, G.A. (1972) Pheromone production by the New Guinea sugarcane weevil. *Environmental Entomology* 1, 476–481.

Coffelt, J.A., Vick, K.W., Sower, L.L. and McClellan, W.T. (1978) Sex pheromone of the sweetpotato weevil, *Cylas formicarius elegantulus*: laboratory bioassay and evidence for a multiple component system. *Environmental Entomology* 7, 756–758.

Coudriet, D.L. and Kishaba, A.N. (1988) Bioassay procedure for an attractant of the pepper weevil (Coleoptera: Curculionidae). *Journal of Economic Entomology* 81, 1499–1502.

Cross, W.H. and Mitchell, H.C. (1966) Mating behavior of the female boll weevil. *Journal of Economic Entomology* 59, 1503–1507.

Dickens, J.C. (1989) Green leaf volatiles enhance aggregation pheromone of boll weevil, *Anthonomus grandis*. *Entomologia Experimentalis et Applicata* 52, 191–203.

Dickens, J.C. and Mori, K. (1989) Receptor chirality and behavioral specificity of the boll weevil, *Anthonomus grandis* Boh. (Coleoptera: Curculionidae), for its pheromone, (+)-grandisol. *Journal of Chemical Ecology* 15, 517–528.

Dickens, J.C., Jang, E.B., Light, D.M. and Alford, A.R. (1990) Enhancement of insect pheromone responses by green leaf volatiles. *Naturwissenschaften* 77, 29–31.

Eller, F.J. and Bartelt, R.J. (1996) Grandisoic acid, a male-produced aggregation pheromone from the plum curculio, *Conotrachelus nenuphar*. *Journal of Natural Products* 59, 451–453.

Eller, F.J., Bartelt, R.J., Shasha, B.S., Schuster, D.J., Riley, D.G., Stansly, P.A., Mueller, T.F., Shuler, K.D., Johnson, B., Davis, J.H. and Sutherland, C.A. (1994) Aggregation pheromone for the pepper weevil, *Anthonomus eugenii* Cano (Coleoptera: Curculionidae): identification and field activity. *Journal of Chemical Ecology* 20, 1537–1555.

Evans, K.A. and Bergeron, J. (1994) Behavioral and electrophysiological response of cabbage seed weevils (*Ceutorhynchus assimilis*) to conspecific odor. *Journal of Chemical Ecology* 20, 979–989.

Faustini, D.L., Giese, W.L., Phillips, J.K. and Burkholder, W.E. (1982) Aggregation pheromone of the male granary weevil, *Sitophilus granarius* (L.). *Journal of Chemical Ecology* 8, 679–687.

Ferguson, A.W. and Williams, I.H. (1991) Deposition and longevity of oviposition-deterring pheromone in the cabbage seed weevil. *Physiological Entomology* 16, 27–33.

Fontaine, M.S. and Foltz, J.L. (1982) Field studies of a male-released aggregation pheromone in *Pissodes nemorensis*. *Environmental Entomology* 11, 881–883.

Giblin-Davis, R.M., Weissling, T.J., Oehlschlager, A.C. and Gonzalez, L.M. (1994) Field response of *Rhynchophorus cruentatus* (Coleoptera: Curculionidae) to its aggregation pheromone and fermenting plant volatiles. *Florida Entomologist* 77, 164–177.

Giblin-Davis, R.M., Oehlschlager, A.C., Perez, A., Gries, G., Gries, R., Weissling, T.J., Chinchilla, C.M., Peña, J.E., Hallett, R.H. and Pierce, H.D. Jr (1996a) Chemical and behavioral ecology of palm weevils (Curculionidae: Rhynchophorinae). *Florida Entomologist* 79, 153–167.

Giblin-Davis, R.M., Peña, J.E., Oehlschlager, A.C. and Perez, A.L. (1996b) Optimization of semiochemical-based trapping of *Metamasius hemipterus sericeus* (Olivier) (Coleoptera: Curculionidae). *Journal of Chemical Ecology* 22, 1389–1410.

Giblin-Davis, R.M., Gries, R., Gries, G., Peña-Rojas, E., Pinzón, I., Peña, J.E., Perez, A.L., Pierce, H.D. Jr and Oehlschlager, A.C. (1997) Aggregation pheromone of palm weevil, *Dynamis borassi*. *Journal of Chemical Ecology* 23, 2287–2297.

Goff, C.C. and Wilson, J.W. (1937) *The Pepper Weevil.* Bulletin 310, University of Florida Agricultural Experiment Station, Gainesville, Florida, 12 pp.

Gries, G., Gries, R., Perez, A.L., Oehlschlager, A.C., Gonzales, L.M., Pierce, H.D. Jr, Kouda-Bonafos, M., Zebeyou, M. and Nanou, N. (1993) Aggregation pheromone of the African palm weevil, *Rhynchophorus phoenicis* F. *Naturwissenschaften* 80, 90–91.

Gries, G., Gries, R., Perez, A.L., Gonzales, L.M., Pierce, H.D. Jr, Oehlschlager, A.C., Rhainds, M., Zebeyou, M. and Kouame, B. (1994) Ethyl propionate: synergistic kairomone for African palm weevil, *Rhynchophorus phoenicis* L. (Coleoptera: Curculionidae). *Journal of Chemical Ecology* 20, 889–897.

Hallett, R.H., Gries, G., Gries, R., Borden, J.H., Czyzewska, E., Oehlschlager, A.C., Pierce, H.D. Jr, Angerilli, N.P.D. and Rauf, A. (1993) Aggregation pheromones of two Asian palm weevils, *Rhynchophorus ferrugineus* and *R. vulneratus*. *Naturwissenschaften* 80, 328–331.

Harari, A.R. and Landolt, P.J. (1997) Orientation of sugarcane rootstalk borer weevil, *Diaprepes abbreviatus*, to weevil, frass, and food odors. *Journal of Chemical Ecology* 23, 857–868.

Hardee, D.D. and Mitchell, E.B. (1997) Boll weevil, *Anthonomus grandis* Boheman (Coleoptera: Curculionidae): a summary of research on behavior as affected by chemical communication. *Southwestern Entomologist* 22, 465–491.

Hardee, D.D., Mitchell, E.B. and Huddleston, P.M. (1967) Procedure for bioassaying the sex attractant of the boll weevil. *Journal of Economic Entomology* 60, 169–171.

Hardee, D.D., McKibben, G.H., Gueldner, R.C., Mitchell, E.B., Tumlinson, J.H. and Cross, W.H. (1972) Boll weevils in nature respond to grandlure, a synthetic pheromone. *Journal of Economic Entomology* 65, 97–100.

Heath, R.R., Coffelt, J.A., Sonnet, P.E., Proshold, F.I., Dueben, B. and Tumlinson, J.H. (1986) Identification of sex pheromone produced by female sweetpotato weevil, *Cylas formicarius elegantulus* (Summers). *Journal of Chemical Ecology* 12, 1489–1503.

Hedin, P.A., McKibben, G.H., Mitchell, E.B. and Johnson, W.L. (1979) Identification and field evaluation of the compounds comprising the sex pheromone of the female boll weevil. *Journal of Chemical Ecology* 5, 617–627.

Hedin, P.A., Ellsbury, M.M. and Dysart, R.J. (1988) Volatiles from clover head *Hypera meles* (Fab.) and alfalfa *H. postica* (Gyllenhal) weevils: search for pheromones. *Experientia* 44, 792–794.

Hedin, P.A., Dollar, D.A., Collins, J.K., DuBois, J.G., Mulder, P.G., Hedger, G.H., Smith, M.W. and Eikenbary, R.D. (1997) Identification of male pecan weevil pheromone. *Journal of Chemical Ecology* 23, 965–977.

Hibbard, B.E. and Webster, F.X. (1993) Enantiomeric composition of grandisol and grandisal produced by *Pissodes strobi* and *P. nemorensis* and their electroantennogram response to pure enantiomers. *Journal of Chemical Ecology* 19, 2129–2141.

Hobbs, P.D. and Magnus, P.D. (1976) Synthesis of optically active grandisol, the boll weevil pheromone. *Journal of the American Chemical Society* 98, 4594–4600.

Jaffé, K., Sánchez, P., Cerda, H., Hernández, J.V., Jaffé, R., Urdaneta, N., Guerra, G., Martínez, R. and Miras, B. (1993) Chemical ecology of the palm weevil *Rhynchophorus palmarum* (L.) (Coleoptera: Curculionidae): attraction to host plants and to a male-produced aggregation pheromone. *Journal of Chemical Ecology* 19, 1703–1720.

Jansson, R.K., Heath, R.R. and Coffelt, J.A. (1989) Temporal and spatial patterns of sweetpotato weevil (Coleoptera: Curculionidae) counts in pheromone-baited traps in white-fleshed sweet potato fields in southern Florida. *Environmental Entomology* 18, 691–697.

Jansson, R.K., Mason, L.J. and Heath, R.R. (1991) Use of sex pheromone for monitoring and managing *Cylas formicarius*. In: Jansson, R.K. and Raman, K.V. (eds) *Sweet Potato Pest Management: a Global Perspective*. Westview Press, Boulder, Colorado, pp. 97–138.

Jansson, R.K., Mason, L.J., Heath, R.R., Sorensen, K.A., Hammond, A.M. and Robinson, J.V. (1992) Pheromone-trap monitoring system for sweetpotato weevil (Coleoptera: Apionidae) in the southern United States: effects of trap type and pheromone dose. *Journal of Economic Entomology* 85, 416–423.

Jansson, R.K., Mason, L.J., Heath, R.R., Lecrone, S.H. and Forey, D.E. (1993) Pheromone-trap monitoring system for sweetpotato weevil (Coleoptera: Apionidae) in the southern United States: effects of lure type, age, and duration in storage. *Journal of Economic Entomology* 86, 1109–1115.

Jayaraman, S., Ndiege, I.O., Oehlschlager, A.C., Gonzalez, L.M., Alpizar, D., Falles, M., Budenberg, W.J. and Ahuya, P. (1997) Synthesis, analysis, and field activity of sordidin, a male-produced aggregation pheromone of the banana weevil, *Cosmopolites sordidus*. *Journal of Chemical Ecology* 23, 1145–1161.

Jones, I.F. and Schroeder, W.J. (1984) Capture of *Diaprepes abbreviatus* (Coleoptera: Curculionidae) in frass extract-baited traps in citrus. *Journal of Economic Entomology* 77, 334–336.

Kalo, P. and Nederström, A. (1986) Female-specific compounds in the ovaries of the large pine weevil, *Hylobius abietis* L. (Coleoptera, Curculionidae). *Annales Entomologici Fennici* 52, 95–101.

Keller, J.C., Mitchell, E.B., McKibben, G. and Davich, T.B. (1964) A sex attractant for female boll weevils from males. *Journal of Economic Entomology* 57, 609–610.

Kozlowski, M.W. (1989) Oviposition and host object marking by the females of *Ceutorhynchus floralis* (Coleoptera: Curculionidae). *Entomologia Generalis* 14, 197–201.

Kozlowski, M.W., Lux, S. and Dmoch, J. (1983) Oviposition behaviour and pod marking in the cabbage seed weevil, *Ceutorhynchus assimilis*. *Entomologia Experimentalis et Applicata* 34, 277–282.

Maclaughlan, L.E. and Borden, J.H. (1995) Discrimination between pine shoots with and without oviposition by *Pissodes terminalis* Hopping (Coleoptera: Curculionidae). *Canadian Entomologist* 127, 267–269.

Maclaughlan, L.E., Borden, J.H. and Price, I. (1993) Life history and pheromone response in *Pissodes schwarzi* Hopk. (Coleoptera: Curculionidae). *Journal of the Entomological Society of British Columbia* 90, 30–35.

Mason, L.J., Jansson, R.K. and Heath, R.R. (1990) Sampling range of male sweetpotato weevils (*Cylas formicarius elegantulus*) (Summers) (Coleoptera: Curculionidae) to pheromone traps: influence of pheromone dosage and lure age. *Journal of Chemical Ecology* 16, 2493–2502.

McKibben, G.H., Hedin, P.A., McGovern, W.L., Wilson, N.M. and Mitchell, E.B. (1977) A sex pheromone for male boll weevils from females. *Journal of Chemical Ecology* 3, 331–335.

Mori, K. (1978) Synthesis of the both enantiomers of grandisol, the boll weevil pheromone. *Tetrahedron* 34, 915–920.

Mori, K., Tamada, S. and Hedin, P.A. (1978) (−)-Grandisol, the antipode of the boll weevil pheromone, is biologically active. *Naturwissenschaften* 65, 653–654.

Mori, K., Kiyota, H., Malosse, C. and Rochat, D. (1993a) Synthesis of the enantiomers of *syn*-4-methyl-5-nonanol to determine the absolute configuration of the naturally occurring (4S,5S) isomer isolated as the male-produced pheromone compound of *Rhynchophorus vulneratus* and *Metamasius hemipterus*. *Liebigs Annalen der Chemie* 1993, 1201–1204.

Mori, K., Kiyota, H. and Rochat, D. (1993b) Synthesis of the stereoisomers of 3-methyl-4-octanol to determine the absolute configuration of the naturally occurring (3S,4S)-isomer isolated as the male-produced aggregation pheromone of the African palm weevil, *Rhynchophorus phoenicis*. *Liebigs Annalen der Chemie* 1993, 865–870.

Mori, K., Nakayama, T. and Takikawa, H. (1996) Synthesis and absolute configuration of sordidin, the male-produced aggregation pheromone of the banana weevil, *Cosmopolites sordidus*. *Tetrahedron Letters* 37, 3741–3744.

Mudd, A., Ferguson, A.W., Blight, M.M., Williams, I.H., Scubla, P., Solinas, M. and Clark, S.J. (1997) Extraction, isolation and composition of oviposition-deterring secretion of cabbage seed weevil *Ceutorhynchus assimilis*. *Journal of Chemical Ecology* 23, 2227–2240.

Ndiege, I.O., Jayaraman, S., Oehlschlager, A.C., Gonzalez, L., Alpizar, D. and Fallas, M. (1996) Convenient synthesis and field activity of a male-produced aggregation pheromone of *Cosmopolites sordidus*. *Naturwissenschaften* 83, 280–282.

Oehlschlager, A.C., Pierce, H.D. Jr, Morgan, B., Wimalaratne, P.D.C., Slessor, K.N., King, G.G.S., Gries, G., Gries, R., Borden, J.H., Jiron, L.F., Chinchilla, C.M. and Mexzan, R.G. (1992) Chirality and field activity of rhynchophorol, the aggregation pheromone of the American palm weevil. *Naturwissenschaften* 79, 134–135.

Oehlschlager, A.C., Chinchilla, C.M., Gonzalez, L.M., Jiron, L.F., Mexzon, R. and Morgan B. (1993) Development of a pheromone-based trapping system for *Rhynchophorus palmarum* (Coleoptera: Curculionidae). *Journal of Economic Entomology* 86, 1381–1392.

Oehlschlager, A.C., McDonald, R.S., Chinchilla, C.M. and Patschke, S.N. (1995a) Influence of a pheromone-based mass-trapping system on the distribution of *Rhynchophorus palmarum* (Coleoptera: Curculionidae) in oil palm. *Environmental Entomology* 24, 1005–1012.

Oehlschlager, A.C., Prior, R.N.B., Perez, A.L., Gries, R., Gries, G., Pierce, H.D. Jr and Laup, S. (1995b) Structure, chirality, and field testing of a male-produced aggregation pheromone of Asian palm weevil *Rhynchophorus bilineatus* (Montr.) (Coleoptera:Curculionidae). *Journal of Chemical Ecology* 21, 1619–1629.

Patrock, R.J., Schuster, D.J. and Mitchell, E.R. (1992) Field evidence for an attractant produced by the male pepper weevil (Coleoptera: Curculionidae). *Florida Entomologist* 75, 138–144.

Perez, A.L., Gries, G., Gries, R., Giblin-Davis, R.M. and Oehlschlager, A.C. (1994) Pheromone chirality of African palm weevil, *Rhynchophorus phoenicis* (F.) and palmetto weevil, *Rhynchophorus cruentatus* (F.) (Coleoptera: Curculionidae). *Journal of Chemical Ecology* 20, 2653–2671.

Perez, A.L., Hallett, R.H., Gries, R., Gries, G., Oehlschlager, A.C. and Borden, J.H. (1996) Pheromone chirality of Asian palm weevils, *Rhynchophorus ferrugineus* (Oliv.) and *R. vulneratus* (Panz.) (Coleoptera: Curculionidae). *Journal of Chemical Ecology* 22, 357–368.

Perez, A.L., Campos, Y., Chinchilla, C.M., Oehlschlager, A.C., Gries, G., Gries, R., Giblin-Davis, R.M., Castrillo, G., Peña, J.E., Duncan, R.E., Gonzalez, L.M., Pierce, H.D. Jr, McDonald, R. and Andrade, R. (1997) Aggregation pheromones and host kairomones of West Indian sugarcane weevil, *Metamasius hemipterus sericeus*. *Journal of Chemical Ecology* 23, 869–888.

Phillips, J.K. and Burkholder, W.E. (1981) Evidence for a male-produced aggregation pheromone in the rice weevil. *Journal of Economic Entomology* 74, 539–542.

Phillips, J.K., Walgenbach, C.A., Klein, J.A., Burkholder, W.E., Schmuff, N.R. and Fales, H.M. (1985) (*R**,*S**)-5-Hydroxy-4-methyl-3-heptanone: male-produced aggregation pheromone of *Sitophilus oryzae* (L.) and *S. zeamais* Motsch. *Journal of Chemical Ecology* 11, 1263–1274.

Phillips, J.K., Miller, S.P.F., Andersen, J.F., Fales, H.M. and Burkholder, W.E. (1987) The chemical identification of the granary weevil aggregation pheromone. *Tetrahedron Letters* 28, 6145–6146.

Phillips, J.K., Chong, J.M., Andersen, J.F. and Burkholder, W.E. (1989) Determination of the enantiomeric composition of (*R**,*S**)-1-ethylpropyl 2-methyl-3-hydroxypentanoate, the male-produced aggregation pheromone of *Sitophilus granarius*. *Entomologia Experimentalis et Applicata* 51, 149–153.

Phillips, T.W. (1990) Attraction of *Hylobius pales* (Herbst) (Coleoptera: Curculionidae) to pheromones of bark beetles (Coleoptera: Scolytidae). *Canadian Entomologist* 122, 423–427.

Phillips, T.W. and Lanier, G.N. (1986) Interspecific activity of semiochemicals among sibling species of *Pissodes* (Coleoptera: Curculionidae). *Journal of Chemical Ecology* 12, 1587–1601.

Phillips, T.W., West, J.R. Foltz, J.L., Silverstein, R.M. and Lanier, G.N. (1984) Aggregation pheromone of the deodar weevil, *Pissodes nemorensis* (Coleoptera: Curculionidae): isolation and activity of grandisol and grandisal. *Journal of Chemical Ecology* 10, 1417–1423.

Phillips, T.W., Jiang, X.-L., Burkholder, W.E., Phillips, J.K. and Tran, H.Q. (1993) Behavioral responses to food volatiles by two species of stored-product Coleoptera, *Sitophilus oryzae* (Curculionidae) and *Tribolium castaneum* (Tenebrionidae). *Journal of Chemical Ecology* 19, 723–734.

Proshold, F.I., Gonzalez, J.L., Asencio, C. and Heath, R.R. (1986) A trap for monitoring the sweetpotato weevil (Coleoptera: Curculionidae) using pheromone or live females as bait. *Journal of Economic Entomology* 79, 641–647.

Ramirez-Lucas, P., Malosse, C., Ducrot, P.H., Lettere, M. and Zagatti, P. (1996a) Chemical identification, electrophysiological and behavioral activities of the pheromone of *Metamasius hemipterus* (Coleoptera: Curculionidae). *Bioorganic and Medicinal Chemistry* 4, 323–330.

Ramirez-Lucas, P., Rochat, D. and Zagatti, P. (1996b) Field trapping of *Metamasius hemipterus* with synthetic aggregation pheromone. *Entomologia Experimentalis et Applicata* 80, 453–460.

Ridgway, R.L., Inscoe, M.N. and Dickerson, W.A. (1990) Role of the boll weevil pheromone in pest management. In: Ridgway, R.L., Silverstein, R.M. and Inscoe, M.N. (eds) *Behavior-Modifying Chemicals for Insect Management*. Marcel Dekker, New York, pp. 437–471.

Rochat, D., Gonzalez-V., A., Mariau, D., Villanueva-G., A. and Zagatti, P. (1991a) Evidence for male-produced aggregation pheromone in American palm weevil, *Rhynchophorus palmarum* (L.) (Coleoptera: Curculionidae). *Journal of Chemical Ecology* 17, 1221–1230.

Rochat, D., Malosse, C., Lettere, M., Ducrot, P.H., Zagatti, P., Renou, M. and Descoins, C. (1991b) Male-produced aggregation pheromone of the American palm weevil, *Rhynchophorus palmarum* (L.) (Coleoptera, Curculionidae): collection, identification, electrophysiological activity, and laboratory bioassay. *Journal of Chemical Ecology* 17, 2127–2141.

Rochat, D., Malosse, C., Lettere, M., Ramirez-Lucas, P., Einhorn, J. and Zagatti, P. (1993) Identification of new pheromone-related compounds from volatiles produced by males of four Rhynchophorinae weevils (Coleoptera, Curculionidae). *Comptes Rendus de l'Académie des Sciences, Série II* 316, 1737–1742.

Rochat, D., Akamou, F., Sangare, Al., Mariau, D. and Mori, K. (1995) Field trapping of *Rhynchophorus phoenicis* (Coleoptera: Curculionidae) with stereoisomers of a synthetic aggregation pheromone. *Comptes Rendus de l'Académie des Sciences, Série III* 318, 183–190.

Roseland, C.R., Bates, M.B. and Oseto, C.Y. (1990) Role of a male-produced pheromone of the red sunflower seed weevil (Coleoptera: Curculionidae) in host finding. *Environmental Entomology* 19, 1675–1680.

Sánchez, P., Cerda, H., Cabrera, A., Caetano, F.H., Materán, M., Sánchez, F. and Jaffé, K. (1996) Secretory mechanisms for the male produced aggregation pheromone of the palm weevil *Rhynchophorus palmarum* L. (Coleoptera: Curculionidae). *Journal of Insect Physiology* 42, 1113–1119.

Schmuff, N.R., Phillips, J.K., Burkholder, W.E., Fales, H.M., Chen, C.-W., Roller, P.P. and Ma, M. (1984) The chemical identification of the rice weevil and maize weevil aggregation pheromone. *Tetrahedron Letters* 25, 1533–1534.

Schroeder, W.J. (1981) Attraction, mating, and oviposition behavior in field populations of *Diaprepes abbreviatus* on citrus. *Environmental Entomology* 10, 898–900.

Selander, J. (1978) Evidence of pheromone-mediated behaviour in the large pine weevil, *Hylobius abietis* (Coleoptera, Curculionidae). *Annales Entomologici Fennici* 44, 105–112.

Smart, L.E., Blight, M.M., Pickett, J.A. and Pye, B.J. (1994) Development of field strategies incorporating semiochemicals for the control of the pea and bean weevil, *Sitona lineatus* L. *Crop Protection* 13, 127–135.

Smit, N.E.J.M., Magenya, O. and Parker, B.L. (1994) Biology and pheromone studies with the sweet potato weevils: *Cylas puncticollis* (Bohe.) and *C. brunneus* (Fabr.). *Acta Horticulturae* 380, 399–404.

Stansly, P. and Cate, J.R. (1984) Discrimination by ovipositing boll weevils (Coleoptera: Curculionidae) against previously infested *Hampea* (Malvaceae) flower buds. *Environmental Entomology* 13, 1361–1365.

Tilles, D.A., Nordlander, G., Nordenhem, H., Eidmann, H.H., Wassgren, A.-B. and Bergström, G. (1986) Increased release of host volatiles from feeding scars: a major cause of field aggregation in the pine weevil *Hylobius abietis* (Coleoptera: Curculionidae). *Environmental Entomology* 15, 1050–1054.

Tilles, D.A., Eidmann, H.H. and Solbreck, B. (1988) Mating stimulant of the pine weevil *Hylobius abietis* (L.). *Journal of Chemical Ecology* 14, 1495–1503.

Tumlinson, J.H., Hardee, D.D., Gueldner, R.C., Thompson, A.C., Hedin, P.A. and Minyard, J.P. (1969) Sex pheromones produced by male boll weevil: isolation, identification, and synthesis. *Science* 166, 1010–1012.

Vanderwel, D. and Oehlschlager, A.C. (1987) Biosynthesis of pheromones and endocrine regulation of pheromone production in Coleoptera. In: Prestwich, G.D. and Blomquist, G.J. (eds) *Pheromone Biochemistry*. Academic Press, Orlando, Florida, pp. 175–215.

Walgenbach, C.A. and Burkholder, W.E. (1986) Factors affecting the response of the maize weevil, *Sitophilus zeamais* (Coleoptera: Curculionidae), to its aggregation pheromone. *Environmental Entomology* 15, 733–738.

Walgenbach, C.A., Phillips, J.K., Faustini, D.L. and Burkholder, W.E. (1983) Male-produced aggregation pheromone of the maize weevil, *Sitophilus zeamais*, and interspecific attraction between three *Sitophilus* species. *Journal of Chemical Ecology* 9, 831–841.

Walgenbach, C.A., Burkholder, W.E., Curtis, M.J. and Khan, Z.A. (1987a) Laboratory trapping studies with *Sitophilus zeamais* (Coleoptera: Curculionidae). *Journal of Economic Entomology* 80, 763–767.

Walgenbach, C.A., Phillips, J.K., Burkholder, W.E., King, G.G.S., Slessor, K.N. and Mori, K. (1987b) Determination of chirality in 5-hydroxy-4-methyl-3-heptanone, the aggregation pheromone of *Sitophilus oryzae* (L.) and *S. zeamais* Motschulsky. *Journal of Chemical Ecology* 13, 2159–2169.

Weissling, T.J., Giblin-Davis, R.M. and Scheffrahn, R.H. (1993) Laboratory and field evidence for male-produced aggregation pheromone in *Rhynchophorus cruentatus* (F.) (Coleoptera: Curculionidae). *Journal of Chemical Ecology* 19, 1195–1203.

Weissling, T.J., Giblin-Davis, R.M., Gries, G., Gries, R., Perez, A.L., Pierce, H.D. Jr and Oehlschlager, A.C. (1994) Aggregation pheromone of palmetto weevil, *Rhynchophorus cruentatus* (F.) (Coleoptera: Curculionidae). *Journal of Chemical Ecology* 20, 505–515.

Yasuda, K. (1995) Mass trapping of the sweet potato weevil *Cylas formicarius* (Fabricius) (Coleoptera: Brentidae) with a synthetic sex pheromone. *Applied Entomology and Zoology* 30, 31–36.

6 Forest Beetles

FREDRIK SCHLYTER[1] AND GÖRAN A. BIRGERSSON[2]

[1]*Chemical Ecology, Department of Plant Protection Sciences, Swedish Agricultural University, PO Box 44, S-230 53 Alnarp, Sweden;* [2]*Chemical Ecology, Department of Botany, Göteborg University, PO Box 461, S-405 30 Göteborg, Sweden*

Introduction

The Coleoptera covered in this chapter are those that regularly infest bark and wood of forest trees and that contain identified compounds in their communication systems: bark beetles (Scolytidae including Platypodidae), longhorn beetles (Cerambycidae) and bark weevils (Curculionidae). These beetles are of great economic and ecological importance, killing millions of cubic metres of conifer trees in production forests during beetle epidemics but also acting as keystone species in natural ecosystems by nutrient cycling of mature trees.

These beetles tunnel through and reproduce in the tree hosts, often depleting the nutritional value for their brood and destroying the breeding patch. Olfactory guidance is important in finding a suitable host at each level of search: habitat, host species, host condition, density of conspecifics, and competing insect species. In all three beetle taxa there is a complexity of signal production and response, including attraction to volatiles produced purely by the host tree (kairomones) and attraction or inhibition by pheromone signals produced *de novo* from both sexes of the insect. Inhibitory signals, from non-hosts or competitors, have so far been found mainly in bark beetles. In relation to many other taxa, there is a high frequency of male-produced sex or aggregation pheromones in forest beetles. This may reflect the size of the food patches, which are far too large to be utilized and defended by an individual insect. They are intially located by pioneering male or female beetles, while later-arriving insects build up the populations.

This chapter is not a comprehensive coverage of the literature, as excellent summaries of earlier bark beetle studies can be found in Wood (1982), Borden (1982, 1985, 1997) and Byers (1989, 1995). Instead, we have attempted to focus more on recent studies, on the Palearctic fauna, and on the function of selected examples of communication systems in forest Coleoptera and their ecological, physiological and evolutionary aspects.

Ecologically, we discuss the three steps in the sequence of host finding: (i) the habitat; (ii) the host species; and (iii) the host quality. In the first two steps, host and non-host plant compounds are used, whereas the pheromone signal is crucial in assessing host quality as well as mate location. Physiologically, we describe chemical signals and their biosynthetic origin. From the evolutionary point of view we attempt to analyse the origin of aggregation pheromones and other male-produced signals. Lastly, we discuss how the knowledge of chemical ecology established in forest Coleoptera can be used in forest protection.

The Appendix to this chapter gives a list of the IUPAC names for various chemicals referred to in the text.

Host Selection Sequence

Habitat location

The first step in finding dispersed host trees is to locate the best habitat for host location. For the conifer bark beetle, *Ips typographus* (L.) (Fig. 6.1), a hypothetical sequence of choices is shown in Fig. 6.2. Conifer bark beetles are repelled by common volatiles from broadleaved plants. The green leaf volatiles (GLVs; mostly C_6 alcohols and aldehydes; Zhang *et al.*, 1999) dominate habitats of non-hosts and indicate the low possibility of finding a suitable (i.e. coniferous) host tree.

The attraction to aggregation pheromones is inhibited by individual GLVs for two *Ips* species in the southern USA (Dickens *et al.*, 1991, 1992), by a GLV blend

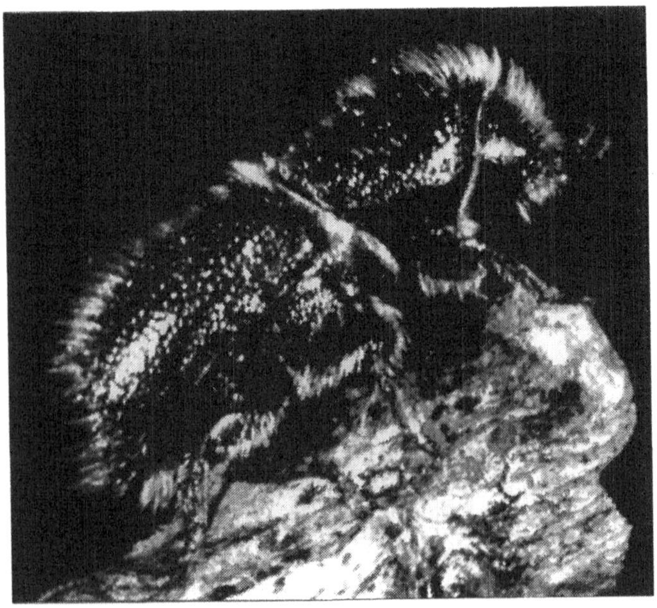

Fig. 6.1. Female bark beetle, *Ips typographus*.

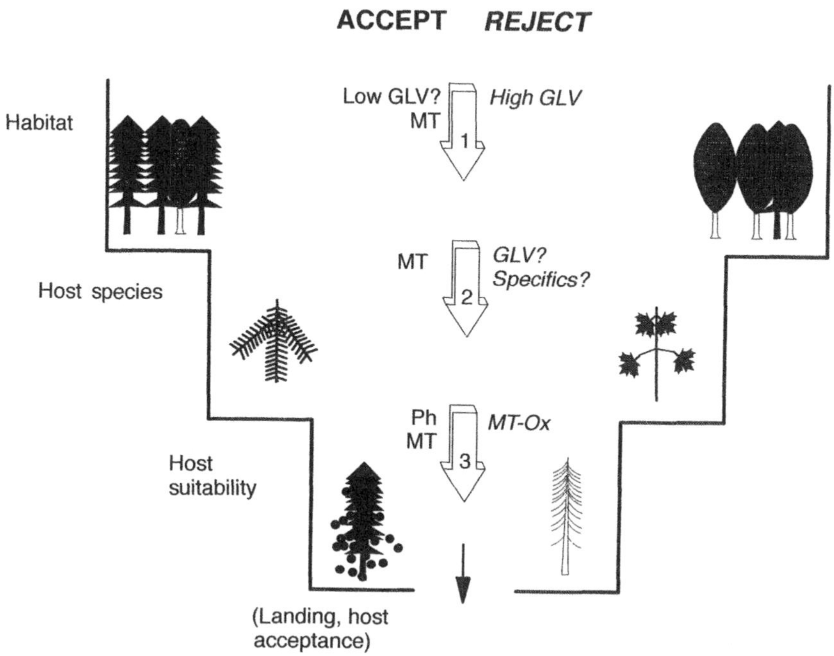

Fig. 6.2. A hypothetical choice sequence for a forest beetle to find (1) the right habitat, (2) the correct host tree species, and (3) a suitable host tree. The chemical signals used for host/non-host and suitable/unsuitable host trees are shown for the conifer bark beetle, *Ips typographus*. Monoterpene hydrocarbons (MT) have been shown to attract other conifer-infesting Coleoptera, but this has not yet been demonstrated in *I. typographus*. Low concentrations of green leaf volatiles (GLV), specific for non-host trees, would assist habitat location. The avoidance of non-hosts could be effected by recognition of high concentrations of green leaf volatiles; while other 'specific' non-host signals, such as (5*S*,7*S*)-conophthorin, and oxygenated monoterpenes (MT-Ox), such as verbenone, would also enhance avoidance of non-hosts. Once insects are established, aggregation pheromones (Ph), in this case a blend of at least two compounds, *cis*-verbenol and 2-methyl-3-buten-2-ol (3-hydroxy-3-methylbutene), also indicate host suitability.

for two Scandinavian *Ips* (Schlyter *et al.*, 1995), and in western Canada it was shown for several *Dendroctonus* and a *Trypodendron* species (Wilson *et al.*, 1996; Borden *et al.*, 1997; Poland *et al.*, 1998). The response to kairomones by *Tomicus piniperda* (L.) was also inhibited but in *Hylurgops palliatus* (Gyll.) it was synergized by a GLV blend (Schlyter *et al.*, 1995). Whether the GLV signal is in fact ecologically relevant at a diffuse habitat scale (as depicted in Fig. 6.2) or only behaviourally active at point sources (as presently demonstrated) has not yet been shown, but is an attractive hypothesis.

Host species choice

The second step – finding a tree of the correct host species in a mixed forest – could be guided by rather general signals, such as low levels of green leaf volatiles, or the presence of a positive signal of conifer host volatiles, such as monoterpene hydrocarbons (Fig. 6.2). However, a specific negative signal exists in volatiles from the bark of deciduous non-host trees that inhibits the attraction to host odours in conifer bark beetles (Schroeder, 1992). Quite recently a component in this signal, (5S,7S)-conophthorin (see Fig. 6.6) has been chemically identified in aerations from bark of *Betula* spp. The bark reduces the attraction to aggregation pheromone in the field (Byers *et al.*, 1998) and synthetic (5S,7S)-conophthorin is synergistic with GLVs and/or verbenone (see Fig. 6.7) in reducing attraction to pheromone in the field (Q.-H. Zhang and F. Schlyter, 1998, unpublished data). It is present in deciduous (angiosperm) trees of North America (*Betula*, *Acer*: Huber *et al.*, 1999) and reduces attraction to pheromone in bark beetles from several genera (*Dendroctonus*, *Ips*, *Dryocoetes*: Borden *et al.*, 1998a). Other angiosperm bark volatiles, such as a blend of four antennaly active compounds, 1-hexanol, benzaldehyde, benzyl alcohol, and nonanol emitted from *Populus tremuloides* Michx. also interrupt the pheromone response of D. *ponderosae* Hopk. in the field (Borden *et al.*, 1998b).

Host suitability recognition

The third step is more complex, as the beetle might encounter a mixture of compounds with negative or positive effects on attraction from the host trees as well as from the beetles on or in the host.

Negative signals

Verbenone is a compound that has been shown to be present in many, if not all, conifers attacked by beetles. It is derived from α-pinene, a major monoterpene hydrocarbon constituent in many conifers (Fig. 6.3), that can be oxygenated directly by exposure to air (autoxidation: Hunt *et al.*, 1989) or by biological agents such as yeasts or bark beetles or by their combined activity (Leufvén *et al.*, 1984; Byers *et al.*, 1989) to form the corresponding ketone, verbenone (see Fig. 6.7). It has been found to inhibit the attraction to kairomones or pheromones in over ten species of bark beetles (as reviewed by Borden, 1997).

It has been established (see above) that GLVs, (5S,7S)-conophthorin and verbenone can all act as inhibitors of the attraction of conifer bark beetles to their respective pheromones. Whether these inhibitory signals also reduce the attraction of conifer-inhabiting bark weevils and cerambycids remains to be shown. However, in the cerambycid *Monochamus alternatus* (Hope) infesting *Pinus densiflora* (Sieb. et Zucc.) the opposite pattern seems to occur with the α-pinene/verbenone responses reported for bark beetles. The stressed host contains three attractive oxygenated terpenes: the monoterpene alcohol (+)-*cis*-3-pinen-2-ol, the sesquiterpene alcohol (+)-juniperol and the diterpene aldehyde (+)-pimaral (Sakai and Yamasaki, 1990, 1991). The healthy host contains none of these oxygenated compounds, but does contain (+)-germacrene D and other terpene hydrocarbons that can behaviourally mask the attrac-

Normal volatiles | **Stress-induced volatiles**

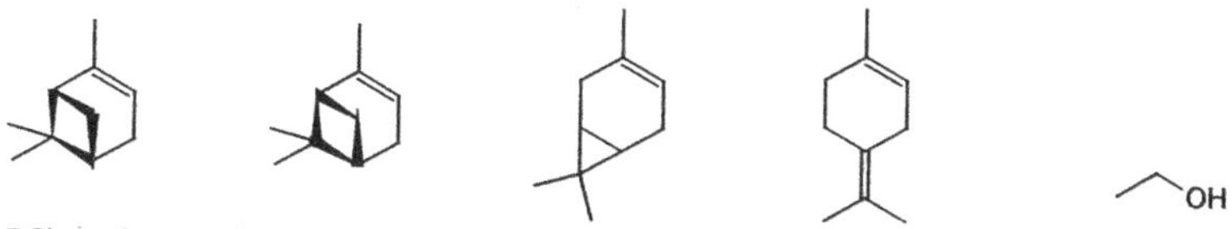

(1*S*,5*S*)-α-pinene	(1*R*,5*R*)-α-pinene	3-Carene	Terpinolene	Ethanol
Species responding:				
T. piniperda	*T. piniperda*	*T. piniperda*	*T. piniperda*	*(T. piniperda)*
H. abietis	*H. abietis*			*H. abietis*
D. micans ?	*D. micans* ?			*D. micans* ?
D. terebrans ?				*D. terebrans* ?
	D. valens	*D. valens*		*D. valens* ?
Cerambycidae	*Cerambycidae*			*Cerambycidae*

Fig. 6.3. Kairomone compounds used in host attraction by the larger pine shoot beetle (*Tomicus piniperda*), the pine weevil (*Hylobius abietis*), the turpentine beetles (*Dendroctonus micans, D. terebrans*, and *D. valens*) and in several longhorn beetles (Cerambycidae). ?, attraction has not been demonstrated unequivocally; ( ), ethanol alone is not attractive. See text for further details.

tiveness of the oxygenated compounds from the stressed host (Yamasaki *et al.*, 1997).

While the two signals result in opposite behaviours, the biological significance in bark beetles and the longhorn beetle is the same. A chemical change in the terpenes (oxygenation from hydrocarbon to alcohol, aldehyde or ketone) gives a signal indicating the status of a tree. For bark beetles demanding a fresh host, the oxygenated compound indicates unsuitability and the signal is inhibitory, whereas for the longhorn beetle, requiring a stressed host, oxygenated compounds identify suitability and provide an attractive signal.

Positive signals

Chemical signals indicating a suitable host could come from several sources, including the host tree itself, the insects, or combinations of these. The following section will give examples.

How Primary and Secondary Attraction Systems Function

Primary attraction in bark beetles is the orientation to host-tree volatiles (kairomones) without the influence of beetle-produced attractants (pheromones). Secondary attraction involves the increase in beetles orienting in response to volatiles produced by conspecifics (Person, 1931).

While common wisdom today holds that all insects use pheromones for mate location, there are some notable exceptions among forest Coleoptera. Here compounds acting as pheromone components in other species may be produced, as in the bark beetle *T. piniperda*, but these may not always be used for long-range mate location or aggregation in that particular species.

From host kairomones to de novo *beetle-produced pheromones*

Four levels of signalling systems, or solutions to the problem of finding a temporary resource scattered in space in a conifer forest, will be described: (i) attraction to host kairomones; (ii) detoxification of host monoterpenes; (iii) pheromones produced *de novo* by one sex; and (iv) pheromones produced *de novo* by both sexes. Although there is increasing chemical and behavioural complexity as we progress through these four levels, we do not imply an evolutionary trend here, mainly because the phylogeny is not known at higher levels for the taxa involved (for a different view see Francke *et al.*, 1995). Even the comparison of the phylogeny and the chemical signals in a well investigated subgeneric group of only seven *Ips* spp. and two outgroups proved to be quite complex (Cognato *et al.*, 1997). We will now discuss the first level: the attraction to host kairomones.

Primary attraction: host attraction and kairomones

The attraction of longhorn beetles (cerambycids), bark weevils (curculionids) and non-aggressive bark beetles (only attacking damaged trees) to combinations of conifer volatiles and ethanol is well established (Moeck, 1970; Klimetzek *et al.*, 1986; Schroeder, 1988; Byers, 1992a; Lindelöw *et al.*, 1992; Erasmus and Chown, 1994) (Fig 6.3, Table 6.1). Attraction to ethanol alone or in combination with odours from stressed hosts occurs also in longhorns and bark beetles living in hardwoods (Montgomery and Wargo, 1983; Dunn and Potter, 1991). In dying stems, roots or logs the release of ethanol (Moeck, 1970) and other low-molecular products due to anaerobic stress (Graham, 1968; Davies, 1980; Kimmerer and Kozolowski, 1982; Kelsey, 1996; Kelsey and Joseph, 1997) or decay fungi (Gara *et al.*, 1993) makes the host material more attractive. Often, a period of storage is required to make logs attractive to ambrosia beetles (Pulliainen, 1983; Kelsey, 1994).

Whether stress will also make host trees more attractive to aggressive bark beetle species (those able to attack and kill living, standing trees: Raffa *et al.*, 1993;

Table 6.1. Summary of stress-induced and normal primary attraction as well as secondary attraction in conifer Coleoptera.

	Primary attraction (plant volatiles)		
Beetle category	Stress-induced (detection of stressed trees)	'Normal' or wide-sense (detection of host species)	Secondary attraction (pheromone)
Aggressive bark beetles[a]	No	Often (host attraction and/or host synergism)	All cases
Non-aggressive bark beetles	Some (ethanol)	Yes	Most
Ambrosia beetles[b]	All (ethanol)	Yes	Many
Longhorns	Many	Yes	Many?
Weevils	Some	Yes	Several

[a] As defined by Raffa *et al.* (1993) and Paine *et al.* (1997).
[b] Scolytidae and Platypodidae with larvae dependent on symbiotic fungi (Francke-Grossmann, 1952).

Paine *et al.*, 1997) is controversial. The stress-induced primary attraction – the attraction to stressed but mechanically intact host trees releasing volatile signals, indicating stress predisposing it to bark beetle attack – is not well established. On the other hand, the attraction due to increased release of host volatiles from ruptured host tissues after mechanical damage or insect attack is well established, not only in bark beetles (Ryker and Oester, 1982; Byers *et al.*, 1985; Campos *et al.*, 1994), but also in bark weevils (Zagatti *et al.*, 1997) and in other insects (Lougrin *et al.*, 1995; Bolter *et al.*, 1997).

However, the attraction of 'aggressive' bark beetles to stressed but living and physically intact hosts is difficult to establish experimentally (for a review of the older literature see Moeck *et al.*, 1981). There are suggestions that stress-induced primary attraction may also make the host more attractive to the more aggressive bark beetle species. Moreover, sometimes the term primary attraction seems to be limited only to stress-induced primary attraction, but unfortunately the terms are not always clearly formulated (Tømmerås, 1989; Führer *et al.*, 1991; Reddeman, 1993; Baier, 1996). Attempts to identify stress-induced compounds from *Pinus sylvestris* L. (Sjödin *et al.*, 1989) or *Picea abies* (L.) have produced chemical candidates (Lindelöw *et al.*, 1992) but the biological activity of compounds remains equivocal (R. Jakus and F. Schlyter, 1995, unpublished results).

Correlative studies implying stress-induced primary attraction exist (Schimitschek and Wienke, 1966; Rutledge *et al.*, 1988; Führer *et al.*, 1991; Baier, 1996) but critical tests are rare. The most conclusive experiment to date was performed by Moeck *et al.* (1981), who freeze-killed stem bases of *Pinus ponderosae* Dougl. ex Laws. by dry ice treatment, and subsequently recorded bark beetle arrival to these maximally stressed (freeze-girdled) trees. Some trees were screened from the base to the beginning of the crown to prevent beetles from attacking the bark and releasing pheromones. Neither *Dendroctonus brevicomis* LeC., *D. ponderosae*, *Ips paraconfusus* Lanier, *I. latidens* (LeC.) nor *Gnathotrichus retusus* LeC. showed increased landing rates on freeze-girdled trees with screened boles over 3 months. Unscreened freeze-girdled trees had more attacks and landings than unscreened controls. Naturally stressed trees, infected by the fungal root pathogen *Verticicladiella wagenerii*, had higher landing rates and were often killed by attacking beetles, but only when they were not protected by a screen. Thus, beetle-produced pheromone attraction was repeatedly demonstrated on unscreened trees (including some screened trees where two species had penetrated the screen) while stress-induced primary attraction was not supported. Conflicting information was obtained in a study of lodgepole pine (*P. contorta* var. *latifolia* Engelmann) where evidence for, at least, wide-sense primary attraction was found for *I. latidens* (Miller *et al.*, 1986). A monoterpene hydrocarbon found in *P. contorta*, β-phellandrene, is an attractive kairomone component for *I. latidens* (Miller and Borden, 1990).

The fact that, during outbreaks of aggressive species, bark beetles land on most trees in a forest each day, forms the basis for the random landing hypothesis (Raffa and Berryman, 1980). This hypothesis holds that only after landing will the decision to colonize be made and pheromone production start, attracting conspecifics in larger numbers. Theoretical arguments, based on computer simulations, both for (Byers, 1996) and against (Gries *et al.*, 1989) random landing in *I.*

typographus have been made. At the moment, the random landing theory seems to be best supported for aggressive species, but the matter remains controversial.

Four examples of attraction directly to the host trees (chosen from all three families) will be discussed below. At this level, only attraction to host kairomones, such as monoterpenes, is shown, while no pheromone activity has been established.

THE LARGER PINE SHOOT BEETLE, *TOMICUS PINIPERDA*. Earlier claims in the literature that *T. piniperda*, breeding in wind-broken or otherwise severely stressed trees, would use pheromone (reviewed by Lanne *et al.*, 1987) were not well supported experimentally. While there was ample evidence of a host attraction, the odd fact remained that in hindguts of *T. piniperda* females, the pioneering sex, there was an array of oxygenated monoterpenes identical to or similar to earlier proven bark beetle pheromone attractants. Given the ideas at the time, that bark beetle pheromones were mostly produced through oxygenation of host monoterpene hydrocarbons, these hindgut volatiles seemed to be strong candidates for pheromone components.

However, a rigorous isolation scheme with GC fractionation and behavioural testing (Silverstein, 1970; Byers, 1992b) failed to show the existence of any pheromonally active compounds in *T. piniperda* (Byers *et al.*, 1985). Four host monoterpene hydrocarbons (Fig. 6.3) were isolated and shown to be active at ecologically relevant release rates. The use of host monoterpenes alone for mate finding was later theoretically justified by a computer model, showing that the sexes aggregating on a host log would have a good chance of physically encountering each other within 1 h by a random search at normal field densities (Byers, 1991). It has been suggested that ethanol is a general synergist of host terpenes for less aggressive bark beetles (Klimetzek *et al.*, 1986), but in the pine shoot beetle it is an attractant only at low concentrations and with a specific ratio to host-tree terpenes (Schroeder, 1988). Ethanol alone is not attractive in *T. piniperda* (Byers, 1992a).

Recently, a new claim for pheromonal activity has been made based on electrophysiological and field trapping evidence for biological activity of female-produced compounds (Czokajlo *et al.*, 1996, 1997). At the time of writing, the published behavioural evidence for this claim appears to be inconclusive. However, the elusive character of the secondary attraction (pheromone) in this species may be caused mainly by the strong activity of the primary attraction (kairomone). A possible parallel to *T. piniperda* is found in another less aggressive bark beetle, the fir engraver *Scolytus ventralis* LeC., which showed a high sensitivity to host odours and the existence of primary attraction, but no evidence of secondary attraction (Macías-Sámano *et al.*, 1998). It is possible that these two species have a weak or short-range pheromone signal, perhaps ecologically relevant and only possible to demonstrate experimentally when the usually dominating kairomone signal is quite weak. Thus, it can be concluded that the primary attractive signal for *T. piniperda* is a kairomone signal of monoterpene hydrocarbons from the host (wide-sense primary attraction), at times combined with a weak signal of host stress (ethanol).

THE PINE WEEVIL, *HYLOBIUS ABIETIS*. The pine weevil, *Hylobius abietis* (L.), breeds in dying roots of conifers, such as those of storm-felled or cut trees, and does not appear to have a long-distance pheromone (Zagatti *et al.*, 1997 and references therein; see also Chapter 5). Like *T. piniperda*, weevils feeding on logs of Scots pine (*Pinus sylvestris*) enhance attraction somewhat over uninfested logs, but no more than can be explained by the increased release of host odours (Zagatti *et al.*, 1997). Uninfested spruce and pine logs are by themselves attractive, proving wide-sense primary attraction (Tunset *et al.*, 1993). A stress-induced primary attraction seems active, as in a study of *H. abietis* in Scandinavia by Lindelöw *et al.* (1992). This was the only species consistently attracted, not only to combinations of host monoterpenes and ethanol, but also to ethanol itself, indicating a preference for, or an acceptance of, strongly stressed host roots which had been damaged for some time. This is a case of stress-induced primary attraction.

THE TURPENTINE BEETLES, *DENDROCTONUS MICANS*, *D. TEREBRANS* AND *D. VALENS*. Some *Dendroctonus* species do not mass attack in large numbers, but the larvae feed in dense groups at the base of trees and the adults seem to avoid the normal terpene-based defence (Grégoire, 1985). The signalling system of these species seems to reflect the need for aggregation; in the European *D. micans* Kug., adults do not appear to use a pheromone system, while their larvae have an aggregation pheromone. Larvae of *D. micans* use several oxygenated monoterpenes and one, (−)-fenchone, is also used in the larval aggregation pheromone of *D. valens* LeC. During feeding the larvae line up side by side (Grégoire *et al.*, 1991). The disadvantage of this larval chemical communication signal is that the predator *Rhizophagus grandis* Gyll. uses the same compound (fenchone) as a kairomone during the localization of its prey (Grégoire *et al.*, 1991, 1992).

The attraction of the Nearctic *D. valens* (red turpentine beetle) and *D. terebrans* (Olivier) (black turpentine beetle) to host volatiles is clear. While *D. valens* does not seem to have a pheromone (Hobson *et al.*, 1993, and references therein), females of *D. terebrans* do produce (−)-frontalin, besides an array of compounds which mainly comprise oxygenated monoterpenes. The males are attracted to the combination of frontalin and host turpentine (sex pheromone activity) while females are not clearly attracted to male produced (+)-*exo*-brevicomin (Phillips *et al.*, 1989).

THE LONGHORN BEETLES, CERAMBYCIDAE. The forest longhorn beetles breed almost exclusively in dead or moribund trees. Many species living in conifers are attracted to combinations of monoterpenes and ethanol (Chenier and Philogène, 1989). Some longhorn beetles living in the oak *Quercus alba* L. also show a synergistic response in attraction to the combination of ethanol and odours from the stressed host tree (Dunn and Potter, 1991).

Most longhorn beetles have not been studied in detail for the existence of a pheromone system, but recent work has indicated that many species breeding in forest trees may have pheromones (Iwabuchi *et al.*, 1986, 1987; Fettköther *et al.*, 1995; Leal *et al.*, 1995). Thus, as with the scolytids, only a few species of cerambycids might finally be found to rely solely on stress-induced primary attraction for location of suitable breeding patches.

Secondary attraction: detoxification of host monoterpenes

Detoxification of host-tree monoterpene hydrocarbon defence compounds through hydroxylation (Vanderwel and Oehlschlager, 1987, and references therein) and subsequently turning these into a signal promoting insect aggregation and ultimately killing the host tree, is an interesting example of plant–insect interaction. The production of bark beetle pheromones has long been typified by this kind of system (Hughes, 1974; Renwick *et al.*, 1976a,b). Today, however, few species are believed to use detoxification as their sole system for pheromone production, the first example here may do so, whereas the other has a combination of production systems.

THE LESSER PINE SHOOT BEETLE, *TOMICUS MINOR*. The ambrosia fungi-dependent *T. minor* (Hart.) uses (−)-*trans*-verbenol (Fig. 6.4) as the main component in a female-produced aggregation pheromone for host and mate location. The pheromone is synthesized from host volatiles and the synthetic kairomone blend of the larger pine shoot beetle is active (Lanne *et al.*, 1987). The 1*S*,4*R*,5*S*-(−)-*trans*-verbenol is biogenetically related to the 1*S*,4*S*-(−)-α-pinene (Fig. 6.3), a major host resin constituent, which then acts both as a component of the host tree kairomone signal and as a precursor for pheromone biosynthesis.

Somewhat surprisingly, the volatile profiles from the hindgut of *T. piniperda* and *T. minor* were almost identical after contacting resin in the host phloem (Lanne *et al.*, 1987), indicating in both cases a simple detoxification process of monoterpene hydrocarbons from the shared host. In addition, the antennal sensitivity by EAG to a range of host- and beetle-produced synthetic compounds was similar (Lanne *et al.*, 1987) in the two species. However, field and laboratory bioassays showed that a clear attraction existed to both natural and synthetic pheromone sources in *T. minor*, while the larger species was strongly attracted only to treatments including host odours (Lanne *et al.*, 1987).

The evolutionary background for the two signalling systems can only be speculated about, but a possible cause is major differences in feeding biology. In later instars, *T. minor* larvae are dependent on symbiotic fungi and function ecologically as ambrosia beetles (Francke-Grosmann, 1952). In contrast, *T. piniperda* is a phloem feeder. Thus, a narrower range of hosts might be suitable for fungal establishment and the aggregation of many parent beetles may be required for

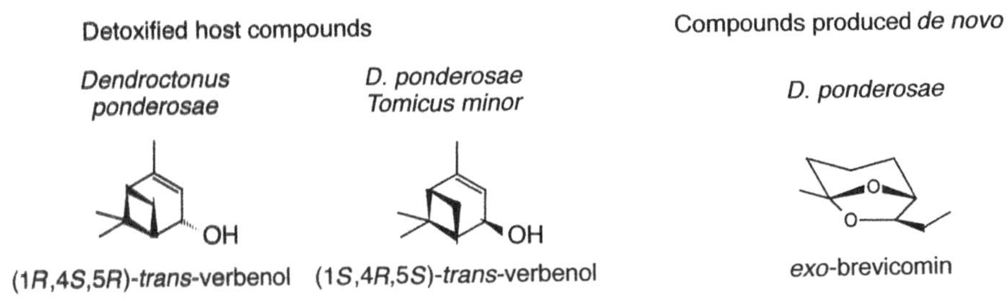

Fig. 6.4. Compounds comprising pheromone components produced by detoxification of host monoterpenes (the smaller pine shoot beetle, *Tomicus minor*) or a combination with probably *de novo* produced compounds (the mountain pine beetle, *D. ponderosae*). In both species there is synergism between beetle-produced compounds and host compounds (not shown).

success of the symbiotic fungi (Paine *et al.*, 1997). Thus, colonization of a good breeding patch may be more dependent on the successful aggregation.

THE MOUNTAIN PINE BEETLE, *DENDROCTONUS PONDEROSAE*. The notorious tree killer *D. ponderosae* Hopk. seems initially to use a simple pheromone signal with female-produced (−)-*trans*-verbenol synthesized from host monoterpenes (Borden *et al.*, 1987, and references therein). This female signal is analogous to that of *T. minor*. Later-arriving males make *exo*-brevicomin *de novo* and contribute to the attraction (Fig. 6.4); however, it could be regarded as multi-functional, as it is inhibitory to males at higher doses. In addition, an array of mostly female-produced compounds, originating from both host and *de novo* sources, have been inferred to be part of the attractive or inhibitory pheromone signals (Conn *et al.*, 1983; Hunt and Borden, 1988; Miller and Lafontaine, 1991; Mori *et al.*, 1997). For tree baiting, a binary bait with only *trans*-verbenol and *exo*-brevicomin is sufficient (Borden *et al.*, 1993).

Secondary attraction: pheromones produced de novo *by one sex*

This is probably the most common communication among wood-living forest beetles, and males are the sex that often produce these *de novo* compounds. In addition to the examples below, the pinhole borers or Platypodidae seem to fit here as well. The platypodids are ambrosia beetles superficially similar in morphology and ecology to the scolytids, but they may not be phylogenetic sister groups. There is biological evidence of a male-produced pheromone in four species of platypodids, attractive together with host volatiles (Renwick *et al.*, 1977; Milligan *et al.*, 1988; Milligan and Ytsma, 1988; Ytsma, 1989). The male pheromone seems to act as an aggregation pheromone (Milligan *et al.*, 1988; Milligan and Ytsma, 1988). The only chemical identification in platypodids so far (Renwick *et al.*, 1977) indicates sulcatol and grandisol (Figs 6.5 and 6.8), compounds that are not suspected to derive from host terpenes.

We have chosen three examples from each of the three taxa, but only the bark beetle compounds have been directly shown to be produced *de novo*.

THE DOUBLE-SPINED SPRUCE BARK BEETLE, *IPS DUPLICATUS*. *Ips duplicatus* (Sahlb.) often breeds in slash or in trees colonized by the larger and more aggressive *I. typographus* (Schlyter and Anderbrant, 1993). It has recently been causing single-species outbreaks in both Inner Mongolia and central Europe (Zhang *et al.*, 1995; Knízek and Zahradník, 1996). Its pheromone consists of two alcohols, ipsdienol and *E*-myrcenol (Fig. 6.5), both seemingly closely related to the host monoterpene hydrocarbon myrcene (Fig. 6.6; Byers *et al.*, 1990). A combination of these two alcohols at 200 male equivalents was better than or equal to a similar number of unpaired males boring in a host log (*P. abies*) (Schlyter *et al.*, 1992). In beetles from Norway the ipsdienol is produced as a racemic mixture [50:50 (4*R*):(4*S*)-enantiomers] and the *E*-myrcenol makes up about 15% of the emitted blend (Schlyter *et al.*, 1992). A blend of between 23 and 47% *E*-myrcenol is optimal for attracting several Palearctic populations, except an isolated population in Inner Mongolia, where optimum response was between 58 and 76% *E*-myrcenol (F. Schlyter *et al.*, 1996, unpublished results). The two pheromone components are both produced *de novo* when males are making their nuptial chamber (Ivarsson *et*

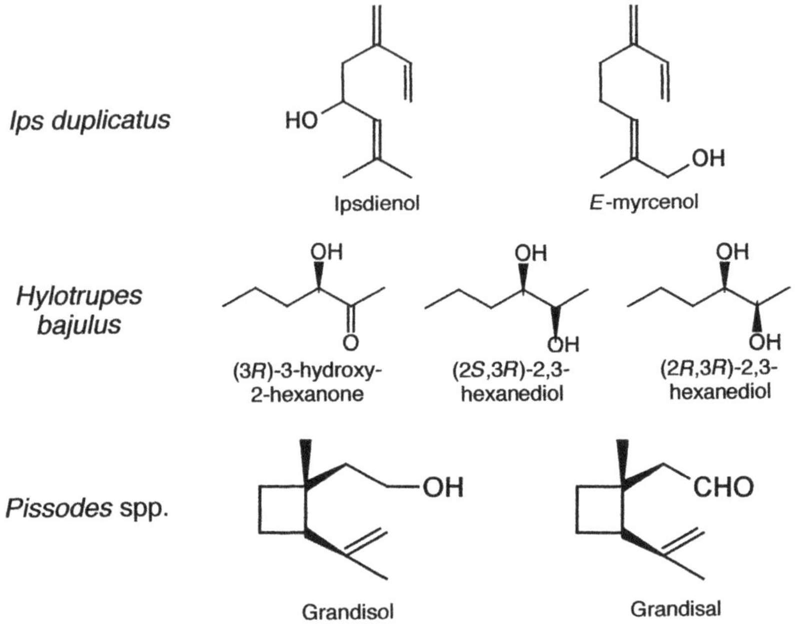

Fig. 6.5. Pheromone components produced *de novo* in one sex by the double-spined spruce bark beetle, *Ips duplicatus*. *De novo* production is also proposed in the old house borer, *Hylotrupes bajulus*, and pine bark weevils, *Pissodes* spp. In the latter genus, host kairomones synergize the attraction to insect-produced pheromones but the compounds have not yet been identified.

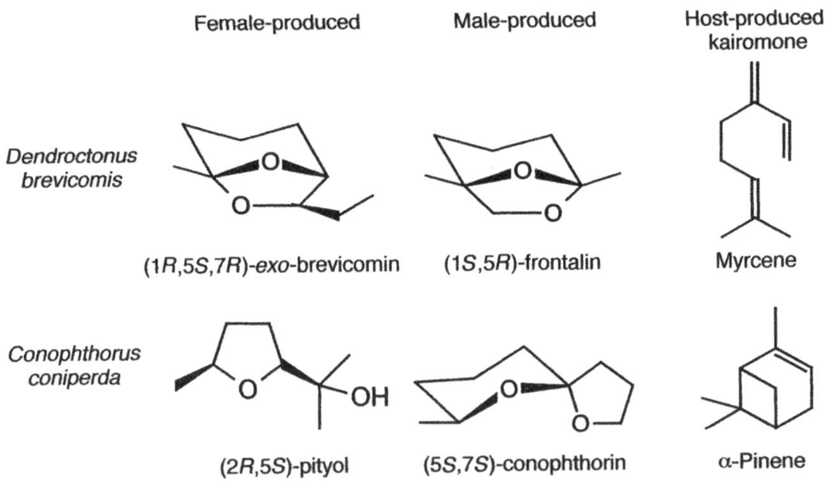

Fig. 6.6. Pheromone components produced *de novo* by both sexes in the western pine beetle, *Dendroctonus brevicomis*, and in the white pine cone beetle, *Conophthorus coniperda*, together with synergistic host kairomones.

al., 1993), while males are able to produce smaller amounts of ipsdienol if exposed to high doses of myrcene (Schlyter *et al.*, 1992). Interestingly, in the Inner Mongolia population both compounds are also found in beetles from maturation feeding aggregations before reproduction (Q.-H. Zhang, 1998, unpublished results).

THE OLD HOUSE BORER, *HYLOTRUPES BAJULUS*. Attraction to host odours is important for many forest cerambycids, but male-produced pheromones have been described for several species (Iwabuchi, 1982). The old house borer of Europe, *Hylotrupes bajulus* (L.), has been shown to have a multi-component male-produced pheromone. Although at present a pest in domestic situations, feeding on dry conifer logs or structural boards in houses, the original habitat is undoubtedly old forest with an abundance of moribund trees.

The male pheromone signal comprises the two diols (2*R*,3*R*)-2,3-hexanediol and (2*S*,3*R*)-2,3-hexanediol and the corresponding ketone, (3*R*)-3-hydroxy-2-hexanone (Fig. 6.5) (Fettköther *et al.*, 1995), as major components of the blend attracting virgin females. These C_6 compounds show a clear affinity with the C_8 compounds found in the grape borer *Xylotrechus pyrrhoderus* Bates and in the mulberry borer *X. chinensis* Chevrolat: (2*S*,3*S*)-octanediol and (2*S*)-hydroxy-3-octanone (Iwabuchi *et al.*, 1986, 1987). Interestingly, the males of the cryptomeria twig borer, *Anaglyptus subfasciatus* Pic, release (*R*)-3-hydroxy-2-octanone and (*R*)-3-hydroxy-2-hexanone (Leal *et al.*, 1995) in a geographically variable ratio (from 1:7 to 1:125) and make up an aggregation pheromone showing a distinct synergism with floral odour (Nakamuta *et al.*, 1997).

There is still a lack of information on the active range of these male signals; it seems to be shorter than that found in bark beetles (Schlyter, 1992), but longer than for a contact pheromone. A female contact sex pheromone seems to exist in some other species, and one component was recently identified from the body surface of *Psacothea hilaris* (Pascoe) as the hydrocarbon (*Z*)-21-methyl-8-pentatriacontene (Fukaya *et al.*, 1996).

THE PINE BARK WEEVILS, *PISSODES* SPP. Two oxygenated monoterpenic compounds have been identified in males of the *Pissodes* bark or pine weevils, which feed on live shoots of *Pinus* spp. These are grandisol and grandisal (Fig. 6.5), which in *P. nemorensis* Germar are 100% (1*R*,2*R*) and 100% (1*S*,2*R*), respectively, while in *P. strobi* (Peck) the enantiomers are 99% (1*R*,2*S*) and 60% (1*R*,2*S*), respectively (Hibbard and Webster, 1993). Combined with host odours, the alcohol and aldehyde were active as aggregation pheromones for populations regarded as *P. nemorensis* Germar and *P. approximatus sensu* Hopkins (Booth *et al.*, 1983; Phillips *et al.*, 1984).

The species separation problems amongst the three sibling species are obvious. They are probably resolved by temporal separation and by the inhibition due to differences in enantiomeric composition (Phillips and Lanier, 1986). Later, *P. approximatus* was synonymized with *P. nemorensis* (Phillips *et al.*, 1987). Possibly, more pheromone components remain to be discovered.

Secondary attraction: pheromones produced de novo *by both sexes*

The fourth level of communication is found in a few species, where the semiochemical systems comprise compounds that are produced *de novo* by both sexes.

These compounds can either show synergism, or have antagonistic effects. Two of the better known examples of this complex, and possibly evolutionarily more advanced category, are discussed in detail below.

THE WESTERN PINE BEETLE, *DENDROCTONUS BREVICOMIS*: SYNERGISM. In the monogamous western pine beetle (*D. brevicomis* LeC.), the female is the pioneering sex which initially attacks the host trees. When a female is tunnelling through the outer bark, she releases a mixture of volatiles including (+)-*exo*-brevicomin (Fig. 6.6; Byers, 1989, and references therein). This bicyclic acetal acts as a contact pheromone, produced by the beetle prior to attack and before contact with the host resin, and is most probably produced *de novo*. When the tunnelling female contacts the resin channels of the tree, terpene-rich resin will exude through the entrance hole. The release of *exo*-brevicomin, together with myrcene (one of the monoterpene hydrocarbons in the host-tree resin), specifically attracts males. To overcome the defences of the host *Pinus* tree, a mass attack is achieved by the later-arriving males, which produce the bicyclic acetal (−)-frontalin (Fig. 6.6). This compound also changes the attraction to the mixture, from being a sex pheromone, attracting mostly males, to a three-component mixture attracting both males and females, and can be called an aggregation pheromone. This efficient aggregation pheromone enables *D. brevicomis* to attack healthy host trees.

Later during the attack, the males also produce verbenone (Fig. 6.7), which inhibits attraction (Byers, 1989, and references therein; Paine and Hanlon, 1991; Bertram and Paine, 1994). Verbenone might also be produced by autoxidation from host α-pinene (Hunt *et al.*, 1989). Besides verbenone, (+)-ipsdienol, naturally produced by competing *Ips* species and by male *D. brevicomis*, has a very strong inhibitory effect in field bioassays, being active at much lower doses than verbenone (Byers, 1989, and references therein; Paine and Hanlon, 1991; Bertram and Paine, 1994).

The system is complex. There is an attractive principle consisting of three components: one from each sex of the beetles, plus one from the host tree. This aggregation signal is later countered by negative signals from at least two sources. One inhibitor, verbenone, is produced by the males, or directly from the tree, due to autoxidation of the monoterpene hydrocarbon α-pinene in the resin. The other inhibitor, (+)-ipsdienol, is produced by competing sympatric bark beetle species and by males.

THE WHITE PINE CONE BEETLE, *CONOPHTHORUS CONIPERDA*: INHIBITION. Not all scolytids reproduce under the bark of their host trees and this is especially true in the tropics. One

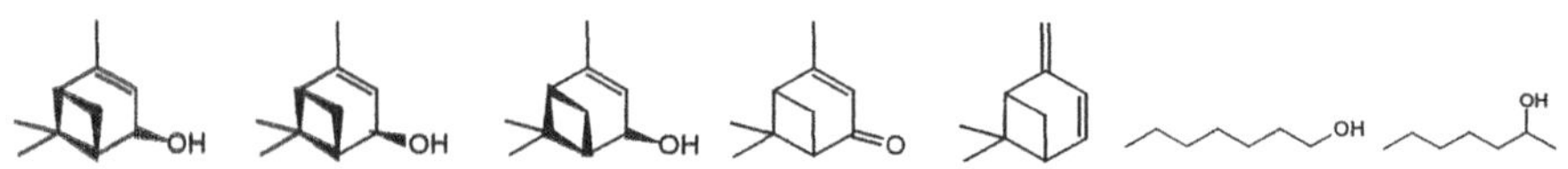

(1*S*,4*S*,5*S*)-*cis*-verbenol (1*S*,4*R*,5*S*)-*trans*-verbenol (1*R*,4*S*,5*R*)-*trans*-verbenol Verbenone Verbenene 1-Heptanol 2-Heptanol

Fig. 6.7. Bark beetle pheromone components known to be derived from host precursors.

temperate species that attacks conifer cones is the monogamous white pine cone beetle, *Conophthorus coniperda* (Schwartz), with the female as the pioneering sex. In the spring, when the cones are elongating, the female attacks the cone and bores into the base. She may mine through a few cones for feeding, before she attracts a male and oviposits (Godwin and Odell, 1965; de Groot *et al.*, 1991). During her tunnelling through a pine cone, the female produces *de novo* a female-specific furanoid compound, (2*R*,5*S*)-pityol (Fig. 6.6; Birgersson *et al.*, 1995). This compound, synergized by host tree monoterpene hydrocarbons (or just α-pinene), is a very strong attractant to conspecific males and acts as a sex pheromone (de Groot *et al.*, 1998). In a laboratory walking bioassay, this mixture even turned the males aggressive, as they cut legs and antennae off each other with their mandibles (G. Birgersson, 1998, unpublished data). The male accepted by a female produces large amounts of a spiroacetal, (5*S*,7*S*)-conophthorin (Fig. 6.6). This compound strongly inhibits the male attraction to the female-produced sex attractant (de Groot *et al.*, 1998).

Pheromone biosynthesis in bark beetles: shift from oxygenation of host monoterpenes to de novo *production*

During the 1970s there was a debate among researchers about the biosynthesis of female-produced sex pheromones in Lepidoptera, especially moths. Some argued for a relationship with diet, i.e. a simple conversion and direct utilization of the diet into the components that comprise the pheromone (Hendry *et al.*, 1975), while others argued for a *de novo* biosynthesis, i.e. without a direct connection to dietary material (Miller *et al.*, 1976). The debate finally reached consensus (Hendry, 1976), and for many years it has been clear that moth sex pheromones are produced *de novo* from acetates via long-chain fatty acids. Over the years, pheromones from several insect species have been investigated, and for most insects it has been concluded that their pheromones are produced *de novo*. The last group of insects for which the diet-related strategy for pheromone production is proposed is scolytid bark beetles.

The production of bark beetle pheromones has long been explained as detoxification of host tree monoterpene hydrocarbons through hydroxylation (Vanderwel and Oehlschlager, 1987, and references therein). However, recent investigations into bark beetle pheromones have questioned this. Biosynthetic research and the identification of new bark beetle pheromone components have revealed a new understanding of the biosynthesis of scolytid pheromones. There has been a shift from ideas of detoxification of host-tree monoterpenes to *de novo* production.

Most bark beetle species have the capacity to hydroxylate toxic host-tree monoterpene hydrocarbons of the host (Hunt *et al.*, 1989), but this is the major biosynthetic pathway for very few compounds used as pheromone components (Fig. 6.7). Currently, only *cis*- and *trans*-verbenol, mainly in *Ips* and *Dendroctonus* species, can be directly derived from monoterpene hydrocarbons, as hydroxylation products from α-pinene (Renwick *et al.*, 1976a,b; Lanne *et al.*, 1989). Further oxidation of these two monoterpene alcohols, by either the beetles themselves, or by associated microorganisms, or through autoxidation (Hunt *et al.*, 1989), gives verbenone: a general inhibitor of bark beetle attraction. Some pine species in western North America, such as Jeffrey pine (*Pinus jeffreyi* Grev. & Balf.), have heptane-based resins, and contain almost no monoterpene hydrocarbons. Parallel to the hydroxylation of α-pinene in many conifer resins, this major resin straight-chain hydrocarbon is detoxified through hydroxylation by the attacking bark beetles. When a female Jeffrey pine beetle (*Dendroctonus jeffreyi* Hopk.) attacks a host tree, she encounters large amounts of heptane-rich resin which she hydroxylates to heptanol and 2-heptanol (Barkawi *et al.*, 1997). These compounds are used in the aggregation pheromone of this species (Renwick and Pitman, 1979).

Besides the oxygenated monoterpenes, other terpenoid compounds have been identified as pheromone components in bark beetles. Isoprenic alcohols, such as methylbutenols (MBs), are found in several bark beetle species (Borden, 1985): the 2,3,2-MB in *I. typographus* and *Orthotomicus erosus* Woll., the 3,3,1-MB and 3,2,1-MB in *Ips cembrae* (Heer) (Francke *et al.*, 1995) and the 3,3,1-MB in *Polygraphus rufipennis* (Kirby) (Bowers *et al.*, 1991). These hemiterpene alcohols (Fig. 6.8) can be produced via the same biosynthetic pathway as the oxygenated monoterpenes. A biosynthetic study by Lanne *et al.* (1989), using radiolabelled precursors, showed that 2-methyl-3-buten-2-ol is produced *de novo* by *I. typographus*.

Parallel to the production of the verbenols, the biosynthesis of the 'myrcene alcohols', ipsdienol, ipsenol and *E*-myrcenol (Fig. 6.8), common pheromone components in *Ips* bark beetles, were regarded as hydroxylation and reduction products of the host tree monoterpene hydrocarbon myrcene (Hughes, 1974; Fish *et al.*, 1979; Hendry *et al.*, 1980; Vanderwel and Oehlschlager, 1987). A quantitative study by Byers and Birgersson (1990), where males of the California five-spined ips (*I. paraconfusus*) were allowed to attack bolts of different pine species, raised doubts about the detoxification theory. Independent of the quantity of myrcene in the host tree resin, the beetles produced the same amount of ipsdienol and ipsenol. In contrast, the amount of *cis*-verbenol produced by *Ips* species reflects the amount of (−)-α-pinene in the host tree resins (Klimetzek and Francke, 1980; Lindström *et al.*, 1989; Byers and Birgersson, 1990).

A breakthrough in the *de novo* hypothesis for the production of monoterpenic pheromone components in bark beetles was a study by Ivarsson *et al.* (1993) that used compactin, an enzyme-specific inhibitor affecting the rate-limiting step of the mevalonic (or isoprenic) pathway. The double-spined bark beetle, *I. duplicatus*, was the model insect. The aggregation pheromone of this species is a 5:1 mixture of ipsdienol and E-myrcenol, as described earlier. When the *de novo* biosynthesis of monoterpenes (and other terpenes) was blocked by compactin, the male production of both pheromone components was significantly reduced. To prove that the inhibition of the pheromone was not an indirect inhibition, by blocking the synthesis of the terpenoid juvenile hormone (JH), the first study was balanced in a 'rescue' experiment, where the blocking effect by compactin was countered by treatment with the JH analogue, methoprene. Food-deprived males of *I. duplicatus* do not produce pheromone as this treatment reduces JH synthesis, but topical application of methoprene restores pheromone production. However, when other males were treated with both compactin and the synthetic JH, their production of pheromone was drastically reduced (Ivarsson and Birgersson, 1995). Work with

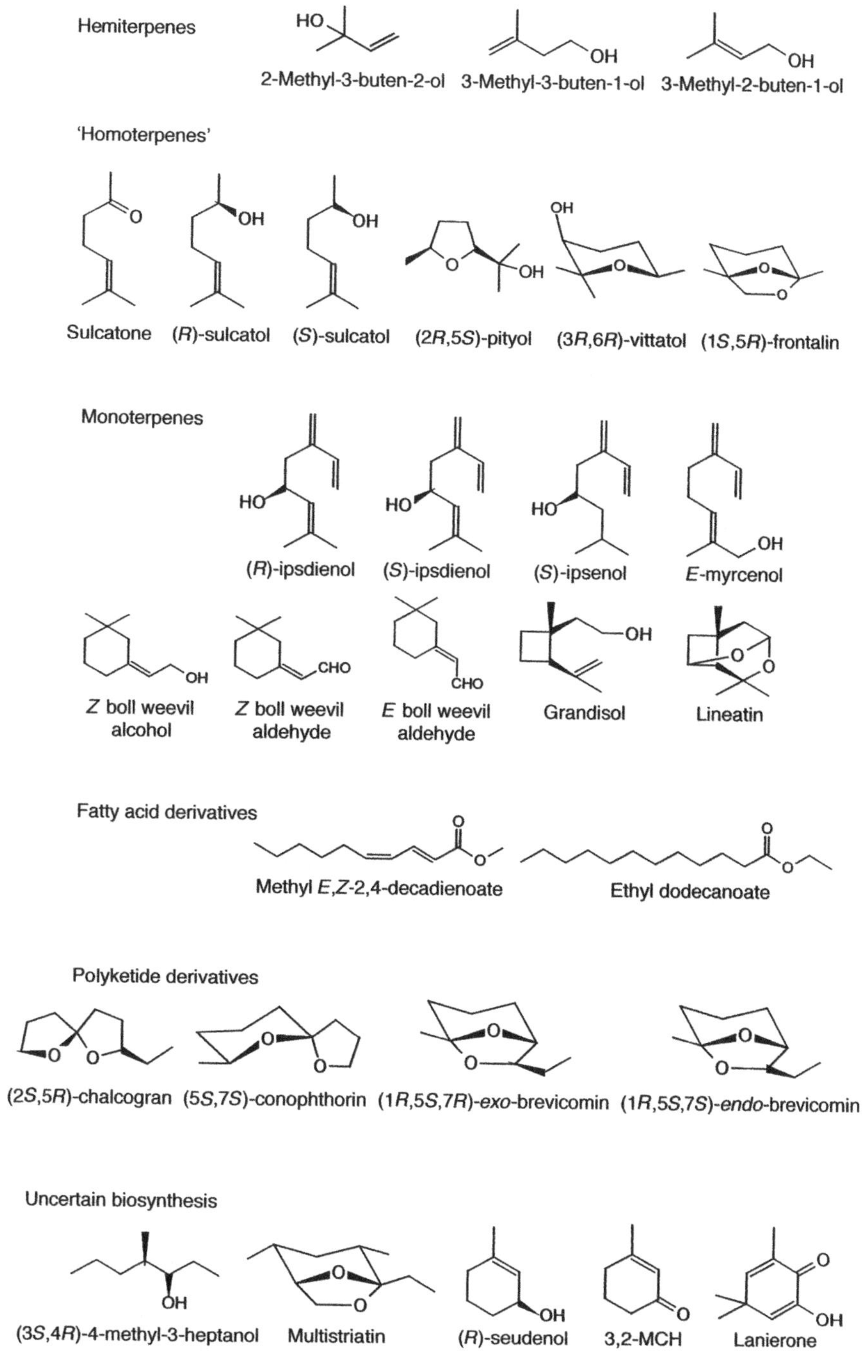

Fig. 6.8. Bark beetle pheromone components where the biosynthesis is known not to involve host precursors. Verbenene, a pheromone component reported from *Dendroctonus rufipennis* (Gries *et al.*, 1992), could possibly also be placed under uncertain biosynthetic pathways.

North American *Ips* species, using radiolabelled precursors, has proved that acyclic monoterpene alcohols in other species of *Ips* are also produced *de novo* by the beetles (Seybold *et al.*, 1995; Ivarsson *et al.*, 1997, 1998; Tillman *et al.*, 1998).However, a small amount might still be produced from preformed myrcene (S. Seybold, Reno, 1998, personal communication).

Grandisol, a cyclobutanoic monoterpene alcohol in the boll weevil, *Anthonomus grandis* Bohem., as well as the cyclohexanoic boll weevil alcohols and boll weevil aldehydes (see Chapter 5) have been identified in forest weevils, such as *Pissodes* spp. (Phillips *et al.*, 1984), and bark beetles, such as *Pityogenes* spp. (Figs 6.5 and 6.8; Francke *et al.*, 1995; J.A. Byers and G. Birgersson, 1998, unpublished data). The biosynthetic pathway was shown either to use host geraniol or nerol (Thompson and Mitlin, 1979), or to be entirely *de novo* (Mitlin and Hedin, 1974). Further cyclization of grandisol gives the tricyclic acetal lineatin (Fig. 6.8), a pheromone component of the striped ambrosia beetle, *Trypodendron lineatum* (Oliv.) (Borden, 1985).

Other bark beetle compounds like the hemiterpenes with reduced terpenic structures are the sulcatols (6-methyl-5-hepten-2-ol) and the corresponding ketone, sulcatone (6-methyl-5-hepten-2-one) (Fig. 6.8). The ketone can be produced as an oxidative cleavage of the 2,3 double bond of the monoterpenic precursor geraniol (β-oxidation; Vanderwel and Oehlschlager, 1987) and further reduced to the alcohols. The two alcohol isomers are pheromone components of the ambrosia beetles *Gnathotrichus sulcatus* (LeC.) and *G. retusus* (LeC.) (Borden *et al.*, 1980), while the ketone is believed to be an anti-aggregant of the Douglas-fir beetle, *Dendroctonus pseudotsugae* Hopk. (Ryker *et al.*, 1979). The sulcatols can, in turn, function as biosynthetic precursors in cyclization reactions. If sulcatol cyclizes, as proposed by Klimetzek *et al.* (1989), this will give either the furanoid, pityol, or the pyranoid, vittatol (Fig. 6.8), in analogy with the cyclization of linalool to the linalool 3,6- and 3,7-oxides, respectively. On the other hand, if the double bond of sulcatol isomerizes, giving a methylene group, followed by an epoxidation and subsequent cyclization, the product will be the bicyclic acetal frontalin (Fig. 6.8), as outlined by Vanderwel and Oehlschlager (1987).

Other major biosynthetic routes for pheromone components are the fatty acid and acetogenic pathways, both producing straight-chain molecules. The fatty acid pathway has historically been the 'normal' production of pheromones in moths. The first moth pheromone-like compound that was identified in a bark beetle was the 'pear ester', methyl E,Z-2,4-decadienoate, in *Pityogenes chalcographus* (L.) by Byers *et al.* (1988), which has been followed by ethyl dodecanoate in *P. hopkinsii* Swaine (Birgersson *et al.*, 1990), *P. bidentatus* (Herbst) and *P. chalcaratus* (Eichh.) (Francke *et al.*, 1995) (Fig. 6.8). These fatty acid esters synergize chalcogran (Fig. 6.8), the spiroacetal identified in several species of *Pityogenes*. The attractivity to chalcogran alone is weak in *P. chalcographus*, while the ester is basically inactive by itself. Together they make a very strong aggregation pheromone, with 25–50 times the potency of chalcogran alone (Byers *et al.*, 1988). The biosynthetic pathway proposed for the spiroacetals chalcogran (in *Pityogenes* spp.) and (5S,7S)-conophthorin (in *Conophthorus* spp.) is through acetogenins, or polyketids (Vanderwel and Oehlschlager, 1987; Francke *et al.*, 1995).

For some compounds used by scolytids as pheromone components, the biosynthetic pathways have not been elucidated. These compounds have not been identified in host trees, and must be produced by the insects. Structural similarities between the bicyclic acetal, multistriatin, and the two brevicomin isomers and frontalin (Fig. 6.8) might indicate similar biosynthetic steps. Seudenol and 3,2-MCH (3-methyl-2-cyclohexenone) in the Douglas-fir beetle, *D. pseudotsugae* (Ryker *et al.*, 1979) and lanierone in *Ips pini* (Teal *et al.*, 1992) all show structural similarities with the host resin monoterpenes, and might be derivatives of α-pinene (Figs. 6.6 and 6.8). However, these compounds could also be produced *de novo*, as are most other scolytid pheromone components.

Why Certain Pheromone Systems are Used

Pheromone production and mating systems

Many long-distance pheromone signals have been described from females but they also arise from males in some taxa and especially in forest beetles. As suggested by Thornhill and Alcock (1983), male pheromones seem to be less common and it would be interesting to examine the cases where they occur, to analyse their ecological and evolutionary background, and to answer three more specific questions:

1. Are female-produced sex pheromones 'the rule', and, if so: why?
2. Under which conditions do male pheromones occur?
3. Are male-produced pheromones, as in many bark beetles, always aggregation pheromones?

Individual species cannot be used as the unit of comparison, as a severe bias would be produced by the many species studied in economically important groups, particularly moths and bark beetles. Instead, we chose to use families as the unit of comparison, as pheromones in an adequate number of families have been described and because different families often represent distinctly different life histories, mating systems and resource utilization patterns. In Table 6.2, pheromone systems are further classified as long- or short-distance systems. Long-distance systems include communication distances as in noctuid moths (Lepidoptera: Noctuidae, hundreds of metres; Schlyter, 1992) and bark beetles (Coleoptera: Scolytidae, tens to hundreds of metres; Schlyter, 1992). Moths also exemplify the typical sex pheromones where only the male, as the sex not producing a long-distance pheromone, is attracted. Bark beetles show the best known examples of aggregation pheromones, where the pheromone-producing sex is also attracted, although the pheromone-producing sex is attracted in smaller numbers. Short-distance systems include 'short-range' types where maximum communication distances are typically not more than a few metres even in relatively large insects (1.5 m in a cerambycid beetle), or 'courtship' types with distances from centimetres (Lepidoptera: Pyralidae) down to physical contact (Dictyoptera: Blattidae). The primary data sources for Table 6.2 are Tamaki (1977) and Cardé and Baker (1984).

To answer the specific questions set out above:

1. Of 53 cases of long-distance systems listed in Table 6.2, female-produced sex pheromones account for 39 cases (74%), and can be called 'the rule'. In most of

Table 6.2. Production of pheromones by insects of either sex from different families with notes on reproduction-limiting resources.

Taxa[a]	Sex[b]	Pheromone type[c] Long-distance	Short-distance	Resource type[d]	Reference[e]
Dictyoptera					
Blattidae	F	X	X	Female	Cardé and Baker (1984)
Hemiptera					
Diaspididae	F	X		Female	Cardé and Baker (1984)
Coccidae	F	X		Female	Cardé and Baker (1984)
Neuroptera					
Mantispidae	M		X?	Female	Redborg (1998)
Myrmeliotidae	M		X	Female?	J. Löfqvist (Alnarp, 1998, personal communication)
Dilaridae	F	X			Chinery (1973)
Mecoptera					
Bittacidae (Panorpidae)	M	X		Nuptial gift *CD*	Thornhill (1992)
Lepidoptera					
Eriocraniidae	F	X			Löfstedt and Kozlov (1997)
Mnesarchaeidae	F	X		Female	Löfstedt and Kozlov (1997)
Hepialidae	F	X		Female	Löfstedt and Kozlov (1997)
Nepticulidae	F	X		Female	Löfstedt and Kozlov (1997)
Prodoxidae	F	X		Female	Löfstedt and Kozlov (1997)
Tischeridae	F	X		Female	Löfstedt and Kozlov (1997)
Pterophoridae	F	X		Female	Tamaki (1977)
Pyralidae	F	X		Female	Tamaki (1977)
	M		X	Female	Cardé and Baker (1984)
Tortricidae	F	X		Female	Tamaki (1977)
	M		X	Female	Cardé and Baker (1984)
Sesiidae	F	X		Female	Tamaki (1977)
Gelechiidae	F	X		Female	Tamaki (1977)
Momphidae (Cosmopterygidae)	F	X		Female	Tamaki (1977)
Yponomeutidae	F	X		Female	Tamaki (1977)
Tineidae	F	X		Female	Tamaki (1977)
Notodontidae	F	X		Female	Tamaki (1977)
Arctiidae	F	X		Female	Tamaki (1977)
	M	A (lek)		Lek	Cardé and Baker (1984)
Noctuidae	F	X		Female	Tamaki (1977)
Lymantriidae	F	X		Female	Tamaki (1977)
Geometridae	F	X		Female	Tamaki (1977)
Lasiocampidae	F	X		Female	Tamaki (1977)
Saturniidae	F	X		Female	Tamaki (1977)
Bombycidae	F	X	(X)	Female	Tamaki (1977)
Trichoptera					
Seriocomatidae	F	X		Female	Löfstedt and Kozlov (1997)
Limnephilidae	F	X		Female	Löfstedt and Kozlov (1997)
Rhyacophilidae	F	X		Female	Löfstedt and Kozlov (1997)
Hydropsychidae	M	A?		Female?	Löfstedt and Kozlov (1997)
	F	A?		Female?	Löfstedt and Kozlov (1997)

Table 6.2 (*Continued*).

Taxa[a]	Sex[b]	Pheromone type[c]: Long-distance	Pheromone type[c]: Short-distance	Resource type[d]	Reference[e]
Diptera					
Mycetophilidae (incl. Sciaridae)	F	X		Female	Cardé and Baker (1984)
Trypetidae	F	X		Female	Cardé and Baker (1984)
(Tephritidae)	M	A (lek)		Female	Cardé and Baker (1984)
Drosophilidae	M	A (lek)		Lek, female	Cardé and Baker (1984)
Muscidae	F		X	Female	Cardé and Baker (1984)
Cecidomyiidae	F	X		Female	Harris and Foster (this volume, Chapter 2)
Hymenoptera					
Diprionidae	F	X		Female	Cardé and Baker (1984)
Pamphiliidae	F	X		Female	Anderbrant (this volume, Chapter 8)
Tenthredinidae	F	X		Female	Anderbrant (this volume, Chapter 8)
Formicidae	F	X		Female	Cardé and Baker (1984)
Apidae	F	X		Female	Cardé and Baker (1984)
Coleoptera					
Silphidae	M	X		Carcass *CD*	Scott (1998)
Scarabaeidae	F	X	?	Female	Leal (this volume, Chapter 3)
Dynastinae	M	A		Female	Leal (this volume, Chapter 3)
Dermestidae	F	X		Female	Cardé and Baker (1984)
Bostrichidae	M	A		Food *C*	Burkholder and Ma (1985)
Anobiidae	F	X?		*C*/female	Burkholder and Ma (1985)
Cucujidae	M	A		*C*	Burkholder and Ma (1985)
Tenebrionidae	M	A		*C*	Burkholder and Ma (1985)
Cerambycidae	M	X	(X)	Tree *C*	This chapter
	F		X	Female	This chapter
Curculionidae	F		X	Female	Cardé and Baker (1984)
	M	A		Host plant *C*	This chapter
Platypodidae	M	A		Tree *C*	This chapter
Scolytidae	F	A		Tree *C*	This chapter
	M	A		Tree *C*	This chapter
Totals					
Female	45				
Male	18				
Sex[f]		42			
Aggregation[f]		11			
Short-distance			9		
All	63	53			

[a] Families according to Chinery (1973).
[b] F = female-produced; M = male-produced.
[c] A = aggregation (attracts both sexes); X = sex pheromone (attracts opposite sex only).
[d] *C* = clumped or strongly patchy food; *CD* = clumped and defendable food for larvae.
[e] Most literature citations include references therein.
[f] Sum excluding cases with '?'. Only these cases are used in Tables 6.3 and 6.4.

these groups the female egg-laying capacity is the critical resource for reproduction and represents the largest parental investment (Trivers, 1972).

2. Long-distance pheromones are usually produced only in one of the sexes, most frequently the female (Table 6.2). Correspondingly, males more often produce short-distance/courtship pheromones. In their egg-laying capacity, females provide a limiting resource for reproduction in both sexes, while males provide a measure of fitness and species-specificity by their courtship. The few cases of long-distance male pheromones, either sex or aggregation pheromones, such as in bark beetles, are dependent on male control over, or information on, clumped food resources for the brood.

3. In 12 groups we found male-produced long-distance pheromones but only one case of a distinct sex pheromone in the predatory scorpionflies (Panorpidae). These provide an illuminating example where the males provide a clearly defined entity, a nuptial gift in the form of a captured prey, which is necessary for the females' subsequent egg-laying (Thornhill, 1992, and references therein). In this example, the male controls the resource almost completely by holding the nuptial gift clasped with his legs while emitting pheromone to attract females. However, other males may also be attracted and may try to steal the gift from the pheromone-emitting male. In the other cases we found aggregation pheromones produced by males; while in one group (the bark beetles), female-produced aggregation pheromones also exist (Table 6.3).

Aggregation pheromones and resource-based signalling

The aggregation pheromones seem to occur in cases where food or host-plant resources have a patchy distribution (Table 6.4). In these cases, male control of the resource is physically impossible due to its size, in contrast to the small nuptial gift held by the male in the Panorpidae. This is exemplified particularly in the bark beetles where susceptible host trees are, as a rule, rare and scattered, but also impossible to fully utilize by a first-arriving beetle. A similar pattern of male pheromone production associated with plant resources was recognized by Landolt (1997) for species of phytophagous insects. The larger individual variation in pheromone content found in bark beetles than in moths could be due to calling in large groups, allowing matings by chance encounters also for low producers because

Table 6.3. Distribution of sexes producing sex and aggregation pheromones, in relation to the spacing of larval food resources (compiled from Table 6.2).

	Sex pheromones		Aggregation pheromones		
Producing sex	Clumped food	Dispersed food	Clumped food	Dispersed food	Total
Male	2	0	7	3[a]	12
Female	0	40	1	0	41
Total	2	40	8	3	53

[a] Lek systems.

Table 6.4. Sex and aggregation pheromones in relation to the spacing of larval food resources (compiled from Table 6.3).

Pheromone type	Clumped food	Dispersed food	Total
Sex pheromone	2	40	42
Aggregation pheromone	8	3[a]	11
Total	10	43	53

[a] Lek systems.
$\chi^2 = 22.1$; $P < 0.001$***.

of the large number of insects attracted by the group signal (Birgersson *et al.*, 1988; Schlyter and Birgersson, 1989). The aggregation pheromones could be seen as 'backfired' sex pheromones, where the sex arriving first on a suitable host would try to attract the opposite sex, but at the same time would announce a clumped resource which also attracts competing individuals of its own sex. Incidentally, these backfired sex pheromones are used by a few aggressive bark beetle species (Raffa *et al.*, 1993) as a mechanism for regularly mass attacking and killing healthy, mature forest trees valuable to humans.

Will Understanding These Systems Help IPM?

The understanding of ecological, physiological and evolutionary aspects of host finding, pheromone production and mating systems is critical for research programmes and the development of semiochemical-based strategies for control of bark and wood-infesting forest beetles.

For an identification programme of attractants in pest species, an understanding of the great diversity of signalling systems is mandatory. To know that: (i) not all pest insects in this group have a similar pheromone system, and that (ii) male, female or both, together with host compounds, might be necessary for full attraction, is essential for success. A genetic programme aimed at increasing host-tree resistance against bark beetles might be focused on changing host monoterpene composition to make trees less suitable for pheromone production. We now realize that such an effort would be futile in many cases, since oxygenation of monoterpenes is a relatively rare mode of pheromone biosynthesis.

Generally, the use of semiochemicals appears to be favourable for controlling forest Coleoptera populations (Borden, 1997), as both attractants and inhibitors of attraction can be used in a variety of ways (Fig. 6.9). This discussion will be limited to bark beetles. In species with aggregation systems the overall population is affected, as both males and females respond. However, in spite of a steady stream of publications on small and large field experiments (Borden, 1997, and references therein), there have been few practical applications on a landscape scale or commercial successes. While monitoring of populations is possible for many species and mass trapping is possible in special cases, the use of inhibitors seems more promising in the long run.

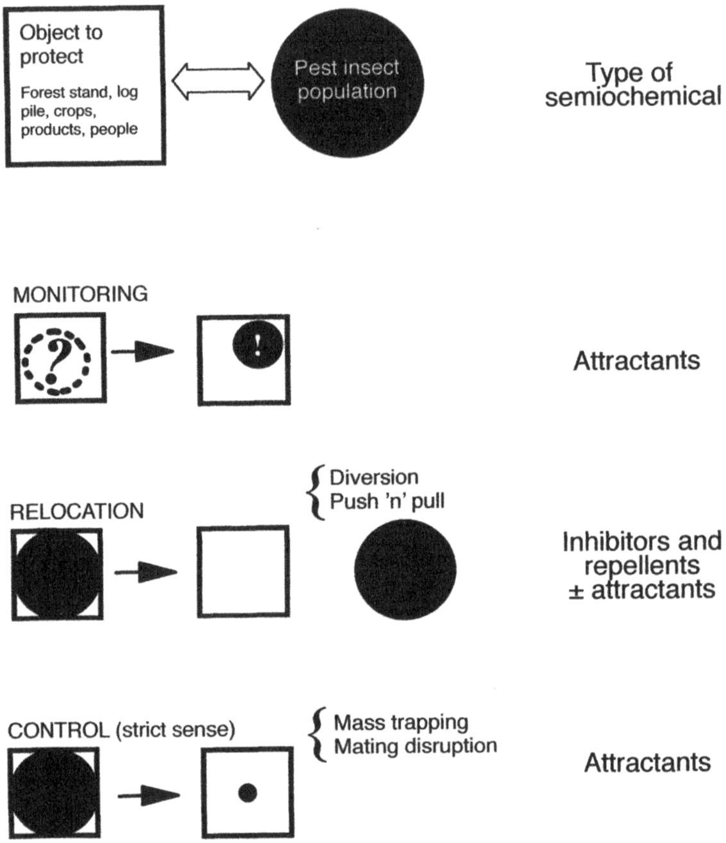

Fig. 6.9. Application strategies for semiochemicals. The box on the left, of fixed size and position, represents the object to be protected, such as a field or forest stand. The circle on the right, of variable size and position, represents the pest insect population to be controlled. For most forest Coleoptera all strategies can be applied, as most species use aggregation pheromones attracting both sexes.

Monitoring, the first strategy in Fig. 6.9, is successful for a handful of species in Europe, and commercial products exist for major species: Pheroprax© and IT-ETOKAP, Linoprax©, Chalcoprax©, for *I. typographus*, *T. lineatum* and *P. chalcographus*, respectively. Development continues in Europe (Vrkoč *et al.*, 1997) and in North America by companies like Phero Tech Inc. In spite of the frequent use of monitoring, quantitative estimates of attraction range (Schlyter, 1992) or the effective sampling area (Turchin and Odendaal, 1996) of pheromone traps for bark beetles are rare.

Relocation, the second strategy in Fig. 6.9, by use of inhibitors or combinations of attractants and inhibitors, seems to work technically in some cases but has not been commercialized. Usually a single compound, verbenone (Fig. 6.7), has been used in various dispensers, but with variable success (Payne *et al.*, 1992; Kostyk *et al.*, 1993; Borden, 1997, and references therein; F. Schlyter *et al.*, 1998, unpublished data). For the Douglas-fir beetle, *D. pseudotsugae*, a beetle-produced inhibitory compound, 3,2-MCH (Fig. 6.8), has been used. Formulations of this compound have given consistent results in reducing attacks on logs or felled trees since the 1970s (Borden, 1997, and references therein) and recently on standing trees (Ross and Daterman, 1995), but are not yet registered for commercial use. When compared with the positive, attractive signal, which is invariably multi-component in forest Coleoptera, it is reasonable that more compounds, like habitat-indicating green leaf volatiles (Dickens *et al.*, 1992; Schlyter *et al.*, 1995; Wilson *et al.*, 1996; Borden *et al.*, 1997), need to be included to give the inhibitory signal a sufficient effect. Also the odours from bark of non-host (angiosperm) trees (Borden *et al.*, 1998; Byers *et al.*, 1998a) and herbs (Guerrero *et al.*, 1997) show promise here.

The third strategy in Fig. 6.9, direct control by reduction of pest population size by **mass-trapping**, is easy to understand but is the most difficult to implement. An advantage in using direct control by mass-trapping of the species in this chapter is that, with few exceptions, they use aggregation pheromones, attracting both sexes. This makes trapping at least theoretically possible, in contrast to the systems with female-produced sex pheromones where > 90% of males must be captured to effect any depletion of the following generation. However, the efficiency of mass-trapping is controversial, and the European experience of mass-trapping with the behaviourally highly efficient synthetic aggregation pheromone of *I. typographus* in Scandinavia and central Europe has not been entirely positive. Several authors (Zumr, 1987; Bakke, 1989; Vité, 1989) have claimed positive effects of the use of mass-trapping as part of integrated pest management. Other authors (Dimitri *et al.*, 1992; Weslien, 1992; Pfeffer and Skuhravì, 1995) have shown unsuccessful results for this method in the field. However, intensive trapping at stand level may locally reduce populations and tree-killing by *I. typographus* (Dimitri *et al.* 1992; Jakus, 1998). The only commercial success for mass-trapping in scolytids is the programme for ambrosia beetles in sawmills in British Columbia (Lindgren and Borden, 1983; Shore and McLean, 1988). High-value timber is protected by intensively managed mass-trapping in well-defined areas that are to a degree isolated from beetle populations in forests.

In the future, probably a combination of the two strategies of relocation and mass-trapping should be used, especially in high-risk areas. This 'push 'n' pull' tactic (Borden, 1997), with deployment of inhibitors inside a stand at risk and trap trees or traps outside, has been successful against the mountain pine beetle (Lindgren and Borden, 1993) and *I. paraconfusus* (Shea and Neustein, 1995).

Acknowledgements

We appreciate the comments from two referees and the editors on earlier drafts. Special thanks go to Q.-H. Zhang for comments and for maintaining the reprint collection. Work on this paper was partly supported by grants from SJFR to FS (No. 23.0521/96) and to GB (No. 23.0433/95).

Appendix 6A: IUPAC Names for Chemicals Referred to in the Text

For geometric isomers, *E* and *Z* are used to describe positions according to a double-bond; *cis* and *trans* indicate the orientation of acyls and functional groups according to the theoretical plane in cyclic molecules.

Chemical	IUPAC name
Z Boll weevil alcohol	(*Z*)-2-(3,3-dimethylcyclohexylidene)-ethanol
E Boll weevil aldehyde	(*E*)-2-(3,3-dimethylcyclohexylidene)-ethanal
Z Boll weevil aldehyde	(*Z*)-2-(3,3-dimethylcyclohexylidene)-ethanal
endo-Brevicomin	*endo*-7-ethyl-5-methyl-6,8-dioxabicyclo[3.2.1]octane
exo-Brevicomin	*exo*-7-ethyl-5-methyl-6,8-dioxabicyclo[3.2.1]octane
3-Carene	3,7,7-trimethylbicyclo[4.1.0]hept-3-ene
(2*S*,5*R*)-Chalcogran	(2*S*,5*R*)-2-ethyl-1,6-dioxaspiro[4.4]nonane
(5*S*,7*S*)-Conophthorin	(5*S*,7*S*)-7-methyl-1,6-dioxaspiro[4.5]decane
Frontalin	1,5-dimethyl-6,8-dioxabicyclo[3.2.1]octane
Grandisal	*cis*-1-methyl-2-(1-ethenyl)cyclobutaneethanal
Grandisol	*cis*-1-methyl-2-(1-ethenyl)cyclobutaneethanol
(*R*)-Ipsdienol	(4*R*)-2-methyl-6-methylene-2,7-octadien-4-ol
(*S*)-Ipsdienol	(4*S*)-2-methyl-6-methylene-2,7-octadien-4-ol
(*R*)-Ipsenol	(4*R*)-2-methyl-6-methylene-7-octen-4-ol
Lanierone	2-hydroxy-4,4,6-trimethyl-2,5-cyclohexadien-1-one
Lineatin	3,3,7-trimethyl-2,9-dioxatricyclo[3.3.1.$0^{4,7}$]nonane
3,2-MCH	3-methyl-2-cyclohexen-1-one
Multistriatin	2,4-dimethyl-5-ethyl-6,8-dioxabicyclo[3.2.1]octane
Myrcene	7-methyl-3-methylene-1,6-octadiene
E-Myrcenol	(2*E*)-2-methyl-6-methylene-2,7-octadien-1-ol
a-Pinene	2,6,6-trimethylbicyclo[3.1.1]hept-2-ene
(1*S*,5*S*)-a-Pinene	(1*S*,5*S*)-2,6,6-trimethylbicyclo[3.1.1]hept-2-ene
(1*R*,5*R*)-a-Pinene	(1*R*,5*R*)-2,6,6-trimethylbicyclo[3.1.1]hept-2-ene
(2*R*,5*S*)-Pityol	(2*R*,5*S*)-2-(1-hydroxy-1-methylethyl)-5-methyl-tetrahydrofuran
(*R*)-Seudenol	(1*R*)-3-methyl-2-cyclohexen-1-ol,
(*R*)-Sulcatol	(2*R*)-6-methylhept-5-en-2-ol
(*S*)-Sulcatol	(2*S*)-6-methylhept-5-en-2-ol
Sulcatone	6-methylhept-5-en-2-one
Terpinolene	1-methyl-4-(1-methylethylidene)cyclohexene
Verbenene	6,6-dimethyl-2-methylenbicyclo[3.1.1]hept-3-ene
(1*S*,4*S*,5*S*)-*cis*-Verbenol	(1*S*,4*S*,5*S*)-*cis*-4,6,6-trimethylbicyclo[3.1.1]hept-3-en-2-ol
(1*R*,4*S*,5*R*)-*trans*-Verbenol	(1*R*,4*S*,5*R*)-*trans*-4,6,6-trimethylbicyclo[3.1.1]hept-3-en-2-ol
(1*S*,4*R*,5*S*)-*trans*-Verbenol	(1*S*,4*R*,5*S*)-*trans*-4,6,6-trimethylbicyclo[3.1.1]hept-3-en-2-ol
Verbenone	4,6,6-trimethylbicyclo[3.1.1]hept-3-en-2-one
(3*R*,6*R*)-Vittatol	(3*R*,6*R*)-*cis*-3-hydroxy-2,2,6-trimethyltetrahydropyran

References

Baier, P. (1996) Inference of vigour and host quality of Norway spruce, *Picea abies*, towards the development of *Ips typographus* (Coleoptera: Scolytidae). *Entomologia Generalis* 21, 27–35.

Bakke, A. (1989) The recent *Ips typographus* outbreak in Norway – experiences from a control program. *Holarctic Ecology* 12, 515–519.

Barkawi, L.S., Peterson, S.L., Riba, J.M., Shea, J.P., Mori, K., Blomquist G.J. and Seybold, S.J. (1997) Biosynthetic studies of the aggregation pheromone of the jeffrey pine beetle, *Dendroctonus jeffreyi*. *International Society of Chemical Ecology 14th Annual Meeting, Vancouver, 12–16 July*, Abstract P-17.

Bertram, S.L. and Paine, T.D. (1994) Response of *Dendroctonus brevicomis* Le Conte (Coleoptera: Scolytidae) to different release rates and ratios of aggregation semiochemicals and the inhibitors verbenone and ipsdienol. *Journal of Chemical Ecology* 20, 2931–2941.

Birgersson, G., Schlyter, F., Bergström, G. and Löfqvist, J. (1988) Individual variation of pheromone content in the bark beetle *Ips typographus*. *Journal of Chemical Ecology* 14, 1735–1759.

Birgersson, G., Dalusky, M.J., Espelie, K.E. and Berisford, C.W. (1990) The aggregation pheromone of the bark beetle *Pityogenes hopkinsi* Swaine. *ESA National Conference, New Orleans, 2–7 December*, Abstract, p. 152.

Birgersson, G., DeBarr, G.L., de Groot, P., Dalusky, M.J., Pierce, H.D. Jr, Borden, J.H., Meyer, H., Francke, W., Espelie, K.E. and Berisford, C.W. (1995) Pheromone in the white pine cone beetle, *Conophthorus coniperda* (Coleoptera; Scolytidae). *Journal of Chemical Ecology* 21, 143–167.

Bolter, C.J., Dicke, M., van Loon, J.J.A., Visser, J.H. and Posthumus, M.A. (1997) Attraction of Colorado beetle to herbivore-damaged plants during herbivory and after its termination. *Journal of Chemical Ecology* 23, 1003–1023.

Booth, D.C., Phillips, T.W., Claesson, A., Silverstein, R.M., Lanier, G.N. and West, J.R. (1983) Aggregation pheromone components of 2 species of *Pissodes* weevils (Coleoptera, Curculionidae). Isolation, identification and field activity. *Journal of Chemical Ecology* 9, 1–12.

Borden, J.H. (1982) Aggregation pheromones. In: Mitton, J.B. and Sturgeon, K.B. (eds) *Bark Beetles in North American Conifers – a System for the Study of Evolutionary Biology*. University of Texas Press, Austin, pp. 74–139.

Borden, J.H. (1985) Aggregation pheromones. In: Kerkut, G.A. and Gilbert, L.I. (eds) *Comprehensive Insect Physiology, Biochemistry and Pharmacology*. Oxford University Press, Oxford, pp. 257–285.

Borden, J.H. (1997) Disruption of semiochemical-mediated aggregation in bark beetles. In: Cardé, R.T. and Minks, A.K. (eds) *Insect Pheromone Research: New Directions*. Chapman & Hall, New York, pp. 421–438.

Borden, J.H., Lindgren, B.S. and Chong, L. (1980) Ethanol and α-pinene as synergists for the aggregation pheromones of two *Gnathotricus* species. *Canadian Journal of Forest Research* 10, 290–292.

Borden, J.H., Ryker, L.C., Chong, L.J., Pierce, H.D. Jr, Johnston, B.D. and Oehlschlager, A.C. (1987) Response of the mountain pine beetle, *Dendroctonus ponderosae* Hopkins (Coleoptera: Scolytidae) to five semiochemicals in British Columbia lodgepole pine forests. *Canadian Journal of Forest Research* 17, 118–128.

Borden, J.H., Chong, L.J., Lindgren, B.S., Begin, E.J., Ebata, T.M., Maclauchlan, L.E. and Hodgkinson, R.S. (1993) A simplified tree bait for the mountain pine beetle. *Canadian Journal of Forest Research* 23, 1108–1113.

Borden, J.H., Chong, L.J., Savoie, A. and Wilson, I.M. (1997) Responses to green leaf volatiles in two biogeoclimatic zones by striped ambrosia beetle, *Trypodendron lineatum*. *Journal of Chemical Ecology* 23, 2479–2491.

Borden, J.H., Poland, T.M., Huber, D.P.W., Gries, R., Gries, G., and Pierce, H.D. Jr (1998a) Two new semiochemical-based tactics for disruption of bark beetle attack. *2nd International Symposium on Insect Pheromones, Wageningen, 30 March–3 April*, Abstract, pp. 108–110.

Borden, J.H., Wilson, I.M., Gries, R., Chong, L.J., Pierce, H.D. Jr, and Gries, G. (1998b) Volatiles from the bark of trembling aspen, *Populus tremuloides* Michx. (Salicaceae) disrupt secondary attraction by the mountain pine beetle, *Dendroctonus ponderosae* Hopkins (Coleoptera: Scolytidae). *Chemoecology* 8, 69–75.

Bowers, W.W., Gries, G., Borden, J.H. and Pierce, H.D. Jr (1991) 3-Methyl-3-buten-1-ol: an aggregation pheromone of the four-eyed spruce beetle, *Polygraphus rufipennis* (Kirby) (Coleoptera: Scolytidae). *Journal of Chemical Ecology* 17, 1989–2002.

Burkholder, W.E. and Ma, M. (1985) Pheromones for monitoring and control of stored-product insects. *Annual Review of Entomology* 30, 257–272.

Byers, J.A. (1989) Chemical ecology of bark beetles. *Experientia* 45, 271–283.

Byers, J.A. (1991) Simulation of mate-finding behaviour of pine shoot beetles, *Tomicus piniperda*. *Animal Behaviour* 41, 649–660.

Byers, J.A. (1992a) Attraction of bark beetles, *Tomicus piniperda, Hylurgops palliatus*, and *Trypodendron domesticum* and other insects to short-chain alcohols and monoterpenes. *Journal of Chemical Ecology* 18, 2385–2402.

Byers, J.A. (1992b) Optimal fractionation and bioassay plans for isolation of synergistic chemicals – the subtractive–combination method. *Journal of Chemical Ecology* 18, 1603–1621.

Byers, J.A. (1995) Host-tree chemistry affecting colonization in bark beetles. In: Cardé, R.T. and Bell, W.J. (eds) *Chemical Ecology of Insects 2*. Chapman & Hall, New York, pp. 154–214.

Byers, J.A. (1996) An encounter rate model of bark beetle populations searching at random for susceptible host trees. *Ecological Modelling* 97, 57–66.

Byers, J.A. and Birgersson, G. (1990) Pheromone production in a bark beetle independent of myrcene precursor in host pine species. *Naturwissenschaften* 77, 385–387.

Byers, J.A., Lanne, B.S., Löfqvist, J., Schlyter, F. and Bergström, G. (1985) Olfactory recognition of host-tree susceptibility by pine shoot beetles. *Naturwissenschaften* 72, 324–326.

Byers J.A., Birgersson, G., Löfqvist, J. and Bergström, G. (1988) Synergistic pheromones and monoterpenes enable aggregation and host recognition by a bark beetle. *Naturwissenschaften* 75, 153–155.

Byers, J.A., Lanne, B.S. and Löfqvist, J. (1989) Host tree unsuitability recognized by pine shoot beetles in flight. *Experientia* 45, 489–492.

Byers, J.A., Schlyter, F., Birgersson, G. and Francke, W. (1990) E-Myrcenol in *Ips duplicatus*: an aggregation pheromone component new for bark beetles. *Experientia* 46, 1209–1211.

Byers, J.A., Zhang, Q.-H., Schlyter, F. and Birgersson, G. (1998) Volatiles from nonhost birch trees inhibit pheromone response in spruce bark beetles. *Naturwissenschaften* 85, 557–561.

Campos, M., Peña, A. and Raya, A.J.S. (1994) Release of ethylene from pruned olive logs: influence on attack by bark beetles (Coleoptera, Scolytidae). *Journal of Chemical Ecology* 20, 2513–2521.

Cardé, R.T. and Baker, T.C. (1984) Sexual communication with pheromones. In: Bell, W.J. and Cardé, R.T. (eds) *Chemical Ecology of Insects*. Chapman & Hall, New York, pp. 356–383.

Chenier, J.V.R. and Philogène, B.J.R. (1989) Field responses of certain forest Coleoptera to conifer monoterpenes and ethanol. *Journal of Chemical Ecology* 15, 1729–1746.

Chinery, M. (1973) *A Field Guide to the Insects of Britain and Northern Europe*. W. Collins and Son, London, 371 pp.

Cognato, A.I., Seybold, S.J., Wood, D.L. and Teale, S.A. (1997) A cladistic analysis of pheromone evolution in *Ips* bark beetles (Coleoptera: Scolytidae). *Evolution* 51, 313–318.

Conn, J.E., Borden, J.H., Scott, B.E., Friskie, L.M., Pierce, H.D. Jr and Oehlschlager, A.C. (1983) Semiochemicals for the mountain pine beetle *Dendroctonus ponderosae* (Coleoptera: Scolytidae) in British-Columbia, Canada: field trapping studies. *Canadian Journal of Forest Research* 13, 320–324.

Czokajlo, D., Webster, F.X. and Teale, S.A. (1996) Behaviour modifying chemicals for the pine shoot beetle. *International Society of Chemical Ecology 13th Annual Meeting, Prague, 18–22 August*, Abstract.

Czokajlo, D., Webster, F.X. and Teale, S.A. (1997) Pheromone complex of *Tomicus piniperda* and its activity on native and European predators, *Thanasimus dubius* and *T. formicarius*. *International Society of Chemical Ecology 14th Annual Meeting, Vancouver, 12–16 July*, Abstract C–1–3.

Davies, D.D. (1980) Anaerobic metabolism and the production of organic acids. In: Stumpf, P.K. and Conn, E.E. (eds) *The Biochemistry of Plants*, Vol. 2. Academic Press, New York, pp. 581–611.

Dickens, J.C., Billings, R.F. and Payne, T.L. (1991) Green leaf volatiles: a ubiquitous chemical signal modifies insect pheromone responses. In: Hrdy, I. (ed.) *Insect Chemical Ecology. Proceedings of a Conference held in Tábor, Czechoslovakia, 12–18 August 1990*. Academia Prague and SPB Academic Publisher, The Hague, pp. 277–280.

Dickens, J.C., Billings, R.F. and Payne, T.L. (1992) Green leaf volatiles interrupt aggregation pheromone response in bark beetles infesting southern pines. *Experientia* 48, 523–524.

Dimitri, L., Gebauer, U., Löserkrug, L. and Vaupel, O. (1992) Influence of mass trapping on the population dynamic and damage-effect of bark beetles. *Journal of Applied Entomology* 114, 103–109.

Dunn, J.P. and Potter, D.A. (1991) Synergistic effects of oak volatiles with ethanol in the capture of saprophagus wood borers. *Journal of Entomological Science* 26, 425–429.

Erasmus, M.J. and Chown, S.L. (1994) Host location and aggregation behaviour in *Hylastes angustatus* (Herbst) (Coleoptera: Scolytidae). *African Entomology* 2, 7–11.

Fettköther, R., Dettner, K., Schröder, F., Meyer, H., Francke, W. and Noldt, U. (1995) The male pheromone of the old house borer *Hylotrupes bajulus* (L.) (Coleoptera: Cerambycidae): identification and female response. *Experientia* 51, 270–277.

Fish, R.H., Browne, L.E., Wood, D.L. and Hendry, L.B. (1979) Conversions of deuterium labelled ipsdienol with sexual and enantio-selectivity in *Ips paraconfusus* Lanier. *Tetrahedron Letters* 17, 1465–1468.

Francke, W., Bartels, J., Meyer, H., Schröder, F., Kohnle, U., Baader, E. and Vité, J.P. (1995) Semiochemicals from bark beetles: new results, remarks and reflections. *Journal of Chemical Ecology* 21, 1043–1063.

Francke-Grosmann, H. (1952) Über die Ambrosiazucht der beiden Kiefernborkenkäfer *Myelophilus minor* Htg. und *Ips acuminatus* Gyll. *Meddelanden från Statens Skogsforskningsinstitut* 41, 1–52.

Führer, E., Hausmann, B. and Wiener, L. (1991) Borkenkäferbefall und Terpenmuster der Fichtenrinde (*Picea abies* Karst) an Fangbäumen. *Journal of Applied Entomology* 112, 113–129.

Fukaya, M., Yasuda, T., Wakamura, S. and Honda, H. (1996) Reproductive biology of the yellow-spotted longicorn beetle, *Psacothea hilaris* (Pascoe) (Coleoptera: Cerambycidae). III. Identification of contact sex pheromone on female body surface. *Journal of Chemical Ecology* 22, 259–270.

Gara, R.I., Littke, W.R. and Rhoades D.F. (1993) Emission of ethanol and monoterpenes by fungal infected lodgepole pine trees. *Phytochemistry* 34, 987–990.

Godwin, P.A. and Odell, T.M. (1965) The life history of the white pine cone beetle, *Conophthorus coniperda*. *Annals of the Entomological Society of America* 58, 213–219.

Graham, K. (1968) Anaerobic induction of primary chemical attractancy for ambrosia beetles. *Canadian Journal of Zoology* 46, 905–908.

Grégoire, J.-C. (1985) Host colonization stragies in *Dendroctonus*: Larval gregariousness or mass attack by adults? In: Safranik, L. (ed.) *The Role of the Host in Population Dynamics of Forest Insects*. Canadian Forestry Service and USDA Forest Service, Victoria, BC, pp. 147–154.

Grégoire, J.-C., Baisier, M., Drumont, A., Dahlsten, D.L., Meyer, H. and Francke, W. (1991) Volatile compounds in the larval frass of *Dendroctonus valens* and *D. micans* in relation to oviposition by the predator *Rhizophagus grandis* (Coleoptera: Rhizophagidae). *Journal of Chemical Ecology* 17, 2003–2020.

Grégoire, J.-C., Coullien, D., Krebber, R., König, W.A., Meyer, H. and Francke, W. (1992) Orientation of *Rhizophagus grandis* (Coleoptera: Rhizophagidae) to oxygenated monoterpenes in a species-specific predator–prey relationship. *Chemoecology* 3, 14–18.

Gries, G., Nolte, R. and Sanders, W. (1989) Computer simulated host selection in *Ips typographus*. *Entomologia Experimentalis et Applicata* 53, 211–218.

Gries, G., Borden, J.H., Gries, R., Lafontaine, J.P., Dixon, E.A., Wieser, H. and Whitehead, A.T. (1992) 4-Methylene-6,6-dimethylbicyclo[3.1.1]hept-2-ene (Verbenene): new aggregation pheromone of the scolytid beetle *Dendroctonus rufipennis*. *Naturwissenschaften* 79, 367–368.

Groot, P. de, DeBarr, G.L., Birgersson, G., Pierce, H.D. Jr, Borden, J.H., Berisford, Y.C. and Berisford, C.W. (1991) Evidence for a female-produced pheromone in the white pine cone beetle, *Conophthorus coniperda* (Schwarz), and in the red pine cone beetle, *C. resinosae* Hopkins (Coleoptera; Scolytidae). *Canadian Entomologist* 123, 1057–1064.

Groot, P. de, DeBarr, G.L. and Birgersson, G. (1998) Field bioassays of synthetic pheromones and host monoterpenes for *Conophthorus coniperda* (Coleoptera: Scolytidae). *Environmental Entomology* 27, 382–387.

Guerrero, A., Feixas, J., Pajares, J., Wadhams, L.J., Pickett, J.A. and Woodcock, C.M. (1997) Semiochemically induced inhibition of behaviour of *Tomicus destruens* (Woll.) (Coleoptera: Scolytidae). *Naturwissenschaften* 84, 155–157.

Hendry, L.B. (1976) Insect pheromones: diet related. *Science* 192, 143–145.

Hendry, L.B., Wichmann, J.K., Hindenlang, D.M., Mumma, R.O. and Anderson, M.E. (1975) Evidence for origin of insect sex pheromones: presence in food plants. *Science* 183, 59–62.

Hendry, L.B., Piatek, B., Browne, L.E., Wood, D.L., Byers, J.A., Fish, R.H. and Hicks, R.A. (1980) *In vivo* conversion of labelled host chemical to pheromones in the bark beetle *Ips paraconfusus*. *Nature* 284, 485.

Hibbard, B.E. and Webster, F.X. (1993) Enantiomeric composition of grandisol and grandisal produced by *Pissodes strobi* and *P. nemorensis* and their electroantennogram response to pure enantiomers. *Journal of Chemical Ecology* 19, 2129–2141.

Hobson, K.R., Wood, D.L., Cool, L.G., White, P.R., Ohtsuka, T., Kubo, I. and Zavarin, E. (1993) Chiral specificity in responses by the bark beetle *Dendroctonus valens* to host kairomones. *Journal of Chemical Ecology* 19, 1837–1846.

Huber, D.P.W., Gries, R., Borden, J.H. and Pierce, H.D. Jr (1999) Two pheromones of coniferophagous bark beetles occur in the bark of non-host angiosperm trees. *Journal of Chemical Ecology* 25, 805–816.

Hughes, P.R. (1974) Myrcene: a precursor of pheromones in *Ips* beetles. *Journal of Insect Physiology* 20, 1271–1275.

Hunt, D.W.A. and Borden, J.H. (1988) Response of mountain pine beetle *Dendroctonus ponderosae* Hopkins and pine engraver *Ips pini* Say to ipsdienol in southwestern British-Columbia, Canada. *Journal of Chemical Ecology* 14, 277–294.

Hunt, D.W.A., Borden, J.H., Lindgren, B.S. and Gries, G. (1989) The role of autoxidation of α-pinene in the production of pheromones of *Dendroctonus ponderosae* (Coleoptera: Scolytidae). *Canadian Journal of Forest Research* 19, 1275–1282.

Ivarsson, P. and Birgersson, G. (1995) Regulation and biosynthesis of pheromone components in the double spined bark beetle *Ips duplicatus* (Coleoptera: Scolytidae). *Journal of Insect Physiology* 41, 843–849.

Ivarsson, P., Schlyter, F. and Birgersson, G. (1993) Demonstration of *de novo* pheromone biosynthesis in *Ips duplicatus* (Coleoptera: Scolytidae): inhibition of ipsdienol and E-myrcenol production by compactin. *Insect Biochemistry and Molecular Biology* 23, 655–662.

Ivarsson, P., Blomquist, G.J. and Seybold, S.J. (1997) *In vitro* production of pheromone intermediates ipsdienone and ipsenone by the bark beetles *Ips pini* Say and *I. paraconfusus* Lanier (Coleoptera: Scolytidae). *Naturwissenschaften* 84, 454–457.

Ivarsson, P., Tittiger, C., Blomquist, C., Borgeson, C.E., Seybold, S.J., Blomquist, G.J., and Högberg, H.-E. (1998) Pheromone precursor synthesis is localized in the metathorax of *I. paraconfusus* Lanier (Coleoptera: Scolytidae). *Naturwissenschaften* 85, 507–511.

Iwabuchi, K. (1982) Mating behavior of the grape borer *Xylotrechus pyrrhoderus* Bates (Coleoptera: Cerambycidae). I. Behavioural sequences and existence of the male sex pheromone. *Applied Entomology and Zoology* 17, 494–500.

Iwabuchi, K., Takahashi, J., Nakagawa, Y. and Sakai, T. (1986) Behavioral response of female grape borer *Xylotrechus pyrrhoderus* (Coleoptera: Cerambycidae) to synthetic male sex pheromone components. *Applied Entomology and Zoology* 21, 21–27.

Iwabuchi, K., Takahashi, J. and Sakai, T. (1987) Occurrence of 2,3-octanediol and 2-hydroxy-3-octane, possible male sex pheromone in *Xylotrechus chinesis* Chevrolat (Coleoptera: Cerambycidae). *Applied Entomology and Zoology* 22, 110–111.

Jakus, R. (1998) A method for the protection of spruce stands against *Ips typographus* by the use of barriers of pheromone traps in north-eastern Slovakia. *Anzeiger für Schädlingskunde, Pflanzenschutz, Umweltschutz* 71, 152–158.

Kelsey, R.G. (1994) Ethanol synthesis in Douglas-fir logs felled in November, January, and March, and its relationship to ambrosia beetle attack. *Canadian Journal of Forest Research* 24, 2096–2104.

Kelsey, R.G. (1996) Anaerobic induced ethanol synthesis in the stems of greenhouse-grown conifer seedlings. *Trees* 10, 183–188.

Kelsey, R.G. and Joseph, G. (1997) Ambrosia beetle host selection among logs of Douglas fir, western hemlock, and western red cedar with different ethanol and α-pinene concentrations. *Journal of Chemical Ecology* 23, 1035–1051.

Kimmerer, T.W. and Kozolowski, T.T. (1982) Ethylene, ethane, acetaldehyde and ethanol production by plants under stress. *Plant Physiology* 69, 840–847.

Klimetzek, D. and Francke, W. (1980) Relationship between the enantiomeric composition of α-pinene in the host trees and the production of verbenols in *Ips* species. *Experientia* 36, 1343–1344.

Klimetzek, D., Köhler, J., Vité, J.P. and Kohnle, U. (1986) Dosage response to ethanol mediates host selection by 'secondary' bark beetles. *Naturwissenschaften* 73, 270–272.

Klimetzek, D., Bartels, J. and Francke, W. (1989) Das Pheromon-System des Bunten Ulmenbastkäfers *Pteleobius vittatus*. *Journal of Applied Entomology* 107, 518–523.

Knízek, M., and Zahradník, P. (1996) Mass outbreak of *Ips duplicatus* Sahlberg (Coleoptera, Scolytidae). In: Doe, J. (ed.) *Proceedings of the XXth International Congress of Entomology, Firenze, Italy, August 25–31, 1996*, p. 527.

Kostyk, B.C., Borden, J.H. and Gries, G. (1993) Photoisomerization of antiaggregation pheromone verbenone: biological and practical implications with respect to the mountain pine beetle, *Dendroctonus ponderosae* Hopkins (Coleoptera: Scolytidae). *Journal of Chemical Ecology* 19, 1749–1759.

Landolt, P.J. (1997) Sex attractant and aggregation pheromones of male phytophagous insects. *American Entomologist* 43, 12–22.

Lanne, B.S., Schlyter, F., Byers, J.A., Löfqvist, J., Leufvén, A., Bergström, G., van der Pers, J.N.C., Unelius, R. and Norin, T. (1987) Differences in attraction to semiochemicals present in the sympatric pine shoot beetles *Tomicus minor* and *T. piniperda*. *Journal of Chemical Ecology* 13, 1045–1067.

Lanne, B.S., Ivarsson, P., Johnsson, P., Bergström, G. and Wassgren, A.-B. (1989) Biosynthesis of 2-methyl-3-buten-2-ol, a pheromone component of *Ips typographus* (Coleoptera: Scolytidae). *Insect Biochemistry* 19, 163–168.

Leal, W.S. (1998) Chemical ecology of phytophagous scarab beetles. *Annual Review of Entomology* 43, 39–61.

Leal, W.S., Shi, X., Nakamuta, K., Ono, M. and Meinwald, J. (1995) Structure, stereochemistry, and thermal isomerization of the male sex pheromone of the longhorn beetle *Anaglyptus subfasciatus*. *Proceedings of the National Academy of Sciences USA* 92, 1038–1042.

Leufvén, A., Bergström, G. and Falsen, E. (1984) Interconversion of verbenols and verbenone by identified yeasts isolated from the spruce bark beetle *Ips typographus*. *Journal of Chemical Ecology* 10, 1349–1361.

Lindelöw, Å., Risberg, B. and Sjödin, K. (1992) Attraction during flight of scolytids and other bark- and wood-dwelling beetles to volatiles from fresh and stored spruce wood. *Canadian Journal of Forest Research* 22, 224–228.

Lindgren, B.S. and Borden, J.H. (1983) Survey and mass trapping of ambrosia beetles (Coleoptera: Scolytidae) in timber processing areas on Vancouver Island, Canada. *Canadian Journal of Forest Research* 13, 481–493.

Lindgren, B.S. and Borden, J.H. (1993) Displacement and aggregation of mountain pine beetles, *Dendroctonus ponderosae* (Coleoptera: Scolytidae), in response to their anti-aggregation and aggregation pheromones. *Canadian Journal of Forest Research* 23, 286–290.

Lindström, M., Norin, T., Birgersson, G. and Schlyter, F. (1989) Variation of enantiomeric composition of α-pinene in Norway spruce *Picea abies*, and its influence on the production of verbenol isomers by *Ips typographus* in the field. *Journal of Chemical Ecology* 15, 541–548.

Löfstedt, C. and Kozlov, M. (1997) A phylogenetic analysis of pheromone communication in primitive moths. In: Cardé, R.T. and Minks, A.K. (eds) *Insect Pheromone Research: New Directions*. Chapman & Hall, New York, pp. 473–498.

Lougrin, J.H., Potter, D.A. and Hamilton-Kemp, T.R. (1995) Volatile compounds induced by herbivory act as aggregation kairomones for the Japanese beetle (*Popillia japonica* Newman). *Journal of Chemical Ecology* 21, 1457–1467.

Macías-Sámano, J.E., Borden, J.H., Gries, R., Pierce, H.D. Jr, Gries, G. and King, G.G.S. (1998) Primary attraction of the fir engraver, *Scolytus ventralis*. *Journal of Chemical Ecology* 24, 1049–1075.

Miller, D.R. and Borden, J.H. (1990) The use of monoterpenes as kairomone for pine engraver *Ips latidens* Leconte (Coleoptera: Scolytidae). *Canadian Entomologist* 122, 301–308.

Miller, D.R. and Lafontaine, J.P. (1991) *cis*-Verbenol: an aggregation pheromone for the mountain pine beetle *Dendroctonus ponderosae* Hopkins (Coleoptera: Scolytidae) in British Columbia, Canada. *Journal of the Entomological Society of British Columbia* 88, 34–38.

Miller, D.R., Madden, J.L. and Borden, J.H. (1986) Primary attraction of *Ips latidens* and *Hylastes gracilis* (Coleoptera: Scolytidae) to high-girdled lodgepole pine *Pinus contorta* var. *latifolia*. *Canadian Entomologist* 118, 85–88.

Miller, J.R., Baker, T.C., Cardé, R.T. and Roelofs, W.L. (1976) Reinvestigation of oak leaf roller sex pheromone components and the hypothesis that they vary with diet. *Science* 192, 140–143.

Milligan, R.H. and Ytsma, G. (1988) Pheromone dissemination by male *Platypus apicalis* White and *Platypus gracilis* Broun (Coleoptera: Platypodidae). *Journal of Applied Entomology* 106, 113–118.

Milligan, R.H., Osborne, G.O. and Ytsma, G. (1988) Evidence for an aggregation pheromone in *Platypus gracilis* Broun (Coleoptera: Platypodidae). *Journal of Applied Entomology* 106, 20–24.

Mitlin, N. and Hedin, P.A. (1974) Biosynthesis of grandlure, the pheromone of the boll weevil, *Anthonomus grandis*, from acetate, mevalonate, and glucose. *Journal of Insect Physiology* 20, 1825–1831.

Moeck, H.A. (1970) Ethanol as the primary attractant for the ambrosia beetle *Trypodendron lineatum* (Coleoptera: Scolytidae). *Canadian Entomologist* 102, 985–995.

Moeck, H.A., Wood, D.L. and Lindahl, K.Q. Jr (1981) Host selection behavior of bark beetles (Coleoptera: Scolytidae) attacking *Pinus ponderosa* with special emphasis on the western pine beetle *Dendroctonus brevicomis*. *Journal of Chemical Ecology* 7, 49–84.

Montgomery, M.E. and Wargo, P.M. (1983) Ethanol and other host-derived volatiles as attractants to beetles that bore into hardwoods. *Journal of Chemical Ecology* 9, 181–190.

Mori, K., Takikawa, H., Nishimura, Y. and Horikiri, H. (1997) Pheromone synthesis, CLXXVII: Synthesis of (−)-*exo*-isobrevicomin and its (−)-*endo* isomer, the components of the male-produced volatiles of the mountain pine beetle, *Dendroctonus ponderosae* (Coleoptera: Scolytidae) in British Columbia, Canada. *Liebigs Annalen-Recueil* 0(2), 327–332.

Nakamuta, K., Leal, W.S., Nakashima, T., Tokoro, M., Ono, M. and Nakanishi, M. (1997) Increase of trap catches by a combination of male sex pheromones and floral attractant in longhorn beetle, *Anaglyptus subfasciatus*. *Journal of Chemical Ecology* 23, 1635–1640.

Paine, T.D. and Hanlon, C.C. (1991) Response of *Dendroctonus brevicomis* and *Ips paraconfusus* (Coleoptera: Scolytidae) to combinations of synthetic pheromone attractants and inhibitors verbenone and ipsdienol. *Journal of Chemical Ecology* 17, 2163–2176.

Paine, T.D., Raffa, K. and Harrington, T.C. (1997) Interactions among scolytid bark beetles, their associated fungi, and live host trees. *Annual Review of Entomology* 42, 179–206.

Payne, T.L., Billings, R.F., Berisford, C.W., Salom, S.M., Grosman, D.M., Dalusky, M.J. and Upton, W.W. (1992) Disruption of *Dendroctonus frontalis* (Col., Scolytidae) infestations with an inhibitor pheromone. *Journal of Applied Entomology* 114, 341–347.

Person, H.L. (1931) Theory in explanation of the selection of certain trees by the western pine beetle. *Journal of Forestry* 29, 696–699.

Pfeffer, A. and Skuhravì, V. (1995) Der Buchdrucker (*Ips typographus* L.) (Col., Scolytidae) und seine Problematik in der Tschechischen Republik. *Anzeiger Schädlingskunde, Pflanzenschutz und Umweltschutz* 68, 151–152.

Phillips, T.W. and Lanier, G.N. (1986) Interspecific activity of semiochemicals among sibling species of *Pissodes* (Coleoptera: Curculionidae). *Journal of Chemical Ecology* 12, 1587–1602.

Phillips, T.W., West, J.R., Foltz, J.L., Silverstein, R.M. and Lanier, G.N. (1984) Aggregation pheromone of the Deodar weevil *Pissodes nemorensis* (Coleoptera: Curculionidae): isolation and activity of grandisol and grandisal. *Journal of Chemical Ecology* 10, 1417–1424.

Phillips, T.W., Teale, S.A. and Lanier, G.N. (1987) Biosystematics of *Pissodes* Germar (Coleoptera: Curculionidae): seasonality morphology and synonym of *Pissodes approximatus* Hopkins and *Pissodes nemorensis* Germar. *Canadian Entomologist* 119, 465–480.

Phillips, T.W., Nation, J.L., Wilkinson, R.C. and Foltz, J.L. (1989) Secondary attraction and field activity of beetle-produced volatiles in *Dendroctonus terebrans*. *Journal of Chemical Ecology* 15, 1513–1534.

Poland, T.M., Borden, J.H., Stock, A.J. and Chong, L.J. (1998) Green leaf volatiles disrupt responses by the spruce beetle, *Dendroctononus rufipennis*, and the western pine beetle, *Dendroctonus brevicomis* (Coleoptera: Scolytidae) to attractant-baited traps. *Journal of the Entomological Society of British Columbia* 95, 17–24.

Pulliainen, E. (1983) Infestation by *Trypodendron lineatum* (Oliv.) (Coleoptera, Scolytidae) of spruce and pine felled in different years and months near the subarctic timberline. *Entomologica Fennica* 49, 45–48.

Raffa, K.F. and Berryman, A.A. (1980) Flight responses and host selection by bark beetles. In: Berryman, A.A. and Safranik, L. (eds) *Dispersal of Forest Insects: Evaluation, Theory, and Management Implications*. Cooperative Extension Service, Washington State University, Pullman, Washington, pp. 213–233.

Raffa, K., Phillips, T.W. and Salom, S. (1993) Strategies and mechanisms of host colonization by bark beetles. In: Schowalter, T.D. and Filip, G.M. (eds) *Beetle–Pathogen Interactions in Conifer Forests*. Academic Press, London, pp. 103–128.

Redborg, K.E. (1998) Biology of the Mantispidae. *Annual Review of Entomology* 43, 175–194.

Reddemann, J. (1993) Monoterpenkohlwasserstoffe in der Aggregation von *Ips typographus* L. (Col., Scolytidae). PhD thesis, Forstliche Fakultät der Ludwig-Maximilians-Universität-München, Germany.

Renwick, J.A.A. and Pitman, G.B. (1979) An attractant isolated from female Jeffrey pine beetles, *Dendroctonus jeffreyi*. *Environmental Entomology* 8, 40–41.

Renwick, J.A.A., Hughes, P.R. and Krull, I.S. (1976a) Selective production of *cis*- and *trans*-verbenol from (−)- and (+)-α-pinene by a bark beetle. *Science* 191, 199–201.

Renwick, J.A.A., Pitman, G.B. and Vité, J.P. (1976b) Oxidation products of terpenes identified from *Dendroctonus* and *Ips* bark beetles. *Journal of Insect Physiology* 22, 725–727.

Renwick, J.A.A., Vité, J.P. and Billings, R.F. (1977) Aggregation pheromones in the ambrosia beetle *Platypus flavicornis*. *Naturwissenschaften* 64, 226.

Ross, D.W. and Daterman, G.E. (1995) Efficacy of an antiaggregation pheromone for reducing Douglas-fir beetle, *Dendroctonus pseudotsugae* Hopkins (Coleoptera: Scolytidae), infestation in high risk stands. *Canadian Entomologist* 127, 805–811.

Rutledge, D.N., Chararas, C. and Ducauze, C.J. (1988) Investigating the hypothesis of primary attraction in conifer-specific Scolytidae using a chemometrical approach. In: Mattson, W.J., Levieux, J. and Bernard-Dagan, C. (eds) *Mechanisms of Woody Plant Defenses Against Insects: Search for Pattern*. Springer Verlag, New York, pp. 391–409.

Ryker, L.C. and Oester, P.T. (1982) *Pseudohylesinus nebulosus* (Coleoptera: Scolytidae) aggregation by primary attraction. *Zeitschrift für Angewandte Entomologie* 94, 377–382.

Ryker, L.C., Libbey, L.M. and Rudinsky, J.A. (1979) Comparison of volatile compounds and stridulation emitted by the Douglas-fir beetle from Idaho and western Oregon populations. *Environmental Entomology* 8, 789–798.

Sakai, M. and Yamasaki, T. (1990) (+)-Juniperol and (+)-pimaral: attractants for the cerambycid beetle, *Monochamus alternatus* (Hope). *Journal of Chemical Ecology* 16, 3383–3392.

Sakai, M. and Yamasaki, T. (1991) (+)-*cis*-3-Pinen-2-ol: attractant for male cerambycid beetle, *Monochamus alternatus* (Hope). *Journal of Chemical Ecology* 17, 757–765.

Schimitschek, E. and Wienke, E. (1966) Untersuchungen über die Befallsbereitschaft von Baumarten für Sekundärschädlinge II. Teil: Untersuchungen über die Befallsbereitschaft der europäischen Fichte. *Zeitschrift für Angewandte Entomologie* 57, 73–139.

Schlyter, F. (1992) Sampling range, attraction range, and effective attraction radius: estimates of trap efficiency and communication distance in coleopteran pheromone and host attractant systems. *Journal of Applied Entomology* 114, 439–454.

Schlyter, F. and Anderbrant, O. (1993) Competition and niche separation between two bark beetles: existence and mechanisms. *Oikos* 68, 437–447.

Schlyter, F. and Birgersson, G. (1989) Individual variation in bark beetle and moth pheromones – a comparison and an evolutionary background. *Holarctic Ecology* 12, 457–465.

Schlyter, F., Birgersson, G., Byers, J.A. and Bakke, A. (1992) The aggregation pheromone of *Ips duplicatus* and its role in competitive interactions with *I. typographus*. *Chemoecology* 3, 103–112.

Schlyter, F., Löfqvist, J. and Jakus, R. (1995) Green leaf volatiles and verbenone modify attraction of European *Tomicus*, *Hylurgops*, and *Ips* bark beetles. In: Hain, F.P., Salom, S.S., Ravlin, W.F., Payne, T.L. and Raffa, K.F. (eds) *Behavior, Population Dynamics, and Control of Forest Insects*. Ohio State University, OARDC, Wooster, pp. 29–44.

Schroeder, L.M. (1988) Attraction of the bark beetle *Tomicus piniperda* and some other bark- and woodliving beetles to the host volatiles α-pinene and ethanol. *Entomologia Experimentalis et Applicata* 46, 203–210.

Schroeder, L.M. (1992) Recognition of nonhosts aspen and birch by conifer bark beetles *Tomicus piniperda* and *Hylurgops palliatus*. *Journal of Chemical Ecology* 18, 1583–1593.

Scott, M.P. (1998) The ecology and behavior of burying beetles. *Annual Review of Entomology* 43, 595–618

Seybold, S.J., Quilici, D.R., Tillman, J.A., Vanderwel, D., Wood, D.L. and Blomquist, G.L. (1995) *De novo* biosynthesis of the aggregation pheromone components ipsenol and ipsdienol by the bark beetles *Ips paraconfusus* Lanier and *Ips pini* (Say) (Coleoptera: Scolytidae). *Proceedings of the National Academy of Sciences USA* 92, 8393–8397.

Shea, P.J. and Neustein, M. (1995) Protection of a rare stand of Torrey pine from *Ips paraconfusus*. In: Salom, S.S. and Hobson, K.R. (eds) *Proceedings of the Symposium on Application of Semiochemicals for Management of Bark Beetle Infestations*. USDA, Forest Service, General Technical Report INT-GTR-318, pp. 123–456.

Shore, T.L. and McLean, J.A. (1988) The use of mark–recapture to evaluate a pheromone-based mass trapping program for ambrosia beetles in a sawmill. *Canadian Journal of Forest Research* 18, 1113–1117.

Silverstein, R.M. (1970) Methodology for isolation and identification of pheromones – examples from Coleoptera. In: Wood, D.L, Silverstein, R.M. and Nakajiama, M. (eds) *Control of Insect Behaviour by Natural Products*. Academic Press, New York, pp. 285–299.

Sjödin, K., Schroeder, L.M., Eidmann, H.H., Norin, T. and Wold, S. (1989) Attack rates of scolytids and composition of volatile wood constituents in healthy and mechanically weakened pine trees. *Scandinavian Journal of Forest Research* 4, 379–391.

Tamaki, Y. (1977) Complexity, diversity, and specificity of behavioral-modifying chemicals in Lepidoptera and Diptera. In: Shorey, H.H. and McKelvey, J.J. Jr (eds) *Chemical Control of Insect Behavior*. John Wiley & Sons, New York, pp. 253–285.

Teal, S.A., Webster, F.X., Zhang, A. and Lanier, G.N. (1992) Lanierone: a new pheromone component from *Ips pini* (Coleoptera: Scolytidae) in New York. *Journal of Chemical Ecology* 17, 1159–1176.

Thompson, A.C. and Mitlin, N. (1979) Biosynthesis of the sex pheromone of the male boll weevil from monoterpene precursors. *Insect Biochemistry* 9, 293–294.

Thornhill, R. (1992) Female preferences for the pheromone of males with low fluctuating asymmetry in the Japanese scorpionfly (*Panorpa japonica*: Mecoptera). *Behavioral Ecology* 3, 277–283.

Thornhill, R. and Alcock, J. (1983) *The Evolution of Insect Mating Systems*. Harvard University Press, Cambridge, Massachusetts, 546 pp.

Tillman, J.A., Holbrook, G.L., Dallara, P.L., Schal, C., Wood, D.L., Blomquist, G.J. and Seybold, S.J. (1998) Endocrine regulation of *de novo* aggregation pheromone biosynthesis in the pine engraver, *Ips pini* (Say) (Coleoptera: Scolytidae). *Insect Biochemistry and Molecular Biology* 28, 705–715.

Tømmerås, B. (1989) Host selection by odorous compounds from host and non-host trees in bark beetles. *Fauna Norvegica Series B* 36, 75–79.

Trivers, R. (1972) Parental investment and sexual selection. In: Campbell, B.G. (ed.) *Sexual Selection and the Descent of Man*. Aldine Press, Chicago, Illinois, pp. 136–179.

Tunset, K., Nilssen, A.C. and Andersen, J. (1993) Primary attraction in host recognition of coniferous bark beetles and bark weevils (Col., Scolytidae and Curculionidae). *Journal of Applied Entomology* 115, 155–169.

Turchin, P. and Odendaal, F.J. (1996) Measuring the effective sampling area of a pheromone trap for monitoring population density of southern pine beetle (Coleoptera: Scolytidae). *Environmental Entomology* 25, 582–588.

Vanderwel, D. and Oehlschlager, A.C. (1987) Biosynthesis of pheromones and endocrine regulation of pheromone production in Coleoptera. In: Prestwich, G.D. and Blomquist, G.J. (eds) *Pheromone Biochemistry*. Academic Press, New York, pp. 175–215.

Vité, J.P. (1989) The European struggle to control *Ips typographus* – past, present and future. *Holarctic Ecology* 12, 520–525.

Vrkoč, J., Zahradnik, P., Knizek, M., and Turcani, M. (1997) Semiochemicals of bark beetles: application for area-wide management of pest populations. *International Society of Chemical Ecology 14th Annual Meeting, Vancouver, 12–16 July*, Abstract P–22.

Weslien, J. (1992) Effects of mass trapping on *Ips typographus* (L.) populations. *Journal of Applied Entomology* 114, 228–232.

Wilson, I.M., Borden, J.H., Gries, R. and Gries, G. (1996) Green leaf volatiles as antiaggregants for the mountain pine beetle, *Dendroctonus ponderosae* Hopkins (Coleoptera: Scolytidae). *Journal of Chemical Ecology* 22, 1861–1875.

Wood, D.L. (1982) The role of pheromones, kairomones and allomones in the host selection and colonization behavior of bark beetles. *Annual Review of Entomology* 27, 411–446.

Yamasaki, T., Sato, M. and Sakoguchi, H. (1997) (−)-Germacrene D: masking substance of attractants for the cerambycid beetle, *Monochamus alternatus* (Hope). *Applied Entomology and Zoology* 32, 423–429.

Ytsma, G. (1989) Colonization of southern beech by *Platypus caviceps* (Coleoptera: Platypodidae). *Journal of Chemical Ecology* 15, 1171–1176.

Zagatti, P., Lemperière, G. and Malosse, C. (1997) Monoterpenes emitted by the large pine weevil, *Hylobius abietis* (L.) feeding on Scots pine, *Pinus sylvestris* L. *Physiological Entomology* 22, 394–400.

Zhang, Q.-H., Schlyter, F. and Liu, G.-T. (1995) Distribution, mortality, and sex-ratio of overwintering *Ips duplicatus* (Coleoptera: Scolytidae) in the soil of a *Picea koraiensis* reserve in Inner Mongolia, China, with a model of diffusion. In: Hain, F.P., Salom, S.S., Ravlin, W.F., Payne, T.L. and Raffa, K.F. (eds) *Behavior, Population Dynamics, and Control of Forest Insects*. Ohio State University, OARDC, Wooster, pp. 109–122.

Zhang, Q.-H., Birgersson, G., Zhu, J.-W., Löftstedt, C., Löfqvist, J. and Schlyter, F. (1999) Leaf volatiles from nonhost deciduous trees: variation by tree species, season, and temperature and electrophysiological activity in *Ips typographus*. *Journal of Chemical Ecology* (in press).

Zumr, V. (1987) Reduction of mass outbreaks in spruce bark beetle, *Ips typographus* (L.) (Coleoptera, Scolytidae), by pheromone traps. *Lesnictví* 33, 49–64 (in Czech, with English summary).

7 Stored-product Beetles

Rudy Plarre[1] and Desiree C. Vanderwel[2]

[1]BAM–Federal Institute of Materials Research and Testing, Lab IV.11, Materials Protection against Biological Agents, Unter den Eichen 87, D-12205 Berlin, Germany; [2]Department of Chemistry, University of Winnipeg, 515 Portage Avenue, Winnipeg, Manitoba, Canada R3B 2E9

Introduction

The custom of storing plant materials, mainly grain and its products, provides a habitat with shelter, feeding grounds and breeding sites to which a group of insects, known as stored-product insects, has successfully adapted. About 100 insect species, a few from the orders Psocoptera and Lepidoptera but mainly from the order Coleoptera, are able to complete their development and entire life cycle on stored products. Worldwide losses of stored products due to post-harvest insect attack are estimated to be 15% annually, and tremendous costs are involved to protect stored commodities against insect infestations (Reichmuth *et al.*, 1997). It is therefore not surprising that, in order to optimize insect control programmes, the biology of stored-product insects is better studied than that of other ecological groups of insects.

Modern stored-product pest control focuses on integrated pest management (IPM). Sampling, including trapping to monitor the presence and abundance of stored-product beetles, is a key element in IPM (Hagstrum and Flinn, 1996). Pheromones usually increase the efficacy of a trap. Thus, the study and the use of sex and aggregation pheromones, as well as other semiochemicals, to monitor and trap stored-product beetles has steadily increased over the past 25–30 years (Burkholder, 1981a; Burkholder and Faustini, 1991; Phillips, 1994). It should be noted, however, that reports of the existence of sex attractants in beetles associated with stored products appeared well before the term 'pheromone' was coined in the late 1950s (Voelkel, 1924; Valentine, 1931).

Today, after detailed chemical and biological studies, the pheromones of nearly all major economically important insect pests of stored products have been at least partly identified and are listed in several review publications (Burkholder, 1982, 1990; Burkholder and Ma, 1985; Pinniger and Chambers, 1987; Chambers, 1990; Levinson and Levinson, 1995; Plarre, 1998). The pheromones produced by Coleoptera associated with stored products are far more diverse than those of

Lepidoptera found on stored commodities. The carbon skeleton of the coleopteran pheromone may be branched or unbranched, cyclic or acyclic, and may be functionalized by alcohol, aldehyde, ketone, carboxylic acid or ester moieties. Double bonds might be *E* or *Z*, isolated, conjugated, allenic or 'skipped'. As an added complication, the absolute configuration about the chiral centres plays a crucial role in defining the activity of the molecule. Alternative stereoisomers of a pheromone may have no or only minor activity or may even inhibit the behavioural response.

The following chapter is organized into three main sections. Firstly, the pheromone chemistry and biology of the economically important stored-product beetles in the families Anobiidae, Dermestidae, Bruchidae, Tenebrionidae, Curculionidae, Bostrichidae, Cucujidae and Silvanidae are reviewed. This is followed by a critical discussion, presenting novel ideas on the biological background of aggregation pheromones in stored-product beetles. Finally, different strategies for the application of pheromones in the control of stored-product beetles in the field are presented.

Species-specific Pheromone Chemistry and Biology

Anobiidae

The drugstore beetle, Stegobium paniceum

Females of the drugstore beetle, *Stegobium paniceum* L., produce a sex pheromone to which only males are attracted (Barratt, 1974). Females start to produce the pheromone immediately after eclosion while still in the pupal cocoon, but secretion is delayed until maturity. Adult females become attractive to males when they are 4–5 days old with maximum attractiveness after about 1 week.

Kuwahara *et al.* (1975) confirmed the presence of a female sex pheromone in the drugstore beetle, and later elucidated the structure as 2,3-dihydro-2,3,5-trimethyl-6-(1′-methyl-2′-oxobutyl)-4*H*-pyran-4-one (Kuwahara *et al.*, 1978) (Fig. 7.1). The first successful synthesis was reported by Ansell *et al.* (1979), who proposed the trivial name stegobinone. As little as 10^{-4} μg of stegobinone is active as a sex attractant in bioassays.

The absolute stereochemical configuration of the active pheromone was determined to be 2*S*,3*R*,1′*R* (Hoffmann and Ladner, 1979; Hoffmann *et al.*, 1981). The diastereomeric isomers 2*S*,3*S*,1′*RS* are inactive, while the 2*S*,3*R*,1′*S*-epimer reduces the male response to natural stegobinone (Kodama *et al.*, 1987a). This inhibitory isomer does not occur in female drugstore beetles, but it can form spontaneously in pure samples of stegobinone held at room temperature (Kodama *et al.*, 1987a). The hydrogen on C_1' is relatively acidic, due to the proximity of the electron withdrawing groups and the potential for resonance stabilization of the carbanion (Fig. 7.2). The *S* configuration about the C_1' is lost when the planar carbanion is formed and, upon reprotonation, both the 2*S*,3*R*,1′*S*- and the 2*S*,3*R*,1′*R*-isomers are formed. The potential to form the inhibitory 2*S*,3*R*,1′*S*-epimer should be considered when using stegobinone in pheromone traps. It either needs to be replaced more often than other lures, or the lure must be designed to

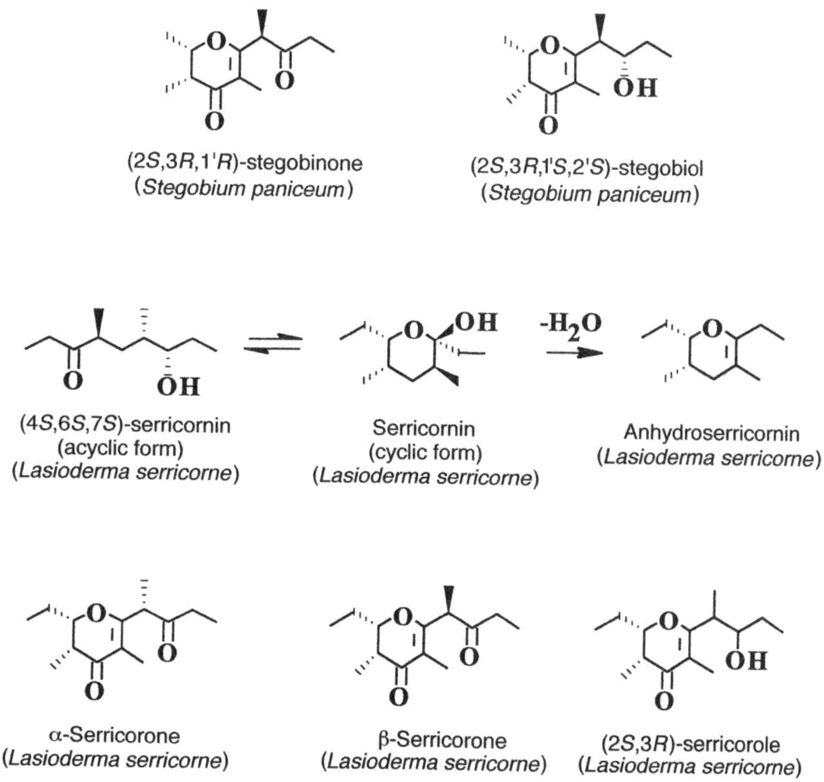

Fig. 7.1. Pheromones of selected anobiid beetles. **Stegobinone** = 2,3-dihydro-2,3,5-trimethyl-6-(1′-methyl-2′-oxobutyl)-4*H*-pyran-4-one; **stegobiol** = 2,3-dihydro-2,3,5-trimethyl-6-(1′-methyl-2′-hydroxybutyl)-4*H*-pyran-4-one; **serricornin** = 4,6-dimethyl-7-hydroxy-3-nonanone; **anhydroserricornin** = 2,6-diethyl-3,5-dimethyl-3,4-dihydro-2*H*-pyran; **serricorone** = 2,3-dihydro-3,5-dimethyl-2-ethyl-6-(1-methyl-2-oxobutyl)-4*H*-pyran-4-one; **serricorole** = 2,3-dihydro-3,5-dimethyl-2-ethyl-6-(1-methyl-2-hydroxybutyl)-4*H*-pyran-4-one. **α-Serricorone** is the (2*S*,3*R*,1′*S*)-epimer of serricorone, while **β-serricorone** is the (2*S*,3*R*,1′*R*)-epimer.

reduce epimerization (by, for example, diluting the stegobinone in a nonpolar, aprotic solvent).

About a decade after the elucidation of the structure of stegobinone, Kodama *et al.* (1987b) identified a second pheromone component, an alcohol analogue of stegobinone called stegobiol [(2*S*,3*R*,1′*S*)-2,3-dihydro-2,3,5-trimethyl-6-(1′-methyl-2′-hydroxybutyl)-4*H*-pyran-4-one] (Fig. 7.1). The configuration about the C_2' carbon was unknown until Mori and Ebata (1986) synthesized the (2*S*,3*R*,1′*S*,2′*S*)-isomer and determined that it was identical to natural stegobiol, through spectroscopic analyses and behavioural bioassays. Like stegobinone, stegobiol can elicit both searching and mating behaviour from male drugstore beetles (Kodama *et al.*, 1987b). The authors also noted that the beetles took a few

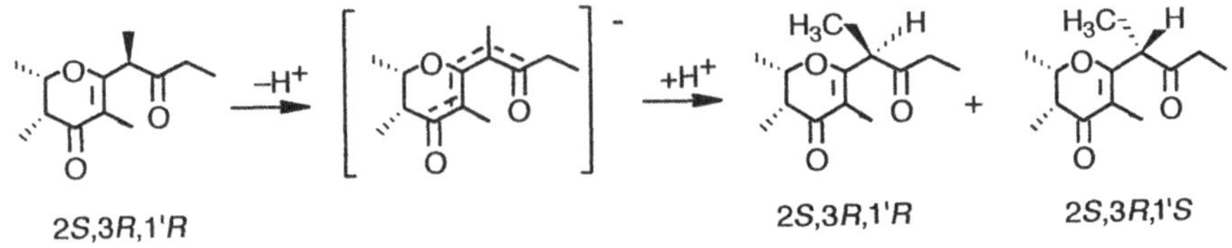

Fig. 7.2. Likely mechanism for the formation of the inhibitory (2*S*,3*R*,1′*S*)-epimer in preparations of (2*S*,3*R*,1′*R*)-stegobinone (see text for explanation).

minutes longer to respond to stegobiol than to stegobinone. The volatility of stegobiol should be depressed compared with stegobinone, due to the hydroxy substituent at C_2' and the fact that the S configuration about C_2' does not easily allow intramolecular hydrogen bonding (which would enhance the volatility) (Mori and Ebata, 1986). Taking these factors into account, Kodama *et al.* (1987b) suggested that stegobiol may operate as a short-range sex attractant, while stegobinone may be a long-range sex attractant. However, pure synthetic stegobiol had no detectable attractant activity (Burkholder and Wang, 1994, personal communication, cited in Matteson *et al.*, 1996).

The cigarette beetle, Lasioderma serricorne

Males of the cigarette beetle, *Lasioderma serricorne* Fabricius (Fig. 7.3a), are strongly attracted to the odour of females (Burkholder, 1970; Coffelt and Burkholder, 1972). Chuman *et al.* (1979) identified the major component of the female-produced sex pheromone as 4,6-dimethyl-7-hydroxy-3-nonanone, which was given the trivial name serricornin (Fig. 7.1). As little as 10 pg elicits searching and mating behaviour in male cigarette beetles. The absolute configurations of serricornin at C_4, C_6 and C_7 were established as 4S, 6S and 7S (Mori *et al.*, 1981a, 1982). With biological tests and electroantennogram (EAG) recordings, Chuman *et al.* (1985) clearly demonstrated that all three chiral centres are involved in pheromone recognition, and that the activity of (4S,6S,7S)-serricornin is much greater than those of the other stereoisomers. Moreover, the *erythro*-diastereoisomers have an antagonistic or at least weakly suppressing effect on the sexual behaviour of males (Levinson and Levinson, 1986, 1987).

(4S,6S,7S)-Serricornin exists in both an acyclic keto-alcohol form and a cyclic hemiacetal form, which are in equilibrium (Mori *et al.*, 1984; Chuman *et al.*, 1985) (Fig. 7.1). Both forms are likely to be present in appreciable quantities under natural conditions, but the active form of the pheromone is not known. Dehydration of the cyclic form of serricornin produces anhydroserricornin (2,6-diethyl-3,5-dimethyl-3,4-dihydro-2*H*-pyran), which is found in trace quantities in female beetles (Levinson *et al.*, 1981a; Chuman *et al.*, 1985). Since only the cyclic hemiacetal form of serricornin can dehydrate to form the less active anhydroserricornin, Chuman *et al.* (1985) speculated that the interconversion of the two forms of serricornin may be involved in regulating pheromone activity.

When reinvestigating the sex pheromone system of *L. serricorne*, Chuman *et al.* (1983, 1985) discovered that female beetles produce several compounds that are

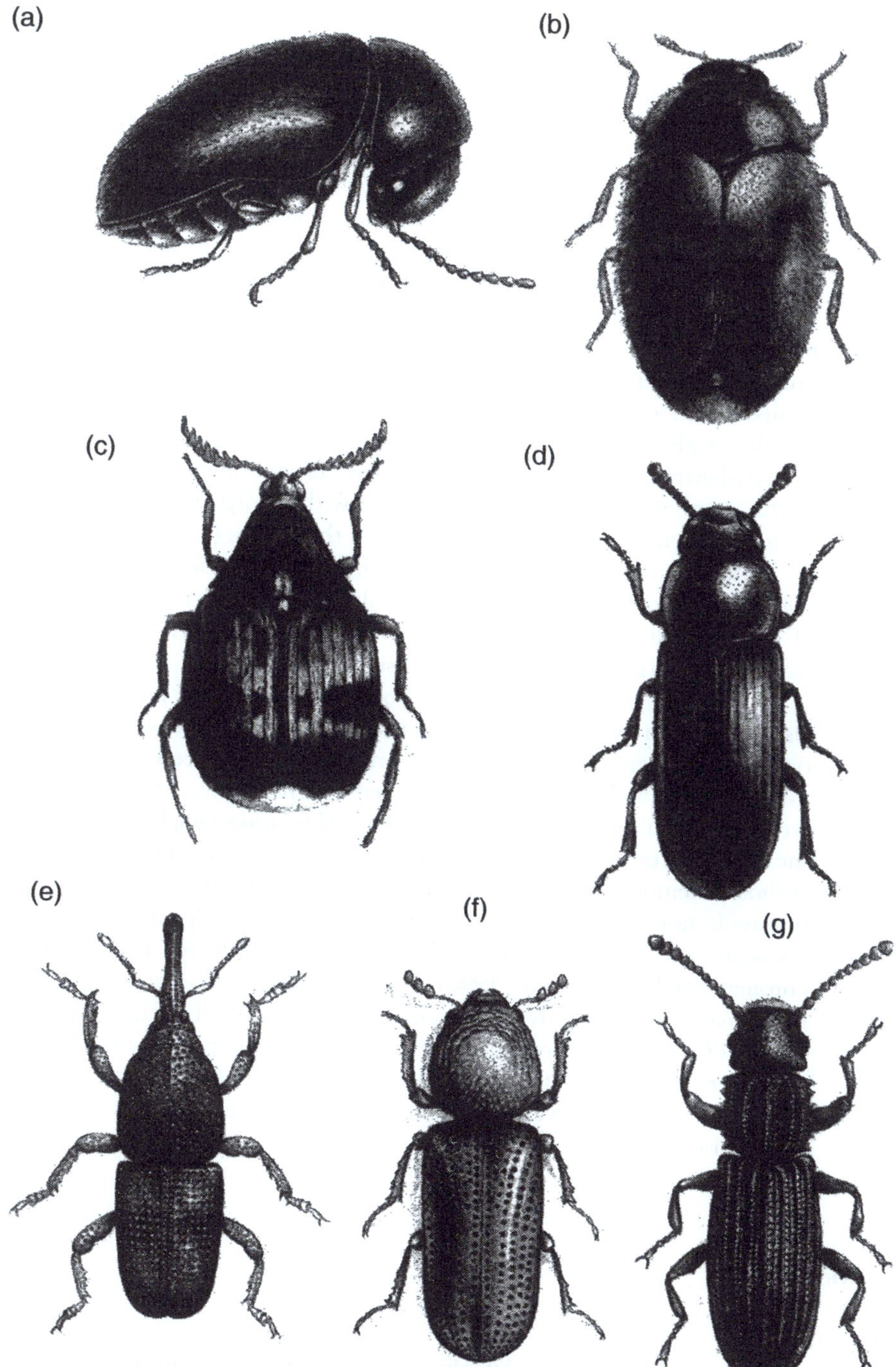

Fig. 7.3. (a) The cigarette beetle; (b) the khapra beetle; (c) the cowpea weevil; (d) the red flour beetle; (e) the maize weevil; (f) the lesser grain borer; (g) the sawtooth grain beetle.

structurally related to serricornin (Fig. 7.1). These compounds include (2*S*,3*R*)-serricorone [2,3-dihydro-3,5-dimethyl-2-ethyl-6-(1-methyl-2-oxobutyl)-4*H*-pyran-4-one], (2*S*,3*R*)-serricorole [2,3-dihydro-3,5-dimethyl-2-ethyl-6-(1-methyl-2-hydroxybutyl)-4*H*-pyran-4-one], 4,6-dimethylnonan-3,7-dione, 4,6-dimethylnonan-3,7-diol and 4,6-dimethyl-7-hydroxy-4-nonen-3-one. Serricorone and serricorole do elicit aggregation and mating behaviour in male cigarette beetles, but serricornin elicits a much greater response at comparable doses. Given that the levels of serricorone and serricorole in the female are also much lower than serricornin (the ratio of serricornin:serricorone:serricorole was determined to be 4.4:1.7:1), serricornin is recognized as the most important component of the sex pheromone. The other structurally related compounds found in the females are only weakly active in bioassay (Chuman *et al.*, 1983). The presence of all these compounds does not automatically indicate a positive contribution to the overall female sex pheromone bouquet. They are believed to play only a slight, if any, role in sexual attraction. However, their occurrence and obvious structural relationship to the sex pheromone were reason to suggest that polyketides were involved in the natural pheromone production pathways of the female (Chuman *et al.*, 1985).

The pheromone-producing gland is located in the female's second abdominal segment (Levinson *et al.*, 1983), with males lacking this structure. Hexane extracts of the pheromone glands are attractive to male *L. serricorne* and induce high receptor potentials in the olfactory sensilla of the antenna.

In addition to its role as a minor component of the sex pheromone, serricorone also serves as an oviposition deterrent in *L. serricorne* (Imai *et al.*, 1990). Surprisingly, two stereoisomers of serricorone are present in adult beetles: (2*S*,3*R*,1′*S*), called α-serricorone, and its 1′*R*-epimer, called β-serricorone. Epimerization at C_1' of serricorone could have occurred spontaneously during sample preparation (through the same process as described earlier for the epimerization of stegobinone), but the authors attempted to minimize this possibility by extracting the serricorone in hexane (a non-polar, aprotic solvent) for only 5 min, evaporating the solvent under nitrogen, and immediately analysing the extract by high pressure liquid chromatography (Imai *et al.* 1990). Both α- and β-serricorone deter oviposition by the females, either singly or in combination. Recall that, in contrast, only one epimer of stegobinone serves as a sex pheromone component in *S. paniceum* while the other actually inhibits pheromonal activity. This difference may be appreciated if one considers the relatively longer time period required for an oviposition deterrent to function (days or weeks) as compared with a sex pheromone (minutes to hours) (Imai *et al.*, 1990).

Dermestidae

Attagenus *spp.*

Mature females of several *Attagenus* species demonstrate diurnal calling behaviour, which is associated with the release of a sex attractant (Barak and Burkholder, 1977, 1978). In *Attagenus brunneus* Fald. (= *A. elongatulus* (Casey)) and *Attagenus rufipennis* LeConte, female calling activity, female pheromone content and male responsiveness are highest during the early hours of the photophase. *Attagenus unicolor* Brahm (formerly known as A. *megatoma* (Fabricius) or A. *piceus* Olivier), *Attagenus schaefferi* LeConte and *Attagenus bicolor* Harold show calling behaviour mainly during the middle of the photophase. Slight deviations in the calling behaviour do occur among subspecies.

The female-produced sex pheromone of the black carpet beetle, A. *unicolor*, was the first sex pheromone of an insect associated with stored products to be identified and synthesized. In a classic study, Silverstein *et al.* (1967) identified the pheromone as (*E*,*Z*)-3,5-tetradecadienoic acid (Fig. 7.4). Synthetic (*E*,*Z*)-3,5-tetradecadienoic acid, named megatomoic acid (Rodin *et al.*, 1970), elicited the natural behaviour of males toward females. The *Z*,*Z*-isomer of 3,5-tetradecadienoic acid was identified as the major sex attracting component in extracts of virgin females of A. *brunneus* (Fukui *et al.*, 1977) (Fig. 7.4).

The site of the pheromone production is believed to be in the female abdominal sternites (Barak and Burkholder, 1977).

Anthrenus *spp.*

Female furniture carpet beetles, *Anthrenus flavipes* LeConte, have an entrainable daily calling rhythm which occurs during the early photophase and is correlated

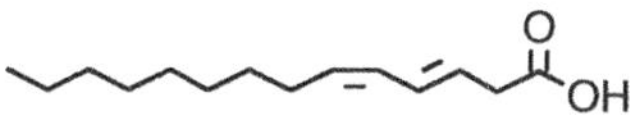

(*E*,*Z*)-3,5-tetradecadienoic acid
(*Attagenus unicolor*)

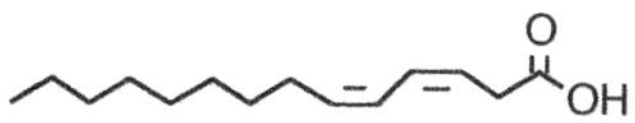

(*Z*,*Z*)-3,5-tetradecadienoic acid
(*Attagenus brunneus*)

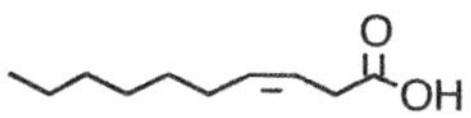

(*Z*)-3-decenoic acid
(*Anthrenus flavipes*)

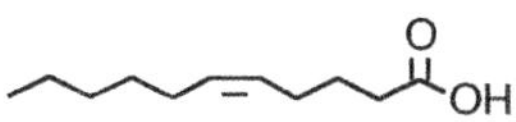

(*Z*)-5-undecenoic acid
(*Anthrenus verbasci*)

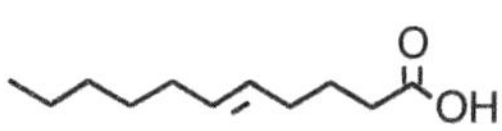

(*E*)-5-undecenoic acid
(*Anthrenus verbasci*)

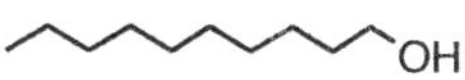

Decanol
(*Anthrenus sarnicus*)

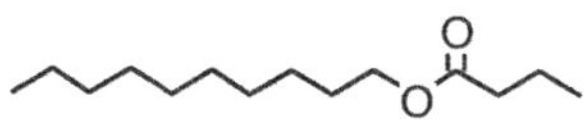

Decyl *n*-butyrate
(*Anthrenus sarnicus*)

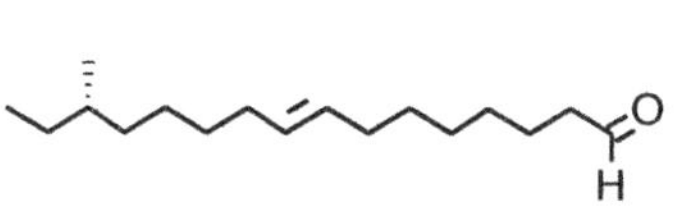

(*R*)-(*E*)-trogodermal
(*Trogoderma glabrum*,
Trogoderma granarium)

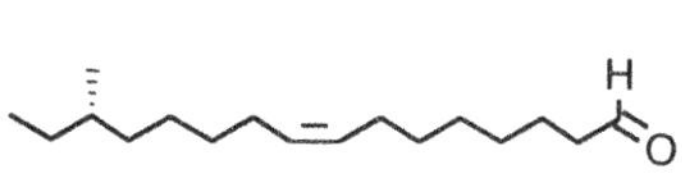

(*R*)-(*Z*)-trogodermal
(*Trogoderma granarium*
Trogoderma inclusum,
Trogoderma variabile)

Fig. 7.4. Pheromones of selected dermestid beetles. **Trogodermal** = 14-methyl-8-hexadecenal.

closely with active pheromone production, maximum male response to the sex attractant, and the peak mating period (Burkholder *et al.*, 1974; Ma and Burkholder, 1978). The pheromone, produced by virgin furniture carpet beetle females, was identified as (Z)-3-decenoic acid (Fukui *et al.*, 1974) (Fig. 7.3). On average a 10-day-old virgin female of *A. flavipes* produced 198.6 ng of the attractant chemical, but only 13.1 ng of the synthetic compound was necessary to elicit response in 50% of bioassayed males (Ma *et al.*, 1980).

Virgin females of the varied carpet beetle, *Anthrenus verbasci* L., also release a sex pheromone that attracts males and stimulates their sexual response. This sex pheromone was concluded to be a mixture of (Z)-5 and (E)-5-undecenoic acid (Kuwahara and Nakamura, 1985) (Fig. 7.4).

The female-produced sex attractant of the Guernsey carpet beetle, *Anthrenus sarnicus* Mroczkowski, contains both decanol and decyl *n*-butyrate (Finnegan and Chambers, 1993) (Fig. 7.4). The components appeared in approximately equal amounts, but decyl butyrate produced a larger EAG response in males. Since decyl butyrate was only slightly less attractive when presented alone, decanol is considered a minor synergist in the pheromone mixture.

Trogoderma *spp.*

Voelkel (1924) was the first to hypothesize that female khapra beetles, *Trogoderma granarium* Everts (Fig. 7.3b), produce an odoriferous chemical substance that attracts males. The existence of a female-produced pheromone that attracts conspecific males and, to some extent, conspecific females was confirmed by Bar Ilan *et al.* (1965). Female-produced sex pheromones were also detected in *Trogoderma inclusum* LeConte and *Trogoderma glabrum* (Herbst) (Burkholder and Dicke, 1966).

Despite intense investigations (Yinon and Shulov, 1967a,b; Levinson and Bar Ilan, 1967, 1970a,b; Ikan *et al.*, 1969; Rodin *et al.*, 1969; Cohen *et al.*, 1971; Levinson and Levinson, 1973), the identity of the sex pheromones of the *Trogoderma* spp. eluded researchers for some time.

Chemical analysis of female body extracts from *T. inclusum* led to the identification of the two compounds, (Z)-14-methyl-8-hexadecenol and methyl (Z)-14-methyl-8-hexadecenoate. Each could elicit a male response very similar to that of live females, and their effects were additive (Rodin *et al.*, 1969). Similarly, body extracts from virgin *T. glabrum* females revealed the major active compounds methyl (E)-14-methyl-8-hexadecenoate, (E)-14-methyl-8-hexadecenol, and additional minor compounds also capable of independently eliciting sexually excitatory responses in conspecific males (Yarger *et al.*, 1975).

Finally, it was found that an airborne component, not found in macerated females, was the primary compound responsible for male attraction. The major female sex pheromone components were identified as (Z)-14-methyl-8-hexadecenal in *T. inclusum* and *T. variabile* Ballion, (E)-14-methyl-8-hexadecenal in *T. glabrum*, and a 92:8 mixture of the (Z):(E) isomers of 14-methyl-8-hexadecenal in *T. granarium* (Cross *et al.*, 1976, 1977; Greenblatt *et al.*, 1976) (Fig. 7.4). The corresponding alcohol of 14-methyl-8-hexadecenal appeared as only a minor constituent in the trapped volatiles, and showed significantly less activity in eliciting male arousal than the aldehyde. Using the EAG technique, Levinson *et al.*

(1978) showed the importance of (*Z*)- and (*E*)-14-methyl-8-hexadecenal in *T. granarium* over the corresponding alcohols and esters, by recording higher potentials in the presence of the aldehydes. The pheromone may be stored in the alcohol or ester forms, for conversion to the more volatile aldehyde immediately before or during release. Given its widespread use as a sex pheromone amongst *Trogoderma* spp., the (*Z*)- and (*E*)-isomers of 14-methyl-8-hexadecenal were given the trivial names (*Z*)- and (*E*)-trogodermal, respectively (Levinson *et al.*, 1978).

There are contradictory reports in the literature regarding the absolute configuration of the chiral centre in trogodermal, C-14. Using synthetic samples of the (*R*)- and (*S*)-enantiomers of both (*E*)- and (*Z*)-trogodermal, Rossi and Niccoli (1978) and Rossi *et al.* (1979) reported that the (*S*)-enantiomers were more active than the (*R*)-enantiomers in bioassays with *T. granarium*. In contrast, Levinson and Mori (1980) and Levinson *et al.* (1981b) concluded that the (*R*)-enantiomers of (*Z*)- and (*E*)-trogodermal were far more active than the (*S*)-enantiomers, after conducting biological tests and EAG recordings. Similarly, the (*R*)-trogodermal was more active than the (*S*)-enantiomer in biological tests with *T. glabrum*, *T. inclusum* and *T. variabile* by about two orders of magnitude (Silverstein *et al.*, 1980). In none of the studies was the enantiomeric composition of the compounds emitted by the insects determined, due to the technical difficulties involved in such a measurement with this particular compound. Trogodermal has a relatively small specific rotation, making it difficult to isolate large enough quantities for an accurate determination of whether it is levorotatory or dextrorotatory. However, (*Z*)-14-methyl-8-hexadecenol and methyl (*Z*)-14-methyl-8-hexadecenoate, isolated from *T. inclusum*, were determined to be levorotatory by Rodin *et al.* (1969). Mori (1974) stereoselectively synthesized the (*S*)-enantiomers of these compounds, and determined that both (*S*)-enantiomers are dextrorotatory. Thus the naturally occurring, levorotatory alcohol and ester derivatives of trogodermal must be of the (*R*)-configuration. Since they are the likely biosynthetic precursors of (*Z*)-trogodermal, this can be taken as evidence that the (*Z*)-trogodermal is of the (*R*)-configuration.

Histological studies of the female abdomen of *T. granarium* suggested that the intersegmental fold between the fifth and the sixth sternite serves as a multicellular gland excreting the pheromone (Stanic *et al.*, 1970). Hammack *et al.* (1973) found an enlarged glandular epithelium in the inner surface of the seventh sternite (the fifth visible sternite) of *T. glabrum*, *T. inclusum*, *T. variabile*, *T. grassmani* Beal and *T. sternale* Jayne. Hexane extracts of intact *T. glabrum*, *T. inclusum* and *T. variabile* were more active to males than extracts of females with the seventh abdominal sternite removed.

Interspecific responses to female extracts of several *Trogoderma* species and cross-attraction of males to non-conspecific females in biological tests were recorded by Vick *et al.* (1970). The cross-attraction was explained by the occurrence of the pheromone precursors (*Z*)-14-methyl-8-hexadecenol and methyl (*Z*)-14-methyl-8-hexadecenoate in female body extracts of *T. inclusum*, *T. variabile* and *T. granarium* that had some attractiveness to males of *T. inclusum* and *T. variabile*, which utilize (*Z*)-trogodermal as their sex pheromone (Greenblatt *et al.*, 1977). *T. glabrum* appeared different because the pheromone precursors are (*E*)-14-methyl-8-hexadecen-ol and (*E*)-14-methyl-8-hexadecenoate. However, since some cross-attraction did occur, other isolating mechanisms must exist to prevent interspecific mating attempts. In fact, different rhythms in the female sex pheromone releasing behaviour and the male responsiveness were found in *Trogoderma* species. Calling behaviour of virgin *T. glabrum* females, which is associated with pheromone release, is largely restricted to an 8 h interval that peaks toward the middle of the 16 h photophase. Male activity and mating increase during the same 8 h interval (Vick *et al.*, 1973a; Hammack *et al.*, 1976; Shapas and Burkholder, 1978a). Pheromone release by *T. variabile* females and male response peak at the beginning of the photophase, whereas female calling behaviour and mating by *T. inclusum* occur mainly during midphotophase. *T. glabrum* calls during the midphotophase as well, but slightly later than *T. inclusum* (Vick *et al.*, 1973a; Shapas and Burkholder, 1978b). Female calling and mating frequency in *T. granarium* are also centred around the middle of the photophase (Hammack and Burkholder, 1981). Thus the two species that rely solely on (*Z*)-trogodermal, *T. variabile* and *T. inclusum*, are sexually most active at different times of the day. *T. inclusum*, *T. glabrum* and *T. granarium*, which have nearly coincident calling and mating times, utilize opposite isomers of trogodermal or a specific ratio of the sex attractant and, thus, avoid cross-attraction to some extent (Shapas and Burkholder, 1978a). Although *T. granarium* and *T. inclusum* rarely occur sympatrically, due to their difference in food preference and global distribution (Weidner, 1983a), *T. granarium* (relying mainly on (*Z*)-trogodermal) and *T. inclusum* (relying exclusively on (*Z*)-trogodermal) must have additional species-isolating mechanisms.

Bruchidae

The dry bean weevil, Acanthoscelides obtectus

Males of the dry bean weevil, *Acanthoscelides obtectus* (Say), produce a sex pheromone that is attractive only to virgin females, while the females apparently do not produce a sex attractant (Hope *et al.*, 1967). Virgin adult female *A. obtectus* of about 1 day old are attracted to both unmated and mated caged males (Halstead, 1973), but after mating the females are no longer attracted to males and remain unresponsive for the rest of their lives. Males are neither attracted nor repelled by females, female extracts, conspecific males or male extracts. In dual-choice tests, virgin females prefer male-contaminated beans over female-contaminated or uncontaminated beans, while mated females did not show any preference. The male scent was extractable in hexane and associated with the outer cuticular surface or a gland opening on to the male's thorax or abdomen (Hope *et al.*, 1967). Abdominal integumentary glands that presumably produce the sex pheromone were described by Biemont *et al.* (1990).

Horler (1970) identified the male-produced sex pheromone as methyl (*E*)-2,4,5-tetradecatrienoate (Fig. 7.5). The structure is obviously of fatty acid origin, as are many pheromones of stored-product insects, but the pheromone of the dry bean weevil is especially interesting due to the allene moiety. The absolute configuration about the chiral allene moiety was deduced by synthesizing both the (*R*)- and the (*S*)-enantiomers, and determining the specific rotations of each. The (*R*)-enantiomer was found to be levorotatory, as is the natural pheromone, allowing the

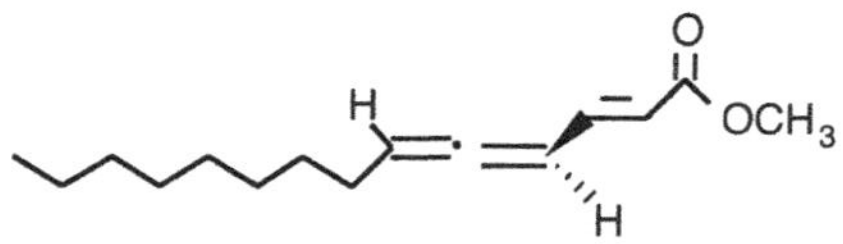

Methyl (*R*)-(*E*)-2,4,5-tetradecatrienoate
(*Acanthoscelides obtectus*)

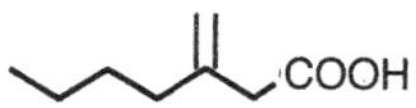

3-Methyleneheptanoic acid
(*Callosobruchus maculatus*)

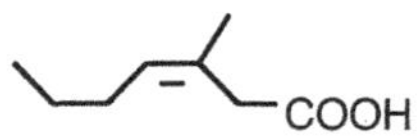

(Z)-3-methyl-3-heptenoic acid
(*Callosobruchus maculatus*)

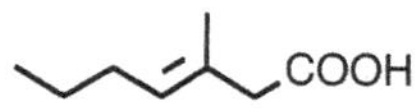

(*E*)-3-methyl-3-heptenoic acid
(*Callosobruchus maculatus*)

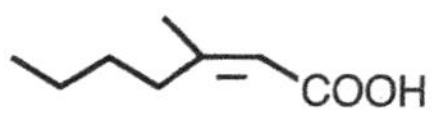

(Z)-3-methyl-2-heptenoic acid
(*Callosobruchus maculatus*)

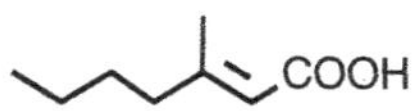

(*E*)-3-methyl-2-heptenoic acid
(*Callosobruchus maculatus*)

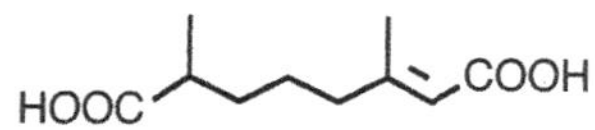

(*R*/*S*)-(*E*)-3,7-dimethyl-2-octene-1,8-dioic acid
(*Callosobruchus chinensis*)

Fig. 7.5. Pheromones of selected bruchid beetles.

conclusion that the pheromone is of the (*R*)-configuration (Pirkle and Boeder, 1978; Mori *et al.*, 1981b.)

Callosobruchus *spp.*

Males of the cowpea weevil, *Callosobruchus maculatus* (Fabricius) (Fig. 7.3c), are attracted to females and show a characteristic courtship behaviour when exposed to female odour (Rup and Sharma, 1978). Female attraction toward males was not observed. The pheromone is released from the abdominal tip while the female assumes a typical calling position (Qi and Burkholder, 1982). Mating reduces pheromone emission by the female, but males remain responsive after mating.

Parameters affecting pheromone release were compared in flightless and flighted strains of *C. maculatus* (Lextrait *et al.*, 1994, 1995). In both strains, emission of the pheromone by the females seems to be correlated with the reproductive status. The flightless-form females, which are sexually mature at emergence, begin to release the pheromone on the first day after eclosion. The flighted females, which exhibit no reproductive activity at emergence, show a delayed emission of pheromone. Furthermore, the presence of seeds of the host plant tends to stimulate pheromone release by flighted females, while the seeds have no apparent effect on pheromone release by females of the flightless form.

The female sex pheromone of *C. maculatus* was revealed in part by Shu *et al.* (1996), and in more detail by Phillips *et al.* (1996), to contain five relatively short-chain carboxylic acids: 3-methyleneheptanoic acid, (*Z*)-3-methyl-3-heptenoic acid, (*E*)-3-methyl-3-heptenoic acid, (*Z*)-3-methyl-2-heptenoic acid and (*E*)-3-methyl-2-heptenoic acid (Fig. 7.5). Each acid is active, but the 3(*Z*) and 3(*E*) compounds elicit the highest response from males. Combinations of two or more compounds have an additive effect. The source of the sex pheromone of *C. maculatus* is believed to be the female's pygidium, because solvent extracts of that region elicited the largest EAG responses from the male antennae (Ramaswamy *et al.*, 1995).

Two pheromones mediate mating activity in the azuki bean weevil, *Callosobruchus chinensis* L.: a female-produced male attractant and a copulation-release pheromone (Tanaka *et al.*, 1981). The existence of a female sex pheromone was demonstrated in biological tests in which males were highly attracted to confined females (Honda and Yamamoto, 1976; Gharib *et al.*, 1992). Attraction of females toward males, or between adults of the same sex, was not observed. However, isolation and identification of this sex pheromone have not been fully successful to date. A female *C. chinensis*, dead or alive, is capable of both attracting and causing copulation in a conspecific male, but the sex attractant alone does not cause copulation. These observations led to the identification of a copulation-release pheromone, which was given the trivial name erectin (Tanaka *et al.*, 1981, 1982). Erectin is distinct from the sex pheromone, and is a complex bouquet containing ten distinct compounds. These can be divided into two synergistically acting fractions, neither having any activity alone. One fraction is simply composed of (*E*)-3,7-dimethyl-2-octene-1,8-dioic acid, named callosobruchusic acid (Fig. 7.5). The (*R*)- and (*S*)-enantiomers of callosobruchusic acid, synthesized by Mori *et al.* (1983), were equally effective in eliciting copulation response in males. The other fraction of erectin is a more complicated mixture of methyl-branched, long-chain (25–33 carbons) hydrocarbons: 3-methylpentacosane, 11-methylheptacosane, 3-methylheptacosane, 11-methylnonacosane, 13-methylnonacosane, 11,15-dimethylnonacosane, 13-methylhentriacontane, 9,13-dimethylhentriacontane and 11,15-dimethyltritriacontane. All compounds found in erectin should have relatively low vapour pressures (due to their extremely large size or, in the case of callosobruchusic acid, the presence of two carboxylic acid moieties), facilitating the role of erectin as a short-range or contact pheromone.

The Mexican bean weevil, Zabrotes subfasciatus

Bioassay data indicate that virgin females of *Zabrotes subfasciatus* Boheman produce a pheromone that is attractive to males, and that the males do not produce a sex attractant (Pimbert, 1987). The chemical identity of this pheromone is as yet unknown, but several interesting factors involved in the regulation of pheromone production have been elucidated. Sex pheromone production and/or release is apparently correlated with sexual maturity, since the attraction of males to virgin females is enhanced as the females age (Pimbert, 1987). The odours of females held in the absence of host plant seeds do not elicit strong EAG responses from the males until the females are about 5 days post-emergence. In contrast, if the females are held for 6 h in contact with crushed bean seeds, the odours of virgin females only 3 days post-emergence can elicit strong EAG responses from the males (Pimbert and Pouzat, 1988). The authors speculated that the host seeds may accelerate the sexual maturation of the females, thereby causing a concomitant stimulation of pheromone production/emission. Lastly, insemination of the females causes a transitory decrease in the attractiveness of the females (Pimbert, 1987).

Oviposition deterrents or 'host-marking pheromones' have been reported in several bruchids (Giga and Smith, 1984/1985; Credland and Wright, 1990; Mbata, 1993; Mbata and Ramaswamy, 1995). Males and females of *A. obtectus* are thought to chemically mark beans, which results in avoidance of these beans by other females for oviposition (Szentesi, 1981). Messina and Renwick (1985) and Messina *et al.* (1987) showed that females of *C. maculatus* avoid oviposition on host seeds already bearing conspecific eggs. This behaviour was presumed to be mediated by an oviposition marker that was deposited with the eggs. Azuki bean weevils, *C. chinensis*, also clearly distinguish beans conditioned with isolates of a presumable oviposition marker from unconditioned beans or even glass beads (Oshima *et al.*, 1973).

Tenebrionidae

Tribolium *spp.*

Male-produced aggregation pheromones are present in the flour beetles *Tribolium confusum* Jacquelin duVal (Ryan and O'Ceallachain, 1976; O'Ceallachain and Ryan, 1977; Suzuki and Sugawara, 1979), *Tribolium castaneum* (Herbst) (Suzuki and Sugawara, 1979), *Tribolium brevicornis* (LeConte), *Tribolium destructor* Uyttenboogaart, *Tribolium audax* Halstead (Faustini *et al.*, 1982a) and *Tribolium freemani* Hinton (Suzuki *et al.*, 1987). The pheromone production site in the red flour beetle, *T. castaneum* (Fig. 7.3d), is believed to be the exocrine glands beneath the setiferous sex patch on the ventral side of the femur (Faustini *et al.*, 1981). The females have neither the patch nor the associated brush of hairs (Hinton, 1942). The aggregation pheromones of *T. castaneum* and *T. confusum* were isolated and both identified as 4,8-dimethyldecanal (Suzuki and Sugawara, 1979; Suzuki, 1980, 1981) (Fig. 7.6). This compound was given the trivial name tribolure, since it was also found to serve as the aggregation pheromone in *T. freemani* (Suzuki *et al.*, 1987).

The *R*,*R*-enantiomer of 4,8-dimethyldecanal elicits the greatest response in bioassays (Suzuki and Mori, 1983; Barak and Burkholder, 1984/1985) and EAGs (Levinson and Mori, 1983). The *R*-configuration about C_4 is apparently essential for biological activity (Suzuki *et al.*, 1984). However, the (4*R*,8*S*)-isomer showed minor attraction to *T. castaneum* in bioassays, and a combination of the (4*R*,8*R*)- and (4*R*,8*S*)-isomers in a ratio of 8:2 was most attractive and more alluring than the (4*R*,8*R*)-isomer alone (Suzuki *et al.*, 1984). It was proposed that the (4*R*,8*S*)-isomer might be a secondary pheromone component acting as a synergist in the red flour beetle's aggregation pheromone. This remains to be clarified.

The average production of the aggregation pheromone by individual male *T. castaneum* is 1.3 ± 0.2 μg over a 2-day interval (Hussain *et al.*, 1994a). Aging has no effect on pheromone production, although a certain rhythmic change in pheromone emission was observed. A dramatic decrease in pheromone production occurred when the males had no physical contact with food, irrespective of exposure to volatiles of cracked wheat (Hussain *et al.*, 1994a). Thus, the production and emission of the aggregation pheromone are dependent on feeding.

Suzuki (1985) described a second aggregation substance in *T. castaneum*. This substance attracted males and females but was less active than tribolure. It is produced by both sexes and believed to be excreted along with faeces.

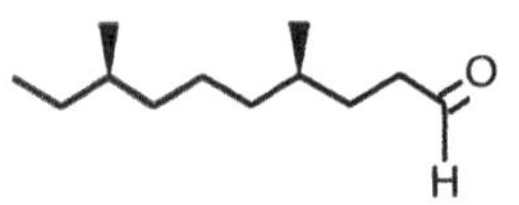

(4*R*,8*R*)-tribolure
(*Tribolium* spp.)

(*R*)-4-methyl-1-nonanol
(*Tenebrio molitor*)

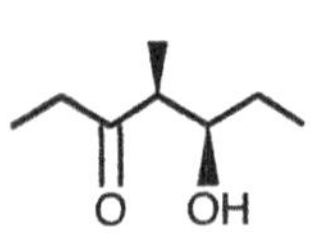

(4*S*,5*R*)-sitophinone
(*Sitophilus oryzae*,
Sitophilus zeamais)

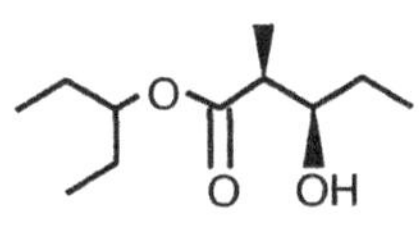

(2*S*,3*R*)-sitophilate
(*Sitophilus granarius*)

Fig. 7.6. Pheromones of selected tenebrionid and curculionid beetles. **Tribolure** = 4,8-dimethyldecenal; **sitophinone** = 5-hydroxy-4-methyl-3-heptanone; **sitophilate** = 1-ethylpropyl 2-methyl-3-hydroxypentanoate.

In addition to aggregation pheromones, adults of many *Tribolium* species also produce defence or spacing chemicals (Roth and Howland, 1941; Roth, 1943, 1945). These compounds are emitted from well-developed odoriferous glands in the prothorax and abdomen, and include quinones, certain alkenes, aromatic ketones and aromatic esters (Alexander and Barton, 1943; Ladisch *et al.*, 1967; Happ, 1968; Howard and Mueller, 1987). Some of them, especially 1-pentadecene, were believed to serve to bring the sexes together and induce copulation because, although the chemicals were produced in both males and females, only males appeared to be affected (Keville and Kannowski, 1975). However, the same compounds were clearly repellent at very low concentrations (Suzuki *et al.*, 1975a). The 14-, 15-, 16- and 17-carbon alkenes, the benzoquinones and the hydroquinones can vary in quantity and quality among and between the species (Suzuki *et al.*, 1975b; Howard, 1987; Pappas and Morrison, 1995; Pappas and Wardrop, 1996). Their function can be of pheromonal or allomonal character. Within a species they can counteract the effect of the aggregation pheromone, depending on the number and duration of beetles in the medium. Under the conditions of overcrowding or lack of food, quinones are believed to act as population-regulating pheromones either by directly repelling conspecifics (Loconti and Roth, 1953; Ghent, 1963; Mondal and Port, 1994) or by dissolving the waxy globules on the male prothoracic femora, where the aggregation pheromone is emitted, and thus suppressing aggregation behaviour (Faustini and Burkholder, 1987). It is interesting to note that not only adult *T. castaneum*, but also the larvae, were repelled by flour conditioned with high densities of male and female red flour beetles (Mondal, 1985). The synomonal character of quinones was shown when parasitoids of stored-product moth larvae were repelled (Press *et al.*, 1986; Press, 1988). This can also be regarded as an allomonal effect because red flour beetles can prey on moth larvae and quinones repel competitors for this kind of food.

The yellow mealworm, Tenebrio molitor

Several pheromones may be involved in mediating the mating behaviour of the yellow mealworm, *Tenebrio molitor* L.: a female-produced male attractant (Valentine, 1931; Tschinkel *et al.*, 1967; Happ, 1969; Happ and Wheeler, 1969; August, 1971; Tanaka *et al.*, 1986); a male-produced female attractant (Valentine, 1931; Happ, 1969; August, 1971); one or more copulation-release pheromones (Tschinkel *et al.*, 1967; Happ and Wheeler, 1969; August, 1971; Tanaka *et al.*, 1986); and a male-produced anti-aphrodisiac (Happ, 1969).

Tanaka *et al.* (1986) identified the female-produced sex pheromone as 4-methylnonanol (Fig. 7.6). Newly eclosed and immature adult females produce relatively little of the attractant, but bioassays indicate that sex pheromone production increases as the female matures and reaches a maximum at 4 or 5 days post-eclosion (Valentine, 1931; Happ and Wheeler, 1969). Coincidentally, male responsiveness to the female-produced sex pheromone increased with age and maturation, reaching a maximum about 5 days post-eclosion (Happ, 1970).

The male-produced female attractant has not been identified. However, bioassay results indicate that only mature males (5 or more days post-eclosion) produce appreciable quantities of the pheromone (Menon, 1976). Thus the male-produced sex pheromone apparently acts in concert with the female-produced sex pheromone, to facilitate the intersection of the two sexes when the insects are mature enough to participate in productive matings. These findings correlate well with the observation that the majority of adult beetles do not copulate until 4–5 days post-eclosion (Gerber, 1973).

Sex pheromone biosynthesis in both male and female *T. molitor* is regulated by juvenile hormone (JH) III (Menon, 1970, 1976). Corpora allata excised from mature male and female adults can produce JH III *in vitro*, whereas those excised from immature females cannot (Judy *et al.*, 1975). These results explain why mature adults produce more sex pheromone than immature adults: as the insects mature, the JH titres rise to a level where sex pheromone production is stimulated.

The adult female also produces at least one copulation-release pheromone (Tschinkel, 1970; August, 1971; Tanaka *et al.*, 1986). The female-produced sex pheromone, 4-methylnonanol, attracts males at low doses (10 ng), and elicits a copulation response from the males at higher doses (20 ng) (Tanaka *et al.*, 1986). Adult females each contain approximately 30 ng of 4-methylnonanol (Tanaka *et al.*, 1986), indicating that 4-methylnonanol can both attract males and induce copulation at biologically relevant concentrations. A non-volatile copulation-release pheromone may also be involved during amplexus. August (1971) observed that male beetles whose mouthparts were covered with soft wax or collodion could become stimulated in the presence of a female, and amplexus could occur, but the males did not extend their genitalia. Tanaka *et al.* (1986) reported that the ether extract of a virgin female from which all volatile compounds (including 4-methylnonanol) had been removed could elicit copulation activity in males. The activity of this extract was synergized with 4-methylnonanol. Hurd and Parry (1991) observed that if female extract was inserted into a hollow glass tube, males would not exhibit copulatory activity, whereas if the same extract was coated on to a solid glass rod, the males would attempt copulation with the rod. Thus, under natural conditions, the copulation-release pheromone may facilitate copulation after the male has located the female by her sex pheromone and physical contact has been made, very similar to the pheromone and mating biology of *Callosobruchus chinensis* (see above).

Male beetles may also contain low concentrations of a copulation-release pheromone (Tschinkel *et al.*, 1967; Tanaka *et al.*, 1986), although this conclusion is disputed (Happ and Wheeler, 1969; August, 1971). Part of the difficulty in resolving this issue is that it is difficult to isolate the effects of the putative copulation-release pheromone from those of the sex attractant and tactile stimuli (which are primarily functional during amplexus). Moreover, while the behavioural sequence of the male may normally result from concerted stimuli, it may be possible to complete the behavioural sequence even in the absence of one or more stimuli (August, 1971).

Curculionidae

Sitophilus *spp.*

Male-produced aggregation pheromones occur in the rice weevil, *Sitophilus oryzae* L. (Phillips and Burkholder, 1981), the granary weevil, *Sitophilus granarius* L.

(Faustini *et al.*, 1982b) and the maize weevil, *Sitophilus zeamais* Motschulsky (Fig. 7.3e; Walgenbach *et al.*, 1983). Laboratory bioassays indicated some interspecific cross-attraction, especially between *S. oryzae* and *S. zeamais* (Walgenbach *et al.*, 1983). Subsequent studies revealed that *S. oryzae* and *S. zeamais* share an aggregation pheromone, while that of *S. granarius* is closely related not only in structure but also in the configuration of the chiral carbons (Fig. 7.6).

The aggregation pheromone of both *S. oryzae* and *S. zeamais* is 5-hydroxy-4-methyl-3-heptanone (Schmuff *et al.*, 1984), which has been given the trivial name sitophinone (Walgenbach and Burkholder, 1986). The (4*S*,5*R*)-stereoisomer of sitophinone elicits the highest response from both male and female adult weevils in biotests (Phillips *et al.*, 1985) and EAG recordings (Levinson *et al.*, 1990). The (4*R*,5*R*)-isomer is produced in only trace quantities (Walgenbach *et al.*, 1987a).

The main pheromone component in *S. granarius* is 1-ethylpropyl 2-methyl-3-hydroxypentanoate, an ester named sitophilate (Phillips *et al.*, 1987). The (2*S*,3*R*)-stereoisomer is the most active isomer (Phillips *et al.*, 1989; Levinson *et al.* 1990). Chambers *et al.* (1996) reported that a second, as yet unidentified, pheromone component may be produced by male *S. granarius*.

The effect of mating and the presence of grain on the response of both male and female, virgin and mated, granary weevils has been examined using live male granary weevils as pheromone donors in olfactometer assays (Plarre, 1991, 1992). Virginal male granary weevils are more attractive than mated males, in the presence of wheat. Unmated females showed stronger positive responses toward male odours than mated females or males in general. In the absence of any grain, *S. granarius* males and females strongly respond to the pheromone in the range 20–2500 ng in small-scale bioassays (Phillips *et al.*, 1989). With access to suitable substrate, unmated females are sensitive to *c.* 600 ng of synthetic sitophilate and aggregate near a pheromone source (Plarre, 1994). Unmated males do not respond to such low levels of pheromone. Higher pheromone levels of 6000 ng attract unmated males and mated weevils of both sexes. High pheromone concentrations reduce attraction (Jiang, 1995) or even show strong repellent effects (Plarre, 1994).

The production site of the pheromone in grain weevils has not yet been determined, but it is believed that the pheromone is confined to the male's thorax or abdomen (Chambers *et al.*, 1996). Although male frass is occasionally attractive to weevils (Plarre, 1991), release of the aggregation pheromone in the frass is unlikely. Rather, slight absorption of pheromone from the males may allow the frass to act as a secondary pheromone source (Chambers *et al.*, 1996).

Bostrichidae

The lesser grain borer, Rhyzopertha dominica

The male-produced aggregation pheromone of the lesser grain borer, *Rhyzopertha dominica* (Fabricius) (Fig. 7.3f), is composed of two compounds, *S*-(+)-1-methylbutyl (*E*)-2-methyl-2-pentenoate (dominicalure 1) and *S*-(+)-1-methylbutyl (*E*)-2,4-dimethyl-2-pentenoate (dominicalure 2) (Khorramshahi and Burkholder, 1981; Williams *et al.*, 1981) (Fig. 7.7). Dominicalures 1 and 2 are as effective individually as they are in various combinations, and attractancy to wild populations

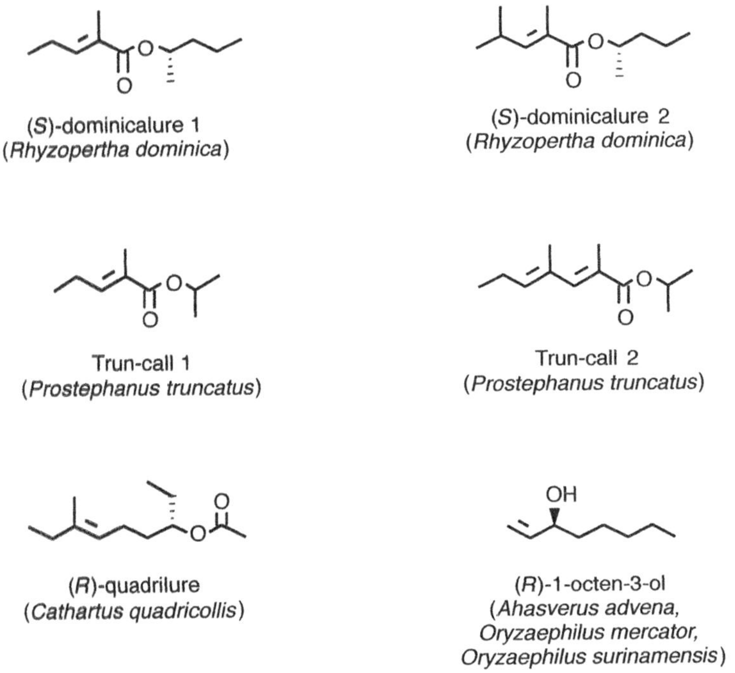

Fig. 7.7. Pheromones of selected bostrychid and silvanid beetles. **Dominicalure 1** = 1-methylbutyl 2-methyl-2-pentenoate; **dominicalure 2** = 1-methylbutyl 2,4-dimethyl-2-pentenoate; **trun-call 1** = 1-methylethyl 2-methyl-2-pentenoate; **trun-call 2** = 1-methylethyl 2,4-dimethyl-2,4-heptadienoate; **quadrilure** = 7-methyl-6-nonen-3-yl acetate.

increased with higher dosages (Cogburn *et al.*, 1984; Selitskaya and Shamshev, 1994). Mayhew and Phillips (1994) showed that the absolute production of dominicalures 1 and 2 and their ratio change continuously over a period of 1 month. The maximum production of each compound is about 1 μg per beetle per day. Mating status has no influence on the levels of pheromone production by males of the same age, and the response to the pheromone is not influenced by age, mating status or sex (Dowdy *et al.*, 1993). Dominicalures 1 and 2 and several related compounds, possibly metabolites of the lesser grain borer's pheromone, are believed to be responsible for the severe off-odour associated with grain heavily infested with *R. dominica* (Seitz and Sauer, 1996).

The larger grain borer, Prostephanus truncatus

The larger grain borer, *Prostephanus truncatus* (Horn), produces an aggregation pheromone that is very similar to that of *R. dominica*. In fact, dominicalures 1 and 2 are highly attractive to *P. truncatus*, and were used to monitor for the larger grain borer in East Africa before its pheromone was identified (Hodges *et al.*, 1983). The pheromone of the larger grain borer was eventually determined to comprise two compounds, 1-methylethyl (*E*)-2-methyl-2-pentenoate (trun-call 1) (Hodges *et al.*,

1984) and 1-methylethyl (*E,E*)-2,4-dimethyl-2,4-heptadienoate (trun-call 2) (Cork *et al.*, 1991) (Fig. 7.7). The structures of trun-calls 1 and 2 are strikingly similar to those of dominicalures 1 and 2, but one important difference in the biological role of the pheromones has been observed. In contrast to the situation with the lesser grain borer, age does affect the response of larger grain borers to their pheromone. Younger larger grain borers respond more quickly and stay longer in the vicinity of the pheromone than older beetles (Boughton and Fadamiro, 1996).

It is believed that trun-call 2 is the more effective and important component of the pheromone (Leos-Martinez *et al.*, 1995). A mixture of trun-calls 1 and 2 attracts ten times more *P. truncatus* than trun-call 1 alone (Dendy *et al.*, 1989a; Leos-Martinez *et al.*, 1995). In the study by Leos-Martinez *et al.* (1995), trun-call 2 alone attracted more larger grain borers than a mixture of both trun-calls (although the difference was not statistically significant). In wind tunnel experiments, the response of male and female larger grain borers to a single pheromone source containing only trun-call 2 did not differ from that containing a mixture of both trun-calls. However, males and females equally and significantly discriminated trun-call 1 from trun-call 2 when presented separately at the same time, preferring trun-call 2 over trun-call 1 (Fadamiro *et al.*, 1996a). Most recent studies by Hodges *et al.* (1998) reconfirmed the importance of trun-call 2 in flight traps for long-range attraction of *P. truncatus* but also clearly demonstrated the importance of trun-call 1 as a short-range attractant when applied in walk-in crevice traps. Thus trun-call 2 appears to be a long-range attractant and trun-call 1 is important for modifying the response at close-range orientation.

For practical applications, a higher proportion of trun-call 2 is recommended (Dendy *et al.*, 1991) because this achieves a more steady release rate, without affecting trap catch.

Cucujidae and Silvanidae

The aggregation pheromones of several economically important species of cucujid and silvanid beetles have been identified (Borden *et al.*, 1979; Wong *et al.*, 1983; Millar *et al.*, 1985a,b; Pierce *et al.*, 1984a,b, 1985). The males produce structurally similar macrocyclic lactones, which have been given the trivial name cucujolides. The aggregation pheromones identified to date include the following: (*E,E*)-4,8-dimethyl-4,8-decadien-10-olide (cucujolide I); (Z)-3-dodecen-11-olide (cucujolide II); (Z)-5-tetradecen-13-olide (cucujolide III); (Z,Z)-3,6-dodecadien-11-olide (cucujolide IV); (Z,Z)-5,8-tetradecadien-13-olide (cucujolide V); (Z)-3-dodecen-12-olide (cucujolide VIII); and (Z,Z)-3,6-dodecadien-12-olide (cucujolide IX) (Table 7.1, numbered according to Oehlschlager *et al.*, 1988). Males and females are attracted to species-specific combinations of these cucujolides (Oehlschlager *et al.*, 1987).

Cryptolestes *spp.*

The male-produced aggregation pheromone of the rusty grain beetle, *Cryptolestes ferrugineus* (Stephens), comprises cucujolides I and II (formerly known as ferrulactone

Table 7.1. Cucujolides produced as aggregation pheromones, and their distribution amongst five species of *Oryzaephilus* and *Cryptolestes* grain beetles.

Structure	Trivial name[a] Chemical name	Genus and species[b]				
		Cryptolestes ferrugineus	*Cryptolestes pusillus*	*Cryptolestes turcicus*	*Oryzaephilus mercator*	*Oryzaephilus surinamensis*
	**Cucujolide I** (*E*,*E*)-4-8-Dimethyl-4,8-decadien-10-olide	**A/Syn**				
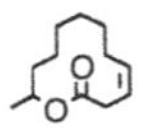	**Cucujolide II** (*Z*)-3-Dodecen-11-olide	**A/Syn** ***S***			**A** ***R***	
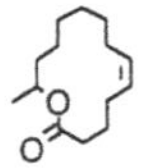	**Cucujolide III** (*Z*)-5-Tetradecen-13-olide		**Syn** ***S***	**Syn** ***R:S* = 33:67**		
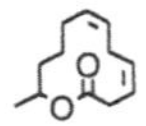	**Cucujolide IV** (*Z*,*Z*)-3,6-Dodeca-dien-11-olide				**A** ***R***	**A/Syn** ***R***
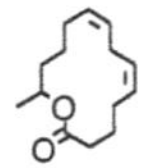	**Cucujolide V** (*Z*,*Z*)-5,8-Tetradeca-dien-13-olide			**A** ***R:S* = 85:15**		**Syn** ***R***
	**Cucujolide VIII** (*Z*)-3-Dodecen-12-olide		**A**			
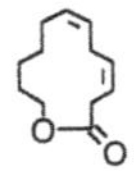	**Cucujolide IX** (*Z*,*Z*)-3,6-Dodeca-dien-12-olide					**A/Syn**

[a] The cucujolides are numbered according to Oehlschlager *et al.* (1988).
[b] A, compound is attractive alone and does not act as a synergist; A/Syn, compound is attractive alone and enhances the activity of other pheromone compounds; Syn, compound alone is not attractive but acts as a synergist; *R* and *S* indicate absolute configuration about chiral carbons.

I and II), which act synergistically (Borden *et al.*, 1979; Wong *et al.*, 1983) (Table 7.1). The males produce only the *S*-enantiomer of cucujolide II, while the *R*-enantiomer neither attracts nor repels *C. ferrugineus* (Wong *et al.*, 1983; Oehlschlager *et al.*, 1987). Approximately 990 pg and 640 pg of cucujolides I and (*S*)-II, respectively, are produced per beetle per hour (Lindgren *et al.*, 1985). EAG studies indicate that the females are more sensitive to the pheromone than the males (Chambers *et al.*, 1990). While the naturally occurring ratio of cucujolide I to (*S*)-II is about 1.6:1 (Lindgren *et al.*, 1985), the ratio that elicits the best response by EAGs (using racemic II) is 9:1 (Chambers *et al.*, 1990).

The aggregation pheromone of the flat grain beetle, *Cryptolestes pusillus* (Schönherr), is composed of cucujolides III and VIII (Quaife, 1980; Millar *et al.*, 1983, 1985a) (Table 7.1). Cucujolide VIII is active alone and is therefore considered to be the main pheromone component. At the relatively low levels at which it is naturally produced, cucujolide III does not attract beetles alone. However, when presented with cucujolide VIII in the natural ratio of about 1:20, the *S*-enantiomer of cucujolide III increases the overall response to cucujolide VIII and thus acts as a synergist. When cucujolide III is presented at higher levels, comparable to those of cucujolide VIII, it is as effective as cucujolide VIII in stimulating activity (Chambers *et al.*, 1990).

The main pheromone component of *Cryptolestes turcicus* Grouvelle is cucujolide V, with cucujolide III acting as a synergist (Millar *et al.*, 1985b) (Table 7.1). Pure *R*- or *S*-enantiomers of either cucujolides V or III are inactive. Rather, a specific ratio of the isomers is required for response to the pheromone (Millar *et al.*, 1985b; Oehlschlager *et al.*, 1987).

Oryzaephilus *spp.*

Male and female frass from the merchant grain beetle, *Oryzaephilus mercator* Fauvel, and from the sawtoothed grain beetle, *Oryzaephilus surinamensis* L. (Fig. 7.3g), contain attractive volatiles in addition to and distinct from food odours (Pierce *et al.*, 1981). Both sexes respond to the frass odours and, although some interspecific attraction was observed, both species are more sensitive to frass produced by conspecifics. The production of the aggregation odours was initially attributed to both sexes, but it was later determined that the macrolide aggregation pheromones are male-produced. The identification of the multiple compound pheromone complex in *Oryzaephilus* was reported by Pierce *et al.* (1984a,b, 1985). The pheromone components of *O. mercator* are cucujolide II and cucujolide IV, both in the *R* configuration (Oehlschlager *et al.*, 1987; Pierce *et al.*, 1987) (Table 7.1). The cucujolides are produced in a ratio of about 1:1.25, but sexes are attracted toward each compound individually, and toward a 1:1 mixture. The males appear to have a lower threshold for a positive response to cucujolide II, while females have a lower threshold for responding to cucujolide IV.

The pheromone system of *O. surinamensis* comprises the three cucujolides IV, V and IX (Table 7.1). Cucujolides IV and IX elicit low aggregation responses from either sex when presented individually, and attraction is somewhat enhanced when the two are offered in a 1:1 combination. Cucujolide V alone elicits no response, but it does enhance the response to either cucujolide IV or cucujolide IX. The combination of all three compounds elicits the strongest aggregation response from both sexes. The approximate natural ratio of cucujolides I:V:IX was reported to be

1:3:1 by Pierce *et al.* (1984a), with cucujolide V synergizing the response to the 1:1 mixture of cucujolides IV and IX. Varying the proportion of the synergist does not alter the aggregation response. A different ratio of the three cucujolides was reported to be 4.4:2:1 (White *et al.*, 1989), but the difference was attributed to strain variation. EAG recordings show no difference between the responses of the two sexes to cucujolides IV and IX (White and Chambers, 1989), but in bioassays more females were attracted when cucujolide IV was present in higher proportions.

The production of an additional pheromone compound, (*R*)-1-octen-3-ol (Fig. 7.7), was reported to occur under certain circumstances in the later stages of adulthood (Pierce *et al.*, 1989). The alcohol is produced by both sexes of *O. mercator* and *O. surinamensis* by the second month after eclosion when the population is maintained at a low density. It is attractive at dosages between 1 ng and 10 μg, but it is highly repellent in larger dosages. Interestingly, 1-octen-3-ol is also associated with food volatiles (White *et al.*, 1989), and is produced by certain fungi associated with stored grain (Pierce *et al.*, 1991a). The 1-octen-3-ol produced by low-density beetle populations is believed to extend the attractive range of the main pheromone components. Thus 1-octen-3-ol may serve as an attractive food volatile, as a beetle-produced pheromone and/or as a fungal-produced allomone (since grain beetles serve as vectors for fungi).

Recently, (*Z*)-13-oxo-9-octadecenoic acid and (*Z*)-15-oxo-11-icosenoic acid were detected in wheat flour infested by *O. surinamensis* (Nakajima *et al.*, 1996). The two compounds were not detected in fresh wheat flour, and apparently cause arrestive behaviour in the sawtoothed grain beetle. However, bioassay data have not yet been reported.

The square-necked grain beetle, Cathartus quadricollis

Pierce *et al.* (1988) identified the male-produced aggregation pheromone of the square-necked grain beetle, *Cathartus quadricollis* (Guerin-Méneville), as (*R*)-(*E*)-7-methyl-6-nonen-3-yl acetate (Fig. 7.7), and proposed the trivial name quadrilure. The pheromone is attractive to both sexes, but has a lower threshold for eliciting behaviour in females. The *S*-enantiomer neither repels nor inhibits the response to the *R*-enantiomer. Its possible role as a synergist was not clearly established.

The foreign grain beetle, Ahasverus advena

The alcohol (*R*)-1-octen-3-ol elicits aggregation of the foreign grain beetle, *Ahasverus advena* (Waltl) (Pierce *et al.*, 1991b) (Fig. 7.7). However, the pheromone is produced mainly at low population densities, and is not produced in significant amounts until 4–6 weeks post-eclosion. Once pheromone production is initiated, it is maintained for at least 11 weeks. Since this compound is produced by both sexes, Pierce *et al.* (1991b) suggested that 1-octen-3-ol may play a similar biological role in *A. advena* as it does in *Oryzaephilus* spp.: that is, to extend the attractive range of the (as yet unidentified) main aggregation pheromone component.

Biological Background of Aggregation Pheromones

The pheromone-mediated behaviours described in the preceding section can be grouped into one of two main strategies (Burkholder, 1982). Species in which the adults usually do not feed, and generally live less than 1 month, tend to use female-produced sex pheromones that attract conspecific males for the purpose of mating (Fig. 7.8a). This strategy is found in species of the Anobiidae, most Bruchidae and most Dermestidae. Species with adults that do feed and have longer life cycles tend to have male-produced aggregation pheromones that attract both sexes to sites suitable for feeding, mating and breeding (Fig 7.8b). Typically this strategy is utilized in the stored-product Bostrichidae, Cucujidae, Curculionidae, Silvanidae and Tenebrionidae. The aggregation pheromone is only produced when the adult males have direct access to food (Borden *et al.*, 1979; Millar *et al.*, 1985a,b; Walgenbach and Burkholder, 1986; Walgenbach *et al.*, 1987b; Pierce *et al.*, 1988, 1990a; Obeng-Ofori and Coaker, 1990a; Phillips *et al.*, 1993; Mayhew and Phillips, 1994; Hussain *et al.*, 1994a; Boughton and Fadamiro, 1996). The mechanism by which this process is regulated is not well understood. The simplest mechanism might be that pheromone biosynthesis is obligatorily dependent on the ingestion of precursors in the diet, but this is unlikely. For example, since some cucujolides are of fatty acid origin (Vanderwel *et al.*, 1990, 1992), and grains are excellent sources of fatty acids, it was originally believed that cucujolide biosynthesis can proceed only in the presence of dietary fatty acids (Pierce *et al.*, 1984b). In fact, however, the cucujolides can be produced through *de novo* biosynthesis, indicating that dietary fatty acids are not required as precursors in pheromone biosynthesis (Vanderwel *et al.*, 1990). Moreover, methoprene (an analogue of JH) stimulates cucujolide production in *O. mercator*, *O. surinamensus* and *C. ferrugineus* (Pierce *et al.*, 1986), indicating that cucujolide production is hormonally regulated. Nevertheless, pheromones are not produced in the absence of food and, in most cases, both beetle-produced pheromone and food volatiles are responsible for aggregations (Levinson and Levinson, 1977, 1979; Barak and Burkholder, 1984/1985; Burkholder, 1985; Burkholder and Phillips, 1988; Barak, 1989; Trematerra and Girgenti, 1989; Pinniger, 1990; Jiang, 1995; Mueller, 1995), but their relative importance varies (Lanier and Burkholder, 1974; Phillips *et al.*, 1991). Thus the males must first locate a food source before conspecifics can be attracted.

It is difficult to understand the evolutionary advantage accrued to males that produce a pheromone which attracts not only potential mates (conspecific females) but also potential competitors (conspecific males). One explanation is that the original primary function of a specifically male-produced pheromone was likely to attract a potential mate (Raffa *et al.*, 1993; Phillips, 1997; Landolt, 1997). In this scenario, the use of the pheromone by other males of the same species must have evolved secondarily and independently in the respective species, presumably to allow them to locate resources and assembled females. However, this explanation is not completely satisfactory. Firstly, one would expect that it would be difficult for intruding males to compete successfully with already established males. Male-produced sex pheromones attracting conspecific females have been reported, for example from cerambycid beetles (Sakai *et al.*, 1984; Iwabuchi *et al.*, 1987; Fettköther *et al.*, 1995) (Fig. 7.8c). Males fight over suitable sites for mating and breeding. In most cases intruding males are fought off by established males calling for females or being already *in copula* with a female (Michelsen, 1966). Secondly, given the lack of isolation mechanisms (such as calling rhythms and female courtship pheromones), one would expect higher specificity in aggregation pheromones if they originally served

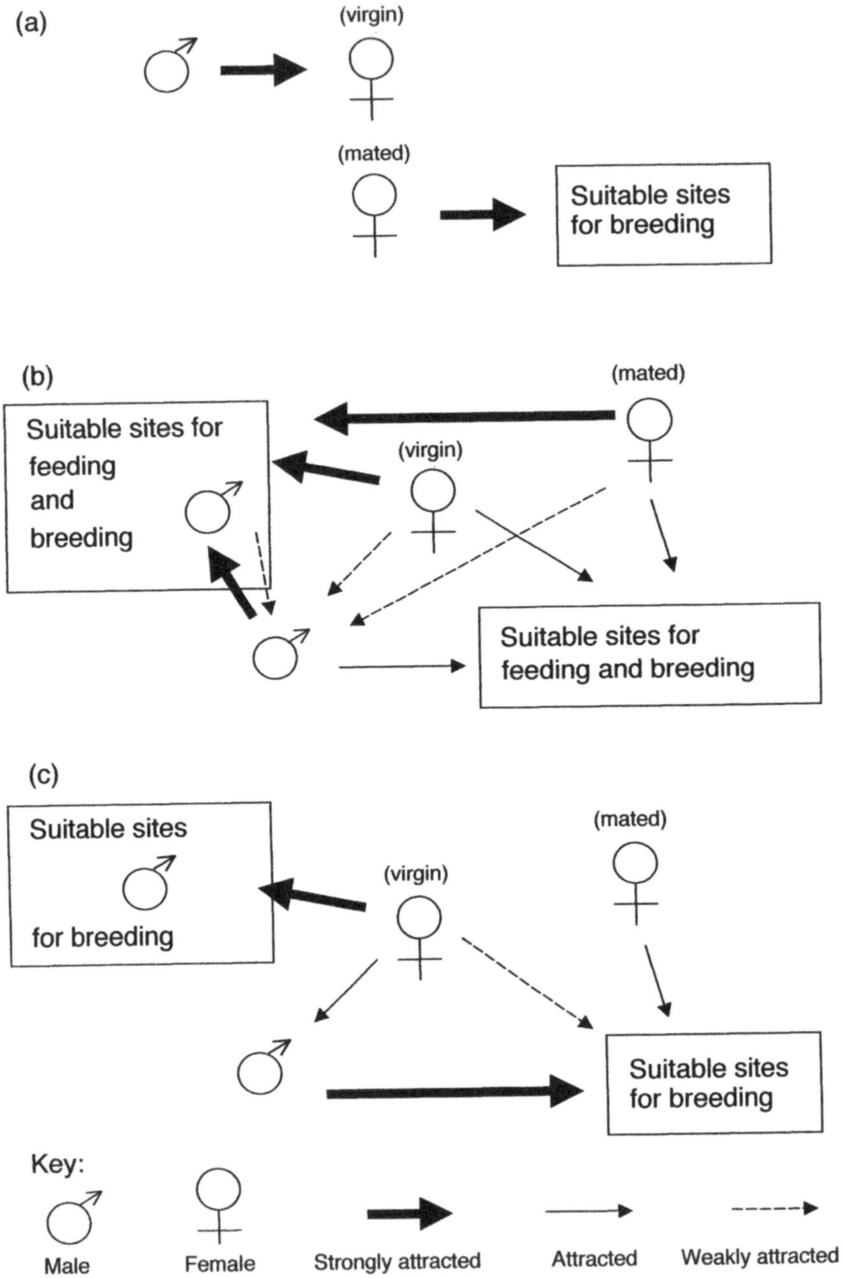

Fig. 7.8. Attraction pattern between the sexes brought about by (a) female-produced sex pheromones found in *Stegobium, Lasioderma, Attagenus, Anthrenus, Trogoderma, Callosobruchus* and *Zabrotes* species; (b) male-produced aggregation pheromones found in *Tribolium, Sitophilus, Rhyzopertha, Prostephanus, Cryptolestes, Oryzaephilus* and *Cathartus* species; (c) male-produced sex pheromones found in *Acanthoscelides* and several Cerambycidae species.

only as mating attractants. In fact, the utilization of the same compounds by multiple species, resulting in cross-attraction, is not uncommon (Faustini *et al.*, 1982a; Walgenbach *et al.*, 1983; Hodges *et al.*, 1983, 1984; Suzuki *et al.*, 1987, 1988; Chambers *et al.*, 1990). Thirdly, beetles habituate to high concentrations of 'true sex pheromones', but are not repelled by them as they are by high concentrations of aggregation pheromones. Fourthly, mating should reduce responsiveness toward a sex attractant. However, mating did not affect the response of *T. confusum*, *P. truncatus* or *R. dominica* females toward the aggregation pheromone (Obeng-Ofori and Coaker, 1990a; Dowdy *et al.*, 1993) (Fig. 7.8b). Moreover, when given a choice between plain wheat, and wheat with conspecific males, in a dual choice olfactometer, mated females of *S. granarius* prefer wheat with males (Plarre, 1991). Even if the male-produced aggregation pheromone is used as a signal for suitable feeding and egg-laying sites by the mated females, the fresh wheat should be preferred to avoid feeding competition by adults. There must be a reason why mated females respond to aggregation pheromones. If the pheromone is a signal for a favourable microclimate (discussed below), mated females that approach a source of aggregation pheromone would ensure favourable abiotic conditions for their brood.

Nevertheless, there is some evidence that females have a slightly different response than males to the aggregation pheromone of conspecific males. Findings of White and Chambers (1989) suggest that the aggregation pheromone of *O. surinamensis* functions partly as a sex pheromone, since an increase in one component causes an increase in female response and males were able to vary the composition of their pheromone. Tribolure elicited a stronger response from females of *T. castaneum* than from males (Faustini *et al.*, 1981), and male granary weevils elicited a stronger response from virgin females than from males (Plarre, 1991). Pheromone traps for *P. truncatus* and *R. dominica* caught significantly more females than males (Hodges *et al.*, 1998; T.W. Phillips, Wageningen, 1998, personal communication). However, these differences might be caused by different behaviour-eliciting thresholds: in general, these should be lower for females since males, as pheromone emitters, are constantly surrounded by pheromone and thus are habituated to low levels of the attractant chemical.

An alternative explanation for the production of aggregation pheromones is that the primary role of the male-produced aggregation pheromone is to attract conspecifics in order to overcome host defences or unsuitable microclimates. Since males or females can serve equally well in this respect, the pheromone emitter benefits from attracting beetles of either sex. Attracting mates with aggregation pheromones is of secondary importance, since adults of the species that use aggregation pheromones have relatively long life cycles. This system is analogous to the well-studied aggregation pheromone systems which are found in the bark beetle family Scolytidae (see also Chapter 6, this volume). During the first phase of an aggregation at a host tree, the attractant pheromone emitted by the pioneer beetles recruits enough conspecifics to overcome host tree resistance and defence (Borden, 1974; Raffa *et al.*, 1993). Only the number of individuals arriving at the host tree in a short time, regardless of their sex, is relevant to accomplish this primary task. In order to infest stored commodities successfully, stored-product beetles certainly do not have to overcome host defence systems. However, stored-product Coleoptera also occur outdoors, away from storage facilities (see below). Non-stored-product species closely related to post-harvest pests in the Tenebrionidae, Curculionidae and Cucujidae feed and breed on physiologically active host-plant parts (Weidner, 1983b; Mayer and McLaughlin, 1991; Levinson and Levinson, 1994; Valamoti and Buckland, 1995). The beetles most likely rely on aggregation pheromones to weaken the host and make the habitat more suitable. In the storage environment, where seeds and other harvested products are usually stored under low temperatures and rather dry conditions, the aggregation pheromones can be regarded as functioning to congregate a large number of individuals for the purpose of locally altering the microclimate. Rising humidities and temperatures have been recorded from hot spots formed by stored-product beetles and weevils (Frey, 1952; Longstaff, 1981; Sinha, 1984). These changes in the microclimate are partly caused by the by-products of insect metabolism (Singh *et al.*, 1976) and more favourable conditions are restricted to the infestation pockets. Heat loss through conduction to the periphery is minimal (Howe, 1962). Depending on the initial grain temperature, the moisture content and the population size, the metabolic heat production of *C. ferrugineus* can range from 0.72 μW to 21.47 μW per insect (Cofie-Agblor *et al.*, 1996a,b). *S. granarius* showed maximum heat production of 81.4 μW (Cofie-Agblor *et al.*, 1995). The importance of higher relative humidities for a faster larval development with fewer larval instars was demonstrated by Pittendrigh *et al.* (1997) for *S. oryzae*. Thus, aggregation pheromones could have been a preadaptive necessity for certain species, which made it possible for them to succeed in infesting, feeding and surviving in stored commodities as adult insects.

Beetles orient differently to an aggregation pheromone than they do to a sex pheromone. The orientation to an aggregation pheromone source can best be described as orthokinesis. An insect that exhibits orthokinesis moves rapidly when exposed to lower concentrations of the odour, but slowly when exposed to higher concentrations, ending up aggregating near the source (Shorey, 1973). In olfactometer studies, Obeng-Ofori (1991) recorded that the response of *T. confusum* and *T. castaneum* to a commercially available rubber septum lure increased with increasing doses of pheromone (from 1.0 to 4.0 mg). Walking speed increased as the beetles walked up the odour plume, decreased as they approached the source, and was lowest when arriving at the odour origin (Obeng-Ofori and Coaker, 1990b). Similarly, *S. granarius* males and females were observed to aggregate in the vicinity of a pheromone dispenser, but not at the actual source (Plarre, 1994). Aggregation pheromones should therefore be regarded not so much as attractants, like sex pheromones, but more as arrestants (Dethier *et al.*, 1960; Evans, 1984). As opposed to locating a mate with a sex pheromone, locating the origin of an aggregation pheromone source is not necessarily required. Rather, assembling in the vicinity of an aggregation is sufficient for an individual to benefit from a better microclimate. In fact, deposition of eggs in the centre of an aggregation might endanger the offspring due to the presence of feeding adults. When selecting oviposition sites, mated females must strike a balance between favourable humidity and temperature, and security for the offspring. The optimal location is on the periphery of 'hot spots'. This may explain why grain samples taken from the surroundings of a trap baited with the aggregation pheromone contained more weevil progeny than grain samples taken from the surroundings of unbaited traps (Plarre, 1996).

The process by which mates are located in aggregations is not well understood. Wojcik (1969) described the mating behaviour of several different stored-product beetles, including those that utilize aggregation pheromones, such as *T. castaneum*, *T. confusum*, *O. mercator*, *C. pusillus* and *S. granarius*. In all cases, the male approached the female and usually mounted her after physical contact, but the male did not appear to perceive the female from a distance. Since males are the more active sex (Surtees, 1964; Cox *et al.*, 1990, 1993; Plarre, 1996), encountering females in aggregation spots is not unlikely. In no cases reported did the female approach the male. For example, mating in *R. dominica* and *S. zeamais* was initiated by the male after physical contact occurred, but no obvious attraction to the female was observed (Khorramshahi and Burkholder, 1981; Walgenbach and Burkholder, 1987; Zernecke, 1993). Males were observed to try to mate with other males as well, making sex differentiation prior to contact unlikely. Volatile female chemical stimuli attractive only to males are reported in just two species which have male-produced aggregation pheromones: *T. confusum* (Ryan and O'Ceallachain, 1976; O'Ceallachain and Ryan, 1977) and *S. oryzae* (Sharma and Deora, 1980). Males of *T. castaneum* are somehow able to distinguish between mated and unmated females. When males of the red flour beetle are provided with equal access to both virgin and mated females, they initiate contact and copulate more frequently with the virgin females (Lewis and Iannini, 1995). Smith *et al.* (1996) found that male *P. truncatus* reduce pheromone production in the presence of females or some chemical residues left by females on the substrate. This unique suppression of pheromone production was discussed as being a mechanism to limit intra-male competition once females were attracted to suitable breeding sites.

Aggregation pheromones also appear to play a role in regulating population density (Pierce *et al.*, 1990b; Mignon *et al.*, 1996). High population densities cause reduced fecundity, depressed oviposition and increased mortality (Crombie, 1942). High concentrations of aggregation pheromone, as produced by high-density populations, are repelling. Moreover, beetles from high-density populations are less strongly attracted to (or may even be repelled by) the aggregation pheromone. For example, young *O. surinamensis* males, maintained at lower population densities, produced about twice as much pheromone as males maintained at higher population densities (Pierce *et al.*, 1989). 1-Octen-3-ol, an additional pheromone component in *Oryzaephilus* spp., is only produced after some time at low population densities, when it probably acts as a long-range attractant. Young male and female beetles reared at high population densities are not attracted as strongly to male volatiles or to synthetic aggregation pheromones, and may even become unresponsive to or repelled by these odours (Pierce *et al.*, 1983, 1984a). For *S. granarius*, high concentrations of synthetic aggregation pheromone offered in rubber septa, mimicking high weevil densities, are repellent to adults of either sex (Plarre, 1994). Similar population density effects are observed in *P. truncatus* (Fadamiro *et al.*, 1996b). Beetles from high-density cultures are more inclined to fly and disperse than beetles from low-density cultures.

Application of Pheromones for Pest Control

The use of pheromones is becoming more and more important in implementing the concept of IPM in the stored product environment. By far the most common application of beetle pheromones is their use in traps to monitor the presence, abundance and distribution of a pest species. Pheromone traps have also been used for mass-trapping to deplete populations, unfortunately with only limited success. The use of pheromones for mating disruption has not been adequately investigated, but may be possible. The relatively new approach of lure (or attract) and kill has shown great promise. Each of these strategies will be examined in more detail below.

Use of pheromones to monitor pest populations

Pheromone traps have a lower detection threshold than any other kind of trap or visual inspection. When no obvious insect activity could be observed in a cargo terminal, only with pheromone traps was it possible to demonstrate the existence of *Attagenus* spp. and *Trogoderma* spp., and their potential for infestation (Barak and Burkholder, 1976). In another example, sex pheromone traps were more sensitive than ultraviolet light traps in detecting, estimating and pinpointing infestations of *L. serricorne* in tobacco stores (Faustini *et al.*, 1990). Pheromone traps were also very effective in monitoring cigarette beetle populations in Australian museum collection storage areas (Gilberg and Roach, 1991). It was possible to identify and locate infested objects throughout the collection storage area at a very early stage. Pierce (1994) designed a fixed grid triangulation technique based on the inverse relationship between the number of captured insects per pheromone trap and the relative proximity of the infestation. This technique was successfully employed to locate hidden infestations of *L. serricorne* in various food facilities. A 3-year trapping study of *L. serricorne* in Greece revealed that the cigarette beetle probably occurs in three to four generations per year, with a peak during the mid to late summer months (Levinson and Buchelos, 1988). The population density was usually very low during winter, and continuous trapping kept the population at a stable low level without the use of insecticides. Similar population dynamics are reported from noodle factories in Japan, where *L. serricorne* and *S. paniceum* were trapped from May to November with four peaks of population outbreaks (Suezawa *et al.*, 1987).

Outdoor trapping and monitoring of stored-product pests with pheromone traps can give valuable information on the relative abundance of the pest in natural environments, its geographical distribution, and areas of retreat when no stored food is available. For example, the spread of the larger grain borer, *P. truncatus*, native to Central and South America and recently accidentally introduced with maize imports to Africa, has been monitored continuously with pheromone traps (Richter and Biliwa, 1991). Mark–release–recapture studies revealed maximum natural dispersion of *P. truncatus* in Mexico to be 250–340 m in 24 h, depending on the wind speed and direction (Rees *et al.*, 1991; Farell and Key, 1992). Peak flight activity occurred mainly at dusk and, to some extent, at dawn (Tigar *et al.*, 1993). *P. truncatus* was also caught far away from agricultural sites, possibly using the natural vegetation as suitable feeding and breeding sites. This insect was recorded in pheromone traps out in the tropical deciduous forest of the Pacific Coast of Mexico (Ramirez-Martinez *et al.*, 1994). The larger grain borer's natural antagonist, the histerid beetle *Teretriosoma nigrescens* Lewis, is a highly specific predator that uses the aggregation pheromone of *P. truncatus* as a kairomonal cue to locate its prey. This makes it possible to monitor the presence, abundance and general distribution of *T. nigrescens* with *P. truncatus* pheromone-baited traps (Rees *et al.*, 1991; Boeye *et al.*, 1992; Key *et al.*, 1994; Tigar *et al.*, 1994). *T. nigrescens* was recently released in Africa as a biological control agent against the larger grain borer. The distribution of this beneficial histerid beetle is now monitored by traps baited with trun-calls 1 and 2 (Helbig *et al.*, 1992). Outdoor trapping of *R. dominica* in Canada revealed that the lesser grain borer can spread to uninfested grain storage facilities by flying long distances (Fields *et al.*, 1993). It was believed that the lesser grain borer is mainly a pest of the warmer regions of the world and would occur in Canada only with infested grain imports. Peak flight activity of *R. dominica* was shown to be mainly at sunset with a smaller activity peak around sunrise (Leos-Martinez *et al.*, 1987).

Physical characteristics like the outer shape and colour of a trap have major influences in catching flying insects. Using identical lures, differences in trap effectiveness were shown to be based on either the traps' visual cues or their accessibility to the pest insects (Dendy *et al.*, 1989b; Javer *et al.*, 1990; Mullen, 1992; Fargo *et al.*, 1994; Hussain *et al.*, 1994b). For aerial trapping of *R. dominica* and *P. truncatus*, a pheromone trap consisting of several funnels emptying into one another and into a collecting bucket was shown to be more catch-effective than other geometrical designs (Leos-Martinez *et al.*, 1986, 1987; Key *et al.*, 1994). The visual attraction to the vertical silhouette of a trap could reflect a behavioural pattern of orienting toward tree trunks, where most stored-product insects breed naturally (Trematerra and Daolio, 1990).

In general, the effectiveness of a pheromone trap is highly dependent on the temperature. Approximately ten times as many *C. ferrugineus* were trapped at higher temperatures than at lower temperatures using a mixture of cucujolides I and II released from hollow polyester fibres (Loschiavo *et al.*, 1986). In areas where the ambient temperatures never exceeded 19°C, trap catches in both the pheromone-baited and untreated control traps were low and their differences were not significant. Higher temperatures promote higher locomotion activity of the insects and a higher release rate of the pheromones (Bradley *et al.*, 1995). In many cases a thorough screening of optimal pheromone concentrations, release rates and type of dispenser might be necessary before a trapping plan can be effective. For *C. ferrugineus* and *T. castaneum*, for example, release rates of 1.25 mg per day and 0.08 mg per day, respectively, were shown to be best for capturing these beetles (Lindgren *et al.*, 1985). Another important consideration for the successful trapping of insects in the field is that different strains of a beetle species may respond differently to various pheromone concentrations. This was observed in *T. castaneum*, when strain-specific females significantly discriminated between males from different populations in two-choice pitfall bioassays, by favouring males from their own strain (Boake and Wade, 1984). Through artificial selection of *T. granarium* in the laboratory, it was possible to breed a population with the male response to the female sex pheromone reduced by 74% (Rahalkar *et al.*, 1985). Compared with the original stock population, a fivefold increase in pheromone concentration was required for maximal attraction.

The commercial use of some pheromones has not been exploited, due to economic factors: pheromones with complex structures may be difficult to synthesize, so their use may not be cost-effective. It may, in some cases, be possible to overcome this problem by reducing the quantity of the expensive pheromone component required in the trap through the judicious use of pheromone synergists and/or other attractive compounds (e.g. food volatiles). For example, Pierce *et al.* (1990c) were able to reduce by 100-fold the quantity of cucujolides required to trap *O. surinamensis* with the addition of the relatively inexpensive synergist, 1-octen-3-ol. A subsequent study revealed that 1-octen-3-ol may, in some instances, be able to replace the cucujolides for use in traps to monitor *C. ferrugineus*, *C. pusillus* and *O. mercator* (Mushobozy *et al.*, 1993). The combination of pheromones and food attractants was shown to improve insect catch for monitoring purposes (Levinson and Levinson, 1977, 1979; Barak and Burkholder, 1984/1985; Burkholder, 1985, 1988a,b; Burkholder and Phillips, 1988; Barak, 1989; Pinniger, 1990; Mueller, 1995; Wakefield, 1999).

Use of pheromones to mass-trap pest populations

Pheromone traps have been used less frequently for mass-trapping stored-product beetles than for monitoring purposes. Nevertheless, attempts to mass-trap *L. serricorne* in tobacco stores in Greece using a multisurface trap baited with anhydroserricornin showed promising results (Buchelos and Levinson, 1993). Over half a million cigarette beetles were caught on pheromone traps during a 15-month trapping period. However, a control of the pest beyond the level of insectistasis (Levinson, 1983) was not achieved, probably due to insufficient storage management, which made continuous reinfestation with cigarette beetles from the outside possible. In another study, 5 years of continuous mass-trapping of *L. serricorne* was found to be a cost-effective alternative to general fogging with insecticides to suppress the pest populations in two food warehouses in Hawaii (Pierce, 1994). Where possible, mass-trapping both sexes of a population using aggregation pheromones should be more effective than mass-trapping only one sex with sex pheromones (Trematerra, 1997). However, due to the different orientation behaviour that is elicited by aggregation pheromones as compared with sex pheromones (see above), effective mass-trapping of *Rhyzopertha* spp., *Sitophilus* spp. and *Tribolium* spp. has gained only limited success (Sinclair and Howitt, 1984). A strategy worthy of consideration might be to surround a pheromone bait with unbaited traps to catch beetles which are attracted to the pheromone but arrest in the vicinity of the pheromone emission source.

Use of pheromones to disrupt mating

Little is known about mating disruption and habituation effects of pheromones in stored-product beetles. In *T. granarium*, continuous exposure of virgin males to the sex pheromone for 2 days prior to the release of females resulted in failure of mating, even though the females were in the close vicinity of the males (Rahalkar *et al.*, 1972). Prior exposure of male *T. inclusum* to the sex pheromone resulted in diminished response for at least 4.5 h to subsequent doses of the pheromone (Vick *et al.*, 1973b). The degree of habituation was dose dependent.

Mating disruption by applying inhibitory compounds would have an advantage over mating disruption with high dosages of a sex attractant, since males from outside the treated facility would not be affected. Permeating a storage facility with sex pheromones can attract males from the surroundings, which would increase the number of males inside the facility. Due to these additional potential mates, mating by chance becomes more likely. Contamination of the commodity with additional individuals from the outside could be avoided with potent inhibitors. Potential mating inhibitory compounds are known for *L. serricorne* (Levinson and Levinson, 1987) and *S. paniceum* (Kodama *et al.*, 1987a).

Use of pheromones to attract and kill

The potential of the 'attract and kill method' was demonstrated by pheromone-baited traps combined with an inoculation device that contained a pathogen (Burkholder, 1980, 1981b). Adult *T. glabrum* males were successfully drawn to the sex pheromone source containing the pathogenic protozoan *Mattesia* spp. and became externally contaminated with the pathogen (Burkholder and Boush, 1974). After leaving the treated attractive site they were able to transmit infective spores to females and subsequently spread the disease throughout 96% of the test population. Under simulated warehouse conditions, following generations of *T. glabrum* were suppressed after a single introduction of *Mattesia trogodermae* Canning into dense, adult male populations via pheromone-baited spore-transfer sites (Shapas *et al.*, 1977). Population growth in the first generation after treatment was reduced to one-sixth compared with the untreated control population, and fell to below pre-treatment level by the second generation. Despite this success, no follow-up experiments or field tests were undertaken. The concern of public health departments regarding the release of pathogens may have limited the development of this practical control technique.

Conclusion

Pheromones have become established as valuable tools in stored-product IPM over the last few decades. Most improvements in different pheromone trapping strategies were based not only on laboratory studies but also on field experiments. Data collected in warehouses, granaries, food manufacturing companies, etc., add valuable knowledge to our understanding of pheromone biology in stored-product beetles. Since many different parameters can make a great difference in the efficacy of a pheromone trap, their precise description should be given when communicating pheromone trapping results in order to compare and obtain valuable qualitative and quantitative research data gathered in field or semi-field situations (Wright and Cogan, 1995). In addition to the use of sex and aggregation pheromones, the use of other semiochemicals – such as oviposition pheromones, allomones and kairomones – will become more and more important in the management of stored-product beetles. Most of these semiochemicals are as yet unknown. Research into the role and use of kairomones in IPM is just beginning and will provide more fundamental insights into insect behaviour (Lewis and Martin, 1990; Metcalf, 1994).

Acknowledgements

The authors would like to thank the staff of Steenbock Library, University of Wisconsin in Madison and of the Central Library of the Federal Biological Research Centre for Agriculture and Forestry in Berlin for their help in researching and providing literature. We would like to thank W.E. Burkholder (USDA–ARS) and J. Chambers (CSL) for careful reviews of this chapter and helpful suggestions.

References

Alexander, P. and Barton, D.H.R. (1943) The excretion of ethylquinone by the flour beetle. *Biochemical Journal* 37, 463–465.

Ansell, J.M., Hassner, A. and Burkholder, W.E. (1979) The synthesis of the drugstore beetle pheromone (stegobinone). *Tetrahedron Letters* 27, 2497–2498.

August, C.J. (1971) The role of male and female pheromones in the mating behaviour of *Tenebrio molitor*. *Journal of Insect Physiology* 17, 739–751.

Bar Ilan, A.R., Stanic, V. and Shulov, A. (1965) Attracting substance (pheromone) produced by virgin females of *Trogoderma granarium* Everts (Col., Dermestidae). *Rivesta Parasitologica* 26, 27–29.

Barak, A.V. (1989) Development of a new trap to detect and monitor khapra beetle (Coleoptera, Dermestidae). *Journal of Economic Entomology* 82, 1470–1477.

Barak, A.V. and Burkholder, W.E. (1976) Trapping studies with dermestid sex pheromones. *Environmental Entomology* 5, 111–114.

Barak, A.V. and Burkholder, W.E. (1977) Behavior and pheromone studies with *Attagenus elongatulus* Casey (Coleoptera, Dermestidea). *Journal of Chemical Ecology* 3, 219–237.

Barak, A.V. and Burkholder, W.E. (1978) Interspecific response to sex pheromones, and calling behavior of several *Attagenus* species (Coleoptera, Dermestidae). *Journal of Chemical Ecology* 4, 451–461.

Barak, A.V. and Burkholder, W.E. (1984/1985) A versatile and effective trap for detecting and monitoring stored-product Coleoptera. *Agriculture, Ecosystems and Environment* 12, 207–218.

Barratt, B.I.P. (1974) Timing of production of a sex pheromone by females of *Stegobium paniceum* (L.) (Coleoptera, Anobiidae) and factors affecting male response. *Bulletin of Entomological Research* 64, 621–628.

Biemont, J.C., Chauvin, G. and Hamon, C. (1990) Morphology and ultrastructure of the abdominal integumentary glands of *Acanthoscelides obtectus* Say (Coleoptera, Bruchidae). *International Journal of Insect Morphology and Embryology* 19, 1–11.

Boake, C.R. and Wade, M.J. (1984) Populations of the red flour beetle *Tribolium castaneum* (Coleoptera: Tenebrionidae) differ in their sensitivity to aggregation pheromones. *Environmental Entomology* 13, 1182–1185.

Boeye, J., Laborius, G.A. and Schulz, F.A. (1992) The response of *Teretriosoma nigrescens* Lewis (Col.: Histeridae) to the pheromone of *Prostephanus truncatus* (Horn) (Col.: Bostrichidae). *Anzeiger für Schädlingskunde, Pflanzenschutz und Umweltschutz* 65, 153–157.

Borden, J.H. (1974) Aggregation pheromones in the Scolytidae. In: Birch, M.C. (ed.) *Pheromones*. North Holland, Amsterdam, pp. 135–160.

Borden, J.H., Dolinski, M.G., Chong, L., Verigin, V., Pierce, H.D. Jr and Oehlschlager, A.C. (1979) Aggregation pheromone in the rusty grain beetle, *Cryptolestes ferrugineus* (Coleoptera, Cucujidae). *Canadian Entomologist* 111, 681–688.

Boughton, A. and Fadamiro, H.Y. (1996) Effect of age and sex on the response of walking *Prostephanus truncatus* (Horn) (Coleoptera, Bostrichidae) to its male-produced aggregation pheromone. *Journal of Stored Products Research* 32, 13–20.

Bradley, S.J., Suckling, D.M., McNaughton, K.G., Wearing, C.H. and Karg, G. (1995) A temperature-dependent model for predicting release rates of pheromone from a polyethylene tubing dispenser. *Journal of Chemical Ecology* 21, 745–760.

Buchelos, C.T. and Levinson, A.R. (1993) Efficacy of multisurface traps and Lasiotraps with and without pheromone addition, for monitoring and mass-trapping of *Lasioderma serricorne* F. (Col., Anobiidae) in insecticide-free tobacco stores. *Journal of Applied Entomology* 116, 440–448.

Burkholder, W.E. (1970) Pheromone research with stored-product Coleoptera. In: Wood, D.L., Silverstein, R.M. and Nakajima, M. (eds) *Control of Stored Product Insects by Natural Products*. Academic Press, New York, pp. 1–20.

Burkholder, W.E. (1980) Linking insect-behavior-modifying chemicals and pathogens. In: *Proceedings of the Workshop on Insect Pest Management with Microbial Agents: Recent Achievements, Deficiencies, and Innovations*. Ithaca, New York, p. 50.

Burkholder, W.E. (1981a) Biomonitoring for stored-product insects. In: Mitchell, E.R. (ed.) *Management of Insect Pests with Semiochemicals*. Plenum Press, New York, pp. 29–40.

Burkholder, W.E. (1981b) Biological suppression of stored-product insect pests. In: Papavizas, G.C. (ed.) *Biological Control in Crop Protection*. BARC Symposium, Allanheld, Osmun, Totowa, pp. 391–399.

Burkholder, W.E. (1982) Reproductive biology and communication among grain storage and warehouse beetles. *Journal of the Georgia Entomological Society* 17, 1–10.

Burkholder, W.E. (1985) Pheromones, food attractants, traps and behavioral information for insect control. *Association of Operative Millers – Bulletin*, November 1985, 4609–4611.

Burkholder, W.E. (1988a) Management of stored product insects with pheromones. In: *Proceedings of the National Conference on Urban Entomology*, Maryland, USA, pp. 77–85.

Burkholder, W.E. (1988b) Some new lures, traps and sampling techniques for monitoring stored-product insects. In: *Proceedings of the International Congress of Entomology*, Vancouver, Canada, p. 444.

Burkholder, W.E. (1990) Practical use of pheromones and other attractants for stored product insects. In: Ridgway, R.L., Silverstein, R.M. and Inscoe, M.N. (eds) *Behavior-Modifying Chemicals for Insect Management Applications of Pheromones and Other Attractants*. Marcel Dekker, New York, pp. 497–516.

Burkholder, W.E. and Boush, G.M. (1974) Pheromones in stored product insect trapping and pathogen dissemination. *Bulletin of the OEPP* 4, 455–461.

Burkholder, W.E. and Dicke, R.J. (1966) Evidence of sex pheromones of several species of Dermestidae. *Journal of Economic Entomology* 59, 540–543.

Burkholder, W.E. and Faustini, D.L. (1991) Biological methods of survey and control. In: Gorham, R.J. (ed.) *Ecology and Management of Food*. AOAC Press, Arlington, Virginia, pp. 361–372.

Burkholder, W.E. and Ma, M. (1985) Pheromones for monitoring and control of stored-product insects. *Annual Review of Entomology* 30, 252–272.

Burkholder, W.E. and Phillips, J.K. (1988) Trapping techniques for dermestid and anobiid beetles. In: Association of Systematics Collections (ed.) A *Guide to Museum Pest Control*. Foundation of the American Institute for Conservation of Historic and Artistic Works and the Association of Systematics Collections, Washington, DC, pp. 109–111.

Burkholder, W.E., Ma, M., Kuwahara, Y. and Matsumura, F. (1974) Sex pheromone of the furniture carpet beetle, *Anthrenus flavipes* (Coleoptera, Dermestidae). *Canadian Entomologist* 106, 835–839.

Chambers, J. (1990) Overview on stored-product insect pheromones and food attractants. *Journal of the Kansas Entomological Society* 63, 490–499.

Chambers, J., Morgan, C.P., White, P.R., Mori, K., Finnegan, D.E. and Pinniger, D.B. (1990) Rust-red grain beetle, *Cryptolestes ferrugineus*, and flat grain beetle, *Cryptolestes pusillus*: antennal and behavioral responses to synthetic components of their aggregation pheromones. *Journal of Chemical Ecology* 16, 3353–3372.

Chambers, J., Van Wyk, C.B., White, P.R., Gerrard, C.M. and Mori, K. (1996) Grain weevil, *Sitophilus granarius* (L.): antennal and behavioral responses to male-produced volatiles. *Journal of Chemical Ecology* 22, 1639–1654.

Chuman, T., Kohno, M., Kato, K. and Noguchi, M. (1979) 4,6-Dimethyl-7-hydroxy-nonan-3-one, a sex pheromone of the cigarette beetle (*Lasioderma serricorne* F.). *Tetrahedron Letters* 25, 2361–2364.

Chuman, T., Mochizuki, K., Kato, K., Ono, M. and Okubo, A. (1983) Serricorone and serricorole, new sex pheromone components of the cigarette beetle. *Journal of Agricultural and Biological Chemistry* 47, 1413–1415.

Chuman, T., Mochizuki, K., Mori, M., Kohno, M., Kato, K. and Noguchi, M. (1985) *Lasioderma* chemistry: sex pheromone of cigarette beetle (*Lasioderma serricorne* F.). *Journal of Chemical Ecology* 11, 417–434.

Coffelt, J.A. and Burkholder, W.E. (1972) Reproductive biology of the cigarette beetle, *Lasioderma serricorne*. 1. Quantitative laboratory bioassay of the female sex pheromone from females of different ages. *Annals of the Entomological Society of America* 65, 447–450.

Cofie-Agblor, R., Muir, W.E. and Sinha, R.N. (1995) Comparative heat of respiration of five grain beetles in stored wheat. *Postharvest Biology and Technology* 5, 167–175.

Cofie-Agblor, R., Muir, W.E., Zhang, Q. and Sinha, R.N. (1996a) Heat of respiration of *Cryptolestes ferrugineus* (Stephens) adults and larvae in stored wheat. *Canadian Agricultural Engineering* 38, 37–44.

Cofie-Agblor, R., Muir, W.E., Sinha, R.N. and Fields, P.G. (1996b) Heat production by adult *Cryptolestes ferrugineus* (Stephens) of different ages and densities. *Postharvest Biology and Technology* 7, 371–380.

Cogburn, R.R., Burkholder, W.E. and Williams, H.J. (1984) Field tests with the aggregation pheromone of the lesser grain borer (Coleoptera, Bostrichidae). *Environmental Entomology* 13, 162–166.

Cohen, E., Ikan, R. and Shulov, A. (1971) Response of the females of *Trogoderma granarium* (Coleoptera, Dermestidae) to cuticular free fatty acids. *Entomologia Experimentalis et Applicata* 14, 315–318.

Cork, A., Hall, D.R., Hodges, R.J. and Pickett, J.A. (1991) Identification of major component of male-produced aggregation pheromone of larger grain borer, *Prostephanus truncatus* (Horn) (Coleoptera, Bostrichidae). *Journal of Chemical Ecology* 17, 789–803.

Cox, P.D., Parish, W.E. and Ledson, M. (1990) Factors affecting the refuge-seeking behaviour of *Cryptolestes ferrugineus* (Stephens) (Coleoptera, Cucujidae). *Journal of Stored Products Research* 26, 169–174.

Cox, P.D., Atkinson, J.E., Bannon, K.L. and Hatfield, J.H. (1993) Studies of refuge-seeking behaviour and locomotory activity in *Cryptolestes ferrugineus* (Stephens) (Coleoptera, Cucujidae) using time-lapse video recordings. *Journal of Stored Products Research* 29, 283–289.

Credland, P.F. and Wright, A.N. (1990) Oviposition deterrents of *Callosobruchus maculatus* (Coleoptera, Bruchidae). *Journal of Physiological Entomology* 15, 285–298.

Crombie, A.C. (1942) The effect of crowding upon the oviposition of grain-infesting insects. *Journal of Experimental Biology* 19, 311–340.

Cross, H.J., Byler, R.C., Cassidy, R.S., Silverstein, R.M., Greenblatt, R.E., Burkholder, W.E., Levinson, A.R. and Levinson, H.Z. (1976) Porapak-Q collection of pheromone components and isolation of (Z)- and (E)-14-methyl-8-hexadecenal, sex pheromone components, from the females of four species of *Trogoderma* (Coleoptera, Dermestidae). *Journal of Chemical Ecology* 2, 457–468.

Cross, H.J., Byler, R.C., Silverstein, R.M., Greenblatt, R.E., Gorman, J.E. and Burkholder, W.E. (1977) Sex pheromone components and calling behavior of female dermestid beetle, *Trogoderma variabile* Ballion (Coleoptera, Dermestidae). *Journal of Chemical Ecology* 3, 115–125.

Dendy, J., Dobie, P., Saidi, J.A., Smith, J. and Uronu, B. (1989a) Trapping the larger grain borer *Prostephanus truncatus* in maize fields using synthetic pheromones. *Entomologia Experimentalis et Applicata* 50, 241–244.

Dendy, J., Dobie, P., Saidi, J.A. and Sherman, C. (1989b) The design of traps for monitoring the presence of *Prostephanus truncatus* (Horn) (Coleoptera: Bostrichidae) in maize fields. *Journal of Stored Product Research* 25, 187–191.

Dendy, J., Dobie, P., Saidi, J.A., Smith, J. and Uronu, B. (1991) Trials to assess the effectiveness of new synthetic pheromone mixtures for trapping *Prostephanus truncatus* (Horn) (Coleoptera, Bostrichidae) in maize stores. *Journal of Stored Products Research* 27, 69–74.

Dethier, V.G., Barton-Browne, L. and Smith, C.N. (1960) The designation of chemicals in terms of the responses they elicit from insects. *Journal of Economic Entomology* 53, 134–136.

Dowdy, A.K., Howard, R.W., Seitz, L.M. and McGaughey, W.H. (1993) Response of *Rhyzopertha dominica* (Coleoptera, Bostrychidae) to its aggregation pheromone and wheat volatiles. *Environmental Entomology* 22, 965–970.

Evans, D.E. (1984) Biological control of stored grain. In: Champ, B.R. and Highley, E. (eds) *Proceedings of the Australian Development Assistance Course on the Preservation of Stored Cereals*. CSIRO Division of Entomology, Canberra, Australia, pp. 574–582.

Fadamiro, H.Y., Wyatt, T.D. and Hall, D.R. (1996a) Behavioural response of *Prostephanus truncatus* (Horn) (Coleoptera, Bostrichidae) to the individual components of its pheromone in the flight tunnel: discrimination between two odour sources. *Journal of Stored Products Research* 32, 163–170.

Fadamiro, H.Y., Wyatt, T.D. and Birch, M.C. (1996b) Flight activity of *Prostephanus truncatus* (Horn) (Coleoptera, Bostrichidae) in relation to population density, resource quality, age, and sex. *Journal of Insect Behavior* 9, 339–351.

Farell, G. and Key, G.E. (1992) Flight behaviour of the larger grain borer *Prostephanus truncatus* in response to synthetic pheromone. *Tropical Science* 32, 163–170.

Fargo, W.S., Cuperus, G.W., Bonjour, E.L., Burkholder, W.E., Clary, B.L. and Payton, M.E. (1994) Influence of probe trap type and attractants on the capture of four stored-grain Coleoptera. *Journal of Stored Products Research* 30, 237–241.

Faustini, D.L. and Burkholder, W.E. (1987) Quinone-aggregation pheromone interaction in the red flour beetle. *Animal Behaviour* 35, 601–603.

Faustini, D.L., Burkholder, W.E. and Laub, R.J. (1981) Sexually dimorphic setiferous sex patch in the male red flour beetle, *Tribolium castaneum* (Herbst) (Coleoptera, Tenebrionidae): site of aggregation pheromone production. *Journal of Chemical Ecology* 7, 465–480.

Faustini, D.L., Rowe, J.R. and Burkholder, W.E. (1982a) A male-produced aggregation pheromone in *Tribolium brevicornis* (Leconte) (Coleoptera, Tenebrionidae) and interspecific responses of several *Tribolium* species. *Journal of Stored Product Research* 18, 153–158.

Faustini, D.L., Giese, W.L., Phillips, J.K. and Burkholder, W.E. (1982b) Aggregation pheromone of the male granary weevil, *Sitophilus granarius* (L.). *Journal of Chemical Ecology* 8, 679–687.

Faustini, D.L., Barak, A.V., Burkholder, W.E. and Leos-Martinez, J. (1990) Combination-type trapping for monitoring stored-product insects – a review. *Journal of the Kansas Entomological Society* 63, 539–547.

Fettköther, R., Dettner, K., Schröder, F., Meyer, H., Francke, W. and Noldt, U. (1995) The male pheromone of the old house borer *Hylotrupes bajulus* (L.) (Coleoptera: Cerambycidae): isolation and female response. *Experientia* 51, 270–277.

Fields, P.G., van Loon, J., Dolinski, M.G., Harris, J.L. and Burkholder, W.E. (1993) The disribution of *Rhyzopertha dominica* (F.) in western Canada. *Canadian Entomologist* 125, 317–328.

Finnegan, D.E. and Chambers, J. (1993) Identification of the sex pheromone of the Guernsey carpet beetle, *Anthrenus sarnicus* Mroczkowski (Coleoptera, Dermestidae). *Journal of Chemical Ecology* 19, 971–983.

Frey, W. (1952) Lebensweise und Schadbild des Kornkäfers und anderer Speicherschädlinge. In: Bundesministerium für Ernährung, Landwirtschaft und Forsten Referat Pflanzenschutz (ed.) *Vorratsschutz*. Verlag Kommentator, Frankfurt, pp. 13–17.

Fukui, H., Matsumura, F., Ma, M.C. and Burkholder, W.E. (1974) Identification of the sex pheromone of the furniture carpet beetle, *Anthrenus flavipes* LeConte. *Tetrahedron Letters* 40, 3563–3566.

Fukui, H., Matsumura, F., Barak, A.V. and Burkholder, W.E. (1977) Isolation and identification of a major sex attractant component of *Attagenus elongatulus* (Casey) (Coleoptera, Dermestidae). *Journal of Chemical Ecology* 3, 539–548.

Gerber, G.H. (1973) Reproductive behaviour and physiology of *Tenebrio molitor* (Coleoptera, Tenebrionidae). I. Initiation of mating in young adults and the effects of adult density. *Canadian Entomologist* 105, 807–811.

Gharib, A.H., Moftah, E.A. and Foud, M.S. (1992) Preliminary tests on the sex attraction response in the pulse beetle, *Callosobruchus chinensis* L. (Coleoptera, Bruchidae). *Minia Journal of Agricultural Research and Development* 14, 1107–1118.

Ghent, A.W. (1963) Studies of behavior of the *Tribolium* flour beetles I. Contrasting responses of *T. castaneum* and *T. confusum* to fresh and conditioned flours. *Ecology* 44, 269–283.

Giga, D.P. and Smith, R.H. (1984/1985) Oviposition markers in *Callosobruchus maculatus* F. and *Callosobruchus rhodesianus* Pic. (Coleoptera, Bruchidae): asymmetry of interspecific responses. *Agriculture, Ecosystems and Environment* 12, 229–233.

Gilberg, M. and Roach, A. (1991) The use of a commercial pheromone trap for monitoring *Lasioderma serricorne* (F.) infestations in museum collections. *Studies in Conservation* 36, 243–247.

Greenblatt, R.E., Burkholder, W.E., Cross, J.C., Byler, C. and Silverstein, R.M. (1976) Chemical communication in the mating behavior of *Trogoderma glabrum* Herbst (Coleoptera, Dermestidae). *Journal of Chemical Ecology* 2, 285–297.

Greenblatt, R.E., Burkholder, W.E., Cross, J.C., Cassidy, R.F., Silverstein, R.M., Levinson, A.R. and Levinson, H.Z. (1977) Chemical basis for interspecific responses to sex pheromones of *Trogoderma* species (Coleoptera, Dermestidae). *Journal of Chemical Ecology* 3, 337–347.

Hagstrum, D.W. and Flinn, P.W. (1996) Integrated Pest Management. In: Subramanyam, B. and Hagstrum, D.W. (eds) *Integrated Management of Insects in Stored Products*. Marcel Dekker, New York, pp. 399–408.

Halstead, D.G.H. (1973) Preliminary biological studies on the pheromone produced by male *Acanthoscelides obtectus* (Say) (Coleoptera, Bruchidae). *Journal of Stored Products Research* 9, 109–117.
Hammack, L. and Burkholder, W.E. (1981) Calling behavior in female *Trogoderma granarium* Everts (Coleoptera, Dermestidae). *Journal of Stored Products Research* 17, 25–29.
Hammack, L., Burkholder, W.E. and Ma, M. (1973) Sex pheromone localization in females of six *Trogoderma* species. *Annals of the Entomological Society of America* 66, 545–550.
Hammack, L., Ma, M. and Burkholder, W.E. (1976) Sex pheromone-releasing behaviour in females of the dermestid beetle, *Trogoderma glabrum*. *Journal of Insect Physiology* 22, 555–561.
Happ, G.M. (1968) Quinone and hydrocarbon production in the defense glands of *Eleodes longicollis* and *Tribolium castaneum* (Coleoptera, Tenebrionidae). *Journal of Insect Physiology* 14, 1232–1237.
Happ, G.M. (1969) Multiple sex pheromones of the mealworm beetle, *Tenebrio molitor* L. *Nature* 222, 180–181.
Happ, G.M. (1970) Maturation of the response of male *Tenebrio molitor* to the female sex pheromone. *Annals of the Entomological Society of America* 63, 1782.
Happ, G.M. and Wheeler, J. (1969) Bioassay, preliminary purification, and effect of age, crowding, and mating on the release of sex pheromone by female *Tenebrio molitor*. *Annals of the Entomological Society of America* 62, 846–851.
Helbig, J., Laborius, G.-A. and Schulz, F.A. (1992) Untersuchungen zur Distanz der Anlockwirkung des synthetischen Pheromones des Großen Kornborers *Prostephanus truncatus* (Horn) (Col., Bostrichidae) 'Trunc-call (1+2)' auf dessen Prädator *Teretriosoma nicrescens* Lewis (Col., Histeridae). *Journal of Applied Entomology* 113, 425–429.
Hinton, H.E. (1942) Secondary sexual characters of *Tribolium*. *Nature* 149, 500–501.
Hodges, R.J., Hall, D.R., Golob, P. and Meik, J. (1983) Responses of *Prostephanus truncatus* to components of the aggregation pheromone of *Rhyzopertha dominica* in the laboratory and field. *Entomologia Experimentalis et Applicata* 34, 266–272.
Hodges, R.J., Cork, A. and Hall, D.R. (1984) Aggregation pheromones for monitoring the greater grain borer *Prostephanus truncatus*. In: *Proceedings of the British Crop Protection Conference – Pests and Diseases*, pp. 255–259.
Hodges, R.J., Hall, D.R., Mbugua, J.N. and Likhayo, P.W. (1998) The responses of *Prostephanus truncatus* (Coleoptera: Bostrichidae) and *Sitophilus zeamais* (Coleoptera: Curculionidae) to pheromone and synthetic maize volatiles as lures in crevice or flight traps. *Bulletin of Entomological Research* 88, 131–139.
Hoffmann, R.W. and Ladner, W. (1979) On the absolute stereochemistry of C-2 and C-3 in stegobinone. *Tetrahedron Letters* 48, 4653–4656.
Hoffmann, R.W., Ladner, W., Steinbach, K., Massa, W., Schmidt, R. and Snatzke, G. (1981) Stereoselektive Synthese von Alkoholen. IX: Absolute Konfiguration von Stegobinon. *Chemische Berichte* 114, 2786–2801.
Honda, H. and Yamamoto, I. (1976) Evidence for and chemical nature of a sex pheromone present in azuki bean weevil, *Callosobruchus chinensis* L. In: Kono, T. and Ishii, S. (eds) *Proceedings, Insect Pheromones and Their Application*, Nagaoka and Tokyo, Japan, p. 164.
Hope, J.A., Horler, D.F. and Rowlands, D.G. (1967) A possible pheromone of the bruchid, *Acanthoscelides obtectus* (Say). *Journal of Stored Products Research* 3, 387–388.
Horler, D.F. (1970) (−)-Methyl-*n*-tetradeca-*trans*-2,4,5-trienoate, an allenic ester produced by the male dried bean beetle, *Acanthoscelides obtectus* (Say). *Journal of the Chemical Society* C 6, 859–862.
Howard, R.W. (1987) Chemosystematic studies of the Triboliini (Coleoptera, Tenebrionidae): phylogenetic inferences from the defense chemicals of eight *Tribolium* spp., *Palorus ratzeburgi* (Wissmann), and *Latheticus oryzae* Waterhouse. *Annals of the Entomological Society of America* 80, 398–405.

Howard, R.W. and Mueller, D.D. (1987) Defense chemistry of the flour beetle *Tribolium brevicornis* (Lec.): presence of known and potential prostaglandium synthetase inhibitors. *Journal of Chemical Ecology* 13, 1707–1723.

Howe, R.W. (1962) A study of the heating of stored grain caused by insects. *Annals of Applied Biology* 50, 137–158.

Hurd, H. and Parry, G. (1991) Metacestode-induced depression of the production of, and response to, sex pheromone in the intermediate host, *Tenebrio molitor*. *Journal of Invertebrate Pathology* 58, 82–87.

Hussain, A., Phillips, T.W. and Aliniazee, M.T. (1994a) Pheromone biology and factors affecting its production in *Tribolium castaneum*. In: Highley, E., Wright, E.J., Banks, H.J. and Champ, B.R. (eds) *Proceedings of an International Working Conference on Stored Product Protection*. CSIRO, Canberra, Australia, pp. 533–536.

Hussain, A., Phillips, T.W. and Aliniazee, M.T. (1994b) Responses of *Tribolium castaneum* to different lures and traps in the laboratory. In: Highley, E., Wright, E.J. Banks, H.J. and Champ, B.R. (eds) *Proceedings of an International Working Conference on Stored Product Protection*. CSIRO, Canberra, Australia, pp. 406–409.

Ikan, R., Bergmann, E.D., Yinon, U. and Shulov, A. (1969) Identification, synthesis and biological activity of an 'assembling scent' from the beetle *Trogoderma granarium*. *Nature* 223, 317.

Imai, T., Kodama, H., Chuman, T. and Masahiro, K. (1990) Female-produced oviposition deterrents of the cigarette beetle, *Lasioderma serricorne* (F.) (Coleoptera: Anobiidae). *Journal of Chemical Ecology* 16, 1237–1247.

Iwabuchi, K., Takahashi, J. and Sakai, T. (1987) Occurrence of 2,3-octanediol and 2-hydroxy-3-octanone, possible male sex pheromone in *Xylotrechus chinensis* Chevrolat. *Journal of Applied Entomology and Zoology* 22, 110.

Javer, A., Borden, J.H., Pierce, H.D. Jr and Pierce, A.M. (1990) Evaluation of pheromone-baited traps for monitoring of cucujid and tenebrionid beetles in stored grain. *Journal of Economic Entomology* 83, 268–272.

Jiang, X.-L. (1995) Studies on the taxis reflection of *Sitophilus granarius* L. to some vegetable oils and its pheromone. *Entomological Knowledge* 32, 23–25.

Judy, K.J., Schooley, D.A., Troetschler, R.G., Jennings, R.C., Bergot, B.J. and Hall, M.S. (1975) Juvenile hormone production by corpora allata of *Tenebrio molitor* in vitro. *Life Sciences* 16, 1059–1066.

Keville, R. and Kannowski, P.B. (1975) Sexual excitation by pheromones of the confused flour beetle. *Journal of Insect Physiology* 21, 81–84.

Key, G.E., Tigar, B.J., Flores-Sanchez, E. and Vazquez-Arista, M. (1994). Response of *Prostephanus truncatus* and *Teretriosoma nigrescens* to pheromone-baited flight traps. In: Highley, E., Wright, E.J., Banks, H.J. and Champ, B.R. (eds) *Proceedings of an International Working Conference on Stored Product Protection*. CSIRO, Canberra, Australia, pp. 410–414.

Khorramshahi, A. and Burkholder, W.E. (1981) Behavior of the lesser grain borer *Rhyzopertha dominica* (Coleoptera, Bostrychidae) male produced aggregation pheromone attracts both sexes. *Journal of Chemical Ecology* 7, 33–38.

Kodama, H., Mochizuki, K., Kohno, M., Ohnishi, A. and Kuwahara, Y. (1987a) Inhibition of the male response of drugstore beetles to stegobinone by its isomers. *Journal of Chemical Ecology* 13, 1859–1869.

Kodama, H., Ono, M., Kohno, M. and Ohnishi, A. (1987b) Stegobiol, a new sex pheromone component of drugstore beetle (*Stegobium paniceum*). *Journal of Chemical Ecology* 13, 1871–1879.

Kuwahara, Y. and Nakamura, S. (1985) (Z)-5- and (E)-5-undecenoic acid: identification of the sex pheromone of the varied carpet beetle, *Anthrenus verbasci* L. (Coleoptera, Dermestidae). *Journal of Applied Entomology and Zoology* 20, 354–356.

Kuwahara, Y., Fukami, H., Ishii, S., Matsumura, F. and Burkholder, W.E. (1975) Studies on the isolation and bioassay of the sex pheromone of the drugstore beetle, *Stegobium paniceum* (Coleoptera, Anobiidae). *Journal of Chemical Ecology* 1, 413–422.
Kuwahara, Y., Fukami, H., Howard, R., Ishii, S., Matsumura, F. and Burkholder, W.E. (1978) Chemical studies on the Anobiidae: sex pheromone of the drugstore beetle, *Stegobium paniceum* (L.) (Coleoptera). *Tetrahedron* 34, 1769–1774.
Ladisch, R.K., Ladisch, S.K. and Howe, P.M. (1967) Quinoid secretions in grain and flour beetles. *Nature* 215, 939–940.
Landolt, P.J. (1997) Sex attractant and aggregation pheromones of male phytophagous insects. *American Entomologist* 43, 12–22.
Lanier, G.N. and Burkholder, W.E. (1974) Pheromones in speciation of Coleoptera. In: Birch, M.C. (ed.) *Pheromones*. North Holland, Amsterdam, pp. 161–189.
Leos-Martinez, J., Granovsky, T.A., Williams, H.J., Vinson, S.B. and Burkholder, W.E. (1986) Estimation of aerial density of the lesser grain borer (Coleoptera: Bostrichidae) in a warehouse using dominicalure traps. *Journal of Economic Entomology* 79, 1134–1138.
Leos-Martinez, J., Granovsky, T.A., Williams, H.J., Vinson, S.B. and Burkholder, W.E. (1987) Pheromonal trapping methods for lesser grain borer, *Rhyzopertha dominica* (Coleoptera: Bostrichidae). *Environmental Entomology* 16, 747–751.
Leos-Martinez, J., Del Socorro Gonzalez-Alonso, M. and Williams, H.J. (1995) Optimization of pheromonal trapping methods for *Prostephanus truncatus* (Horn) (Coleoptera, Bostrichidae). *Journal of Stored Product Research* 31, 103–109.
Levinson, H.Z. (1983) The scientific background of pest manipulation involving insectistasis. *Annals of the Institute for Phytopathology Benaki* 14, 69–79.
Levinson, H.Z. and Bar Ilan, A.R. (1967) Function and properties of an assembling scent in the khapra beetle *Trogoderma granarium*. *Rivista di Parassitologia* 28, 27–42.
Levinson, H.Z. and Bar Ilan, A.R. (1970a) Olfactory and tactile behaviour of the khapra beetle, *Trogoderma granarium*, with special reference to its assembling scent. *Journal of Insect Physiology* 16, 561–572.
Levinson, H.Z. and Bar Ilan, A.R. (1970b) Behaviour of the khapra beetle *Trogoderma granarium* toward the assembling scent released by the female. *Experientia* 26, 846–847.
Levinson, H.Z. and Buchelos, C.T. (1988) Population dynamics of *Lasioderma serricorne* F. (Col., Anobiidae) in tobacco stores with and without insecticidal treatments: a three-year survey by pheromone and unbaited traps. *Journal of Applied Entomology* 106, 201–211.
Levinson, H.Z. and Levinson, A.R. (1973) The dual function of the assembling scent of the female khapra beetle *Trogoderma granarium*. *Naturwissenschaften* 60, 352–353.
Levinson, H.Z. and Levinson, A.R. (1977) Integrated manipulation of storage insects by pheromones and food attractants – a proposal. *Zeitschrift für Angewandte Entomologie* 84, 337–343.
Levinson, H.Z. and Levinson, A.R. (1979) Trapping of storage insects by sex and food attractants as a tool of integrated control. In: Ritter, F.J. (ed.) *Chemical Ecology, Odour Communication in Animals*. Elsevier/North-Holland Biomedical Press, Amsterdam, pp. 327–341.
Levinson, H.Z. and Levinson, A.R. (1986) Antagonized pheromone responses of male tobacco beetles (*Lasioderma serricorne* F.) due to *erythro*-diastereoisomers of 4*S*,6*S*,7*S*-serricornin. *Naturwissenschaften* 73, 36–37.
Levinson, H.Z. and Levinson, A.R. (1987) Pheromone biology of the tobacco beetle (*Lasioderma serricorne* F., Anobiidae) with notes on the pheromone antagonism between 4*S*,6*S*,7*S* and 4*S*,6*S*,7*R* serricornin. *Journal of Applied Entomology* 103, 217–240.

Levinson, H.Z. and Levinson, A.R. (1994) Origin of grain storage and insect species consuming desiccated food. *Anzeiger für Schädlingskunde, Pflanzenschutz und Umweltschutz* 67, 47–60.

Levinson, H.Z. and Levinson, A.R. (1995) Reflection on structure and function of pheromone glands in storage insect species. *Anzeiger für Schädlingskunde, Pflanzenschutz und Umweltschutz* 68, 99–118.

Levinson, H.Z. and Mori, K. (1980) The pheromone activity of chiral isomers of trogodermal for male khapra beetle. *Naturwissenschaften* 67, 148–149.

Levinson, H.Z. and Mori, K.. (1983) Chirality determines pheromone activity for flour beetles. *Naturwissenschaften* 70, 190–192.

Levinson, H.Z., Levinson, A.R., Schweiger, H., Cassidy, R.F. and Silverstein, R.M. (1978) Olfactory behavior and receptor potentials of the khapra beetle *Trogoderma granarium* (Coleoptera, Dermestidae) induced by the major components of its sex pheromone, certain analogues, and fatty acid esters. *Journal of Chemical Ecology* 4, 95–108.

Levinson, H.Z., Levinson, A.R., Francke, W., Mackenroth, W. and Heemann, V. (1981a) The pheromone activity of anhydroserricornin and serricornin for male cigarette beetle (*Lasioderma serricorne* F.). *Naturwissenschaften* 68, 148–149.

Levinson, H.Z., Levinson, A.R. and Mori, K. (1981b) Olfactory behaviour and receptor potentials of two khapra beetle strains induced by enantiomers of trogodermal. *Naturwissenschaften* 67, 480–481.

Levinson, H.Z., Levinson, A.R., Kahn, G.E. and Schäfer, K. (1983) Occurrence of a pheromone-producing gland in female tobacco beetles. *Experientia* 39, 1095–1097.

Levinson, H.Z., Levinson, A.R., Ren, Z. and Mori, K. (1990) Comparative olfactory perception of the aggregation pheromones of *Sitophilus oryzae* (L.), *S. zeamais* (Motsch.) and *S. granarius* (L.), as well as the stereoisomers of these pheromones. *Journal of Applied Entomology* 110, 203–213.

Lewis, S.M. and Iannini, J. (1995) Fitness consequences of differences in male mating behaviour in relation to female reproductive status in flour beetles. *Animal Behaviour* 50, 1157–1160.

Lewis, W.J. and Martin, W.R. (1990) Semiochemicals for use with parasitoids: status and future. *Journal of Chemical Ecology* 16, 3067–3089.

Lextrait, P., Biemont, J.-C. and Pouzat, J. (1994) Comparison of walking locomotory reactions of two forms of *Callosobruchus maculatus* males subjected to female sex pheromone stimulation (Coleoptera, Bruchidae). *Journal of Chemical Ecology* 20, 2917–2930.

Lextrait, P., Biemont, J.-C. and Pouzat, J. (1995) Pheromone release by the two forms of *Callosobruchus maculatus* females, effects of age, temperature and host plant. *Physiological Entomology* 20, 309–317.

Lindgren, B.S., Borden, J.H., Pierce, A.M., Pierce, H.D. Jr, Oehlschlager, A.C. and Wong, J.W. (1985) A potential method for simultaneous semiochemical-based monitoring of *Cryptolestes ferrugineus* and *Tribolium castaneum* (Coleoptera, Cucujidae and Tenebrionidae). *Journal of Stored Products Research* 21, 83–87.

Loconti, J.D. and Roth, L.M. (1953) Composition of the odorous secretion of *Tribolium castaneum*. *Annals of the Entomological Society of America* 46, 281–289.

Longstaff, B.C. (1981) Biology of the grain pest species of the genus *Sitophilus* (Coleoptera, Curculionidae): a critical review. *Protection Ecology* 2, 83–130.

Loschiavo, S.R., Wong, J., White, N.D.G., Pierce, H.D., Borden, J.H. and Oehlschlager, A.C. (1986) Field evaluation of a pheromone to detect adult rusty grain beetles, *Cryptolestes ferrugineus* (Coleoptera: Cucujidae), in stored grain. *Canadian Entomologist* 118, 1–8.

Ma, M. and Burkholder, W.E. (1978) Sex pheromone releasing behavior of *Anthrenus flavipes* (furniture carpet beetles) females (Coleoptera, Dermestidae). *Annals of the Entomological Society of America* 71, 129–133.

Ma, M., Hummel, H.E. and Burkholder, W.E. (1980) Estimation of single furniture carpet beetle (*Anthrenus flavipes* LeConte) sex pheromone release by dose-response curve and chromatographic analysis of pentafluorobenzylic derivate of (Z)-3-decenoic acid. *Journal of Chemical Ecology* 6, 597–607.

Matteson, D.S., Man, H.W. and Ho, O.C. (1996) Asymmetric synthesis of stegobinone via boronic ester chemistry. *Journal of the American Chemical Society* 118, 4560–4566.

Mayer, M.S. and McLaughlin, J.R. (1991) *Handbook of Insect Pheromones and Sex Attractants*. CRC Press, Boca Raton, Florida, 1083 pp.

Mayhew, T.J. and Phillips, T.W. (1994) Pheromone biology of the lesser grain borer, *Rhyzopertha dominica* (Coleoptera, Bostrichidae). In: Highley, E., Wright, E.J., Banks, H.J. and Champ, B.R. (eds) *Proceedings of the International Working Conference on Stored Product Protection*, CSIRO, Canberra, Australia, pp. 541–544.

Mbata, G.N. (1993) Some factors affecting oviposition and development of *Callosobruchus subinnotatus* (Pic) (Coleoptera, Bruchidae). *Zeitschrift für Pflanzenkrankheiten und Pflanzenschutz* 100, 155–164.

Mbata, G.N. and Ramaswamy, S.B. (1995) Factors affecting the stability and recognition of the oviposition marker pheromone of *Callosobruchus subinnotatus* (Pic). *Journal of Stored Products Research* 31, 157–163.

Menon, M. (1970) Hormone–pheromone relationships in the beetle, *Tenebrio molitor*. *Journal of Insect Physiology* 16, 1123–1139.

Menon, M. (1976) Hormone–pheromone relationships of male *Tenebrio molitor*. *Journal of Insect Physiology* 22, 1021–1023.

Messina, F.J. and Renwick, J.A.A. (1985) Mechanism of egg recognition by the cowpea weevil *Callosobruchus maculatus*. *Entomologia Experimentalis et Applicata* 37, 241–245.

Messina, F.J., Barmore, J.L. and Renwick, J.A.A. (1987) Oviposition deterrent from eggs of *Callosobruchus maculatus*: spacing mechanism or artifact? *Journal of Chemical Ecology* 13, 219–226.

Metcalf, R.E. (1994) Role of kairomones in integrated pest management. *Phytoparasitica* 22, 275–279.

Michelsen, A. (1966) The sexual behaviour of some longhorned beetles (Col., Cerambycidae). *Entomologiske Meddelelser* 34, 329–355.

Mignon, J., Haubruge, E., Lienard, V., Gaspar, Ch. and Lognay, G. (1996) Mortality in *Oryzaephilus surinamensis* following short-term exposure to conditioned kernels by high-density culture. *Entomologia Experimentalis et Applicata* 80, 555–557.

Millar, J.G., Oehlschlager, A.C. and Wong, J.W. (1983) Synthesis of two macrolide aggregation pheromones from the flat grain beetle *Cryptolestes pusillus* (Schönherr). *Journal of Organic Chemistry* 48, 4404–4407.

Millar, J.G., Pierce, H.D. Jr, Pierce, A.M., Oehlschlager, A.C., Borden, J.H. and Barak, A.V. (1985a) Aggregation pheromones of the flat grain beetle, *Cryptolestes pusillus* (Coleoptera, Cucujidae). *Journal of Chemical Ecology* 11, 1053–1070.

Millar, J.G., Pierce, H.D. Jr, Pierce, A.M., Oehlschlager, A.C. and Borden, J.H. (1985b) Aggregation pheromones of the grain beetle *Cryptolestes turcicus* (Coleoptera, Cucujidae). *Journal of Chemical Ecology* 11, 1071–1081.

Mondal, K.A.M.S.H. (1985) Response of *T. castaneum* larvae to aggregation pheromone and quinones produced by adult conspecifics. *International Journal of Pest Control* 27, 64–66.

Mondal, K.A.M.S.H. and Port, G.R. (1994) Pheromones of *Tribolium* spp. (Coleoptera, Tenebrionidae) and their potential in pest management. *Agricultural Zoology Reviews* 6, 121–148.

Mori, K. (1974) Absolute configuration of (−)-14-methylhexadec-8-*cis*-en-1-ol and methyl (−)-14-methylhexadec-8-*cis*-enoate, the sex pheromone of female dermestid beetles. *Tetrahedron* 30, 3817–3820.

Mori, K. and Ebata, T. (1986) Synthesis of (2*S*,3*R*,1′*S*,2′*S*)-stegobiol, a new component of the drugstore beetle pheromone. *Tetrahedron Letters* 42, 4685–4689.

Mori, K., Nomi, H., Chuman, T., Kohno, M., Kato, K. and Nogushi, M. (1981a) Determination of the absolute configuration at C-6 and C-7 of serricornin (4,6-dimethyl-7-hydroxy-3-nonanone), the sex pheromone of the cigarette beetle. *Tetrahedron Letters* 22, 1127–1130.

Mori, K., Nukada, T. and Ebata, T. (1981b) Synthesis of optically active forms of methyl (*E*)-2,4,5-tetradecatrienoate, the pheromone of the male dried bean weevil. *Tetrahedron* 37, 1343–1347.

Mori, M., Chuman, T., Kohno, M., Kato, K., Nogushi, M., Nomi, H. and Mori, K. (1982) Absolute stereochemistry of serricornin, the sex pheromone of cigarette beetle, as determined by the synthesis of its (4*S*,6*R*,7*R*)-isomer. *Tetrahedron Letters* 23, 667–670.

Mori, K., Ito, T., Tanaka, K., Honda, H. and Yamamoto, I. (1983) Synthesis and biological activity of optically active forms of (*E*)-3,7-dimethyl-2-octene-1,8-dioic acid (callosobruchusic acid). A component of the copulation release pheromone (erectin) of the azuki bean weevil. *Tetrahedron* 39, 2303–2306.

Mori, M., Chuman, T. and Kato, K. (1984) Cyclic hemiacetal and acyclic chain – the two forms of serricornin. *Tetrahedron Letters* 25, 2553–2556.

Mueller, D. (1995) All pheromone traps were not created equal. *Pest Control* June 1995, 134.

Mullen, M.A. (1992) Development of a pheromone trap for monitoring *Tribolium castaneum*. *Journal of Stored Products Research* 28, 245–249.

Mushobozy, D.K., Pierce, H.D. Jr and Borden, J.H. (1993) Evaluation of 1-octen-3-ol and nonanal as adjuvants for aggregation pheromones for three species of cucujid beetles (Coleoptera: Cucujidae). *Journal of Economic Entomology* 86, 1835–1845.

Nakajima, S., Sugawara, K., Takeda, T., Tateishi, M., Okamura, J.I. and Baba, N. (1996) Arrestants to *Oryzaephilus surinamensis* L. from wheat flour infested by the same weevil. *Bioscience Biotechnology Biochemistry* 60, 1546–1547.

Obeng-Ofori, D. (1991) Analysis of orientation behaviour of *Tribolium castaneum* and *T. confusum* to synthetic aggregation pheromone. *Entomologia Experimentalis et Applicata* 60, 125–133.

Obeng-Ofori, D. and Coaker, T.H. (1990a) Some factors affecting responses of four stored product beetles (Coleoptera, Tenebrionidae & Bostrichidae) to pheromones. *Bulletin of Entomological Research* 80, 433–441.

Obeng-Ofori, D. and Coaker,T.H. (1990b) *Tribolium aggregation* pheromone: monitoring, range of attraction and orientation behaviour of *T. castaneum* (Coleoptera, Tenebrionidae). *Bulletin of Entomological Research* 80, 443–451.

O'Ceallachain, D.P. and Ryan, M.F. (1977) Production and perception of pheromones by the beetle *Tribolium confusum*. *Journal of Insect Physiology* 23, 1303–1309.

Oehlschlager, A.C., King, G.G.S., Pierce, H.D. Jr, Pierce, A.M., Slessor, K.N., Millar, J.G. and Borden, J.H. (1987) Chirality of macrolide pheromones of grain beetles in the genera *Oryzaephilus* and *Cryptolestes* and its implications for species specificity. *Journal of Chemical Ecology* 13, 1543–1554.

Oehlschlager, A.C., Pierce, A.M., Pierce, H.D. Jr and Borden, J.H. (1988) Chemical communication in cucujid grain beetles. *Journal of Chemical Ecology* 14, 2071–2098.

Oshima, K., Honda, H. and Yamamoto, I. (1973) Isolation of an oviposition marker from azuki bean weevil, *Callosobruchus chinensis* (L.). *Journal of Agricultural and Biological Chemistry* 37, 2679–2680.

Pappas, P.W. and Morrison, S.E. (1995) Benzoquinones of the beetles, *Tribolium castaneum* and *Tribolium confusum*. *Preparative Biochemistry* 25, 155–168.

Pappas, P.W. and Wardrop, S.M. (1996) Quantification of benzoquinones in the flour beetles, *Tribolium castaneum* and *Tribolium confusum*. *Preparative Biochemistry and Biotechnology* 26, 53–66.

Phillips, J.K. and Burkholder, W.E. (1981) Evidence for a male produced aggregation pheromone in the rice weevil *Sitophilus oryzae*. *Journal of Economic Entomology* 74, 539–542.

Phillips, J.K., Walgenbach, C.A., Klein, J.A., Burkholder, W.E., Schmuff, N.R. and Fales, H.M. (1985) (*R**,*S**)5-Hydroxy-4-methyl-3-heptanone male-produced aggregation pheromone of *Sitophilus oryzae* (L.) and *S. zeamais* Motsch. *Journal of Chemical Ecology* 11, 1263–1274.

Phillips, J.K., Miller, P.F., Andersen, J.F., Fales, H.M. and Burkholder, W.E. (1987) The chemical identification of the granary weevil aggregation pheromone. *Tetrahedron Letters* 49, 6145–6146.

Phillips, J.K., Chong, J.M., Andersen, J.F. and Burkholder, W.E. (1989) Determination of the enantiomeric composition of (*R**,*S**)-1-ethylpropyl-2-methyl-3-hydroxy-pentanoate, the male-produced aggregation pheromone of *Sitophilus granarius*. *Entomologia Experimentalis et Applicata* 51, 149–153.

Phillips, T.W. (1994) Pheromones of stored-product insects: current status and future perspectives. In: Highley, E., Wright, E.J., Banks, H.J. and Champ, B.R. (eds) *Proceedings of an International Working Conference on Stored Product Protection*. CSIRO, Canberra, Australia, pp. 479–486.

Phillips, T.W. (1997) Semiochemicals of stored-product insects: research and applications. *Journal of Stored Products Research* 33, 17–30.

Phillips, T.W., Burkholder, W.E., Jiang, X.-L., Phillips, J.K. and Quoc, H.T. (1991) Food odor attractants and pheromone synergists for stored-product Coleoptera. In: *Abstracts 8th Annual Meeting of the International Society of Chemical Ecology*, Dijon France, p. 22.

Phillips, T.W., Jiang, X.-L., Burkholder, W.E., Phillips, J.K. and Tran, H.Q. (1993) Behavioral responses to food volatiles by two species of stored-product Coleoptera, *Sitophilus oryzae* (Curculionidae) and *Tribolium castaneum* (Tenebrionidae). *Journal of Chemical Ecology* 19, 723–734.

Phillips, T.W., Phillips, J.K., Webster, F.X., Tang, R.T. and Burkholder, W.E. (1996) Identification of sex pheromones from the cowpea weevil, *Callosobruchus maculatus*, and related studies with *C. analis* (Coleoptera, Bruchidae). *Journal of Chemical Ecology* 22, 2233–2249.

Pierce, A.M., Borden, J.H. and Oehlschlager, A.C. (1981) Olfactory response to beetle-produced volatiles and host-food attractants by *Oryzaephilus surinamensis* and *O. mercator*. *Canadian Journal of Zoology* 59, 1980–1990.

Pierce, A.M., Borden, J.H. and Oehlschlager, A.C. (1983) Effects of age and population density on response to beetle and food volatiles by *Oryzaephilus surinamensis* and *O. mercator* (Coleoptera: Cucujidae). *Environmental Entomology* 12, 1367–1374.

Pierce, A.M., Pierce, H.D. Jr, Millar, J.G., Borden, J.H. and Oehlschlager, A.C. (1984a) Aggregation pheromones in the genus *Oryzaephilus* (Coleoptera, Cucujidae). In: *Proceedings of the International Working Conference on Stored Product Entomology*, Manhattan, Kansas, pp. 107–120.

Pierce, A.M., Pierce, H.D. Jr, Millar, J.G., Wong, J.W., Verigin, V.G., Oehlschlager, A.C. and Borden, J.H. (1984b) Methodology for isolation and analysis of aggregation pheromones in the genera *Cryptolestes* and *Oryzaephilus* (Coleoptera, Cucujidae). In: *Proceedings of the International Working Conference on Stored Product Entomology*, Manhattan, Kansas, pp. 121–137.

Pierce, A.M., Pierce, H.D. Jr, Oehlschlager, A.C. and Borden, J.H. (1985) Macrolide aggregation pheromones in *Oryzaephilus surinamensis* and *Oryzaephilus mercator*. *Journal of Agricultural and Food Chemistry* 33, 848–852.

Pierce, A.M., Pierce, H.D. Jr, Borden, J.H. and Oehlschlager, A.C. (1986) Enhanced production of aggregation pheromones in four stored-product coleopterans feeding on methoprene-treated oats. *Experientia* 42, 164–165.

Pierce, A.M., Pierce, H.D. Jr, Oehlschlager, A.C., Czyzewska, E. and Borden, J.H. (1987) Influence of pheromone chirality on response by *Oryzaephilus surinamensis* and *Oryzaephilus mercator* (Coleoptera, Cucujidae). *Journal of Chemical Ecology* 13, 1525–1542.

Pierce, A.M., Pierce, H.D. Jr, Johnston, B.D., Oehlschlager, A.C. and Borden, J.H. (1988) Aggregation pheromone of square necked grain beetle, *Cathartus quadricollis* (Guer.). *Journal of Chemical Ecology* 14, 2169–2184.

Pierce, A.M., Pierce, H.D. Jr, Borden, J.H. and Oehlschlager, A.C. (1989) Production dynamics of cucujolide pheromones and identification of 1-octen-3-ol as a new aggregation pheromone for *Oryzaephilus surinamensis* and *O. mercator* (Coleoptera, Cucujidae). *Environmental Entomology* 18, 747–755.

Pierce, A.M., Pierce, H.D. Jr, Oehlschlager, A.C. and Borden, J.H. (1990a) Attraction of *Oryzaephilus surinamensis* (L.) and *Oryzaephilus mercator* (Fauvel) (Coleoptera, Cucujidae) to some common volatiles of food. *Journal of Chemical Ecology* 16, 465–475.

Pierce, A.M., Borden, J.H. and Oehlschlager, A.C. (1990b) Suppression of oviposition in *Oryzaephilus surinamensis* (L.) (Coleoptera, Cucujidae) following prolonged retention in high-density cultures or short-term exposure to larval volatiles. *Journal of Chemical Ecology* 16, 595–601.

Pierce, A.M., Pierce, H.D. Jr, Borden, J.H. and Oehlschlager, A.C. (1990c) Effect of aggregation pheromones on efficacy of cardboard traps for sawtooth grain beetle (Coleoptera: Cucujidae). *Journal of Economic Entomology* 83, 273–276.

Pierce, A.M., Pierce, H.D. Jr, Oehlschlager, A.C. and Borden, J.H. (1991a) Fungal volatiles, semiochemicals for stored-product beetles (Coleoptera, Cucujidae). *Journal of Chemical Ecology* 17, 581–597.

Pierce, A.M., Pierce, H.D. Jr, Oehlschlager, A.C. and Borden, J.H. (1991b) 1-Octen-3-ol, attractive semiochemical for foreign grain beetle, *Ahasverus advena* (Waltl) (Coleoptera, Cucujidae). *Journal of Chemical Ecology* 17, 567–580.

Pierce, L.H. (1994) Using pheromones for location and suppression of phycitid moth and cigarette beetles in Hawaii – a five-year summary. In: Highley, E., Wright, E.J., Banks, H.J. and Champ, B.R. (eds) *Proceedings of an International Working Conference on Stored Product Protection*. CSIRO, Canberra, Australia, pp. 439–443.

Pimbert, M. (1987) Analyse, en olfactomètre tubulaire, de l'influence de différents stimulus olfactifs dans la recherche du partenaire sexuel par *Zabrotes subfasciatus*. *Entomologia Experimentalis et Applicata* 45, 23–27.

Pimbert, M. and Pouzat, J. (1988) Electroantennogram responses of *Zabrotes subfasciatus* to odours of the sexual partner. *Entomologia Experimentalis et Applicata* 47, 49–53.

Pinniger, D.B. (1990) Food-baited traps: past, present and future. *Journal of the Kansas Entomological Society* 63, 533–538.

Pinniger, D.B. and Chambers, J. (1987) The use of pheromones in stored product protection: a UK view. In: Donahaye, E. and Navarro, S. (eds) *Proceedings of the International Working Conference on Stored Product Protection*, Tel Aviv (Israel), pp. 24–34.

Pirkle, W.H. and Boeder, C.W. (1978) Synthesis and absolute configuration of (−)-methyl (E)-2,4,5-tetradecatrienoate, the sex attractant of the male dried bean weevil. *Journal of Organic Chemistry* 43, 2091–2093.

Pittendrigh, B.R., Huesing, J.E., Shade, R.E. and Murdock, L.L. (1997) Monitoring of rice weevil, *Sitophilus oryzae*, feeding behavior in maize seeds and occurrence of supernumerary molts in low humidity conditions. *Entomologia Experimentalis et Applicata* 83, 225–231.

Plarre, R. (1991) Pheromonal communication and orientation of the granary weevil, *Sitophilus granarius* L. (Coleoptera, Curculionidae). In: Fleurat-Lessard, F. and Ducom, P. (eds) *Proceedings of the International Working Conference on Stored Product Protection*, Bordeaux, France, pp. 181–190.

Plarre, R. (1992) Orientierung des Kornkäfers *Sitophilus granarius* (L.) (Coleoptera, Curculionidae) auf Duft- und Pheromonquellen in Abhängigkeit vom physiologischen Zustand. *Mitteilungen der Deutschen Gesellschaft für Allgemeine und Angewandte Entomologie* 8, 257–264.

Plarre, R. (1994) Influence of synthetic sitophilate, the aggregation pheromone of *Sitophilus granarius* (L.) (Col., Curculionidae) on dispersion and aggregation behavior of the granary weevil. In: Highley, E., Wright, E.J., Banks, H.J. and Champ, B.R. (eds) *Proceedings of an International Working Conference on Stored Product Protection*. CSIRO, Canberra, Australia, pp. 570–582.

Plarre, R. (1996) Three dimensional distribution of *Sitophilus granarius* (L.) (Coleoptera, Curculionidae) in wheat influenced by the synthetic aggregation pheromone. *Journal of Stored Products Research* 32, 275–283.

Plarre, R. (1998) Pheromones and other semiochemicals of stored product insects – a historical review, current applications, and prospective needs. *Mitteilungen aus der Biologischen Bundesanstalt für Land- und Forstwirtschaft Berlin* 342, 13–83.

Press, J.W. (1988) Avoidance of the red flour beetle, *Tribolium castaneum* (Coleoptera, Tenebrionidae), by *Bracon hebetor* Say (Hymenopytera, Braconidae). *Journal of the Kansas Entomological Society* 61, 228–230.

Press, J.W., Cline, L.D. and Flaherty, B.R. (1986) Impact of the red flour beetle, *Tribolium castaneum* (Herbst), on suppression of the almond moth, *Cadra cautella* (Walker), by the parasitic wasp, *Bracon hebetor* Say and *Venturia canescens* (Gravenhorst). *Journal of Entomological Science* 21, 271–275.

Qi, Y.T. and Burkholder, W.E. (1982) Sex pheromone biology and behavior of the cowpea weevil *Callosobruchus maculatus* (Coleoptera, Bruchidae). *Journal of Chemical Ecology* 8, 527–534.

Quaife, J.T. (1980) Aggregation and sex pheromones in the flat grain beetle, *Cryptolestes pusillus* (Schonherr) (Coleoptera, Cucujidae). *USDFDA Science Advisor Research Associate Program* No. 96–78, 177–205.

Raffa, K.F., Phillips, T.W. and Salom, S.M. (1993) Strategies and mechanisms of host colonization by bark beetles. In: Schowalter, T.D. and Filip, G.M. (eds) *Beetle–Pathogen Interactions in Conifer Forests*. Academic Press, London, pp. 103–128.

Rahalkar, G.W., Tamhankar, A.J. and Adeesan, C. (1972) Influence of prior exposure of male khapra beetle (*Trogoderma granarium*) to female sex pheromone on their mating ability. *Experientia* 28, 1123–1124.

Rahalkar, G.W., Tamhankar, A.J. and Gothi, K.K. (1985) Selective breeding for reduced male response to female sex pheromone in *Trogoderma granarium* Everts (Coleoptera: Dermestidae). *Journal of Stored Products Research* 21, 123–126.

Ramaswamy, S.B., Shu, S., Monroe, W.A. and Mbata, G.N. (1995) Ultrastructure and potential role of integumentary glandular cells in adult male and female *Callosobruchus subinnotatus* (Pic) and *C. maculatus* (Fabricius) (Coleoptera, Bruchidae). *International Journal of Insect Morphology and Embryology* 24, 51–61.

Ramirez-Martinez, M., Alba-Avila, A. de and Ramirez-Zurbia, R. (1994) Discovery of the larger grain borer in a tropical deciduous forest in Mexico. *Journal of Applied Entomology* 118, 354–360.

Rees, D., Rodriguez, R., Herrera, F. and Ofosu, A. (1991) Advances in monitoring *Prostephanus truncatus* (Horn) (Col.: Bostrichidae) and *Teretriosoma nigrescens* Lewis (Col.: Histeridae) populations. In: Fleurat-Lessard, F. and Ducom, P. (eds). *Proceedings of an International Working Conference on Stored Product Protection*, Bordeaux, France, pp. 1417–1423.

Reichmuth, Ch., Schöller, M. and Ulrichs, Ch. (1997) *Vorratsschädlinge im Getreide*. Verlag Th. Mann, Gelsenkirchen, 119 pp.

Richter, J. and Biliwa, A. (1991) Landesweite Erhebung mittels Pheromonfallen zur Verbreitung von *Prostephanus truncatus* (Horn) (Col., Bostrichidae) in Togo. *Anzeiger für Schädlingskunde, Pflanzenschutz und Umweltschutz* 64, 89–92.

Rodin, J.O., Silverstein, R.M., Burkholder, W.E. and Gorman, J.E. (1969) Sex attractant of the female dermestid beetle *Trogoderma inclusum* Leconte. *Science* 165, 904–906.

Rodin, J.O., Leaffer, M.A. and Silverstein, R.M. (1970) Synthesis of *trans*-3, *cis*-5-tetradecadienoic acid (magatomoic acid), the sex attractant of the black carpet beetle, and its geometric isomers. *Journal of Organic Chemistry* 35, 3152.

Rossi, R. and Niccoli, A. (1978) Relationship between chirality and biological activity. Biological response of *Trogoderma granarium* to optically active synthetic sex attractants. *Naturwissenschaften* 65, 259–260.

Rossi, R., Salvadori, P.A., Carpita, A. and Niccoli, A. (1979) Chirality influences the biological activity of sex pheromones of the khapra beetle. *Naturwissenschaften* 66, 211.

Roth, L.M. (1943) Studies on the gaseous secretion of *Tribolium confusum* Duval. II. The odoriferous glands of *Tribolium confusum*. *Annals of the Entomological Society of America* 36, 397–424.

Roth, L.M. (1945) The odoriferous glands in the Tenebrionidae. *Annals of the Entomological Society of America* 38, 77–87.

Roth, L.M. and Howland, R.B. (1941) Studies on the gaseous secretion of *Tribolium confusum* Duval. I. Abnormalities produced in *Tribolium confusum* Duval by exposure to a secretion given off by the adults. *Annals of the Entomological Society of America* 34, 151–175.

Rup, P.J. and Sharma, S.P. (1978) Behavioural response of male and female of *Callosobruchus maculatus* (F.) to the sex pheromones. *Indian Journal of Ecology* 5, 72–76.

Ryan, M.F. and O'Ceallachain, D.P. (1976) Aggregation and sex pheromones in the beetle *Tribolium confusum*. *Journal of Insect Physiology* 22, 1501–1503.

Sakai, T., Nakagawa, Y., Takahashi, J., Iwabuchi, K. and Ishii, K. (1984) Isolation and identification of the male sex pheromone of the grape borer *Xylotrechus pyrrhoderus* Bates. *Chemical Letters*, 263.

Schmuff, N.R., Phillips, J.K., Burkholder, W.E., Fales, H.M., Chen, C.W., Roller, P.P. and Ma, M. (1984) The chemical identification of the rice weevil and the maize weevil aggregation pheromone. *Tetrahedron Letters* 25, 1533–1534.

Seitz, L.M. and Sauer, D.B. (1996) Volatile compounds and odors in grain sorghum infested with common storage insects. *Cereal Chemistry* 73, 744–750.

Selitskaya, O.G. and Shamshev, I.V. (1994) Response of lesser grain borer (Coleoptera, Bostrichidae) to components of aggregation pheromone. *Zoologicheskii Zhurnal* 73, 43–48.

Shapas, T.J. and Burkholder, W.E. (1978a) Diel and age-dependent behavioral patterns of exposure–concealment in three species of *Trogoderma*. Simple mechanisms for enhancing reproductive isolation in chemically mediated mating systems. *Journal of Chemical Ecology* 4, 409–423.

Shapas, T.J. and Burkholder, W.E. (1978b) Patterns of sex pheromone release from adult females, and effects of air velocity and pheromone release rates on theoretical communication distances in *Trogoderma glabrum*. *Journal of Chemical Ecology* 4, 395–408.

Shapas, T.J., Burkholder, W.E. and Boush, G.M. (1977) Population suppression of *Trogoderma glabrum* by using pheromone luring for protozoan pathogen dissemination. *Journal of Economic Entomology* 70, 469–474.

Sharma, S.P. and Deora, R.K. (1980) Factors affecting production, release and response to female sex pheromone in *Sitophilus oryzae* (L.) (Coleoptera, Curculionidae). *Indian Journal of Experimental Biology* 18, 463–465.

Shorey, H.H. (1973) Behavioral response to insect pheromones. *Annual Review of Entomology* 18, 349–380.

Shu, S., Koepnick, W.L., Mbata, G.N., Cork, A. and Ramaswamy, S.B. (1996) Sex pheromone production in *Callosobruchus maculatus* (Coleoptera, Bruchidae): electroantennographic and behavioral responses. *Journal of Stored Products Research* 32, 21–30.

Silverstein, R.M., Rodin, J.O., Burkholder, W.E. and Gorman, J.E. (1967) Sex attractant of the black carpet beetle. *Science* 157, 85–87.

Silverstein, R.M., Cassidy, R.F., Burkholder, W.E., Shapas, T.J., Levinson, H.Z., Levinson, A.R. and Mori, K. (1980) Perception by *Trogoderma* species of chirality and methyl branching at a site far removed from a functional group in a pheromone component. *Journal of Chemical Ecology* 6, 911–917.

Sinclair, E.R. and Howitt, C. (1984) Tests of pheromones for monitoring infestations of stored products insects. In: *Proceedings of the 4th Australian Conference on Applied Entomological Research*, pp. 482–487.

Singh, N.B., Campbell, A. and Sinha, R.N. (1976) An energy budget of *Sitophilus oryzae* (Coleoptera, Curculionidae). *Annals of the Entomological Society of America* 69, 503–512.

Sinha, R.N. (1984) Effects of weevil (Coleoptera, Curculionidae) infestation on abiotic and biotic quality of stored wheat. *Journal of Economic Entomology* 77, 1483–1488.

Smith, J.L., Cork, A., Hall, D.R. and Hodges, R.J. (1996) Investigation of the effect of female larger grain borer, *Prostephanus truncatus* (Horn) (Coleoptera, Bostrichidae), and their residues on the production of male aggregation pheromone by males. *Journal of Stored Product Research* 32, 171–181.

Stanic, V., Zlotkin, E. and Shulov, A. (1970) Localization of pheromone excretion in the female of *Trogoderma granarium* (Dermestidae). *Entomologia Experimentalis et Applicata* 13, 342–351.

Suezawa, Y., Miyashiro, R. and Tamura, A. (1987) Estimation of population of cigarette beetle, *Lasioderma serricorne* (F), by a synthetic sex pheromone and food attractant in factories of hand-extended fine noodles. *Nippon Shokuhin Kogyo Gakkaishi* 34, 635–639.

Surtees, G. (1964) Laboratory studies on dispersion behaviour of adult beetles in grain: VIII: Spontaneous activity in three species and a new approach to analysis of kinesis mechanisms. *Animal Behaviour* 12, 374–377.

Suzuki, T. (1980) 4,8-Dimethyldecanal: the aggregation pheromone of the flour beetles, *Tribolium castaneum* and *T. confusum* (Coleoptera, Tenebrionidae). *Journal of Agricultural and Biological Chemistry* 44, 2519–2520.

Suzuki, T. (1981) Identification of the aggregation pheromone of the flour beetles *Tribolium castaneum* and *T. confusum* (Coleoptera, Tenebriondae). *Journal of Agricultural and Biological Chemistry* 45, 1357–1363.

Suzuki, T. (1985) Presence of another aggregation substance(s) in the frass of the red flour beetles, *Tribolium castaneum* (Coleoptera, Tenebrionidae). *Applied Entomology and Zoology* 20, 90–91.

Suzuki, T. and Mori, K. (1983) (4*R*,8*R*)-(−)-4,8-Dimethyldecanal: the natural aggregation pheromone of the red flour beetle, *Tribolium castaneum* (Coleoptera, Tenebrionidae). *Applied Entomology and Zoology* 18, 134–136.

Suzuki, T. and Sugawara, R. (1979) Isolation of an aggregation pheromone from the flour beetles *Tribolium castaneum* and *T. confusum* (Coleoptera, Tenebrionidae). *Applied Entomology and Zoology* 14, 228–230.

Suzuki, T., Suzuki, T., Huynh, V.M. and Muto, T. (1975a) Hydrocarbon repellents isolated from *Tribolium castaneum* and *T. confusum* (Coleoptera, Tenebrionidae). *Journal of Agricultural and Biological Chemistry* 39, 2207–2211.

Suzuki, T., Suzuki, T., Huynh, V.M. and Muto, T. (1975b) Isolation of 2-hydroxy-4-methoxy-propiophenone from *Tribolium castaneum* and *T. confusum*. *Journal of Agricultural and Biological Chemistry* 39, 1687–1688.

Suzuki, T., Kozaki, J., Sugawara, R. and Mori, K. (1984) Biological activities of the analogs of the aggregation pheromone of *Tribolium castaneum* (Coleoptera, Tenebrionidae). *Applied Entomology and Zoology* 19, 15–20.

Suzuki, T., Nakakita, H. and Kuwahara, Y. (1987) Aggregation pheromone of *Tribolium freemani* Hinton (Coleoptera, Tenebrionidae). I. Identification of the aggregation pheromone. *Applied Entomology and Zoology* 22, 340–347.

Suzuki, T., Nakakita, H. and Kuwahara,Y. (1988) Cross-attraction among three *Tribolium* species and their hybrids (Coleoptera, Tenebrionidae). *Applied Entomology and Zoology* 23, 477–480.

Szentesi, A. (1981) Pheromone-like substances affecting host-related behaviour of larvae and adults in the dry bean weevil, *Acanthoscelides obtectus*. *Entomologia Experimentalis et Applicata* 30, 219–226.

Tanaka, K., Ohsawa, K., Honda, H. and Yamamoto, I. (1981) Copulation release pheromone, erectin, from the azuki bean weevil (*Callosobruchus chinensis* L.). *Journal of Pesticide Science* 6, 75–82.

Tanaka, K., Ohsawa, K., Honda, H. and Yamamoto, I. (1982) Synthesis of erectin, a copulation release pheromone of the azuki bean weevil, *Callosobruchus chinensis* L. *Journal of Pesticide Science* 7, 535–537.

Tanaka, Y., Honda, H., Ohsawa, K. and Yamamoto, I. (1986) A sex attractant of the yellow mealworm, *Tenebrio molitor* L., and its role in the mating behavior. *Journal of Pesticide Science* 11, 49–55.

Tigar, B.J., Key, G.E., Flores-Sanchez, M.E. and Vazquez-Arista, M. (1993) Flight periodicity of *Prostephanus truncatus* and longevity of attraction to synthetic pheromone. *Entomologia Experimentalis et Applicata* 66, 91–97.

Tigar, B.J., Osborne, P.E., Key, G.E., Flores-Sanchez, M.E. and Vazquez-Arista, M. (1994) Distribution and abundance of *Prostephanus truncatus* (Coleoptera: Bostrichidae) and its predator *Teretriosoma nigrescens* (Coleoptera: Histeridae) in Mexico. *Bulletin of Entomological Research* 84, 555–565.

Trematerra, P. (1997) Integrated pest management of stored-product insects: practical utilization of pheromones. *Anzeiger für Schädlingskunde, Pflanzenschutz und Umweltschutz* 70, 41–44.

Trematerra, P. and Daolio, E. (1990) Capture of *Rhyzopertha dominica* (F.) (Col., Bostrichidae) with dominicalure multifunnel traps and considerations on trapping of non target spiecies: *Sitophilus oryzae*, *Cryptolestes ferrugineus* and *Colydium castaneum*. *Journal of Applied Entomology* 100, 275–280.

Trematerra, P. and Girgenti, P. (1989) Influence of pheromone and food attractants on trapping of *Sitophilus oryzae* (L.) (Col., Curculionidae): a new trap. *Journal of Applied Entomology* 108, 12–20.

Tschinkel, W.R. (1970) Chemical studies on the sex pheromones of *Tenebrio molitor* (Coleoptera, Tenebrionidae). *Annals of the Entomological Society of America* 63, 626–627.

Tschinkel, W.R., Willson, C. and Bern, H.A. (1967) Sex pheromone of the mealworm beetle (*Tenebrio molitor*). *Journal of Experimental Zoology* 164, 81–86.

Valamoti, S.M. and Buckland, P.C. (1995) An early find of *Oryzaephilus surinamensis* (L.) (Coleoptera, Silvanidae) from final neolithic Mandalo, Macedonia, Greece. *Journal of Stored Products Research* 31, 307–309.

Valentine, J.M. (1931) The olfactory sense of the adult mealworm beetle *Tenebrio molitor* (Linn.). *Journal of Experimental Zoology* 58, 165–227.

Vanderwel, D., Pierce, H.D. Jr, Oehlschlager, A.C., Borden, J.H. and Pierce, A.M. (1990) Macrolide (cucujolide) biosynthesis in the rusty grain beetle, *Cryptolestes ferrugineus*. *Insect Biochemistry* 20, 567–572.

Vanderwel, D., Johnston, B. and Oehlschlager, A.C. (1992) Cucujolide biosynthesis in the merchant and rusty grain beetles. *Insect Biochemistry and Molecular Biology* 22, 875–883.

Vick, K.W., Burkholder, W.E. and Gorman, J.E. (1970) Interspecific response to sex pheromones of *Trogoderma* species (Coleoptera, Dermestidae). *Annals of the Entomological Society of America* 63, 379–381.

Vick, K.W., Drummond, P.C. and Coffelt, J.A. (1973a) *Trogoderma inclusum* and *T. glabrum*: effects of time of day on production of female pheromone, male responsiveness, and mating. *Annals of the Entomological Society of America* 66, 1001–1004.

Vick, K.W., Drummond, P.C., Sower, L.L. and Coffelt, J.A. (1973b) Sex pheromone habituation: the effects of habituation on the pheromone response level of *Trogoderma inclusum* (Coleoptera: Dermestidae). *Annals of the Entomological Society of America* 66, 667–670.

Voelkel, H. (1924) Zur Biologie und Bekämpfung des Khaprakäfers, *Trogoderma granarium* Everts. *Arbeiten aus der Biologischen Reichsanstalt für Land- und Fortswirtschaft Berlin* 13, 129–171.

Wakefield, M.E. (1999) The effect of insect age on the response of three species of *Sitophilus* to 4*S*,5*R*-sitophilure and food volatiles. In: *Proceedings of an International Working Conference on Stored Product Protection*, Beijing, PR China (in press).

Walgenbach, C.A. and Burkholder, W.E. (1986) Factors affecting the response of the maize-weevil, *Sitophilus zeamais* (Coleoptera, Curculionidae), to its aggregation pheromone. *Environmental Entomology* 15, 733–738.

Walgenbach, C.A. and Burkholder, W.E. (1987) Mating behavior of the maize weevil, *Sitophilus zeamais* (Coleoptera, Curculionidae). *Annals of the Entomological Society of America* 80, 578–583.

Walgenbach, C.A., Phillips, J.K., Faustini, D.L. and Burkholder, W.E. (1983) Male-produced aggregation pheromone of the maize weevil, *Sitophilus zeamais*, and inter-specific attraction between three *Sitophilus* species. *Journal of Chemical Ecology* 9, 831–841.

Walgenbach, C.A., Phillips, J.K., Burkholder, W.E., King, G.G.S., Slessor, K.N. and Mori, K. (1987a) Determination of chirality in 5-hydroxy-4-methyl-3-heptanone, the aggregation pheromone of *Sitophilus oryzae* (L.) and *S. zeamais* Motschulsky. *Journal of Chemical Ecology* 13, 2159–2169.

Walgenbach, C.A., Burkholder, W.E., Curtis, M.J. and Khan, Z.A. (1987b) Laboratory trapping studies with *Sitophilus zeamais* (Coleoptera, Curculionidae). *Journal of Economic Entomology* 80, 763–767.

Weidner, H. (1983a) Vorratsschädlinge. In: Heinze, K. (ed.) *Leitfaden der Schädlingsbekämpfung*. Wissenschaftliche Verlagsgesellschaft, Stuttgart, pp. 9–226.

Weidner, H. (1983b) Herkunft einiger in Mitteleuropa vorkommender Vorratsschädlinge, 1. Die Sitophilus Arten. (Coleoptera, Cuculionidae). *Mitteilungen des Internationalen Entomologischen Vereins* 8, 1–17.

White, P.R. and Chambers, J. (1989) Saw-toothed grain beetle *Oryzaephilus surinamensis* (L.) (Coleoptera, Silvanidae) antennal and behavioral responses to individual components and blends of aggregation pheromone. *Journal of Chemical Ecology* 15, 1015–1031.

White, P.R., Chambers, J., Walter, C.M., Wilkins, J.P.G. and Millar, J.G. (1989) Saw-toothed grain beetle *Oryzaephilus surinamensis* (L.) (Coleoptera, Silvanidae) collection, identification, and bioassay of attractive volatiles from beetles and oats. *Journal of Chemical Ecology* 15, 999–1013.

Williams, H.J., Silverstein, R.M., Burkholder, W.E. and Khorramshahi, A. (1981) Dominicalure 1 and 2: Components of aggregation pheromone from male lesser grain borer *Rhyzopertha dominica* (F.) (Coleoptera, Bostrichidae). *Journal of Chemical Ecology* 7, 759–780.

Wojcik, D.P. (1969) Mating behavior of 8 stored-product beetles (Coleoptera, Dermestidae, Tenebrionidae, Cucujidae, and Curculionidae). *Florida Entomologist* 52, 171–197.

Wong, J.W., Verigin, V., Oehlschlager, A.C., Borden, J.H., Pierce, H.D. Jr, Pierce, A.M. and Chong, L. (1983) Isolation and identification of two macrolide pheromones from the frass of *Cryptolestes ferrugineus* (Coleoptera, Cucujidae). *Journal of Chemical Ecology* 9, 451–474.

Wright, E.J. and Cogan, P.M. (1995) Trapping and sampling: a proposal for standards in reporting. *Journal of Stored Products Research* 31, 175–176.

Yarger, R.G., Silverstein, R.M. and Burkholder, W.E. (1975) Sex pheromone of the female dermestid beetle *Trogoderma glabrum* (Herbst). *Journal of Chemical Ecology* 1, 323–334.

Yinon, U. and Shulov, A. (1967a) New findings concerning pheromones by *Trogoderma granarium* (Everts) (Coleoptera, Dermestidae). *Journal of Stored Products Research* 3, 251–254.

Yinon, U. and Shulov, A. (1967b) A bioassay of the pheromone of *Trogoderma granarium* males as an attractant for both sexes of the species. *Entomologia Experimentalis et Applicata* 10, 453–462.

Zernecke, R. (1993) Beobachtungen zum Paarungs- und Kampfverhalten an drei *Sitophilus*-Arten. *Nachrichtenblatt der Bayrischen Entomologen* 42, 65–80.

8 Sawflies and Seed Wasps

OLLE ANDERBRANT

Department of Ecology, Ecology Building, Lund University, S-223 62 Lund, Sweden

Introduction

The Hymenoptera is one of the largest and most diverse insect orders: Mason and Huber (1993) estimated 300,000 species as a conservative world total. The order is divided into two suborders: Symphyta, or sawflies; and Apocrita, i.e. wasps, ants and bees. The Symphyta is the most ancient, with fossil records dating back to 200 million years ago (Mason and Huber, 1993), and almost entirely phytophagous. The Apocrita, containing over 90% of the species, is primarily carnivorous, but several groups have independently returned to plant-feeding habits. A general overview of classification, morphology, feeding habits and other aspects of the Hymenoptera is given by Goulet and Huber (1993), and a similar description of families occurring in Britain is provided by Gauld and Bolton (1988).

Several of the carnivorous groups contain species beneficial to agriculture or forestry, as they are natural enemies of various harmful insects. In this volume predators are treated in Chapter 14 and the parasitoids in Chapters 15 and 16. One special phytophagous group, the bees, is treated separately in Chapter 17. The present chapter deals with the presence, or potential occurrence, of pheromones in sawflies (Symphyta), including the parasitic family Orussidae, and in the phytophagous groups within the Apocrita subdivision traditionally called Parasitica or Terebrantia. The families covered by this review are listed in Table 8.1 together with the number of species known and main feeding habits. Thus, the phytophagous hymenopterans not dealt with in this chapter are those belonging to the apocrite subdivision Aculeata, namely bees, ants and members of the vespid subfamily Masarinae.

After a thorough review of the literature until 1997, it is quite evident that knowledge about pheromones in the groups treated here is sparse, with the exception of the sawfly family Diprionidae. Many aspects, including pheromones, of this and other sawfly families were reviewed in a recent book edited by Wagner and Raffa (1993). For a historical résumé and additional details about sawfly

Table 8.1. Hymenopteran families covered by this chapter, their number of species and feeding habits (compiled from Goulet and Huber, 1993).

	Family			
Suborder	Name	No. of species	Larval food source or feeding habit	Pheromone[a]
Symphyta				
	Cephidae	100	Twigs or grass stems	
	Megalodontidae	40	Herbaceous plants	
	Pamphiliidae	200	Foliage	++[b]
	Orussidae	70	Parasites	
	Siricidae	93	Wood	
	Argidae	800	Foliage	
	Blasticotomidae	9	Stems of ferns	
	Cimbicidae	130	Foliage	
	Diprionidae	90	Conifer needles	+++
	Pergidae	> 400	Foliage	
	Tenthredinidae	> 6000	Mainly foliage, some borers in stems and fruits, gall formers, miners	++
	Xyelidae	46	Cones, buds, foliage	
	Anaxyelidae	1	Wood	
	Xiphydriidae	80	Wood	
Apocrita (Parasitica)				
	Agaonidae	650	Figs	
	Eulophidae	3900	Parasites, predators, phytophagous	
	Eurytomidae	1425	Parasites, predators, stem miners, gall formers, seed eaters	+
	Pteromalidae	4115	Parasites, predators, gall inquilines	
	Tanaostigmatidae	90	Parasites, gall formers	
	Torymidae	1150	Parasites, gall formers or inquilines, seed eaters	+
	Cynipidae	1200	Gall formers or inquilines	+
	Mymarommatidae	9	Fungus ?	

[a] The extent to which the family's pheromone biology is known is indicated by: +, behavioural indication only; ++, pheromone of one species identified; +++, pheromone of several species identified.
[b] Only one potential pheromone component identified.

pheromones, refer to Anderbrant (1993). After a brief summary of sex determination and mating systems in Hymenoptera, the different families will be treated with respect to present knowledge about pheromones and relevant aspects of life history and mating systems. Lastly, the use of pheromones for plant protection and possible future research directions will be discussed.

Sex Determination and Reproductive Systems

Hymenopterans are characterized by haplodiploid sex determination. Generally males are haploid and females diploid; unfertilized eggs develop into males and fertilized eggs into females. This asexual production of males is called arrhenotokous parthenogenesis, or arrhenotoky. Several hymenopterans are thelytokous, which means that females (diploid) are produced parthenogenetically from unfertilized eggs, and the occurrence of males is very rare. A third mechanism is deuterotoky, in which either sex can develop from unfertilized eggs. The possible role of males in the latter two systems is not yet clear. A more thorough description of the different systems and their evolutionary implications is given by Kerr (1962), Engelmann (1970) and Gauld and Bolton (1988). During rearing, occasionally diploid or triploid males and triploid females can occur (e.g. Smith and Wallace, 1971; Naito and Suzuki, 1991), but it is not known how frequent these are in natural populations. There also exist gender anomalies – gynandromorphs, exhibiting characteristics of both sexes; these have, for instance, been reported from the diprionid species *Diprion similis* (Hartig) and *Neodiprion sertifer* (Geoffroy) (Mertins and Coppel, 1971; Heliövaara *et al.*, 1992).

Different modes of reproduction often occur within the same family or even the same genus. Thus, no general conclusions about the importance of sex pheromones within different groups can be drawn from the mating systems. Arrhenotokous species can be expected to have advanced mate-finding mechanisms, e.g. by means of pheromones, whereas thelytokous species are less likely to have such a system. However, other kinds of pheromones, especially host-marking pheromones (or ODP, oviposition-deterring pheromones: Prokopy *et al.*, 1984; Anderson, 1988), can be expected to occur in any species irrespective of their reproductive system. Pheromones of this kind emanate from the egg or developing larvae or are deposited by the female during oviposition, and serve as a repellence signal. Females arriving later avoid an occupied site and the competition experienced by the present larvae is minimized (also described for fruitflies in Chapter 1 this volume).

Symphyta: Diprionidae (Conifer Sawflies)

Species and life history

This small and relatively homogeneous family is the only one covered by this chapter that has received appreciable attention from pheromone scientists. The reason is that the family contains several of the most harmful defoliators on conifers (Smith, 1993). The main hosts are *Pinus* spp. and therefore the whole family is sometimes called pine sawflies. However, some species feed on other trees within *Pinaceae*, *Picea* spp. and *Thuja* spp., and also on trees within *Cupressaceae* (Smith, 1974). Diprionids place their eggs in the needles individually or in clusters, depending on species, and consequently the larvae may feed alone or gregariously. They overwinter in either the egg or prepupal stage in cocoons and may have one or more generations per year, depending on species and climate. Most species are arrhenotokous, but with a slightly female-biased sex ratio (Craig and Mopper, 1993), and probably all of these use pheromone communication to locate their mates. Males, but not females, have strongly branched, feather-like antennae (Fig. 8.1).

Various aspects of their biology and economic importance are covered by chapters in Wagner and Raffa (1993). The investigation of the diprionid sex pheromones was started by H.C. Coppel and co-workers in the late 1950s (Coppel *et al.*, 1960) and the first pheromone was identified 16 years later (Jewett *et al.*, 1976). A detailed description of methods used and a thorough review of the knowledge accumulated up to October 1991 was given by Anderbrant (1993). Thus, the presentation below concentrates on the progress achieved since then, but it will also gives a complete list of the species investigated from the pheromone point of view.

According to Goulet and Huber (1993), the number of diprionid species is around 90, but based on Smith (1974, 1988), Zhang and Zhou (1996) and Xu (1997) the present known number of diprionids seems to be around 126. Most of them (114) are found only in one of six distinct areas: (i) western North America (16 species); (ii) Central America and West Indies (10); (iii) eastern North America (23); (iv) Europe, North Africa, northern and central Asia (21); (v) eastern Asia, mainly China (31); and (vi) Japan (13).

Female-produced sex pheromone

The pheromones of the species investigated during the first three decades of research showed unusually little variation. Only one significant substance was found in the females: 3,7-dimethyl-2-pentadecanol (diprionol, Fig. 8.2, **7**).

Fig. 8.1. Copulating pair of *Diprion pini*. Note the strongly branched antennae of the male. (Photograph by Martti Varama.)

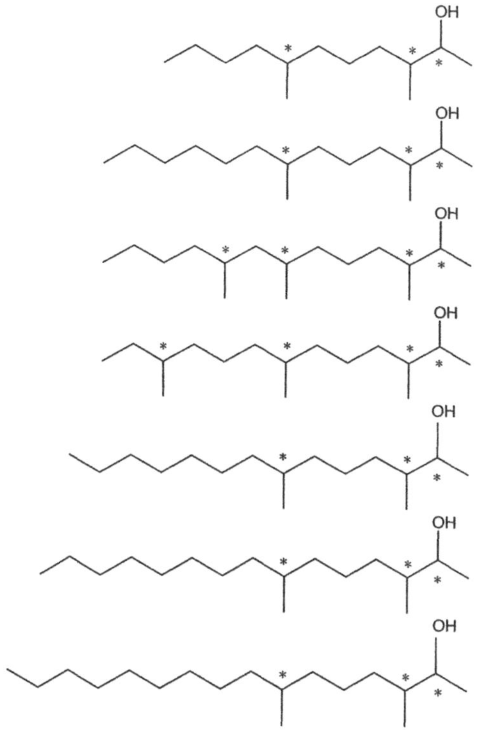

1 3,7-Dimethyl-2-undecanol

2 3,7-Dimethyl-2-tridecanol

3 3,7,9-Trimethyl-2-tridecanol

4 3,7,11-Trimethyl-2-tridecanol

5 3,7-Dimethyl-2-tetradecanol

6 3,7-Dimethyl-2-pentadecanol (diprionol)

7 3,7-Dimethyl-2-hexadecanol

Fig. 8.2. Structures of diprionol and related alcohols found in females of different species of pine sawflies (Diprionidae) (see Table 8.2), in some cases acting as pheromones, when esterified (see Tables 8.3 and 8.4). * Indicates the asymmetric centres. Numbers to the left are used in these tables for reference.

However, since this molecule has three asymmetric centres, eight stereoisomers can exist, allowing for a large number of species-specific pheromone compositions. It was also found that the alcohol is not behaviourally active, but has to be in an ester form to elicit male response. Only in the first pheromone identification (Jewett *et al.*, 1976), in which over 17,000 females were used, was the presence of esters (acetate and propionate) reported. In other studies the female extracts have been esterified before testing their activity. The production site, biochemical pathway and mode of esterification are still unknown. The only hint is that diprionol occurs in all parts of the body, even at the pupal stage (Wassgren *et al.*, 1992), indicating either that it is produced over the whole surface or that it is transported from the production site to other body parts. In the same study it was found that female *Neodiprion sertifer* contained not only diprionol but also homologous alcohols with carbon chains of different length (Fig. 8.2).

The first species that was shown to use a non-diprionol based pheromone was *Diprion pini* (L.). Its principal pheromone component is a homologue with two carbons less in the chain: 3,7-dimethyl-2-tridecanol (Fig. 2.2, 2) (Anderbrant *et al.*, 1995a; Bergström *et al.*, 1995). As with *N. sertifer*, several homologous alcohols were found in the females of *D. pini* (Bergström *et al.*, 1995; Fig. 8.2). 3,7-Dimethyl-2-tridecanol was found to be the principal pheromone component also in the Chinese species *D. jingyuanensis* Xiao and Zhang (Chen *et al.*, 1997), and *D. nipponica* Rohwer contains 3,7-dimethyl-2-undecanol (**1**) (Tai *et al.*, 1998). Recent studies have shown that species within other genera contain similar, but stereochemically more complicated, compounds. *Microdiprion pallipes* (Fallén) contains 3,7,11-trimethyl-2-tridecanol (**4**) (Bergström *et al.*, 1998) and *Macrodiprion nemoralis* (Enslin) contains 3,7,9-trimethyl-2-tridecanol (**3**) (E. Hedenström *et al.*, 1998, unpublished results), both substances with four stereogenic centres, making 16 isomers possible. In Table 8.2 diprionol-related substances found in diprionids are listed together with the species in which they have been found. It should be noted that probably none of the North American species was analysed with methods sensitive enough to detect minor homologues; alternatively they were not looked for. Thus, it is possible that the homologues reported from *N. sertifer* and *D. pini* also occur in some or all of the other species. With today's gas chromatographic techniques it is possible to separate stereoisomers occurring in quantities comparable to those found in a single female (Wassgren and Bergström, 1995). From the analyses of *N. sertifer* and *D. pini* it seems quite clear that the principal pheromone component, i.e. the compound essential for attracting males, is also the compound present in the largest quantity.

In addition to the diprionol-like substances, some 'odd' compounds have been identified from *N. sertifer* and other species. (*E,E*)-2,4-Heptadienal, (*E,E*)-2,4-decadienal, 1,6-germacradienol (Fig. 8.3) and *E*-perillenal (Fig. 8.3) were found in *N. sertifer* females (Wassgren *et al.*, 1992). 1,6-Germacradienol was also found in large quantities in larvae of this species; they apparently accumulate this substance from the pine needle diet, where its relative concentration is much lower (Bergström *et al.*, 1994). *E*-Perillenal is mainly found in the front part of the abdomen (Wassgren *et al.*, 1992), where a pair of lateral glands is situated (Mertins and Coppel, 1972; Hallberg and Löfqvist, 1981). However, so far no behavioural or other role has been found for this or any of the other 'odd' substances.

Female diprionids do not exhibit any distinct calling behaviour. Most of the time they sit quietly with the head towards the base of the needle. Occasionally they move their legs along the sides of the abdomen, as if preening (e.g. Mertins and Coppel, 1972), but it is not known if this behaviour is related to the release of pheromones.

Male response to the sex pheromone

The pheromone molecules elicit a response from receptors situated on long single-walled sensilla on the male antennae (Hansson *et al.*, 1991; Anderbrant *et al.*, 1995a). These sensilla are innervated by 8–12 neurones (Hallberg, 1979; Anderbrant *et al.*, 1995a), which may be compared with the two or three neurones normally occurring in moths. In the two species so far investigated by electrophysiological single-sensillum recordings, *Neodiprion sertifer* and *Diprion pini*, each neurone seems to be isomer specific (Hansson *et al.*, 1991; Anderbrant *et al.*, 1995a). However, one of the isomers – the attractive one – elicits response in several neurones, as indicated by highly variable spike amplitudes. In *D. pini* it

Table 8.2. Saturated methyl-branched alcohols found in female diprionid sawflies (main pheromone precursor indicated by ++).

Compound					Species[a]								
Carbon chain length	Methyl groups	Isomers	Name	Structure in Fig. 8.2	Dn	Dp	Ds	Mp	Mn	Nl	Nnn	Np	Ns
11	2	*threo*	3,7-Dimethyl-2-undecanol	**1**	++								
13	2	*erythro*	3,7-Dimethyl-2-tridecanol	**2**									+
13	2	*threo*	3,7-Dimethyl-2-tridecanol	**2**		++							
13	3	–	3,7,9-Trimethyl-2-tridecanol	**3**					++				
13	3	*erythro*	3,7,11-Trimethyl-2-tridecanol	**4**				++					
14	2	*erythro*	3,7-Dimethyl-2-tetradecanol	**5**		+							+
14	2	*threo*	3,7-Dimethyl-2-tetradecanol	**5**		+							
15	2	*erythro*	3,7-Dimethyl-2-pentadecanol	**6**		+	+			++	++	++	++
15	2	*threo*	3,7-Dimethyl-2-pentadecanol	**6**		+	++					+	+
16	2	*erythro*	3,7-Dimethyl-2-hexadecanol	**7**									+
16	2	*threo*	3,7-Dimethyl-2-hexadecanol	**7**		+							

[a] Dn, *Diprion nipponica* (Tai *et al.*, 1998); Dp, *Diprion pini* (Bergström *et al.*, 1995); Ds, *Diprion similis* (Olaifa *et al.*, 1988); Mp, *Microdiprion pallipes* (Bergström *et al.*, 1998); Mn, *Macrodiprion nemoralis* (E. Hedenström *et al.*, 1998, unpublished results); Nl, *Neodiprion lecontei* (Jewett *et al.*, 1976); Nnn, *Neodiprion nanulus nanulus* (Olaifa *et al.*, 1987); Np, *Neodiprion pinetum* (Olaifa *et al.*, 1988); Ns, *Neodiprion sertifer* (Wassgren *et al.*, 1992). Authors of species names are given in Tables 8.3 and 8.4.

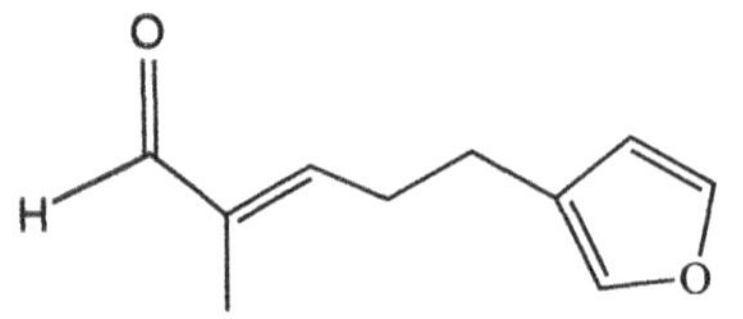

Perillenal

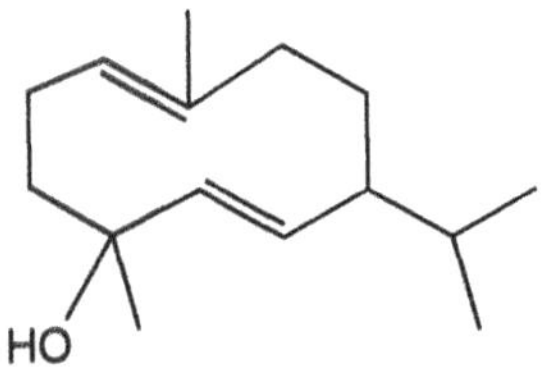

1,6-Germacradien-5-ol

Fig. 8.3. Structures of two compounds found in diprionid sawflies, but with as yet unknown function.

seems as if all neurones are stimulated by the same isomer, the attractive (2*S*,3*R*,7*R*) isomer (Anderbrant *et al.*, 1995a). In *N. sertifer* all neurones, except the one with the largest spikes, responded to the attractive (2*S*,3*S*,7*S*) isomer. The neurone exhibiting the largest spikes is sensitive to the (2*S*,3*R*,7*R*) isomer, which is a behavioural antagonist (Hansson *et al.*, 1991; Anderbrant *et al.*, 1992b). This is different from the situation in moths, in which the largest spikes are usually produced by neurones sensitive to the main pheromone component.

In several of the earlier identification studies, a less specific electrophysiological method, electroantennography (EAG), was used to screen the response to different esters and isomers (Jewett *et al.*, 1976; Kraemer *et al.*, 1979, 1981, 1983, 1984). However, the results were often difficult to interpret, and for most species our knowledge about the pheromone is mainly based on tests of male response to different diprionyl esters and isomers in the field. The male response for the species that have been subjected to tests with different isomers and esters are listed in Table 8.3. The data presented in the table should not be considered as definitive, since the results for most species rely on tests with synthetic pheromone preparations consisting of two or more isomers, sometimes with small amounts of contaminating isomers. If experiments are repeated with pure isomers (Högberg *et al.*, 1990) some of the activity/non-activity indications may have to be changed. For simplicity, the carbon numbers are omitted in connection with the stereogenic identity of the isomers, e.g. (2*S*,3*S*,7*S*)-diprionol is written as *SSS*-diprionol.

It is obvious that most diprionid species rely on one substance only for sexual attraction, which is quite different from the majority of other insects investigated. In a few cases other isomers can have an antagonistic or synergistic effect (Olaifa *et al.*, 1984, 1988; O. Anderbrant *et al.*, 1998, unpublished results). The latter study also revealed a strong intraspecific geographical variation of the male pheromone response in *N. sertifer*. The species for which no thorough tests of isomeric specificity have been reported are listed in Table 8.4, with indications of male responses to different esters.

Table 8.3. The response of male diprionid sawflies in the field to acetates and propionates of different (2*S*)-isomers of their pheromone (in all of the listed species, except *Diprion jingyuanensis* and *Neodiprion dailingensis*, both esters of all or some of the (2*R*)-isomers have been tested, but without showing any notable activity).

Species	Alcohol precursor[a]	Acetates				Propionates				References[b]
		SSS	*SSR*	*SRR*	*SRS*	*SSS*	*SSR*	*SRR*	*SRS*	
Diprion jingyuanensis Xiao and Zhang	**2**			0				++		1
D. nipponica Rohwer	**1**					0	0	++		2
D. pini L.	**2**	0	0	++	0	0	0	++	0	3, 4
D. similis (Hartig)	**6**	0	0	0		+/−	+/−	++	+	5, 6
Gilpinia frutetorum (Fabricius)	**6**	0	0	+		0	0	0		5
Neodiprion dailingensis Xiao and Zhou	**6**	++		0		++				7
N. dubiosus Schedl	**6**	0	0	0		0	0	++	+	8
N. lecontei (Fitch)	**6**	++	0	−		0		0		9, 10
N. nanulus nanulus Schedl	**6**	++	++	0		++	0	0		11, 12
N. nigroscutum Middleton[c]	**6**								++	8
N. pinetum (Norton)	**6**	++	0	+	+?	0	0			6, 13
N. pratti banksianae Rohwer	**6**	++	0	+	+	++	−?	+?		14
N. pratti pratti (Dyar)	**6**	0	0	0	0	0	0	++	0	11
N. rugifrons Middleton	**6**	?	0	0		?	0	++		8
N. sertifer (Geoffroy)	**6**	++	0	+/−	+/−	++	0	+?/−	+?/−	11, 12, 15, 16
N. swainei Middleton	**6**	0		0		++		?		8
N. taeda linearis Ross	**6**	++	0	0		++	0	0		11

[a] See Table 8.2 and Fig. 8.2.

[b] References: 1, Chen *et al.* (1997); 2, Tai *et al.* (1998); 3, Bergström *et al.* (1995); 4, Anderbrant *et al.* (1995a); 5, Kikukawa *et al.* (1982); 6, Olaifa *et al.* (1988); 7, Anderbrant *et al.* (1997); 8, Kraemer *et al.* (1984); 9, Matsumura *et al.* (1979); 10, Kraemer *et al.* (1981); 11, Kraemer *et al.* (1983); 12, Olaifa *et al.* (1987); 13, Kraemer *et al.* (1979); 14, Olaifa *et al.* (1984); 15, Kikukawa *et al.* (1983); 16, Anderbrant *et al.* (1992a,b).

[c] No explicit description of the baits used in tests with this species were given by Kraemer *et al.* (1984), but the *SRS*-propionate was superior to all other isomers or mixtures tested.

Indications: ++, attractive by itself; +, synergistic; −, antagonistic; +/−, synergistic at low ratios (in relation to the principal component) and antagonistic at higher ratios; 0, no activity; ?, interpretation less reliable. When indications fall between columns for isomers *SRR* and *SRS*, only the mixture of these two isomers has been investigated.

Table 8.4. Response of male diprionid sawflies to their pheromones, for species in which different isomers have not yet been tested in the field.

Species	Pheromone[a]	Ester specificity	Reference
Macrodiprion nemoralis (Enslin)	**3** ac	Propionate not active in field	E. Hedenström *et al.* (1998, unpublished results)
Microdiprion pallipes (Fallén)	erythro-**4** pr	Acetate not active in field	Bergström *et al.* (1998)
Neodiprion abbottii (Leach)	**6** pr	EAG showed stronger activity of pr than ac	Jewett *et al.* (1976)
N. abietis (Harris)	**6** pr	EAG showed stronger activity of pr than ac	Jewett *et al.* (1976)

[a] Number indicates the alcohol (see Fig. 8.2 and Table 8.2); ac, acetate; pr, propionate.

The 2*S*-configuration seems to be conserved in all species. The only study in which a 2*R*-diprionyl ester (*RRR*) has been reported to be active was one that attempted to trap *D. similis* in Europe (Longhurst *et al.*, 1980). However, the number of insects caught was low, despite strong pheromone sources, and thus these findings have to be interpreted with caution. It is also clear that all species are very specific with respect to the stereoisomer, whereas the ester group, acetate or propionate, is sometimes less important. Considering all species investigated to date (Tables 8.3 and 8.4), four species react only to the acetate, 11 species are propionate specific and six species are attracted by both esters. Looking closer at the two most studied genera, *Diprion* and *Neodiprion*, some differences appear. All *Diprion* species use the *SRR* propionate, but the carbon chain length differs between species (11, 13 or 15 carbons). In *Neodiprion*, all species use diprionol (15 carbons), whereas the stereoisomer may be *SSS* or *SRR/S* and the ester may be acetate or propionate.

When comparing a synthetic pheromone mixture (*SSS* + *SRR* diprionyl acetate) with a mixture of the analogues, void of the 7-position methyl group (*SS* + *SR*-3-methyl-2-pentadecyl acetate), Tai *et al.* (1992) found the analogue blend to be slightly attractive to male *N. sertifer* in Japan. This might indicate that the methyl group at carbon 3 is more important than that at the 7-position.

The diprionol homologues present in *N. sertifer* were acetylated and showed electrophysiological activity, but when tested in the field, either alone or together with diprionyl acetate, no behavioural activity was detected (Wassgren *et al.*, 1992). Similarly, no effect was observed when several of the minor compounds were added to the principal pheromone component of *D. pini* (Bergström *et al.*, 1995).

A detailed description of pheromone-induced behaviour in male diprionids is still lacking. Flight activity usually takes place from before noon until early evening, and is (at least in *N. sertifer*) to a large extent determined by the temperature (Jönsson and Anderbrant, 1993). Sunny conditions also promote flight, and moderate wind speed (2–3 m s^{-1}) seems to facilitate orientation to a pheromone source (Jönsson and Anderbrant, 1993; Wedding *et al.*, 1995). When approaching a pheromone trap the males exhibit a typical zigzagging flight (O. Anderbrant, 1998, unpublished results), often observed in moths (Cardé and Mafra-Neto, 1997). A preference for intense light during flight is the most probable reason for the lack of a good laboratory bioassay for this insect group.

Other diprionid pheromones

In at least two diprionid species the presence of a host-marking or oviposition-deterrent pheromone has been demonstrated. Tisdale and Wagner (1991) showed that egg washes of *Neodiprion fulviceps* (Cresson) contained one or more substances that deterred egg-laying females. Hilker and Weitzel (1991) and Blümke and Anderbrant (1997) found a similar effect of larval secretion in *D. pini*, whereas no such effect was found in *N. sertifer* (Blümke and Anderbrant, 1997). A reason for this difference could be that the probability of encountering larval colonies during oviposition depends on the life cycle. *D. pini* has overlapping generations in many areas, whereas *N. sertifer* never has more than one generation per year. However, in neither *D. pini* nor *N. sertifer* did the eggs seem to have any deterrent effect on other ovipositing females. Egg clusters are sometimes much larger than the egg number produced by a single female (Juutinen and Varama, 1986; Blümke and Anderbrant, 1997; M. Varama, Vantaa, 1998, personal communication).

In an extensive study of larval defence mechanisms, Prop (1960) noted that larvae of the gregarious species *D. pini* and *N. sertifer* sometimes had to leave a defoliated host plant and move to a new one. Observations of the larval behaviour indicated that the first individuals deposited a scent mark that then guided their siblings, but no further studies have been done.

Pheromones in Other Families of the Symphyta

The Tenthredinidae is by far the largest sawfly family. Most larvae are free-feeding on foliage of woody or herbaceous plants; some are leaf-miners or gall-makers or bore into twigs or fruits. Even mosses (*Bryopsida*) are reported as food plants (Vikberg and Nuorteva, 1997). Host species are from a large number of plant families and both Gauld and Bolton (1988) and Smith (1993) give many examples of genera or species associated with various plants. Several tenthredinids are known to be pests of agriculture and forestry as well as of ornamental plants. Reports about mating behaviour are rare in the literature, but for a few species observations indicated the presence of a female-produced sex pheromone. This is true for *Pristiphora conjugata* (Dahlbom) (Delmas, 1926), *Pristiphora geniculata* (Hartig) (Forbes and Daviault, 1964), *Nematus ribesii* (Scopoli) (Longhurst and Baker, 1980) and *Pikonema alaskensis* (Rowher) (Bartelt *et al.*, 1982a). However, it is only the last species (Fig. 8.4) that has been subject to thorough investigations of the pheromone. Bartelt *et al.* (1982b, 1983, 1984) and Bartelt and Jones (1983) found that the principal pheromone component was (Z)-10-nonadecenal (Fig. 8.5b), an air oxidation product of cuticular (Z,Z)-9,19 hydrocarbon dienes with 28–37 carbons. A second compound, (Z)-5-tetradecenol (Fig. 8.5b), acted as a synergist when released in relatively low quantities and induced a larger proportion of the males to land on the trap. When applied in larger amounts, this compound showed an antagonistic effect on males (Bartelt *et al.*, 1983). Experiments with *Athalia lugens infumata* (Marlatt) indicated that this species does not have a sex pheromone, as males often mounted other males, and that sex recognition was made after contact, possibly aided by female cuticular waxes (Kitano, 1988).

All species of Pamphiliidae (web-spinning sawflies) are associated with woody plants. The larvae often spin webs or roll leaves, and some feed together (Smith, 1993). Several *Cephalcia* species are considered severe pests on Norway spruce, *Picea abies* Karst., in central Europe (Pschorn-Walcher, 1982; Martinek, 1991, 1992; Marchisio *et al.*, 1994), and *C. lariciphila* (Wachtl) feeds on larch, *Larix* spp. (Billany and Brown, 1980; Pschorn-Walcher, 1982). A couple of *Cephalcia* species have been shown to use a female-produced sex pheromone for mate location. Borden *et al.* (1978) found that virgin larch web-spinning sawflies, *C. lariciphila*, attracted males, and observed wing-fanning behaviour among the males. Later, Baker *et al.* (1983) identified *ortho*-aminoacetophenone (Fig. 8.5a) from female

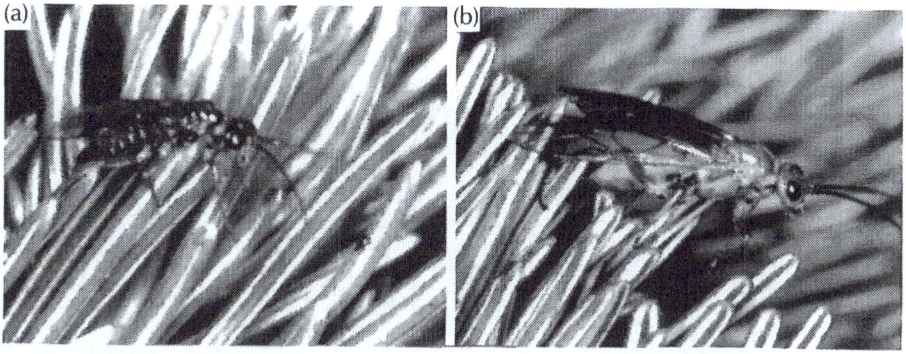

Fig. 8.4. (a) Male and (b) female yellowheaded spruce sawfly, *Pikonema alaskensis*. (Photographs by Robert J. Bartelt.)

extracts and found that it increased male activity in a wind tunnel experiment. However, no upwind flight towards the odour source could be demonstrated, whereas virgin females induced such behaviour among the males. No further investigations of the pheromone of this species have been reported. Battisti (1993) observed that males of the spruce web-spinning sawfly, *C. arvensis* (Panzer), were swarming near the ground (adults emerge from pupal chambers in the soil) and concentrated around females (Fig. 8.6). Males of *C. abietis* (L.), also feeding on spruce, were strongly attracted to traps containing virgin females, but only weakly to traps with males or mated females (Gruppe and Nisslein, 1996).

Within the remaining 11 families of Symphyta (Table 8.1), no studies have been done that indicate the presence of pheromones. Pschorn-Walcher (1982),

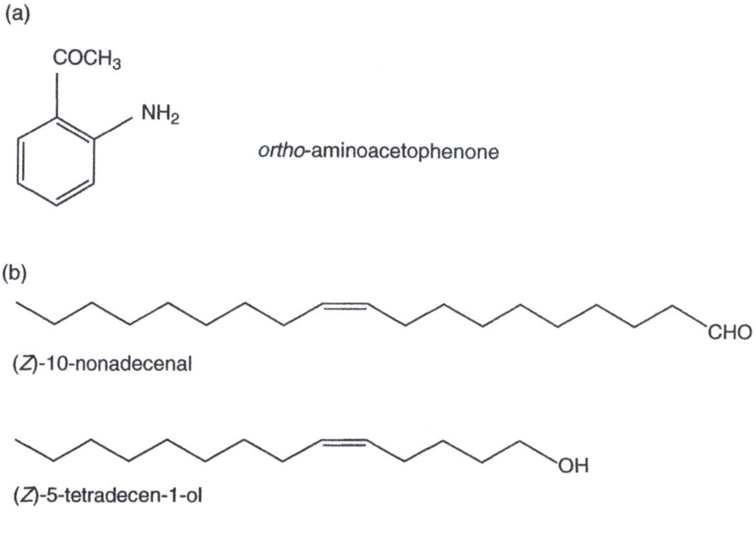

Fig. 8.5. Pheromone components of (a) the larch web-spinning sawfly, *Cephalcia lariciphila* and (b) the yellowheaded spruce sawfly, *Pikonema alaskensis*.

Fig. 8.6. Copulating pair of the spruce web-spinning sawfly, *Cephalcia arvensis*, sitting on *Lonicera* sp. in the undergrowth of a spruce forest. Most adults mate immediately after emergence from the soil. (Photograph by Andrea Battisti.)

Gauld and Bolton (1988) and Smith (1993) provided examples of host plants and species harmful to crops. Notable species are the wheat stem sawfly, *Cephus cinctus* Norton, which is an important pest of wheat, *Triticum aestivum* L., in the northern Great Plains of North America (Morrill and Kushnak, 1996), and the Australian eucalypt leaf-blister sawflies, *Phylacteophaga* spp., which have recently spread to new areas.

Pheromones in Apocrita (Parasitica)

Within this group there are eight partly or entirely phytophagous families (Table 8.1), but in only three of them have indications of pheromones been found.

Eurytomidae contains several phytophagous groups, of which the seed-feeding species are well studied. Damage may in some cases be considerable. For instance, Skrzypczynska (1975) reported 97% damage to larch seeds in Poland caused by *Eurytoma bouceki* Skrzypczynska, and Zerova and Fursov (1991) referred to losses exceeding 70% in almonds and plum caused by *Eurytoma amygdali* Enderlein and *E. schreineri* Schreiner, respectively. *Systole coriandri* (Gussakovsky) destroyed over 60% of coriander seeds in Chile (Lamborot *et al.*, 1993) and *Bruchophagus roddi* (Gussakovsky) can destroy up to 80% of the lucerne (alfalfa) crop (Thoenes and Moffett, 1987).

The reproductive system differs among the phytophagous species both between and within genera. Some are known to be thelytokous, e.g. *Harmelita secale* (Fitch) (Holmes and Blakeley, 1971) and *Bephratelloides cubensis* (Ashmead) (Grissell and Schauff, 1990; Nadel and Peña, 1991), and a number of species are arrhenotokous and obviously use a female-produced sex pheromone. In the lucerne seed chalcid *B. roddi* caged virgin females, but not males or mated females, attracted males (Erdélyi and Manninger, 1978); and similarly in *B. platypterus* (Walker), the trefoil seed chalcid, traps containing virgin females caught males, and in one test also some females (Peterson, 1991).

Much research has been devoted to one of the major pests of almonds, the almond wasp, *E. amygdali* (Fig. 8.7). One- to 12-day-old virgin females attract males, whereas mated females do not (Plaut, 1971; Pittara and Katsoyannos, 1985). Often several males court one female (Plaut, 1971) and copulation takes place on the fruit (Kouloussis and Katsoyannos, 1995). Females of this species prefer to oviposit on uninfested seeds (Plaut, 1971), and it was later shown that the female deposits a water-soluble host-marking pheromone (Kouloussis and Katsoyannos, 1991). The chemical is perceived at contact and, when the concentration is high, also from a distance, indicating an antennal function (Kouloussis and Katsoyannos, 1991). This host-marking or oviposition-deterring pheromone (ODP) could possibly be of applied significance, but when seed availability is reduced the female discrimination between ODP almonds and 'clean' almonds decreases (Kouloussis and Katsoyannos, 1993). There are no reports on the kinds of substances that are involved in the host-marking or sex pheromone of this or any other eurytomid.

In Torymidae, phytophagous species are limited to a few genera, of which

Fig. 8.7. Oviposition of the almond seed wasp, *Eurytoma amygdali*: (a) bringing the tip of the ovipositor in contact with the almond; (b) penetrating the pericarp with the ovipositor; and (c) ovipositing. (Photographs by Nikos Kouloussis.)

Megastigmus is the most studied. The phytophagous *Megastigmus* (there are also parasitoids) are often called seed chalcids as the larvae feed on seeds from a wide variety of plants (Lessmann, 1962). Some species have a very female-biased sex ratio, indicating thelytoky, whereas others have a nearly equal sex ratio and thus probably have arrhenotokous sex determination. Several species are considered severe pests, e.g. in coniferous seed plantations (Skrzypczynska, 1978; Blake *et al.*, 1989) or pistachio nut orchards (Rice and Michailides, 1988; Tzanakakis *et al.*, 1992). On the other hand, M. *aculeatus* (Swederus), which feeds on rose seeds, has been considered for use in the biological control of roses when these occur as weeds in pastures, recreational areas and elsewhere (Syrett, 1990). In addition to a nearly equal sex ratio, there are a few reports that might indicate the presence of a sex pheromone in some species. In M. *bipunctatus* Swederus, which feeds on seeds of juniper, *Juniperus communis* L., the males hatch first and locate seeds where females will emerge. Often the males aggregate, but after getting excited one male chases away the others and mates with the female (Dolgin, 1994). In M. *specularis* Walley, Hedlin (1956) observed males vibrating the antennae and moving the body from side to side when approaching the female, behaviours often recorded in connection with close-range odour communication in other insects. Orr and Borden (1983) described in detail the close-range mating behaviour of M. *pinus* Parfitt. Hussey (1955) similarly described mating in M. *spermotrophus* Wachtl, but did not mention how pairs located each other.

All species in the Cynipidae either form galls or are inquilines in galls of other insects. The vast majority of the species make galls on oaks, *Quercus* spp. (e.g. Evans, 1972; Abe, 1991; Eckberg and Cranshaw, 1994). Askew (1984) provides a general overview of the biology of gall wasps. Several species live on other plants, where they occasionally may be harmful (Ôtake, 1980; Jones, 1983; Baumann, 1992; Shorthouse, 1994). Most species have alternating bisexual and agamic generations. An individual agamic female produces either only haploid eggs developing into males or diploid eggs that give rise to sexual females (Hails and Crawley, 1991). However, several species seem to be entirely parthenogenetic (Ôtake, 1980; Stille, 1984, 1985; Shorthouse and Ritchie, 1984; Manojlovic and Tosevski, 1990). Only a few studies exist that treat the mating behaviour of cynipids. Abe (1991) made detailed observations of the courtship behaviour of *Synergus japonicus* Walker. It was noted that females mated only once, while males mated several times without discriminating between virgin and mated or young and old females. *Andricus quercuscalicis* (Burgsdorf) has a sexual generation that forms galls on Turkey oak, *Quercus cerris* (L.); males hatch first and battle for position over female galls (Hails and Crawley, 1991). Most likely the males are guided by a female-emitted or induced odour. Finally, in *Callirhytis eldoradensis* (Beutenmuller) and C. *flora* Weld, males were attracted to caged couples, indicating the presence of a sex pheromone (Dailey *et al.*, 1974).

Four of the remaining five families contain species harmful to agricultural, ornamental or forestry crops. Examples are found in Lateef *et al.* (1985) for Tanaostigmatidae, in Galil and Copland (1981) for Agaonidae, in Redak and Bethke (1995) for Eulophidae and in West and Shorthouse (1989) for Pteromalidae.

All species within the Agaonidae are associated with figs, *Ficus* spp., either as pollinators of the internal flowers (subfamily Agaoininae) or as phytophages competing with the pollinating wasps for food and space (Goulet and Huber, 1993). Males of pollinating fig wasps are apterous and mating takes place inside the fig before the females leave (Frank, 1985; Murray, 1985). Thus, only very short-distance (if any) chemical communication for mate finding should be expected among these species, although host finding is at least partly aided by olfaction (Hossaert-McKey *et al.*, 1994). Among the non-pollinating species, there are both wingless and winged males and for the latter group one could in theory expect sex pheromones to occur. However, no observations in this direction have been reported.

There are three reports indicating that species within the Pteromalidae are thelytokous, and thus that the presence of a sex pheromone is unlikely: a strongly female-biased sex ratio (5:1) was found in *Trichilogaster acaciaelongifoliae* Froggatt (Dennill, 1987); males of *Trichilogaster* sp. did not improve galling activity (Dennill and Gordon, 1991); and *Hemadas nubilipennis* Ashmead was reported to oviposit immediately after emergence (West and Shorthouse, 1989). In none of the families Eulophidae, Tanaostigmatidae or Mymarommatidae have any notations related to reproductive systems been recorded.

Use in Plant Protection

Detection and monitoring

Considering that comparatively little is known about pheromones in these phytophagous hymenopteran groups, it is not surprising that few attempts have been made to use pheromones for practical purposes. In at least two studies, the natural pheromone has been used for monitoring. Thomas *et al.* (1982) used esterified female extract of *D. similis* to monitor its occurrence in parts of eastern North America, and Katsoyannos *et al.* (1992) used traps with virgin female almond seed wasps, to optimize the timing of chemical control.

Monitoring is much easier once the pheromone of a species has been identified and synthetic samples become available. Various aspects of the biology can then be elucidated with the aid of pheromone traps. For instance, the flight period can be determined (Simandl, 1993; Schedl, 1994) and the abiotic factors important for flight can be investigated (Jönsson and Anderbrant, 1993). Likewise the flight activity at different heights can be measured (Wilkinson *et al.*, 1987; Anderbrant *et al.*, 1989). The importance of trap position was demonstrated when pheromone traps for *N. sertifer* were placed at different heights and at variously sun-exposed sites in a pine plantation (Simandl and Anderbrant, 1995). Catches at the edge were two to five times higher than inside the stand in traps 2.5 m above ground. Moreover, traps installed in the tree crown (> 9 m) caught three to 15 times as many males as did traps suspended at 2.5 m. It is well known that diprionids tend to fly and oviposit at sunny sites (Blümke and Anderbrant, 1997, and references therein). All of these factors are important for a correct interpretation of catches when using pheromone traps for more quantitative estimates.

The aim of using pheromone monitoring traps is usually to indicate when and where a population increase can be expected. To make such a method successful there has to be a relationship between trap catch and population density or population trend. Only a few attempts to arrive at such a relationship have been made with the insects covered by this chapter.

Morse and Kulman (1985) used pheromone traps to monitor flight activity of the yellowheaded spruce sawfly, *P. alaskensis*, and its parasitoid, *Syndipnus rubiginosus* Walley (Ichneumonidae) (see also Chapter 15 this volume). The aim was to study the relationships between catches of the two insects and the degree of defoliation caused by the sawfly. They did not find any significant relationship. However, when comparing the population trend (measured as change in defoliation since the previous year) with trap catches, a significant relation could be formulated. The best fit was achieved when catches of both the sawfly and its parasitoid were included together with their interaction coefficient and the tree height. They concluded that monitoring of both the pest and its natural enemy may give information about the extent of natural control of the pest.

A long-term monitoring study of *N. sertifer* in young Scots pine plantations in Sweden has so far shown relatively weak relationships (Lyytikäinen-Saarenmaa *et al.*, 1999). In general, trap catches varied little compared with population density in the study plots, but increased trap catch for more than one consecutive year coincided with increased population levels.

In species without a known pheromone, attempts have been made to use other methods to catch species regarded as pests. In particular, the use of traps of certain attractive colours should be mentioned. Methods using such traps are under development for several sawflies (Wildbolz and Staub, 1986; Barker *et al.*, 1997; Battisti and Rodeghiero, 1998). For the gall wasp, *Oncastichus goughi* Headrick & LaSalle, Redak and Bethke (1995) showed that sticky traps of several colours were suitable for monitoring. For many species, practical methods are not available and the development of pheromone traps would certainly contribute to more effective pest management (e.g. Peterson, 1991).

Direct control

Coppel *et al.* (1960) pointed out the usefulness of pine sawfly pheromones for population control, and actually predicted what now is called an attracticide, i.e. a combined use of pheromone and insecticide (Lanier, 1990). No attempts either in that direction or by mass trapping have been performed. Mating disruption, so far the most successful sex pheromone-based control method, in which synthetic pheromone is released in relatively large quantities to disturb the males in their mate location, has been attempted on *N. sertifer*. In a series of studies, Anderbrant *et al.* (1995b,c, 1998) showed that trap catches can be reduced almost completely when applying the attractive pheromone, *SSS* diprionyl acetate (pure or in *erythro* mixture), in amounts of approximately 3 mg ha^{-1} day^{-1}. Lower amounts resulted in smaller catch reduction, as did application of the antagonistic *SRR* isomer alone or in combination with the *SSS* isomer. No effect on the sex ratio could be demonstrated the following year (as *N. sertifer* is arrhenotokous, a male-biased ratio would be expected in the absence of matings), and the population density remained unchanged when compared with untreated areas (Anderbrant *et al.*, 1995c). Several hypotheses were put forward to explain these results. One supposition – that females, especially mated ones, fly more than was believed earlier (see references in Anderbrant, 1993) – was tested by releasing and observing virgin females inside and outside pheromone-treated areas (Östrand *et al.*, 1999). It was found that mating was more frequent outside than inside the treated area and that females often left their original shoot, before as well as after copulation. This suggests that females have a wider dispersal ability than was thought, which might contribute to the lack of effects of the pheromone treatment on the population. Also, prolonged diapause may have a great influence on the population dynamics and densities. Additional experiments are needed in order to evaluate whether or not mating disruption is a feasible way to protect pine forests against diprionids.

Conclusions

Sex pheromones occur in several of the hymenopteran families covered by this chapter, and it is likely that thorough investigations would reveal their presence in many more taxonomic groups. This takes us to one area of future research that should be encouraged, namely more pheromone identifications to reveal pheromone diversity and make applied use possible. The substances used as pheromones within the Diprionidae are very similar (Fig. 8.1). Is it the same in other families and can any links between different taxonomic groups be detected by pheromone structures? This task is not as straightforward as it might seem at first glance. It is likely that different groups use different chemicals, many of which will probably be new to science. Thus, challenges are offered not only to biologists but also to chemists.

Since pest species are present in many families in addition to the Diprionidae, such as the Tenthredinidae, Eurytomidae and Torymidae, pheromone identification work would provide the basis for the use of pheromones in plant protection. As soon as the active substances have been identified and synthesized, they can be used in traps to determine distribution and phenology and in studies aimed at uncovering various aspects of a species' biology. After additional development, traps may be used in monitoring programmes.

Even more exciting is the in-depth examination of various aspects of the pheromone communication systems. The following points can be mentioned.

- *Production site and biosynthesis*. Are pheromones in the groups listed here associated with the cuticular substances as in *P. alaskensis* and possibly in the diprionids, or do specific glands occur in other taxa? Experiments with labelled precursors and attempts to determine what is actually released by the female should be encouraged. Godfray and Hardy (1993) suggested that haplodiploid species should use pheromones that are energetically relatively cheap to produce, compared with pheromones in diplodiploid insects. Since females of haplodiploid species are able to reproduce without mating, a weaker selection pressure on the mate-finding mechanisms can be expected.

- *Receiver response*. The only thorough electrophysiological investigation was done on *N. sertifer* (Hansson *et al*., 1991), and it revealed several discrepancies, at the receptor neurone level, from the pattern found in, for example, Lepidoptera (see above and Anderbrant, 1993). When it comes to knowledge about central nervous system organization and signal processing we know absolutely nothing. Even characteristics of the pheromone-induced behaviour are poorly known, and access to a reliable laboratory bioassay would be highly valuable for future research.
- *Genetics and evolution*. Knowledge as to how pheromone production and response are inherited is just starting to accumulate for moths (Phelan, 1997). However, even without such knowledge one could probably gain insight into the role of pheromone communication in various evolutionary processes by a careful selection of study species. One such example might be the closely related *Neodiprion* species in North America, several of which share the same host trees and have similar phenology. Investigations have indicated that many of these *Neodiprion* species use the same substance as attractant, namely the SSS-diprionyl acetate or propionate. Are other cues than pheromones playing an even more important role in mate finding and recognition? Are these species reproductively isolated or are they in the process of speciation by, for instance, reinforcement (Butlin, 1987)? Also, studies of widely distributed species, occurring together with different 'guilds' of diprionids in various parts of its range, would be interesting. Is the pheromone system of such a species adapted to the differences in pheromones among the guilds?

Acknowledgements

Firstly, I would like to thank Robert J. Bartelt at the USDA National Center for Agriculture Utilization Research, Peoria, Illinois; Gunnar Bergström at Göteborg University; Hans-Erik Högberg at Mid-Sweden University, Sundsvall; Byron I. Katsoyannos at the University of Thessaloniki; David R. Smith at Systematic Entomology Laboratory, USDA, Washington, DC; Andreas Taeger at Deutsches Entomologisches Institute, Eberswalde; Martti Varama at the Finnish Forest Research Institute, Vantaa; and the editors of this volume for valuable comments on the manuscript. Photographic illustrations were kindly provided by Andrea Battisti, University of Padova; Nikos Kouloussis, University of Thessaloniki; Robert J. Bartelt and Martti Varama. The extensive literature review was obtained with the kind help of Nicholaus Voin and the staff at the Ecology Library in Lund. This chapter was written with financial support from the Commission of the European Communities, Agriculture and Fisheries (FAIR) specific RTD programme, CT95–0339, 'Pine sawfly pheromones for sustainable management of European forests (PHERODIP)'. It does not necessarily reflect the Commission's view and in no way anticipates its future policy in this area.

References

Abe, Y. (1991) Notices on courtship behavior and copulation of *Synergus japonicus* Walker (Hym., Cynipidae). *Journal of Applied Entomology* 111, 478–483.

Anderbrant, O. (1993) Pheromone biology of sawflies. In: Wagner, M.R. and Raffa, K.F. (eds) *Sawfly Life History Adaptations to Woody Plants*. Academic Press, San Diego, pp. 119–154.
Anderbrant, O., Löfqvist, J., Jönsson, J. and Marling, E. (1989) Effects of pheromone trap type, position and colour on the catch of the pine sawfly *Neodiprion sertifer* (Geoff.) (Hym., Diprionidae). *Journal of Applied Entomology* 107, 365–369.
Anderbrant, O., Bengtsson, M., Löfqvist, J. and Baeckström, P. (1992a) Field response of the pine sawfly *Neodiprion sertifer* to controlled release of diprionyl acetate, diprionyl propionate and *trans*-perillenal. *Journal of Chemical Ecology* 18, 1707–1725.
Anderbrant, O., Löfqvist, J., Högberg, H.-E., Hedenström, E., Wassgren, A.-B., Bergström, G., Bengtsson, M. and Magnuson, G. (1992b) Field response of the pine sawfly *Neodiprion sertifer* to the pheromone (2*S*,3*S*,7*S*)-diprionyl acetate and its stereoisomers. *Entomologia Experimentalis et Applicata* 62, 169–181.
Anderbrant, O., Hansson, B.S., Hallberg, E., Geri, C., Varama, M., Hedenström, E., Högberg, H.-E., Fägerhag, J., Edlund, H., Wassgren, A.-B., Bergström, G. and Löfqvist, J. (1995a) Electrophysiological and morphological characteristics of pheromone receptors in male pine sawflies, *Diprion pini* (Hymenoptera: Diprionidae), and behavioural response to some compounds. *Journal of Insect Physiology* 41, 395–401.
Anderbrant, O., Löfqvist, J., Högberg, H.-E. and Hedenström, E. (1995b) Development of mating disruption for control of pine sawfly populations. *Entomologia Experimentalis et Applicata* 74, 83–90.
Anderbrant, O., Löfqvist, J., Hedenström, E. and Högberg, H.-E. (1995c) Mating disruption of the pine sawfly *Neodiprion sertifer* (Hymenoptera: Diprionidae): effects on pheromone trap catch, sawfly sex ratio and density and on tree damage. In: Hain, F.P., Salom, S.M., Ravlin, W.F., Payne, T.L. and Raffa, K.F. (eds) *Behavior, Population Dynamics, and Control of Forest Insects: Proceedings of the International Union of Forestry Research Organizations Joint Conference, February 1994*. The Ohio State University Press, pp. 415–427.
Anderbrant, O., Zhang, Q.-H. and Chu, D. (1997) Diprionyl esters attractive to males of the Dailing pine sawfly *Neodiprion dailingensis* Xiao et Zhou (Hym., Diprionidae) in north-eastern China. *Journal of Applied Entomology* 121, 281–283.
Anderbrant, O., Hedenström, E., Högberg, H.-E. and Löfqvist, J. (1998) Towards use of pine sawfly pheromones in forest protection: evaluation of behavioral antagonist for mating disruption of *Neodiprion sertifer*. In: McManus, M.L. and Liebhold, A.M. (eds) *Proceedings: Population Dynamics, Impacts, and Integrated Management of Forest Defoliating Insects*. US Department of Agriculture, Forest Service, Northeastern Forest Experiment Station, General Technical Report NE-247, pp. 53–63.
Anderson, P. (1988) Oviposition deterring pheromones in insects. *Entomologisk Tidskrift* 109, 14–18. (In Swedish with English abstract.)
Askew, R.R. (1984) The biology of gall wasps. In: Ananthakrishnan, T.N. (ed.) *The Biology of Gall Insects*. Arnold, London, pp. 223–271.
Baker, R., Longhurst, C., Selwood, D. and Billany, D. (1983) Ortho-aminoacetophenone: a component of the sex pheromone system of the web spinning larch sawfly, *Cephalcia lariciphila* Wachtl. *Experientia* 39, 993–994.
Barker, A.M., Sanbrooke, K.J. and Aebischer, N.J. (1997) The water trap colour preferences of farmland sawflies. *Entomologia Experimentalis et Applicata* 85, 83–86.
Bartelt, R.J. and Jones, R.L. (1983) (*Z*)-10-Nonadecenal: a pheromonally active air oxidation product of (*Z*,*Z*)-9,19 dienes in yellow-headed spruce sawfly. *Journal of Chemical Ecology* 9, 1333–1342.
Bartelt, R.J., Jones, R.L. and Kulman, H.M. (1982a) Evidence for a multicomponent sex pheromone in the yellow-headed spruce sawfly *Pikonema alaskensis*. *Journal of Chemical Ecology* 8, 83–94.

Bartelt, R.J., Jones, R.L. and Kulman, J.M. (1982b) Hydrocarbon components of the yellowheaded spruce sawfly *Pikonema alaskensis* sex pheromone: a series of (Z,Z)-9,19 dienes. *Journal of Chemical Ecology* 8, 95–114.

Bartelt, R.J., Krick, T.P. and Jones, R.L. (1983) (Z)-5-Tetradecen-1-ol: a secondary pheromone of the yellowheaded spruce sawfly and its relationship to (Z)-10-nonadecenal. *Journal of Chemical Ecology* 9, 1343–1352.

Bartelt, R.J., Krick, T.P. and Jones, R.L. (1984) Cuticular hydrocarbons of the yellowheaded spruce sawfly, *Pikonema alaskensis*. *Insect Biochemistry* 14, 209–214.

Battisti, A. (1993) Bionomics of the spruce web-spinning sawfly *Cephalcia arvensis* Panzer (Hym., Pamphiliidae) in northeastern Italy. *Journal of Applied Entomology* 115, 52–61.

Battisti, A. and Rodeghiero, M. (1998) Monitoring spruce web-spinning sawflies *Cephalcia* spp.: the correlation between trap catches and soil sampling. *Entomologia Experimentalis et Applicata* 88, 211–217.

Baumann, T. (1992) Description of the male and a new host record of the gall wasp *Isocolus centaureae* (Hymenoptera: Cynipidae). *Entomologia Generalis* 17, 49–51.

Bergström, G., Wassgren, A.-B. and Birgersson, G. (1994) 1,6-Germacradien-5-ol identified in the larval discharge of pine sawfly *Neodiprion sertifer*. *Acta Chemica Scandinavica* 48, 187–188.

Bergström, G., Wassgren, A.-B., Anderbrant, O., Fägerhag, J., Edlund, H., Hedenström, E., Högberg, H.-E., Geri, C., Auger, M.A., Varama, M., Hansson, B.S. and Löfqvist, J. (1995) Sex pheromone of the pine sawfly *Diprion pini* (Hymenoptera: Diprionidae): chemical identification, synthesis and biological activity. *Experientia* 51, 370–380.

Bergström, G., Wassgren, A.-B., Anderbrant, O., Ochieng, S.A., Östrand, F., Hansson, B.S., Hedenström, E. and Högberg, H.-E. (1998) The sex pheromone of the pine sawfly *Microdiprion pallipes* (Hymenoptera: Diprionidae). *Naturwissenschaften* 85, 244–248.

Billany, D.J. and Brown, R.M. (1980) The web spinning larch sawfly, *Cephalcia lariciphila* Wachtl. (Hymenoptera: Pamphiliidae). A new pest of *Larix* in England and Wales. *Forestry* 53, 71–80.

Blake, E.A., Wagner, M.R. and Koerber, T.W. (1989) Relative effect of seed and cone insects on ponderosa pine in northern Arizona. *Journal of Economic Entomology* 82, 1691–1694.

Blümke, A. and Anderbrant, O. (1997) Oviposition pattern and behaviour of the pine sawfly *Neodiprion sertifer* (Hymenoptera: Diprionidae). *Bulletin of Entomological Research* 87, 231–238.

Borden, J.H., Billany, D.J., Bradshaw, J.W.S., Edwards, M., Baker, R. and Evans, D.A. (1978) Pheromone response and sexual behavior of *Cephalcia lariciphila* Wachtl. (Hymenoptera: Pamphiliidae). *Ecological Entomology* 3, 13–24.

Butlin, R. (1987) Speciation by reinforcement. *Trends in Ecology and Evolution* 2, 8–13.

Cardé, R.T. and Mafra-Neto, A. (1997) Mechanisms of flight of male moths to pheromone. In: Cardé, R.T. and Minks, A.K. (eds) *Insect Pheromone Research, New Directions*. Chapman & Hall, New York, pp. 275–290.

Chen, G.-F., Zhang, Q.-H., Li, Z.-Y., Miao. Z.-W., Wang, L.-Z. and Anderbrant, O. (1997) A preliminary study on the sex pheromone of *Diprion jingyuanensis* Xiao et Zhang. *Chinese Journal of Biological Control* 13, 61–64. (In Chinese with English abstract.)

Coppel, H.C., Casida, J.E. and Dauterman, W.C. (1960) Evidence for a potent sex attractant in the introduced pine sawfly, *Diprion similis* (Hymenoptera: Diprionidae). *Annals of the Entomological Society of America* 53, 510–512.

Craig, T.P. and Mopper, S. (1993) Sex ratio variation in sawflies. In: Wagner, M.R. and Raffa, K.F. (eds) *Sawfly Life History Adaptation to Woody Plants*. Academic Press, San Diego, pp. 61–92.

Dailey, D.C., Perry, T. and Sprenger, C.M. (1974) Biology of three *Callirhytis* gall wasps from Pacific slope *Erythrobalanus* oaks (Hymenoptera: Cynipidae). *Pan-Pacific Entomologist* 50, 60–67.

Delmas, C.S. (1926) Notes sur la biologie de *Pristiphora conjugata* Dahlbom. *Bulletin Biologique de la France et de la Belgique* 60, 448–472.

Dennill, G.B. (1987) Establishment of the gall wasp *Trichilogaster acaciaelongifoliae* (Pteromalidae) for the biological control of *Acacia longifolia* in South Africa. *Agriculture, Ecosystems and Environment* 19, 155–168.

Dennill, G.B. and Gordon, A.J. (1991) *Trichilogaster* sp. (Hymenoptera: Pteromalidae) a potential biocontrol agent for the weed *Acacia pycnantha* (Fabaceae). *Entomophaga* 36, 295–302.

Dolgin, M.N. (1994) Biology of *Megastigmus bipunctatus* Swed. (Hymenoptera, Torymidae) in Komi Republic. *Entomological Review* 73, 30–35.

Eckberg, T.B. and Cranshaw, W.S. (1994) Occurrence of the oak rough bulletgall wasp, *Disholcaspis quercusmamma* (Walsh) (Hymenoptera: Cynipidae), as a street tree pest in Colorado. *Journal of the Kansas Entomological Society* 67, 290–293.

Engelmann, F. (1970) *The Physiology of Insect Reproduction*. Pergamon Press, Oxford, 307 pp.

Erdélyi, C. and Manninger, S. (1978) Evidence for a female sex pheromone in the lucerne seed chalcid *Bruchophagus roddi* (Hymenoptera: Eurytomidae). *Acta Phytopathologica Academiae Scientiarum Hungaricae* 13, 219–222.

Evans, D. (1972) Alternate generations of gall cynipids (Hymenoptera: Cynipidae) on Garry oak. *Canadian Entomologist* 104, 1805–1818.

Forbes, R.S. and Daviault, L. (1964) The biology of the mountain-ash sawfly, *Pristiphora geniculata* (Htg.) (Hymenoptera: Tenthredinidae), in eastern Canada. *Canadian Entomologist* 96, 1117–1133.

Frank, S.A. (1985) Hierarchical selection theory and sex ratios. II. On applying the theory and a test with fig wasps. *Evolution* 39, 949–964.

Galil, J. and Copland, J.W. (1981) *Odontofroggatia galili* in Israel: a primary fig wasp of *Ficus microcarpa* with a unique ovipositor mechanism (Epichrysomallinae: Chalcidoidea). *Proceedings of the Koninklijke Nederlandse Akademie van Wetenschappen Series C, Biological and Medical Sciences* 84, 183–196.

Gauld I. and Bolton B. (1988) *The Hymenoptera*. Oxford University Press, Oxford, 329 pp.

Godfray, H.C.J. and Hardy, I.C.W. (1993) Sex ratio and virginity in haplodiploid insects. In: Wrensch, D.L. and Ebbert, M.A. (eds) *Evolution and Diversity of Sex Ratios in Insects and Mites*. Chapman & Hall, New York, pp. 402–417.

Goulet H. and Huber J.T. (1993) *Hymenoptera of the World*. Research Branch, Agriculture Canada, Publication 1894/E, Ontario, 668 pp.

Grissell, E.E. and Schauff, M.E. (1990) A synopsis of the seed-feeding genus *Bephratelloides* (Chalcidoidea: Eurytomidae). *Proceedings of the Entomological Society of Washington* 92, 177–187.

Gruppe, A. and Nisslein, R. (1996) Bedeutung von Sexualpheromonen und Licht für Flugaktivität und Attraktion männlicher Fichtengespinstblattwespen *Cephalcia abietis* L. (Hym., Pamphiliidae). *Journal of Applied Entomology* 120, 107–109.

Hails, R.S. and Crawley, M.J. (1991) The population dynamics of an alien insect: *Andricus quercuscalicis* (Hymenoptera: Cynipidae). *Journal of Animal Ecology* 60, 545–562.

Hallberg, E. (1979) The fine structure of the antennal sensilla of the pine sawfly *Neodiprion sertifer* (Insecta: Hymenoptera). *Protoplasma* 101, 111–126.

Hallberg, E. and Löfqvist, J. (1981) Morphology and ultrastructure of an intertergal pheromone gland in the abdomen of the pine sawfly *Neodiprion sertifer* (Insecta, Hymenoptera): a potential source of sex pheromones. *Canadian Journal of Zoology* 59, 47–53.

Hansson, B.S., van der Pers, J.N.C., Högberg, H.-E., Hedenström, E., Anderbrant, O. and Löfqvist, J. (1991) Sex pheromone perception in male pine sawflies, *Neodiprion sertifer* (Hymenoptera; Diprionidae). *Journal of Comparative Physiology* A 168, 533–538.

Hedlin, A.F. (1956) Studies on the balsam-fir seed chalcid, *Megastigmus specularis* Walley (Hymenoptera, Chalcididae). *Canadian Entomologist* 88, 691–697.

Heliövaara, K., Väisänen, R., Varama, M. and Viitasaari, M. (1992) Gynandromorphic individuals of *Neodiprion sertifer* (Hymenoptera, Diprionidae). *Entomologica Fennica* 3, 149–153.

Hilker, M. and Weitzel, C. (1991) Oviposition deterrence by chemical signals of conspecific larvae in *Diprion pini* (Hymenoptera: Diprionidae) and *Phyllodecta vulgatissima* (Coleoptera: Chrysomelidae). *Entomologia Generalis* 15, 293–301.

Högberg, H.-E., Hedenström, E., Wassgren, A.-B., Hjalmarsson, M., Bergström, G., Löfqvist, J. and Norin, T. (1990) Synthesis and gas chromatographic separation of the eight stereoisomers of diprionol and their acetates, components of the sex pheromone of pine sawflies. *Tetrahedron* 46, 3007–3018.

Holmes, N.D. and Blakeley, P.E. (1971) The rye jointworm (Hymenoptera: Eurytomidae), a new insect pest in western Canada. *Canadian Entomologist* 103, 277–280.

Hossaert-McKey, M., Gibernau, M. and Frey, J.E. (1994) Chemosensory attraction of fig wasps to substances produced by receptive figs. *Entomologia Experimentalis et Applicata* 70, 185–191.

Hussey, N.W. (1955) The life-histories of *Megastigmus spermotrophus* Wachtl (Hymenoptera: Chalcidoidea) and its principal parasite, with descriptions of the developmental stages. *Transactions of the Royal Entomological Society of London* 106, 133–151.

Jewett, D.M., Matsumura, F. and Coppel, H.C. (1976) Sex pheromone specificity in the pine sawflies: interchange of acid moieties in an ester. *Science* 192, 51–53.

Jones, D. (1983) The influence of host density and gall shape on the survivorship of *Diastrophus kincaidii* Gill. (Hymenoptera: Cynipidae). *Canadian Journal of Zoology* 61, 2138–2142.

Jönsson, P. and Anderbrant, O. (1993) Weather factors influencing catch of *Neodiprion sertifer* (Hymenoptera: Diprionidae) in pheromone traps. *Environmental Entomology* 22, 445–452.

Juutinen, P. and Varama, M. (1986) Occurrence of the European pine sawfly (*Neodiprion sertifer*) in Finland during 1966–83. *Folia Forestalia* 662, 1–39.

Katsoyannos, B.I., Kouloussis, N.A. and Bassiliou, A. (1992) Monitoring populations of the almond seed wasp, *Eurytoma amygdali*, with sex pheromone traps and other means, and optimal timing of chemical control. *Entomologia Experimentalis et Applicata* 62, 9–16.

Kerr, W.E. (1962) Genetics and sex determination. *Annual Review of Entomology* 7, 157–176.

Kikukawa, T., Matsumura, F., Kraemer, M., Coppel, H.C. and Tai, A. (1982) Field attractiveness of chirally defined synthetic attractants to males of *Diprion similis* and *Gilpinia frutetorum*. *Journal of Chemical Ecology* 8, 301–314.

Kikukawa, T., Matsumura, F., Olaifa, J., Kraemer, M., Coppel, H.C. and Tai, A. (1983) Field evaluation of chiral isomers of the sex pheromone of the European pine sawfly *Neodiprion sertifer*. *Journal of Chemical Ecology* 9, 673–694.

Kitano, H. (1988) Experimental studies on the mating behavior of *Athalia lugens infumata*. *Kontyu* 56, 180–188.

Kouloussis, N.A. and Katsoyannos, B.I. (1991) Host discrimination and evidence for a host marking pheromone in the almond seed wasp, *Eurytoma amygdali*. *Entomologia Experimentalis et Applicata* 58, 165–174.

Kouloussis, N.A. and Katsoyannos, B.I. (1993) Egg distribution patterns in the almond seed wasp, *Eurytoma amygdali*. *Entomologia Experimentalis et Applicata* 66, 31–38.

Kouloussis, N.A. and Katsoyannos, B.I. (1995) Distribution and activities of *Eurytoma amygdali* (Hymenoptera: Eurytomidae) wasps on almond trees. *Annals of the Entomological Society of America* 88, 547–553.

Kraemer, M., Coppel, H.C., Matsumura, F., Kikukawa, T. and Mori, K. (1979) Field responses of the white pine sawfly, *Neodiprion pinetum*, to optical isomers of sawfly sex pheromones. *Environmental Entomology* 8, 519–520.

Kraemer, M.E., Coppel, H.C., Matsumura, F., Wilkinson, R.C. and Kikukawa, T. (1981) Field and electroantennogram responses of the red-headed pine sawfly, *Neodiprion lecontei* (Fitch), to optical isomers of sawfly sex pheromones. *Journal of Chemical Ecology* 7, 1063–1072.

Kraemer, M.E., Coppel, H., Kikukawa, T., Matsumura, F., Thomas, H.A., Thompson, L.C. and Mori, K. (1983) Field and electroantennogram responses to sex pheromone optical isomers by four fall flying sawfly species (Hymenoptera: Diprionidae, *Neodiprion*). *Environmental Entomology* 12, 1592–1596.

Kraemer, M.E., Coppel, H.C., Matsumura, F., Kikukawa, T. and Benoit, P. (1984) Field and electroantennogram responses to sex pheromone optical isomers by monophagous jack pine sawflies (Hymenoptera: Diprionidae). *Journal of Chemical Ecology* 10, 983–996.

Lamborot, L., Paredes, F.R., Arretz, P., Guerrero, M.A. and Araya, J.E. (1993) Distribution of *Systole coriandri* (Gussakovski) (Hymenoptera: Eurytomidae) and evaluation of damage to the seeds of coriander, *Coriandrum sativum* L. (Umbelliferae), in Chile. *Acta Entomologica Chilena* 18, 97–100.

Lanier, G.N. (1990) Principles of attraction–annihilation: mass trapping and other means. In: Ridgway, R.L., Silverstein, R.M.and Inscoe, M.N. (eds) *Behavior-modifying Chemicals for Insect Management. Applications of Pheromones and Other Attractants*. Marcel Dekker, New York, pp. 25–45.

Lateef, S.S., Reed, W. and LaSalle, J. (1985) *Tanaostigmodes cajaninae* new species (Hymenoptera: Tanaostigmatidae): a potential pest of pigeon pea, *Cajanus cajan*, in India. *Bulletin of Entomological Research* 75, 305–314.

Lessmann, D. (1962) Übersicht der bisher bekannten *Megastigmus*-Arten und ihrer Wirtspflanzen. *Zeitschrift für Angewandte Entomologie* 49, 233–238.

Longhurst, C. and Baker, R. (1980) Sex pheromones in the gooseberry sawfly *Nematus ribesii*. *Naturwissenschaften* 67, 146.

Longhurst, C., Baker, R. and Mori, K. (1980) Response of the sawfly *Diprion similis* to chiral sex pheromones. *Experientia* 36, 946–947.

Lyytikäinen-Saarenmaa, P., Anderbrant, O., Löfqvist, J., Hedenström, E. and Höberg, H.-E. (1999) Monitoring European pine sawfly population densities with pheromone traps in young pine plantations. *Forest Ecology and Management* (in press).

Manojlovic, B. and Tosevski, I. (1990) Contribution to the knowledge of *Isocolus jaceae* Sch. (Hymenoptera: Cynipidae) influence on the population density of *Centaurea solstitialis* L. (Asteracea). *Zastita Bilja* 41, 141–150.

Marchisio, C., Cescatti, A. and Battisti, A. (1994) Climate, soils and *Cephalcia arvensis* outbreaks on *Picea abies* in the Italian Alps. *Forest Ecology and Management* 68, 375–384.

Martinek, V. (1991) An outbreak of the web-spinning sawfly *Cephalcia arvensis* Panz. (Hym., Pamphiliidae) in eastern Bohemia. *Lesnictví* 37, 543–570. (In Czech with English abstract.)

Martinek, V. (1992) The sawfly *Cephalcia falleni* (Dalm.) (Hym., Pamphilidae) as a new pest on spruce on the Orlické hory Mts. *Lesnictví* 38, 205–220. (In Czech with English abstract.)

Mason, W.R.M. and Huber, J.T. (1993) Order Hymenoptera. In: Goulet, H. and Huber, J.T. (eds) *Hymenoptera of the World*. Research Branch, Agriculture Canada, Publication 1894/E, Ontario, pp. 4–12.

Matsumura, F., Tai, A., Coppel, H.C. and Imaida, M. (1979) Chiral specificity of the sex pheromone of the red-headed pine sawfly, *Neodiprion lecontei*. *Journal of Chemical Ecology* 5, 237–250.

Mertins, J.W. and Coppel, H.C. (1971) Sexual behavior in gynandromorphs of *Diprion similis* (Hymenoptera: Diprionidae). *Annals of the Entomological Society of America* 64, 1191–1192.

Mertins, J.W. and Coppel, H.C. (1972) Previously undescribed abdominal glands in the female introduced pine sawfly, *Diprion similis* (Hymenoptera: Diprionidae). *Annals of the Entomological Society of America* 65, 33–38.

Morrill, W.L. and Kushnak, G.D. (1996) Wheat stem sawfly (Hymenoptera: Cephidae) adaptation to winter wheat. *Environmental Entomology* 25, 1128–1132.

Morse, B.W. and Kulman, H.M. (1985) Monitoring damage by yellowheaded spruce sawflies with sawfly and parasitoid pheromones. *Environmental Entomology* 14, 131–133.

Murray, M.G. (1985) Figs (*Ficus* spp.) and fig wasps (Chalcidoidea, Agaonidae): hypotheses for an ancient symbiosis. *Biological Journal of the Linnean Society* 26, 69–82.

Nadel, H. and Peña, J.E. (1991) Seasonal oviposition and emergence activity of *Bephratelloides cubensis* (Hymenoptera: Eurytomidae), a pest of *Annona* species in Florida. *Environmental Entomology* 20, 1053–1057.

Naito, T. and Suzuki, H. (1991) Sex determination in the sawfly *Athalia rosae-ruficornis* (Hymenoptera): occurrence of triploid males. *Journal of Heredity* 82, 101–104.

Olaifa, J.I., Kikukawa, T., Matsumura, F. and Coppel, H.C. (1984) Response of male jack pine sawfly *Neodiprion pratti banksianae* (Hymenoptera: Diprionidae) to mixtures of optical isomers of the sex pheromone 3,7-dimethylpentadecan-2-ol. *Environmental Entomology* 13, 1274–1277.

Olaifa, J.I., Matsumura, F. and Coppel, H.C. (1987) Field response and gas-liquid chromatograph separation of optically active synthetic and natural pheromones in two sympatric diprionid sawflies, *Neodiprion nanulus nanulus* and *Neodiprion sertifer* (Hymenoptera: Diprionidae). *Journal of Chemical Ecology* 13, 1395–1408.

Olaifa, J.I., Matsumura, F., Kikukawa, T. and Coppel, H.C. (1988) Pheromone-dependent species recognition mechanisms between *Neodiprion pinetum* and *Diprion similis* on white pine. *Journal of Chemical Ecology* 14, 1131–1144.

Orr, D.B. and Borden, J.H. (1983) Courtship and mating behavior of *Megastigmus pinus* Parfitt (Hymenoptera: Torymidae). *Journal of the Entomological Society of British Columbia* 80, 20–24.

Östrand, F., Wedding, R., Jirle, E. and Anderbrant, O. (1999) Effect of mating disruption on reproductive behavior in the European pine sawfly, *Neodiprion sertifer* (Hymenoptera: Diprionidae). *Journal of Insect Behavior* 12, 233–243.

Ôtake, A. (1980) Chestnut gall wasp, *Dryocosmus kuriphilus* Yasumatsu (Hymenoptera: Cynipidae): a preliminary study on trend of adult emergence and some other ecological aspects related to the final stage of its life cycle. *Applied Entomology and Zoology* 15, 96–105.

Peterson, S.S. (1991) Bionomics and management of the insect pests of birdsfoot trefoil seed production. PhD Thesis, University of Wisconsin, Madison.

Phelan, P.L. (1997) Genetics and phylogenetics in the evolution of sex pheromones. In: Cardé, R.T. and Minks, A.K. (eds) *Insect Pheromone Research, New Directions*. Chapman & Hall, New York, pp. 563–579.

Pittara, I.S. and Katsoyannos, B.I. (1985) Male attraction to virgin females in the almond seed wasp *Eurytoma amygdali* (Hymenoptera: Eurytomidae). *Entomologia Hellenica* 3, 43–46.

Plaut, H.N. (1971) On the biology of the adult of the almond wasp, *Eurytoma amygdali* End. (Hymenoptera: Eurytomidae, in Israel. *Bulletin of Entomological Research* 61, 275–281.

Prokopy, P.J., Roitberg, B.D. and Averill, A.L. (1984) Resource partitioning. In: Bell, W.J. and Cardé, R.T. (eds) *Chemical Ecology of Insects*. Chapman & Hall, London, pp. 301–320.

Prop, N. (1960) Protection against birds and parasites in some species of tenthredinid larvae. *Archives Neerlandaises de Zoologie* 13, 380–447.

Pschorn-Walcher, H. (1982) Unterordnung Symphyta, Pflanzenwespen. In: Schwenke, W. (ed.) *Die Forstschädlinge Europas, 4. Band.* Verlag Paul Parey, Hamburg, pp. 4–234.

Redak, R.A. and Bethke, J.A. (1995) Detection and seasonal occurrence of gall-forming wasps (Hymenoptera: Eulophidae) on Geraldton wax plant. *Journal of Economic Entomology* 88, 387–392.

Rice, R.E. and Michailides, T.J. (1988) Pistachio seed chalcid, *Megastigmus pistaciae* Walker (Hymenoptera: Torymidae), in California. *Journal of Economic Entomology* 81, 1446–1449.

Schedl, W. (1994) First field experiments with pheromone sticky traps for sawfly control in Austria (Hymenoptera: Symphyta: Diprionidae). *Entomologia Generalis* 18, 235–239.

Shorthouse, J.D. (1994) Host shift of the leaf galler *Diplolepis polita* (Hymenoptera: Cynipidae) to the domestic shrub rose *Rosa rugosa. Canadian Entomologist* 126, 1499–1503.

Shorthouse, J.D. and Ritchie, A.J. (1984) Description and biology of *Diplolepis* Fourcroy (Hymenoptera: Cynipidae) inducing galls on the stems of *Rosa acicularis. Canadian Entomologist* 116, 1623–1636.

Simandl, J. (1993) Flight dynamics of males in a non-outbreaking population of *Neodiprion sertifer* (Geoff.) (Hymenoptera, Diprionidae). *Lesnictví* 39, 241–243.

Simandl, J. and Anderbrant, O. (1995) Spatial distribution of flying *Neodiprion sertifer* (Hymenoptera, Diprionidae) males in a mature *Pinus sylvestris* stand as determined by pheromone trap catch. *Scandinavian Journal of Forest Research* 10, 51–55.

Skrzypczynska, M. (1975) *Eurytoma bouceki* n. sp. (Hymenoptera, Eurytomidae), its stages of development, biology and economic importance. *Zeitschrift für Angewandte Entomologie* 79, 204–213.

Skrzypczynska, M. (1978) *Megastigmus suspectus* Borries, 1895 (Hymenoptera, Torymidae), its morphology, biology and economic significance. *Zeitschrift für Angewandte Entomologie* 85, 204–215.

Smith, D.R. (1974) Conifer sawflies, Diprionidae: key to North American genera, checklist of world species, and new species from Mexico (Hymenoptera). *Proceedings of the Entomological Society of Washington* 76, 409–418.

Smith, D.R. (1988) A synopsis of the sawflies (Hymenoptera: Symphyta) of America south of the United States: introduction, Xyelidae, Pamphiliidae, Cimbicidae, Diprionidae, Xiphydriidae, Siricidae, Orussidae, Cephidae. *Systematic Entomology* 13, 205–261.

Smith, D.R. (1993) Systematics, life history, and distribution of sawflies. In: Wagner, M.R. and Raffa, K.F. (eds) *Sawfly Life History Adaptations to Woody Plants.* Academic Press, San Diego, pp. 3–32.

Smith, S.G. and Wallace, D.R. (1971) Allelic sex determination in a lower hymenopteran *Neodiprion nigroscutum. Canadian Journal of Genetics and Cytology* 13, 617–621.

Stille, B. (1984) The effect of host plant and parasitoids on the reproductive success of the parthenogenetic gall wasp *Diplolepis rosae* (Hymenoptera, Cynipidae). *Oecologia* 63, 364–369.

Stille, B. (1985) Population genetics of the parthenogenetic gall wasp *Diplolepis rosae* (Hymenoptera: Cynipidae). *Genetica* 67, 145–152.

Syrett, P. (1990) The rose seed chalcid *Megastigmus aculeatus* Swederus (Hymenoptera: Torymidae) on sweet brier, *Rosa rubiginosa*, in the South Island tussock country. *New Zealand Entomologist* 13, 34–38.

Tai, A., Sugimura, T., Kikukawa, T., Naito, C., Nishimoto, Y. and Morimoto, N. (1992) Synthesis and field test of a pheromone mimic of the pine sawfly (*Neodiprion sertifer*). *Bioscience Biotechnology Biochemistry* 56, 1711–1714.

Tai, A., Higashiura, Y., Kakizaki, M., Naito, T., Tanaka, K., Fujita, M., Sugimura, T., Hara, H. and Hayashi, N. (1998) Field and electroantennogram responses of the pine sawfly, *Diprion nipponica,* to chiral synthetic pheromone candidates. *Bioscience Biotechnology Biochemistry* 62, 607–608.

Thoenes, S.C. and Moffett, J.O. (1987) Emergence of alfalfa seed chalcid, *Bruchophagus roddi* (Hymenoptera: Eurytomidae), from overwintered seeds in Oklahoma. *Environmental Entomology* 16, 774–778.

Thomas, H.A., Ghent, J.H. and Stone, C.G. (1982) Use of pheromone traps as a new technique in a large scale survey of the introduced pine sawfly (Hymenoptera: Diprionidae). *Bulletin of the Entomological Society of America* 28, 303–304.

Tisdale, R.A. and Wagner, M.R. (1991) Oviposition behavior of *Neodiprion fulviceps* (Cresson) (Hymenoptera: Diprionidae) on ponderosa pine. *Journal of Insect Behavior* 4, 609–618.

Tzanakakis, M.E., Veenendaal, R.L. and Veerman, A. (1992) Effects of photoperiod and temperature on the termination of diapause in the univoltine seed wasp *Eurytoma plotnikovi. Physiological Entomology* 17, 176–182.

Vikberg, V. and Nuorteva, M. (1997) On the rearing of *Nesoselandria morio* (Fabricius) and *Birka cinereipes* (Klug) (Hymenoptera, Tenthredinidae), with descriptions of their larvae. *Entomologica Fennica* 8, 27–38.

Wagner, M.R. and Raffa, K.F. (1993) *Sawfly Life History Adaptations to Woody Plants.* Academic Press, San Diego, 581 pp.

Wassgren, A.-B. and Bergström, G. (1995) Quantitative high-resolution gas chromatographic determination of stereoisomeric composition of chiral volatile compounds in the picogram range by ec-detection. *Journal of Chemical Ecology* 21, 987–994.

Wassgren, A.-B., Anderbrant, O., Löfqvist, J., Hansson, B.S., Bergström, G., Hedenström, E. and Högberg, H.-E. (1992) Pheromone related compounds in pupal and adult female pine sawflies, *Neodiprion sertifer,* of different age and in different parts of the body. *Journal of Insect Physiology* 38, 885–893.

Wedding, R., Anderbrant, O. and Jönsson, P. (1995) Influence of wind conditions and intertrap spacing on pheromone trap catches of male European pine sawfly, *Neodiprion sertifer. Entomologia Experimentalis et Applicata* 77, 223–232.

West, A. and Shorthouse, J.D. (1989) Initiation and development of the stem gall induced by *Hemadas nubilipennis* (Hymenoptera: Pteromalidae) on lowbush blueberry, *Vaccinium angustifolium* (Ericaceae). *Canadian Journal of Botany* 67, 2187–2198.

Wildbolz, T. and Staub, A. (1986) Fang der Pflaumensägewespen *Hoplocampa minuta* und *H. flava* und der Apfelsägewespe *H. testudinea* mit weissen Fallen – Einfluss von Temperatur, Blütezeit und Fallenposition. *Mitteilungen der Schweizerischen Entomologischen Gesellschaft* 59, 289–296.

Wilkinson, R.C., Chappelka III, A.H., Kraemer, M.E., Coppel, H.C. and Matsumura, F. (1987) Effect of height on responses of redheaded pine sawfly (Hymenoptera: Diprionidae) males to synthetic pheromone and virgin females. *Environmental Entomology* 16, 1152–1156.

Xu, Z. (1997) Two new species and one new combination of the conifer sawfly family Diprionidae (Hymenoptera: Symphyta) from Yunnan, China. *Zoological Research* 18, 171–176.

Zerova, M.D. and Fursov, V.N. (1991) The Palaearctic species of *Eurytoma* (Hymenoptera: Eurytomidae) developing in stone fruits (Rosacae: Prunoideae). *Bulletin of Entomological Research* 81, 209–219.

Zhang, Z. and Zhou, S. (1996) Chinese diprionids I: distribution, host species, damage and life cycle. *Journal of Beijing Forestry University (English edn)* 5, 83–89.

9 Aphids

JIM HARDIE[1], JOHN A. PICKETT[2], ELEANOR M. POW[2] AND DIANE W.M. SMILEY[2]

[1]*Aphid Biology Group, Department of Biology, Imperial College at Silwood Park, Ascot, Berks SL5 7PY, UK;* [2]*Department of Biological and Ecological Chemistry, IACR-Rothamsted, Harpenden, Herts AL5 2JQ, UK*

Introduction

Aphids and their life cycles

Aphids belong to the order Hemiptera, superfamily Aphidoidea, and this chapter will concentrate on the family Aphididae. The more primitive Adelgidae and Phylloxeridae are excluded as little is known of their pheromone biology. Aphids exhibit a variety of life cycles ranging from species showing solely asexual reproduction (anholocyclic) on a single host plant species, usually in warmer climates, to those that alternate between asexual and sexual reproduction (holocyclic) on different host plants in temperate regions. There are no extant aphids that reproduce exclusively by sexual means, and all subfamilies of the Aphididae contain genera that can reproduce both holocyclically and anholocyclically.

Within the family Aphididae, species belonging to the major subfamilies Lachninae, Chaitophorinae, Drepanosiphinae and Greenideinae generally remain on the same host plant species throughout the year (autoecious), while species within the Aphidinae, Anoeciinae, Pemphiginae and Hormaphidinae may alternate between different species of host plants during the annual life cycle (heteroecious) (Blackman and Eastop, 1985). A heteroecious life cycle is associated with alternation between a primary (or winter) host, which is often a tree or shrub, and a secondary (or summer) herbaceous host, with specialist winged morphs moving between the two hosts (Moran, 1992). Heteroecious aphids generally reproduce asexually on the summer host and sexually on the winter host. The asexual generations give birth to live young while mated sexual females lay cold-hardy, overwintering eggs. The switch from asexual to sexual reproduction is most frequently associated with exposure to the short days of autumn (Lees, 1966).

Aphids with holocyclic life cycles show remarkable phenotypic plasticity, with asexual forms that may be winged or wingless, depending primarily upon the population density and/or the nutritional state of the host, and sexual forms with a distinct sexual female (wingless except in the Greenideinae) and a smaller male which is winged in many species but wingless in others. The form of the male is related to the life-cycle strategy. In the host-alternating Aphidinae, winged females (gynoparae) flying to the primary host in autumn produce only sexual females and the winged males fly separately. Wingless males are found in some non-host-alternating species, and in some host-alternating species where the winged females that move from secondary to the primary host give birth to both males and sexual females (i.e. Anoeciinae, Pemphiginae and Hormaphidinae).

Pheromones

Three types of aphid pheromone have been studied: sex pheromones, alarm pheromones and aggregation pheromones.

During the earlier part of this century, there were a number of suggestions that sexual female aphids (oviparae) might produce a sex attractant, i.e. a pheromone (Flögel, 1905; Weber, 1935; Smith, 1936; Stroyan, 1958; van Emden *et al.*, 1969). Ideas on the existence of sex pheromones derived from consideration of the problems of mate location, particularly in host-alternating species where the winged males must leave the natal plant and locate females on an unrelated host, and especially in years when numbers of sexual females are low. Stroyan (1958) also noted that males became excited when within 10–20 mm of a sexual female, whereas others recorded that the distribution of males tended to reflect that of the sexual females. The latter observation led Tamaki *et al.* (1970) on a relatively unsuccessful search for a sex pheromone in the peach–potato aphid, *Myzus persicae* (Sulzer), although they concluded that the sexual females may release a sex pheromone. In the same year, Pettersson (1970), studying *Schizaphis phlei (arrhenatheri)* (Orlob), published the first definitive evidence for an aphid sex pheromone and showed that males in an olfactometer responded to volatiles produced by sexual females. Other studies followed and initially it was suggested that sex pheromones were not species-specific (Pettersson, 1971; Marsh, 1975) and that males were only responsive to the pheromone over a short distance – tens of millimetres (Stroyan, 1958; Eisenbach and Mittler, 1980; Steffan, 1987) to hundreds of millimetres (Pettersson, 1971). Both of these ideas have been challenged by more recent findings, which will be reviewed here.

Nymphal stages and wingless adult morphs are relatively sedentary and the risk of attack by predators and parasitoids is high. In many species, a volatile alarm pheromone is released from lipid droplets secreted at the tip of paired abdominal cornicles when individuals are attacked, for example, by a predator or parasitoid (Strong, 1967; Nault *et al.*, 1973). The pheromone generally initiates an avoidance reaction in neighbouring aphids. In some species, responding aphids move away or drop to the ground (Kislow and Edwards, 1972; Nault and Phelan, 1984); in other species, aphids can adopt an aggressive stance (Arakaki, 1989).

Field observations have also led to the suggestion that accumulations of aphids newly arrived on host plants might result from aggregation pheromones (Broadbent, 1949). More detailed evidence has been produced in laboratory experiments, but no chemical confirmation and characterizations have been made of specific aggregation pheromones.

Biology and Chemistry of Pheromones

Sex pheromones

Sex pheromones released by the sexual females excite males and increase the mating success (Fig. 9.1). The pheromones are produced in glandular epidermal cells lying beneath scent plaques on the tibiae of the hind legs of the sexual females and released through the porous domed cuticle above the plaques (Pettersson, 1971; Marsh, 1972; Roberti, 1979) (Fig. 9.2). The olfactory receptors for sex pheromones are placoid sensilla (the secondary rhinaria) on the antennae of male aphids (Pettersson, 1973; Marsh, 1972, 1975; Dawson *et al.*, 1987a). Male antennae are well endowed with secondary rhinaria, on the third, fourth and sometimes fifth antennal segments, and are particularly sensitive to sex pheromone components (Hardie *et al.*, 1994b). There are separate receptor cells for individual pheromone components (Dawson *et al.*, 1990; Lilley *et al.*, 1994/5). Pettersson (1971) also suggested that other sensilla acting as contact chemoreceptors may play a role in mating.

Chemistry

Sex pheromones have been identified from a number of species, all of which are in the subfamily Aphidinae. The first chemical identification of an aphid sex

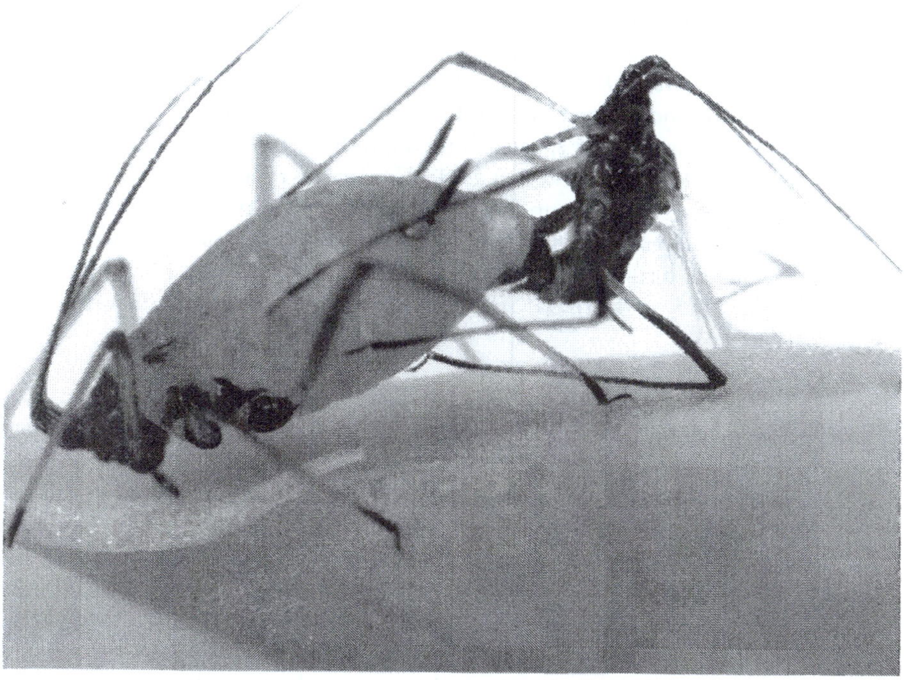

Fig. 9.1. The winged male vetch aphid, *Megoura viciae* Buckton, mating with the wingless sexual female. (Photograph by P. Christiansen-Weniger.)

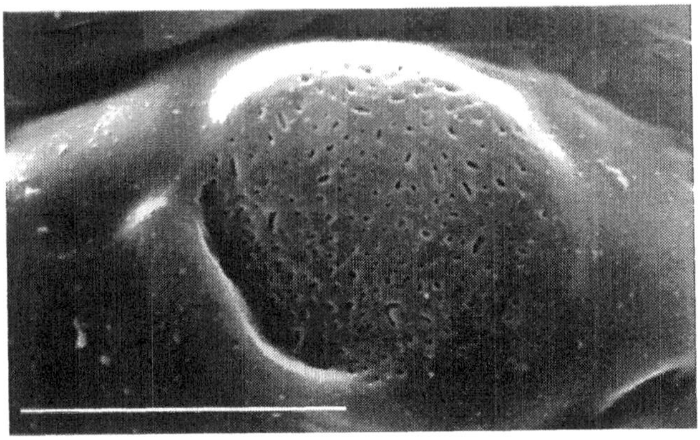

Fig. 9.2. Scanning electronmicrograph of a scent plaque on the hind tibia of a sexual female black bean aphid, *Aphis fabae* Scopoli. Scale bar = 10 μm. (Photograph by Jim Hardie and J.H. Visser.)

pheromone was for the vetch aphid, *Megoura viciae* Buckton (Dawson *et al.*, 1987a). Innovative use of electrophysiological recording from single olfactory cells in the rhinaria directly coupled to gas chromatography (GC) pinpointed two compounds from solvent elutions of hindlegs. Tentative identification by GC–mass spectrometry (GC–MS) and confirmation by comparison with synthetic materials authenticated by ^{1}H and ^{13}C nuclear magnetic resonance (NMR) spectroscopy gave (4aS,7*S*,7a*R*)-nepetalactone (**1** in Fig. 9.3) and, with further X-ray crystallography studies (Dawson *et al.*, 1989), (1*R*,4aS,7*S*,7a*R*)-nepetalactol (**2** in Fig. 9.3) which are monoterpenoids in the cyclopentanoid series. The sex pheromones for a further 15 aphid pest species in the Aphidinae have been identified and principally comprise **1** and **2** (Pickett *et al.*, 1992; Dawson *et al.*, 1996; L.J. Wadhams and C.M. Woodcock, 1998, unpublished results). However, an investigation of the damson–hop aphid, *Phorodon humuli* (Schrank), showed that the pheromone of this species comprised neither compound **1** nor **2**, but a mixture of the two diastereoisomeric (1*S*)- and (1*R*,4a*R*,7*S*,7aS)-nepetalactols, **3** and **4** (Campbell *et al.*, 1990) (Fig. 9.3).

Biosynthesis

The nepetalactone **1** has long been known as a compound from the catmint plant, *Nepeta cataria* L. in the family Lamiaceae (= Labiatae), and can be chemically reduced to the nepetalactol **2**. Also, the nepetalactone related to the nepetalactols **3** and **4** is found in the related *Nepeta racemosa* Lamarck. The biosynthesis of the lactones in plants has received much attention, although the pathway has not been fully elucidated. The isoprenoid precursor has yet to be established but is suggested to be geraniol and/or the (*Z*)-isomer nerol (Uesato *et al.*, 1987), or citronellol (Bellesia *et al.*, 1984). No investigative work has been done to establish whether these nepetalactones are derived from mevalonic acid or 1-deoxyxylulose 5-phosphate.

As nature is often parsimonious with respect to the production of secondary

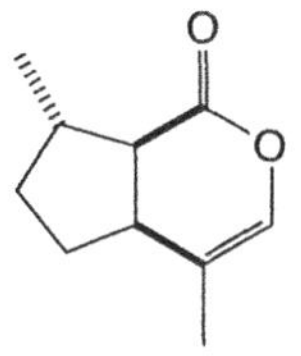

1

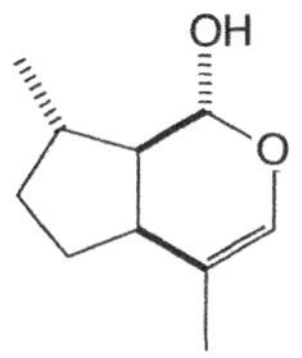

2

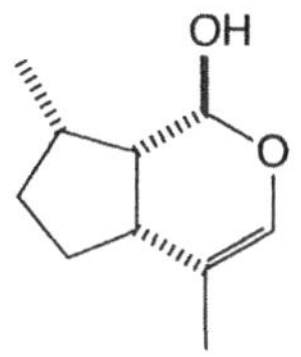

3

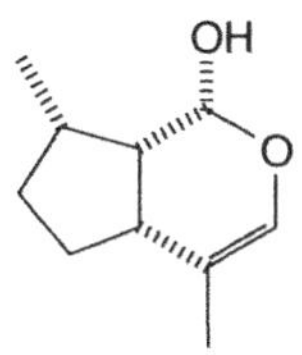

4

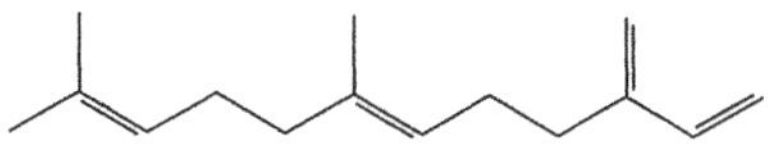

5

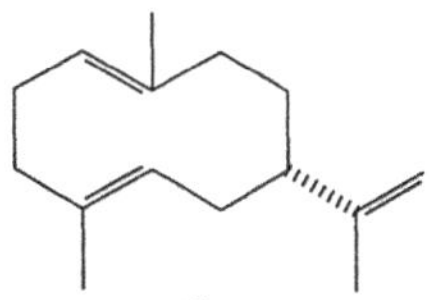

6

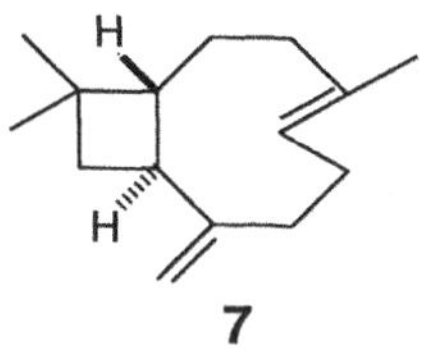

7

Fig. 9.3. Aphid semiochemicals: **1** (4a*S*,7*S*,7a*R*)-nepetalactone; **2** (1*R*,4a*S*,7*S*,7a*R*)-nepetalactol; **3** (1*S*,4a*R*,7*S*,7a*S*)-nepetalactol; **4** (1*R*,4a*R*,7*S*,7a*S*)-nepetalactol; **5** (*E*)-β-farnesene; **6** (−)-germacrene-A; **7** (−)-β-caryophyllene.

metabolites, the biosynthesis of these lactones in plants and insects is likely to proceed via similar pathways. A detailed analytical study of the aphid sex pheromone has revealed that citronellol (with undetermined stereochemistry) is a common, but minor, component of volatiles associated with sex pheromone release in various aphid species. It has, however, shown no electrophysiological or behavioural activity, certainly in connection with mate location, and may therefore be the precursor for the cyclopentanoid biosynthesis of aphid sex pheromones (Dawson *et al.*, 1996).

The sex pheromone components **1** and **2** can be synthesized in the laboratory on a multi-gram scale from (*S*)-citronellol, which is commercially available in a range of enantiomeric purities (Dawson *et al.*, 1996) and which provides the key

(7*S*)-stereochemistry in the pheromone. These synthetic materials have been tested in the field and compared with the plant-derived components (Hardie *et al.*, 1997). The synthetic compounds were significantly more attractive to male aphids than the corresponding plant-derived components, and the compounds produced from high enantiomeric purity (*S*)-citronellol generally had higher trap catches than those produced from the lower enantiomeric purity (*S*)-citronellol.

Composition and species specificity

Although most aphids examined so far employ a limited range of pheromone components, there are differences in relative and absolute compositions. Thus only *P. humuli* produces **3** and **4** while a number of other species produce blends of **1** and **2** or the individual components (Table 9.1).

Although some species utilize blends, there does not seem to be sufficient variability to allow species-specificity for the *c.* 2200 aphid species in the Aphidinae and a number of species appear to use the same single component, e.g. the grain aphid *Sitobion avenae* (Fabricius), the blackberry–cereal aphid *Sitobion fragariae* (Walker) and the cabbage aphid *Brevicoryne brassicae* (L.) (Table 9.1). Even so, laboratory males responded maximally to the pheromone blend which mimicked that released by the conspecific sexual females (Hardie *et al.*, 1990) and field trap-catch data show a remarkable specificity. Water traps releasing only **1** recorded 21 species of male aphid but only one species, *S. fragariae*, was caught in significant numbers (Hardie *et al.*, 1992). The release of various blends of **1** and **2** showed that male bird-cherry–oat aphids, *Rhopalosiphum padi* (L.), were caught only in traps releasing **2**. Even the addition of minimal quantities of **1** reduced catches down to control levels (Hardie *et al.*, 1994a). In similar experiments with traps releasing **1**, **2** or a 1:1 blend of **1** and **2**, 66% of male *Aphis* spp. and 85% of male *Dysaphis* spp., including *Dysaphis plantaginea* (Passerini), were caught in the blend while 88% of *Rhopalomyzus lonicerae* (Siebold) were found in traps releasing

Table 9.1. Composition of aphid sex pheromones established using air-entrainment techniques (see Fig. 9.3 for compound numbers).

Sex pheromone blend	Aphid species	Reference
High **1**:Low **2**	*Aphis fabae*	Dawson *et al.* (1990), Isaacs (1994)
	Megoura viciae	Dawson *et al.* (1990), Hardie *et al.* (1990)
High **2**:Low **1**	*Schizaphis graminum*	Dawson *et al.* (1988)
	Cryptomyzus spp.	Guldemond *et al.* (1993)
	Dysaphis plantaginea	L.J. Wadhams (1998, unpublished results)
1 and **2** equal	*Acyrthosiphon pisum*	Dawson *et al.* (1990)
	Myzus persicae	Dawson *et al.* (1990)
1 only	*Sitobion avenae*	Lilley *et al.* (1994/95)
	Sitobion fragariae	Hardie *et al.* (1992)
	Brevicoryne brassicae	Gabrys *et al.* (1997)
2 only	*Rhopalosiphum padi*	L.J. Wadhams (1998, unpublished results)
3 and **4**	*Phorodon humuli*	Campbell *et al.* (1990)

1 (J. Hardie, R. Harrington and L.J. Wadhams, 1998, unpublished results). Laboratory work with *Cryptomyzus* spp., however, failed to reveal any species differences in pheromone blend but males responded specifically to conspecific females (Guldemond *et al.*, 1993).

Pheromone release by sexual females

Sex pheromone release appears to be discontinuous, coinciding with the light phase, or a part thereof. In at least two species, the rhythm of release has an endogenous circadian component (Pettersson, 1971; Marsh, 1972; Eisenbach and Mittler, 1980). In some species there appear to be particular periods during the day when sex pheromone is released. Thus, a further mechanism for species specificity may be observed in closely related *Cryptomyzus* species, which mate on the same host plant but call maximally during different parts of the day (Guldemond and Dixon, 1994). Similar observations were made on species/subspecies of the *Aphis fabae* species complex (Thieme and Dixon, 1996). In this complex, it appears that male activity corresponds to the periods when females call but there is overlap between species/subspecies. In other aphids, while female calling periods are restricted to within the photophase and pheromone release is age-dependent, males are continuously receptive, day and night, from shortly after the adult moult to just before death (Marsh, 1975; Eisenbach and Mittler, 1980; Hardie *et al.*, 1990). The release rate from female M. *viciae* is relatively high and can achieve 210×10^{-9} g day^{-1} of the **2/1** mixture (Hardie *et al.*, 1990). In a number of species it has been noted that females may raise their hind legs when releasing the sex pheromone (Marsh, 1975; Steffan, 1989). This is considered to be part of the 'calling' behaviour of sexual females and may assist pheromone release and distribution.

The above examples used behavioural assays to establish release times, relying on male responsiveness, which may not be consistent throughout (Guldemond and Dixon, 1994). However, chemical methods have also demonstrated minimal release of the pheromone in the greenbug *Schizaphis graminum* (Rondani) during the scotophase and early photophase (Dawson *et al.*, 1990). In the black bean aphid, A. *fabae*, sexual females call throughout the day (Isaacs, 1994). In S. *graminum*, males do not respond to females that have mated (Eisenbach and Mittler, 1987), possibly because pheromone production ceases, but this is not the case in M. *viciae* (Marsh, 1975).

Location by males

It has recently been shown that male P. *humuli* will fly towards a sex pheromone source at wind speeds of up to 0.7 m s^{-1} and can unequivocally detect the pheromone 3–4 m downwind of the source (Hardie *et al.*, 1996). The latter measure is likely to be an underestimate but certainly requires that the original suggestions that the sex pheromones act over short distances need revision. Precisely how males locate females in the field is still not clear but odour-mediated anemotaxis, as used by male moths to locate sex pheromone sources (Baker and Vickers, 1996; Cardé and Mafra-Neto, 1997), is most likely, although aphid males are restricted to flying at lower wind speeds.

Alarm pheromones

When disturbed, many aphids release an alarm pheromone, which has been studied exclusively in the asexual forms and most often in the wingless females. Aphids nearby exhibit a variety of behaviours ranging from removal of mouthparts from the plant and moving away, to running, dropping off the plant and even attacking the predator with frontal horns (Nault *et al.*, 1973; Roitberg and Myers, 1978; Arakaki, 1989), although often not all aphids in a group respond. For this reason, bioassays frequently record the percentage of aphids responding. Variation in response to alarm pheromone is seen both intra- and inter-specifically. Dill *et al.* (1990) have hypothesized that the relative risks of predation and costs of escape, e.g. cessation of feeding and risk of desiccation, affect the likelihood of any particular response. In studies of M. *persicae*, the pea aphid *Acyrthosiphon pisum* (Harris) and the rose–grain aphid *Metopolophium dirhodum* (Walker), early stadia were found to be less sensitive to alarm pheromone than later ones (Montgomery and Nault, 1978; Roitberg and Myers, 1978; Larsen, 1988). However, older wingless M. *persicae* required the greatest stimulation of alarm pheromone before responding. Winged M. *persicae*, particularly those not feeding, were extremely sensitive to alarm pheromone. The lack of response from the early stadia suggests that the risk of predation to these larvae is lower than the risk involved in ceasing to feed and dropping from the plant. Larsen (1988) found that when first-stadium M. *dirhodum* responded to alarm pheromone they did so by moving to another part of the plant rather than by dropping. Winged adults, on the other hand, are more responsive to alarm pheromone, perhaps because they more readily move off the natal host. In contrast to these results, first-stadium larvae of the sugarcane woolly aphid, *Ceratovacuna lanigera* Zehntner show attack behaviour in response to alarm pheromone while older nymphs and adults disperse (Arakaki, 1989). In this species, the first-stadium larvae have long frontal horns and are able to defend effectively against predators.

Considerable variation is seen amongst aphid species in their sensitivity to alarm pheromones, in both the speed and the form of the response, which can often be related to the ecology of the species (Montgomery and Nault, 1977a). For example, some aphids, particularly those tended by ants, respond by walking or 'waggling' their abdomens rather than falling off the plant (Nault *et al.*, 1976; H.F. van Emden, Reading, 1998, personal communication). They appear to depend more on protection afforded by their ant attenders than their own defensive mechanisms. Aphids that walk away from alarm pheromone tend to form new clusters a short distance from the original site, thus ensuring continued ant attendance (Montgomery and Nault, 1977b). Host plant density and weather also affect tendency to drop from the plant. Among the species in the subfamily Aphidinae, tribe Aphidini, which Montgomery and Nault (1977a) tested for response to alarm pheromone, those that dropped had grasses as their host plant. Since grasses occur in dense stands, it is likely that an aphid on the ground would quickly be able to locate a new host plant. Dixon and Stewart (1975) also noted that the winged tree-dwelling sycamore aphid, *Drepanosiphum platanoidis* (Schrank), which is an agile mover, required additional stimuli, such as vibrations caused by a predator, before responding to alarm pheromone presented c. 5 mm away, whilst the less agile wingless M. *persicae* and M. *viciae* scattered in response to alarm pheromone alone. These differences in behaviour may again be related to the likelihood of locating a new host plant after dropping from the current host. In an illustration of climatic effects, Roitberg and Myers (1978) found that A. *pisum* on lucerne (alfalfa; *Medicago sativa* L.) in a hot, dry region were less likely to drop in response to alarm pheromone than colonies in a cooler, wetter region where the risk of desiccation was lower. Similarly, a strain of A. *pisum* on *Lotus uliginosus* Schkuker, a plant that often grows in ponds and water-filled ditches, showed a lower dropping rate than a strain found on lucerne (Müller, 1983).

Wingless adults from colonies where aphid numbers are high, where winged adults are present and the condition of the host plant is declining, move faster, show less 'searching' behaviour and orient to vertical objects less frequently on mechanical dislodgement than wingless forms from less dense colonies on fresh hosts (Niku, 1975; Phelan *et al.*, 1976; Rice *et al.*, 1983). This results in aphids from colonies enduring less suitable conditions, dispersing further from the original host plant than those from more suitable conditions (Phelan *et al.*, 1976). In the turnip aphid, *Lipaphis erysimi* (Kaltenbach), isothiocyanates, derived from glucosinolates acquired from brassicaceous (= cruciferous) host plants synergize the effect of the alarm pheromone (Dawson *et al.*, 1986, 1987b). These compounds are likely to be released from aphid honeydew. Thus a high number of other aphids in the immediate vicinity releasing isothiocyanates increases the percentage of aphids responding to alarm pheromone and therefore dispersal.

Susceptibility to insecticide has also been found to correlate with responses to alarm pheromone. Susceptible strains produce more pheromone and respond more quickly and in higher numbers than insecticide-resistant strains (Dawson *et al.*, 1983; Rice *et al.*, 1983). This may be due to physiological effects associated with resistance which may affect mobility or sensitivity of the nervous system to stimuli (Foster *et al.*, 1996; Field *et al.*, 1997).

Aphids that have dropped from a plant may recolonize or may move to another host plant further away. Both L. *erysimi* and M. *persicae* dislodged by alarm pheromone are less likely to return to the original host plant than when mechanically dislodged (which may happen in wind or rain) (Phelan *et al.*, 1976). Niku (1975) and Wohlers (1981) found similar patterns of behaviour in A. *pisum*. Aphids dislodged by a predator or experimentally with synthetic alarm pheromone spent longer 'running' (moving in a straight line without moving their antennae or responding to visual stimuli) before 'search' (slow moving with frequent changes in direction and showing responses to visual stimuli) for a host plant began. Aphids dislodged mechanically were more likely to begin to search for a host plant immediately (Niku, 1975).

Chemistry

Alarm pheromones have been identified in aphids from a number of subfamilies in the Aphididae (Nault and Montgomery, 1979). The main component of the alarm pheromone of many aphids is the sesquiterpene hydrocarbon (*E*)-β-farnesene (**5** in Fig. 9.3) (Bowers *et al.*, 1972; Edwards *et al.*, 1973; Wientjens *et al.*, 1973; Pickett and Griffiths, 1980). Other components may also be present. For example, the alarm pheromone of M. *viciae* contains the monoterpenes (−)-α-pinene, (−)-β-pinene, (*Z*,*E*)-α-farnesene and (*E*,*E*)-α-farnesene in addition to **5**, and these can synergize the activity of **5** (Pickett and Griffiths, 1980). There is a high degree of cross-activity of both natural alarm pheromone and **5** among species within the Aphidinae and Chaitophorinae (Bowers *et al.*, 1972; Kislow and Edwards, 1972; Nault and Bowers, 1974). This is typical of insect alarm pheromones in general since such cross-activity does not reduce their evolutionary value (Butler, 1967). The main component of the alarm pheromone of the spotted alfalfa aphid *Therioaphis maculata* Buckton and the sweet clover aphid *Therioaphis riehmi* (Börner) in the Prepanosiphinae however, is the cyclic sesquiterpene (−)-germacrene-A (**6** in Fig. 9.3) (Bowers *et al.*, 1977; Nishino *et al.*, 1977).

Gut and van Oosten (1985) found that, in M. *persicae*, the quantity and composition of the alarm pheromone varied with morph and age of aphid, although **5** was always the main component and (Z,E)-α-farnesene and (E,E)-α-farnesene were minor components. Wingless adults produced a greater amount of alarm pheromone than other morphs, which may be expected since they live gregariously and the chance of predation is high. Some juvenile hormone activity has been found in **5** and related compounds, presumably because of the similarity in their structures (Mauchamp and Pickett, 1987). At unnaturally high levels in the field, **5** is toxic to aphids (Gut *et al.*, 1987).

Both **5** and **6** are volatile, highly labile compounds and produce a short-lived but, for some species, dramatic alarm signal (Nault and Phelan, 1984). Nault *et al.* (1973) estimated that the effective distance of the alarm pheromone in the foxglove aphid *Aulacorthum solani* (Kaltenbach) and in M. *persicae* was approximately 10–30 mm and the duration around 20 min. The proportion of aphids dispersing generally decreased with increasing distance from the source (Montgomery and Nault, 1977b). Alarm pheromone is detected by the primary rhinaria on the fifth and sixth antennal segments (excision studies: Nault *et al.*, 1973; electrophysiological recordings: Wohlers and Tjallingii, 1983, in which segments were covered to locate the position of **5** sensitive cells; Dawson *et al.*, 1987b, recordings from single olfactory cells).

Biosynthesis

Although the biosynthesis of (*E*)-β-farnesene has not been studied in aphids, the gene for an enzyme from peppermint (*Mentha* × *piperita*, L.) that effects conversion of farnesyl pyrophosphate to this sesquiterpene has been cloned and functionally expressed in *Escherichia coli* (Crock *et al.*, 1997). However, many other sesquiterpenes are also produced which could cause inhibition (see below) of the alarm response when attempts are made to clone this gene into crop plants, as proposed originally by Pickett (1985).

The role of dietary components in the production of and responses to alarm pheromone is an area of interest, producing results that may lead to a better understanding of biochemical processes involved. H.F. van Emden and co-workers have found that M. *persicae* reared on artificial diet produces an alarm pheromone containing **5** but which does not stimulate other aphids reared on artificial diet or on host plants (H.F. van Emden and L.J. Wadhams, 1998, unpublished results).

Release

Droplets secreted from the cornicles comprise two types of material: a minor volatile, rapidly vaporizing fraction that comprises the alarm pheromone; and a waxy fraction, consisting mainly of triglycerides, that crystallizes on contact with foreign particles outside of the aphid's body (Strong, 1967; Callow *et al.*, 1973; Greenway and Griffiths, 1973). The waxy component appears to function as a sticky irritant to predators and parasitoids (Dixon, 1958; Edwards, 1966; Nault *et al.*, 1973; Dixon and Stewart, 1975) and can interfere with moulting (Campbell, 1977), but some suggest it may be an excretory product (Hottes, 1928; Lindsay, 1969; Wynn and Boudreaux, 1972) or a releasing substrate for the alarm pheromone fraction (Nault and Phelan, 1984).

The production of cornicle droplets by aphids depends on a number of factors. Mechanical stimulation of the head or abdomen of *A. pisum*, *A. solani* and *M. persicae*, and the abdomen of *D. platanoidis*, can produce a high frequency of cornicle secretion, whereas stimulation of the antennae or legs can give a low frequency of release (Nault *et al.*, 1973; Dixon and Stewart, 1975). Thus, general movement of other aphids in the colony will not stimulate alarm pheromone production. Mechanical stimulation, before or together with the perception of alarm pheromone, increases the proportion of aphids that respond (Dixon and Stewart, 1975; Roitberg and Myers, 1978; Clegg and Barlow, 1982). The degree of vibration produced by a foraging predator may influence the type of alarm behaviour elicited. Thus pea aphids (*A. pisum*) are more likely to drop off a plant when confronted with a coccinellid beetle (*Adalia bipunctata* L.) than with a lighter and slower moving syrphid larva (*Metasyrphus corollae* (Fabricius)) (Brodsky and Barlow, 1986). The type and size of predator also appears to affect the percentage of aphids that produce alarm pheromone and the responses elicited. Goff and Nault (1974) recorded low percentages (less than 25%) of aphids producing cornicle droplets after attack by a parasitoid, though there is evidence that aphid defensive behaviour increases following attack (Gardner *et al.*, 1984). Alarm pheromone production in response to predators can also vary with the type and size of predator (Dixon, 1958; Nault *et al.*, 1973; Arakaki, 1992). Dixon (1958) reported that nettle aphids, *Microlophium carnosum* (Buckton), most frequently 'waxed' coccinellid larvae (*Adalia decempunctata* (L.)) with their cornicle secretions when both aphid and larvae were of a similar size, but total numbers of aphids producing cornicle droplets without waxing were not recorded. Eighty-two per cent of sugarcane woolly aphids (*C. lanigera*) produced cornicle droplets when attacked by adult ladybird beetles (*Pseudoscymnus kurohime*, Miyatake) compared with only 7–12% when attacked by larval beetles (Arakaki, 1992).

There is a correlation between the presence of well developed cornicles and aphid apparency within the habitat, suggesting that the possession of longer cornicles enables aphids to survive in less protected habitats (Nault and Phelan, 1984). The more primitive Aphidoidea (Adelgidae and Phylloxeridae), which lack or have poorly developed cornicles, tend to live in protected environments such as in galls or in soil and are therefore less exposed to predators. The elongation and articulation of the cornicles may increase an aphid's ability to smear wax on or to mark a predator with alarm pheromone and the predator will then be more easily detected as it moves about an aphid colony (Nault and Phelan, 1984).

Aggregation pheromones

Individuals of many aphid species aggregate on the host plant. In the case of larvae and wingless adults, the original aggregation may well result from the larviposition site selected by the mother, but there is evidence that winged adults of *R. padi* also aggregate in the field (Pettersson, 1993). Such behaviour may be due to host plant characteristics which determine suitable feeding sites but evidence is accumulating that specific insect-produced odours may act as aggregation pheromones. Kay (1976) reported such a pheromone produced by *A. fabae* in the laboratory and more recent work has supported this idea. Winged *L. erysimi* and *B. brassicae* were attracted to volatiles produced by conspecific winged forms, but not wingless forms (Pettersson and Stephansson, 1991). The wingless forms were repelled by either morph of the other species but lacked any aggregation response to conspecifics. In the case of the winged morph interactions, the behaviour-modifying odours appeared to derive directly from the insect rather than the plant or the plant/insect combination. These aphid species are non-host-alternating but similar results have been reported from the host-alternating bird-cherry–oat aphid, *R. padi*, on the poaceous summer host (Pettersson *et al.*, 1995). Again, winged forms on the host plant were attractive to other winged forms. Aggregation pheromone may be produced by the spring migrants but only when they are feeding on the summer host, since aggregation on the natal winter host would, of course, conflict with host alternation (Pettersson, 1994). Similar observations of *P. humuli* have shown that spring migrants feeding upon hops release volatiles that attract other migrants (Campbell *et al.*, 1993). In this case, however, although aphid-induced, the active volatiles were plant-derived and are thus kairomones, not pheromones.

Autumn movement from summer to winter host plants also results in aggregation of aphids in the field and it is possible that the autumn winged gynoparae produce an aggregation pheromone, although again this has not been demonstrated unequivocally (Broadbent, 1949; Pettersson, 1993). Sex pheromones produced by the sexual females, however, also appear to act as aggregation pheromones. Field experiments using water traps releasing the specific sex pheromones of *P. humuli*, *R. padi* and *S. fragariae* caught not only males but also significant numbers of winged asexual females (Hardie *et al.*, 1996). These females were most likely to be, and in the case of *P. humuli* were definitely, gynoparae and it seems that the sex pheromone released by sexual females born to earlier gynoparae indicates a suitable host on which later-appearing individuals can successfully deposit their offspring.

Pheromone Interactions with Other Semiochemicals

Sex pheromones

In the 1960s, aphid sex pheromones were believed to act over a very short range and it was proposed that mate location would involve males firstly finding the host plant. This 'rendezvous' hypothesis was most attractive for host-alternating species where the winter hosts comprise one or very few plant species (Steffan, 1987; Ward, 1991). There is now evidence for an interaction between host plant volatiles and sex pheromone in some aphid species. More *P. humuli* males were caught in traps releasing volatiles extracted from winter host plants, *Prunus* spp., along with **3**/**4** than in traps releasing **3**/**4** alone (Campbell *et al.*, 1990; Lösel *et al.*, 1996a,b). Similar results were found with *R. padi* and volatile leaf extracts of bird-cherry, *Prunus padus* L. (Hardie *et al.*, 1994a). Analysis of these leaf volatiles showed large amounts of benzaldehyde which, when released along with the sex pheromone, also increased catches of males (Hardie *et al.*, 1994a). Catches of male *R. padi* were also enhanced by placing pheromone traps in the winter host rather than the non-host tree, *Malus sylvestris* (Mill), from the same family (J.R. Storer and J. Hardie, 1998, unpublished results). Laboratory experiments have produced contradictory results. Guldemond *et al.* (1993) concluded that males of *Cryptomyzus* spp. did not respond to winter-host volatiles and similar conclusions were made for males of *S. fragariae* and *S. avenae* (Lilley and Hardie, 1996). In the latter case, if males could discriminate between the respective host plants – blackberry, *Rubus fruticosus* L., and barley, *Hordeum vulgarae* (L.) – this would provide a reliable cue for locating conspecific sexual females since both species release the same sex pheromone, nepetalactone **1** (Hardie *et al.*, 1992; Lilley *et al.*, 1994/95). This area of aphid chemical ecology requires further investigation.

Chemical cues can interact with visual cues, the latter also affecting the behaviour of male aphids. For example, surface colour is crucial and catches of males of three species in water traps were dramatically affected by trap colour (Hardie *et al.*, 1996). Maximum catches were made in transparent, light green, yellow and orange traps releasing pheromone, while black, blue, white and red caught few males. The response to colour appears to vary with species. Few *P. humuli* males were found in yellow traps unless pheromone was present, while *R. padi* males were found in significant numbers in yellow traps without pheromone, although significantly higher numbers were recorded with pheromone present.

Alarm pheromones

Sesquiterpene hydrocarbons including **5** are commonly produced by plants. Gibson and Pickett (1983) identified **5** in the volatiles produced by leaf hairs of the aphid-resistant wild potato, *Solanum berthaultii* Hawkes; and in bioassays, leaf volatiles from this plant caused alarm behaviour in adult wingless *M. persicae*. Volatiles collected from the cultivated potato *S. tuberosum* L., which is susceptible to aphids, did not affect *M. persicae*. The essential oil from hops (*Humulus lupulus* L.) also contains **5**, but this plant is subject to aphid attack (Dawson *et al.*, 1984). In this case, the sesquiterpene, (−)-β-caryophyllene (**7** in Fig. 9.3) together with other sesquiterpenes, was found to inhibit the activity of **5**. The detection of additional plant volatiles in conjunction with **5** enables the aphid to distinguish between **5** in an alarm pheromone and from a plant source. Increasing the ratio of **5** to **7** increased the level of dispersal. Thus aphids detecting an increase of **5** but not **7** when alarm pheromone is released will respond. Olfactory cells have been found which are sensitive to either **5** or **7**, but not to both, and the two cell types are located in rhinaria on different antennal segments (C.M. Woodcock, 1998, unpublished results), thereby providing the neurophysiological basis for the inhibition of activity of **5** by **7**.

The ratio of **5** to **7** affects the susceptibility to aphid attack of lucerne (Mostafavi *et al.*, 1996). Variation was found in the amount of **5** in extracts from five entries and two germplasms, but the amount of **7** was approximately the same in each. In olfactometer studies, plant material with higher **5** to **7** ratios were chosen less often by blue alfalfa aphids (*Acyrthosiphon kondoi* Shinji) and pea aphids (*A. pisum*). Such information may be useful in the development of aphid-resistant plant varieties.

As mentioned previously, plant-derived isothiocyanates synergize the response to **5** in *L. erysimi* (Dawson *et al.*, 1986, 1987b). Synergism between plant volatiles and **5** also occurs with *A. fabae* but not with the polyphagous *M. persicae* (Dawson *et al.*, 1986)

Aphid Pheromones as Kairomones

Sex pheromones

The aphid sex pheromone component **1** attracts not only males of certain aphid species but also female parasitoid wasps, particularly in the genus *Praon* (see also Chapter 16). Field trials in the UK and Germany have shown significantly higher catches of three *Praon* species in water traps releasing **1** than in control traps. Nepetalactols **2** and **3** were not effective, while the presence of **2** in combination with **1** reduced catches (Hardie *et al.*, 1991; Powell *et al.*, 1993; Hardie *et al.*, 1994c). In laboratory bioassays, naïve virgin *Praon volucre* (Haliday) females responded to the sexual females of *S. fragariae*, which release purely **1** as their sex pheromone (Lilley *et al.*, 1994/95). Male parasitoids did not respond, nor did male or female parasitoids respond to other aphid morphs (Lilley *et al.*, 1994a). It was also of interest that female parasitoids in the field responded to **1** throughout the year and not only in autumn, when sexual female aphids occur (Lilley *et al.*, 1994a). From the viewpoint of possible control measures, it was also shown that when **1** was released close to aphid colonies, there was an increase in the levels of parasitism compared with controls (Lilley *et al.*, 1994b, 1997). In the above field studies, male parasitoids were rarely found in the water traps, even though suction traps showed that they were present. Parasitoid females of other genera, in particular *Aphidius*, were also present but were not attracted to the traps releasing **1**. More recent work has indicated that *Aphidius* females may be initially attracted to the pheromone source, and forage more intensively around it, without flying directly to it (Glinwood *et al.*, 1998). Gabrys *et al.* (1997) also found that *Diaeretiella rapae* (McIntosh), a specialist parasitoid of the cabbage aphid *B. brassicae*, was attracted in significant numbers to water traps releasing **1** in autumn brassicaceous crops.

The lacewing, *Chrysopa cognata* Wesmael, is a common aphid predator in Korea and has antennal receptors for **1** and **2** (Boo *et al.*, 1998). Olfactometer studies in the laboratory also showed an attraction to both pheromone components although direct comparison between **1** and **2** showed a significant preference for **2**. Furthermore, field work also demonstrated that water trap catches were increased significantly above control levels when both **1** and **2** were released together, but the individual components released alone did not enhance trap catches.

Alarm pheromones

Alarm pheromone may indicate the presence of aphids in the local area and stimulate 'searching' behaviour in predators and parasitoids. However, contact with **5** did not influence the behaviour of the predatory lady beetle *Coccinella septempunctata bruckii* Mulsant (Nakamuta, 1991). In contrast, predatory carabids were attracted to **5** in olfactometer studies (Kielty *et al.*, 1996). For parasitoids, the waxy component of aphid cornicle secretions, but not **5**, stimulated oviposition attacks (Grasswitz and Paine, 1992; Battaglia *et al.*, 1993, 1995; see also Chapters 14 and 16) and may be involved in host acceptance. There is also evidence that this kairomone may be present in the host cuticle (Powell *et al.*, 1998; D. Battaglia *et al.*, 1998, unpublished results). Attraction to **5** by the cereal aphid parasitoid, *Aphidius uzbekistanicus* Luzhetski, has been demonstrated in the olfactometer (Micha, 1995).

Possibilities for Aphid Control

The study of aphid chemical ecology, particularly pheromones but including other semiochemicals that influence aphid behaviour, promises alternative methods of pest control to the exclusive use of aphicides that often act as broad-spectrum insect toxicants. In general, semiochemicals, when employed alone, give ineffective or insufficiently reliable pest control and should therefore be combined with other approaches in integrated pest management strategies. The main components of such strategies are pest monitoring, to allow accurate timing of control treatments; combined use of semiochemicals, host plant resistance and trap crops, to manipulate pest behaviour; and selective insecticides or biological control agents, to reduce pest populations. The objective is to draw together these approaches into a push–pull or stimulo-deterrent diversionary strategy. In a push–pull strategy, the harvestable crop is protected by host-masking agents, repellents, anti-feedants or oviposition deterrents. At the same time, aggregative semiochemicals, including sex pheromones and host plant attractants, stimulate colonization of pests on trap crops or entry into traps where pathogens can be deployed. Because the individual components of the push–pull are not in themselves highly efficient, and influence complex behavioural responses, they do not select for resistance as strongly as conventional toxicant pesticides, thereby making the push–pull approach intrinsically more sustainable.

For aphids, attempts to provide semiochemical-based control methods have lagged behind those for some other insect families – for example, the noctuid moths – but semiochemical tools are now available for most aspects of push–pull strategies for certain aphid pest species. Synthetic sex pheromones are becoming more widely available for attracting aphids, with large-scale production being developed by the introduction of new industrial crops in the *Nepeta* genus that yield pheromone components or their precursors (Hick *et al.*, 1997). The sex pheromones specifically attract the males, but they can also attract the autumn asexual migrants (gynoparae) (Hardie *et al.*, 1996; Lösel *et al.*, 1996a,b; Gabrys *et al.*, 1997). The kairomonal role of the sex pheromone in attracting parasitoids is also showing excellent field results

(Glinwood *et al.*, 1998) and may allow improved exploitation of natural populations of beneficial insects (see also Chapter 16, this volume). Alternative aphid attractants generally relate to host plant finding, but further work on aphid aggregation pheromones is expected to contribute new attractants. No pheromonal semiochemicals are yet available for repelling aphids, but non-host plant-derived repellents have proved successful. When the plant-stress-related signal, methyl salicylate, is deployed as slow-release formulations in the crop, it has consistently given a reduction in field cereal aphid populations of around 40% (Pettersson *et al.*, 1994). Anti-feedants have also been investigated experimentally: polygodial, a dialdehydic sesquiterpenoid in the drimanoid series, can be used in the field against cereal aphid-mediated damage (Pickett *et al.*, 1987). The alarm pheromone (*E*)-β-farnesene (**5**) has been used at very high field rates (Gut *et al.*, 1987), where some toxicity, relating presumably to juvenile hormone effects, is observed (Mauchamp and Pickett, 1987). The main role to date for (*E*)-β-farnesene has been in causing aphid mobility and thereby more efficient pick-up of an aphicide or infection with fungal pathogens (Pickett *et al.*, 1994). Deployment of the sex pheromone in the field may potentially improve infection of males with fungal pathogens and cause epizootics in the autumn migrants (Lösel *et al.*, 1996a) and some success has been encountered here (C.A.M. Campbell and C.J. Hartfield, East Malling, 1998, personal communication). As aphid pheromones and other semiochemicals become more readily available, field demonstrations combining the various components of push–pull strategies could facilitate relatively large penetration into horticultural and even agricultural aphid control. In addition, the molecular biological developments in this area, such as the cloning of the sesquiterpene synthase gene from peppermint discussed above, may provide new opportunities.

Future Developments

Aphid aggregation pheromones must be a prime target for future research. Further studies on the interactions between aphids and plants that have already been colonized by conspecifics may extend our knowledge of host-location mechanisms. Recent studies on an oligophagous coleopteran show that, in addition to olfactory cells detecting characteristic host semiochemicals, there are neuronal cell pairs responding specifically to ubiquitous plant volatiles (Blight *et al.*, 1995). Such cell pairing may allow insects to recognize ratios of volatiles produced at different stages of plant growth or enable location of host plants for which no species-specific volatiles exist. A similar mechanism may be important in other phytophagous insects, including aphids.

Perhaps the most exciting area of work is in understanding the mechanisms by which different aphid species induce the release of different compounds from a common host (Du *et al.*, 1998). This work could provide new opportunities for inducing plant defence before aphid colonization or virus transmission. Finally, with the exception of those studies already mentioned, and some preliminary work on other Drepanosiphinae which show considerable chemical diversity (J.A. Pickett, 1998, unpublished results), the semiochemistry of aphids in subfamilies other than the Aphidinae remain largely uninvestigated. With growing concern for losses of species diversity, these multitudes of aphids, often embodying highly specific interactions with plants (including many endangered tree species), invite much more research effort. In addition, many of these aphids have maintained exclusive use of trees, particularly gymnosperms, as hosts, and can be serious pests of economic and amenity tree cultures where aphicide deployment is already unacceptable.

Acknowledgements

Jim Hardie appreciates support from the Biotechnology and Biological Sciences Research Council. IACR-Rothamsted receives grant-aided support from the Biotechnology and Biological Sciences Research Council of the United Kingdom. Part of this work was supported by the UK Ministry of Agriculture, Fisheries and Food (MAFF).

References

Arakaki, N. (1989) Alarm pheromone eliciting attack and escape responses in the sugar cane woolly aphid, *Ceratovacuna lanigera* (Homoptera, Pemphigidae). *Journal of Ethology* 7, 83–90.

Arakaki, N. (1992) Feeding behaviour of the ladybeetle *Pseudoscymnus kurohime*. *Journal of Ethology* 10, 7–13.

Baker, T.C. and Vickers, N.J. (1997) Pheromone-mediated flight in moths. In: Cardé, R.T. and Minks, A.K. (eds) *Insect Pheromone Research, New Directions*. Chapman & Hall, New York, pp. 248–264.

Battaglia, D., Pennacchio, F., Marincola, G. and Tranfaglia, A. (1993) Cornicle secretion of *Acyrthosiphon pisum* (Homoptera: Aphididae) as a contact kairomone for the parasitoid *Aphidius ervi* (Hymenoptera: Braconidae). *European Journal of Entomology* 90, 423–428.

Battaglia, D., Pennacchio, F., Romano, A. and Tranfaglia, A. (1995) The role of physical cues in the regulation of host recognition and acceptance behaviour of *Aphidius ervi* Haliday (Hymenoptera, Braconidae). *Journal of Insect Behavior* 8, 739–750.

Bellesia, F., Grandi, R., Pagnoni, U.M., Pinetti, A. and Trave, R. (1984) Biosynthesis of nepetalactone in *Nepeta cataria*. *Phytochemistry* 23, 83–87.

Blackman, R.L. and Eastop, V.F. (1985) *Aphids on the World's Crops: an Identification Guide*. John Wiley & Sons, Chichester, 466 pp.

Blight, M.M., Pickett, J.A., Wadhams, L.J. and Woodcock, C.M. (1995) Antennal perception of oilseed rape, *Brassica napus* (Brassicaceae), volatiles by the cabbage seed weevil *Ceutorhynchus assimilis* (Coleoptera, Curculionidae). *Journal of Chemical Ecology* 21, 1649–1664.

Boo, K.S., Chung, B., Han, K.S., Pickett, J.A. and Wadhams, L.J. (1998) Response of the lacewing *Chrysopa cognata* Wesmael (Neuroptera: Chrysopidae) to pheromones of its aphid (Homoptera: Aphididae) prey. *Journal of Chemical Ecology* 24, 631–643.

Bowers, W.S., Nault, L.R., Webb, R.E. and Dutky, S.R. (1972) Aphid alarm pheromone: isolation, identification, synthesis. *Science* 177, 1121–1122.

Bowers, W.S., Nishino, C., Montgomery, M.E., Nault, L.R. and Nielson, M.W. (1977) Sesquiterpene progenitor, Germacrene A: an alarm pheromone in aphids. *Science* 196, 680–681.

Broadbent, L. (1949) The grouping and overwintering of *Myzus persicae* on *Prunus* species.

Annals of Applied Biology 36, 334–340.

Brodsky, L.M. and Barlow, C.A. (1986) Escape responses of the pea aphid, *Acyrthosiphon pisum* (Harris) (Homoptera: Aphididae): influence of predator type and temperature. *Canadian Journal of Zoology* 64, 937–939.

Butler, C.G. (1967) Insect pheromones. *Biological Reviews* 42, 42–87.

Callow, R.K., Greenway, A.R. and Griffiths, D.C. (1973) Chemistry of the secretion from the cornicles of various species of aphids. *Journal of Insect Physiology* 19, 737–748.

Campbell, C.A.M. (1977) A laboratory evaluation of *Anthocoris nemorum* and *A. nemoralis* (Hem.: Anthocoridae) as predators of *Phorodon humuli* (Hom.: Aphididae). *Entomophaga* 22, 309–314.

Campbell, C.A.M., Dawson, G.W., Griffiths, D.C., Pettersson, J., Pickett, J.A., Wadhams, L.J. and Woodcock, C.M. (1990) The sex attractant of the damson–hop aphid *Phorodon humuli* (Homoptera, Aphididae). *Journal of Chemical Ecology* 16, 3455–3464.

Campbell, C.A.M., Pettersson, J., Pickett, J.A., Wadhams, L.J. and Woodcock, C.M. (1993) Spring migration of the damson–hop aphid, *Phorodon humuli* (Homoptera, Aphididae), and summer host plant-derived semiochemicals released on feeding. *Journal of Chemical Ecology* 19, 1569–1576.

Cardé, R.T and Mafra-Neto, A. (1997) Mechanisms of flight of male moths to pheromone. In: Cardé, R.T. and Minks, A.K. (eds) *Insect Pheromone Research, New Directions*. Chapman & Hall, New York, pp. 275–290.

Clegg, J.M. and Barlow, C.A. (1982) Escape behaviour of the pea aphid *Acyrthosiphon pisum* (Harris) in response to alarm pheromone and vibration. *Canadian Journal of Zoology* 60, 2245–2252.

Crock, J., Wildung, M. and Croteau, R. (1997) Isolation and bacterial expression of a sesquiterpene synthase cDNA clone from peppermint (*Mentha* × *piperita*, L.) that produces the aphid alarm pheromone (*E*)-β-farnesene. *Proceedings of the National Academy of Sciences USA* 94, 12833–12838.

Dawson, G.W., Griffiths, D.C., Pickett, J.A. and Woodcock, C.M. (1983) Decreased response to alarm pheromone by insecticide-resistant aphids. *Naturwissenschaften* 70, 254–255.

Dawson, G.W., Griffiths, D.C., Pickett, J.A., Smith, M.C. and Woodcock, C.M. (1984) Natural inhibition of the aphid alarm pheromone. *Entomologia Experimentalis et Applicata* 36, 197–199.

Dawson, G.W., Griffiths, D.C., Pickett, J.A., Wadhams, L.J. and Woodcock, C.M. (1986) Plant compounds that synergise activity of the aphid alarm pheromone. In: *Proceedings of the Brighton Crop Protection Conference – Pests and Diseases*, Vol. 2, pp. 829–834.

Dawson, G.W., Griffiths, D.C., Janes, N.F., Mudd, A., Pickett, J.A., Wadhams, L.J. and Woodcock, C.M. (1987a) Identification of an aphid sex pheromone. *Nature* 325, 614–616.

Dawson, G.W., Griffiths, D.C., Pickett, J.A., Wadhams, L.J. and Woodcock, C.M. (1987b) Plant-derived synergists of alarm pheromone from turnip aphid, *Lipaphis (Hyadaphis) erysimi* (Homoptera, Aphididae). *Journal of Chemical Ecology* 13, 1663–1671.

Dawson, G.W., Griffiths, D.C., Merritt, L.A., Mudd, A., Pickett, J.A., Wadhams, L.J. and Woodcock, C.M. (1988) The sex pheromone of the greenbug, *Schizaphis graminum*. *Entomologia Experimentalis et Applicata* 48, 91–93.

Dawson, G.W., Janes, N.F., Mudd, A., Pickett, J.A., Slawin, A.M.Z., Wadhams, L.J. and Williams, D.J. (1989) The aphid sex pheromone. *Pure and Applied Chemistry* 61, 555–558.

Dawson, G.W., Griffiths, D.C., Merritt, L.A., Mudd, A., Pickett, J.A., Wadhams, L.J. and Woodcock, C.M. (1990) Aphid semiochemicals – a review, and recent advances on the sex pheromone. *Journal of Chemical Ecology* 16, 3019–3030.

Dawson, G.W., Pickett, J.A., and Smiley, D.W.M. (1996). The aphid sex pheromone cyclopentanoids: synthesis in the elucidation of structure and biosynthetic pathways.

Bio-organic and Medicinal Chemistry 4, 351–361.

Dill, L.M., Fraser, A.H.G. and Roitberg, B.D. (1990) The economics of escape behaviour in the pea aphid, *Acyrthosiphon pisum*. *Oecologia* 83, 473–478.

Dixon, A.F.G. (1958) The escape responses shown by certain aphids to the presence of the coccinellid *Adalia decempunctata* (L.) *Transactions of the Royal Entomological Society, London* 110, 319–334.

Dixon, A.F.G. and Stewart, W.A. (1975) Function of the siphunculi in aphids with particular reference to the sycamore aphid, *Drepanosiphum platanoides*. *Journal of Zoology, London* 175, 279–289.

Du, Y., Poppy, G.M., Powell, W., Pickett, J.A., Wadhams, L.J. and Woodcock, C.M. (1998) Identification of semiochemicals, released during aphid (Homoptera: Aphididae) feeding, that attract the parasitoid *Aphidius ervi* (Hymenoptera: Aphidiidae). *Journal of Chemical Ecology* 24, 1355–1368.

Edwards, J.S. (1966) Defence by smear: supercooling in the cornicle wax of aphids. *Nature* 211, 73–74.

Edwards, L.J., Siddall, J.B., Dunham, L.L., Uden, P. and Kislow, C.J. (1973) *Trans-β-*farnesene, alarm pheromone of the green peach aphid, *Myzus persicae* (Sulzer). *Nature* 241, 126–127.

Eisenbach, J. and Mittler, T.E. (1980) An aphid circadian rhythm: factors affecting the release of sex pheromone by oviparae of the greenbug, *Schizaphis graminum*. *Journal of Insect Physiology* 26, 511–515.

Eisenbach, J. and Mittler, T.E. (1987) Effects of photoperiod and mating on sex pheromone production and release by oviparae of the aphid *Schizaphis graminum*. *Physiological Entomology* 12, 293–296.

Emden, H.F. van, Eastop, V.F., Hughes, R.D. and Way, M.J. (1969) The ecology of *Myzus persicae*. *Annual Review of Entomology* 14, 197–270.

Field, L.M., Anderson, A.P., Denholm, I., Foster, S.P., Harling, Z.K., Javed, N., Martinez-Torres, D., Moores, G.D., Williamson, M.S. and Devonshire, A.L. (1997) Use of biochemical and DNA diagnostics for characterising multiple mechanisms of insecticide resistance in the peach–potato aphid, *Myzus persicae* (Sulzer). *Pesticide Science* 51, 283–289.

Flögel, J.H.L. (1905) Monographie der Johannisbeeren-Blattlaus, *Aphis ribri* L. *Zeitschrift für Wissenschaftliche Insektenbiologie* 1, 145–155.

Foster, S.P., Harrington, R., Devonshire, A.L., Denholm, I., Devine, G.J. and Kenward, M.G. (1996) Comparative survival of insecticide-susceptible and resistant peach–potato aphids, *Myzus persicae* (Sulzer) (Hemiptera; Aphididae), in low temperature field trials. *Bulletin of Entomological Research* 86, 17–27.

Gabrys, B.J., Gadomski, H.J., Klukowski, Z., Pickett, J.A., Sobota, G.T., Wadhams, L.J. and Woodcock, C.M. (1997) Sex pheromone of cabbage aphid *Brevicoryne brassicae*: identification and field trapping of male aphids and parasitoids. *Journal of Chemical Ecology* 23, 1881–1890.

Gardner, S.M., Ward, S.A. and Dixon, A.F.G. (1984) Limitation of superparasitism by *Aphidius rhopalosiphi*: a consequence of aphid defensive behaviour. *Ecological Entomology* 9, 149–155.

Gibson, R.W. and Pickett, J.A. (1983) Wild potato repels aphids by release of aphid alarm pheromone. *Nature* 302, 608–609.

Glinwood, R.T., Powell, W. and Tripathi, C.P.M. (1998) Increased parasitisation of aphid parasitoids (Hymenoptera: Aphidiinae) on trap plants baited with aphid sex pheromone, and effective range of aphid sex pheromone lures. *Biocontrol Science and Technology* 8, 607–614.

Goff, A.M. and Nault, L.R. (1974) Aphid cornicle secretions ineffective against attack by parasitoid wasps. *Environmental Entomology* 3, 565–566.

Grasswitz, T.R. and Paine, T.D. (1992) Kairomonal effect of an aphid cornicle secretion on *Lysiphlebus testaceipes* (Cresson) (Hymenoptera: Aphidiidae). *Journal of Insect Behavior* 5, 447–457.

Greenway, A.R. and Griffiths, D.C. (1973) A comparison of triglycerides from aphids and their cornicle secretions. *Journal of Insect Physiology* 19, 1649–1655.

Guldemond, J.A. and Dixon, A.F.G. (1994) Specificity and daily cycle of release of sex pheromones in aphids: a case of reinforcement? *Biological Journal of the Linnean Society* 52, 287–303.

Guldemond, J.A., Dixon, A.F.G., Pickett, J.A., Wadhams, L.J. and Woodcock, C.M. (1993) Specificity of sex pheromones, the role of host plant odour in the olfactory attraction of males, and mate recognition in the aphid *Cryptomyzus*. *Physiological Entomology* 18, 137–143.

Gut, J. and van Oosten, A.M. (1985) Functional significance of the alarm pheromone composition in various morphs of the green peach aphid, *Myzus persicae*. *Entomologia Experimentalis et Applicata* 37, 199–204.

Gut, J., Harrewijn, P., van Oosten, A.M. and van Rheenen, B. (1987) Additional function of alarm pheromones in development processes of aphids. *Mededelingen Faculteit LandbouwWetenschappen Rijksuniversiteit Gent* 52, 371–378.

Hardie, J., Holyoak, M., Nicholas, J., Nottingham, S.F., Pickett, J.A., Wadhams, L.J. and Woodcock, C.M. (1990) Aphid sex pheromone components: age dependent release by females and species-specific male response. *Chemoecology* 1, 63–68.

Hardie, J., Nottingham, S.F., Powell, W. and Wadhams, L.J. (1991) Synthetic aphid sex pheromone lures female parasitoids. *Entomologia Experimentalis et Applicata* 61, 97–99.

Hardie, J., Nottingham, S.F., Dawson, G.W., Harrington, R., Merritt, L.A., Pickett, J.A. and Wadhams, L.J. (1992) Attraction of field-flying aphid males to synthetic sex pheromone. *Chemoecology* 3, 113–117.

Hardie, J., Storer, J.R., Nottingham, S.F., Peace, L., Harrington, R., Merritt, L.A., Wadhams, L.J. and Wood, D.K. (1994a) The interaction of sex pheromone and plant volatiles for field attraction of male bird-cherry aphid, *Rhopalosiphum padi*. In: *Proceedings of the Brighton Crop Protection Conference – Pests and Diseases*, Vol. 3, pp. 1223–1230.

Hardie, J., Visser, J.H. and Piron, P.G.M. (1994b) Perception of volatiles associated with sex and food by different adult forms of the black bean aphid, *Aphis fabae*. *Physiological Entomology* 19, 278–284.

Hardie, J., Hick, A.J., Höller, C., Mann, J., Merritt, L.A., Nottingham, S.F., Powell, W., Wadhams, L.J., Witthinrich, J. and Wright, A (1994c) The responses of *Praon* spp. parasitoid to aphid sex pheromone components in the field. *Entomologia Experimentalis et Applicata* 71, 95–99.

Hardie, J., Storer, J.R., Nottingham, S.F., Cook, F.J., Campbell, C.A.M., Wadhams, L.J., Lilley, R. and Peace, L. (1996) Sex pheromone and visual trap interactions in mate location strategies and aggregation by host-alternating aphids in the field. *Physiological Entomology* 21, 97–106.

Hardie, J., Peace, L., Pickett, J.A., Smiley, D.W.M., Storer, J.R. and Wadhams, L.J. (1997) Sex pheromone stereochemistry and purity affect field catches of male aphids. *Journal of Chemical Ecology* 23, 2547–2554.

Hick, A.J., Pickett, J.A., Smiley, D.W.M., Wadhams, L.J. and Woodcock, C.M. (1997) Higher plants as a clean source of semiochemicals and genes for their biotechnological production. In: Wrigley, S., Hayes, M., Thomas, R. and Chrystal, E. (eds) *Phytochemical Diversity: a Source of New Industrial Products*. The Royal Society of Chemistry, Cambridge, pp. 220–236.

Hottes, F.C. (1928) Concerning the structure, function and origin of the cornicles of the family Aphididae. *Proceedings of the Biological Society of Washington* 41, 71–84.

Isaacs, R. (1994) Studies on the chemical ecology of the black bean aphid, *Aphis fabae*. PhD thesis, University of London, UK.

Kay, R.H. (1976) Behavioural components of pheromonal aggregation in *Aphis fabae* Scopoli. *Physiological Entomology* 1, 249–254.

Kielty, J.P., Allen-Williams, L.J., Underwood, N. and Eastwood, E.A. (1996) Behavioral responses of three species of ground beetle (Coleoptera: Carabidae) to olfactory cues associated with prey and habitat. *Journal of Insect Behavior* 9, 237–250.

Kislow, C.J. and Edwards, L.J. (1972) Repellent odour in aphids. *Nature* 235, 108–109.

Larsen, K.S. (1988) Responses of different age classes of the rose-grain aphid, *Metopolophium dirhodum* (Wlk.) to attack by a simulated predator. *Journal of Applied Entomology* 105, 455–459.

Lees, A.D. (1966) The control of polymorphism in aphids. *Advances in Insect Physiology* 3, 207–277.

Lilley, R. and Hardie, J. (1996) Cereal aphid responses to sex pheromones and host-plant odours in the laboratory. *Physiological Entomology* 21, 304–308.

Lilley, R., Hardie, J., Powell, W. and Wadhams, L.J. (1994a) The aphid sex pheromone: a novel host location cue for the parasitoid *Praon volucre*. *In: Proceedings of the Brighton Crop Protection Conference – Pests and Diseases*, Vol. 3, pp. 1157–1162.

Lilley, R., Hardie, J. and Wadhams, L.J. (1994b) Field manipulation of *Praon* populations using semiochemicals. *Norwegian Journal of Agricultural Sciences Supplement* 16, 221–226.

Lilley, R., Hardie, J., Merritt, L.A., Pickett, J.A., Wadhams, L.J. and Woodcock, C.M. (1994/95) The sex pheromone of the grain aphid, *Sitobion avenae* (Fab.) (Homoptera, Aphididae). *Chemoecology* 5/6, 43–46.

Lilley, R., Hardie, J. and Wadhams, L.J. (1997) Manipulation of *Praon* populations with synthetic aphid sex pheromones for the control of cereal aphids. *Boletin de la Asociacion Española de Entomologia* 21, 23–29.

Lindsay, K.L. (1969) Cornicles of the pea aphid, *Acyrthosiphon pisum*; their structure and function. A light and electron microscope study. *Annals of the Entomological Society of America* 62, 1015–1021.

Lösel, P.M., Lindemann, M., Scherkenbeck, J., Campbell, C.A.M., Hardie, J., Pickett, J.A. and Wadhams, L.J. (1996a) Effect of primary-host kairomones on the attractiveness of the hop-aphid sex pheromone to *Phorodon humuli* males and gynoparae. *Entomologia Experimentalis et Applicata* 80, 79–82.

Lösel, P.M., Lindemann, M., Scherkenbeck, J., Maier, J., Engelhard, B., Campbell, C.A.M., Hardie, J., Pickett, J.A., Wadhams, L.J., Elbert, A. and Theilking, G. (1996b) The potential of semiochemicals for control of *Phorodon humuli* (Homoptera: Aphididae). *Pesticide Science* 48, 293–303.

Marsh, D. (1972) Sex pheromone in the aphid *Megoura viciae*. *Nature* 238, 31–32.

Marsh, D. (1975) Responses of male aphids to the female sex pheromone in *Megoura viciae* Buckton. *Journal of Entomology* (A) 50, 43–64.

Mauchamp, B. and Pickett, J.A. (1987) Juvenile hormone-like activity of (*E*)-β-farnesene derivatives. *Agronomie* 7, 523–529.

Micha, S.G. (1995) Die Bedeutung von Semiochemikalien im tritrophischen System aus Hafer (*Avena sativa*), Große Getreideblattlaus (*Sitobion avenae*) und Blattlausparasitoid (*Aphidius uzbekistanicus*). PhD thesis, Christian-Albrechts-Universität, Kiel, Germany.

Montgomery, M.E. and Nault, L.R. (1977a) Comparative response of aphids to the alarm pheromone, (*E*)-β-farnesene. *Entomologia Experimentalis et Applicata* 22, 236–242.

Montgomery, M.E. and Nault, L.R. (1977b) Aphid alarm pheromones: dispersion of *Hyadaphis erysimi* and *Myzus persicae*. *Annals of the Entomological Society of America* 70, 669–672.

Montgomery, M.E. and Nault, L.R. (1978) Effects of age and wing polymorphism on the

sensitivity of *Myzus persicae* to alarm pheromone. *Annals of the Entomological Society of America* 71, 788–790.
Moran, N.A. (1992) The evolution of aphid life cycles. *Annual Review of Entomology* 37, 321–348.
Mostafavi, R., Henning, J.A., Gardea-Torresday, J. and Ray, I.M. (1996) Variation in aphid alarm pheromone content among glandular and eglandular-haired *Medicago* accessions. *Journal of Chemical Ecology* 22, 1629–1638.
Müller, F.P. (1983) Differential alarm pheromone responses between strains of the aphid *Acyrthosiphon pisum*. *Entomologia Experimentalis et Applicata* 34, 347–348.
Nakamuta, K. (1991) Aphid alarm pheromone component, (*E*)-β-farnesene, and local search by a predatory lady beetle, *Coccinella septempunctata bruckii* Mulsant (Coleoptera: Coccinellidae). *Applied Entomology and Zoology* 26, 1–7.
Nault, L.R. and Bowers, W.S. (1974) Multiple alarm pheromones in aphids. *Entomologia Experimentalis et Applicata* 17, 455–457.
Nault, L.R. and Montgomery, M.E. (1979) Aphid alarm pheromones. *Miscellaneous Publications of the Entomological Society of America* 11, 23–31.
Nault, L.R. and Phelan, P.L. (1984) Alarm pheromones and sociality in pre-social insects. In: Bell, W.J. and Cardé, R.T. (eds) *Chemical Ecology of Insects*. Chapman & Hall, London, pp. 237–256.
Nault, L.R., Edwards, L.J. and Styer, W.E. (1973) Aphid alarm pheromones: secretion and reception. *Environmental Entomology* 2, 101–105.
Nault, L.R., Montgomery, M.E. and Bowers, W.S. (1976) Ant–aphid association: role of aphid alarm pheromones. *Science* 192, 1349–1351.
Niku, B. (1975) Verhalten und Fruchtbarkeit ungeflügelter Erbsenläuse (*Acyrthosiphon pisum*) nach einer Fallreaktion. *Entomologia Experimentalis et Applicata* 18, 17–30.
Nishino, C., Bowers, W.S., Montgomery, M.E., Nault, L.R. and Nielson, M.W. (1977) Alarm pheromone of the spotted alfalfa aphid, *Therioaphis maculata* Buckton (Homoptera: Aphididae). *Journal of Chemical Ecology* 3, 349–357.
Pettersson, J. (1970) An aphid sex attractant I. Biological studies. *Entomologica Scandinavica* 1, 63–73.
Pettersson, J. (1971) An aphid sex attractant II. Histological, ethological and comparative studies. *Entomologica Scandinavica* 2, 81–93.
Pettersson, J. (1973) Olfactory reactions of *Brevicoryne brassicae* (L.) (Hom.: Aph.). *Swedish Journal of Agricultural Research* 3, 95–103.
Pettersson, J. (1993) Odour stimuli affecting autumn migration of *Rhopalosiphum padi* (L.) (Hemiptera: Homoptera). *Annals of Applied Biology* 122, 417–425.
Pettersson, J. (1994) The bird cherry–oat aphid, *Rhopalosiphum padi*, (L.) (Homoptera: Aphididae) and odours. In: Leather, S.R., Watt, A.D., Mills, N. and Walters, K.F.A. (eds) *Individuals, Populations and Patterns in Ecology*. Intercept Ltd, Andover, pp. 3–12.
Pettersson, J. and Stephansson, D. (1991) Odour communication between two Brassica feeding aphid species. *Entomologia Generalis* 16, 241–247.
Pettersson, J., Pickett, J.A., Pye, B.J., Quiroz, A., Smart, L.E., Wadhams, L.J. and Woodcock, C.M. (1994) Winter host component reduces colonisation by bird-cherry–oat aphid *Rhopalosiphum padi* (L.) (Homoptera, Aphididae) and other aphids in cereal fields. *Journal of Chemical Ecology* 20, 2565–2574.
Pettersson, J., Quiroz, A. Stephansson, D. and Niemeyer, H.M. (1995) Odour communication of *Rhopalosiphum padi* on grasses. *Entomologia Experimentalis et Applicata* 76, 325–328.
Phelan, P.L., Montgomery, M.E. and Nault, L.R. (1976) Orientation and locomotion of apterous aphids dislodged from their host by alarm pheromone. *Annals of the Entomological Society of America* 69, 1153–1156.

Pickett, J.A. (1985) Production of behaviour controlling chemicals by crop plants. *Philosophical Transactions of the Royal Society of London B* 310, 235–239.

Pickett, J.A. and Griffiths, D.C. (1980) Composition of aphid alarm pheromones. *Journal of Chemical Ecology* 6, 349–360.

Pickett, J.A., Dawson, G.W., Griffiths, D.C., Hassanali, A., Merritt, L.A., Mudd, A., Smith, M.C., Wadhams, L.J., Woodcock, C.M. and Zhang, Z.-N. (1987) Development of plant-derived antifeedants for crop protection. In: Greenhalgh, R. and Roberts, T.R. (eds) *Pesticide Science and Biotechnology*. Blackwell Scientific Publications, Oxford, pp. 125–128.

Pickett, J.A., Wadhams, L.J., Woodcock, C.M. and Hardie, J. (1992) The chemical ecology of aphids. *Annual Review of Entomology* 37, 67–90.

Pickett, J.A., Wadhams, L.J. and Woodcock, C.M. (1994) Attempts to control aphid pests by integrated use of semiochemicals. In: *Proceedings of the Brighton Crop Protection Conference – Pests and Diseases*, Vol. 3, pp. 1239–1246.

Powell, W., Hardie, J., Hick, A.J., Höller, C., Mann, J., Merritt, L.A., Nottingham, S.F., Wadhams, L.J., Witthinrich, J. and Wright, A.F. (1993) Responses of the parasitoid *Praon volucre* (Hymenoptera: Braconidae) to aphid sex pheromone lures in cereal fields in autumn: implications for parasitoid manipulation. *European Journal of Entomology* 90, 435–438.

Powell, W., Pennachio, F., Poppy, G.M. and Tremblay, E. (1998) Strategies involved in the location of hosts by the parasitoid *Aphidius ervi* Haliday (Hymenoptera: Braconidae: Aphidiinae). *Biological Control* 11, 104–112.

Rice, A.D., Gibson, R.W. and Stribley, M.F. (1983) Alarm pheromone secretion by insecticide-susceptible and -resistant *Myzus persicae* treated with demeton-*S*-methyl; aphid dispersal and transfer of plant viruses. *Annals of Applied Biology* 103, 375–381.

Roberti, D. (1979) Sulla attrattivita sessuale degli afida mediante feromoni. *Entomologica* 15, 279–285.

Roitberg, B.D. and Myers, J.H. (1978) Adaptation of alarm pheromone responses of the pea aphid *Acyrthosiphon pisum* (Harris). *Canadian Journal of Zoology* 56, 103–108.

Smith, L.M. (1936) Biology of the mealy plum aphid, *Hyalopterus pruni* (Geoffroy). *Hilgardia* 10, 167–211.

Steffan, A.W. (1987) Fern- und Nahorientierung geflügelter Gynoparae und Sexualis-Männchen bei Blattläusen (Homoptera: Aphidinea: Aphididae). *Entomologia Generalis* 12, 235–258.

Steffan, A.W. (1989) Courtship behaviour and possible pheromone spread by hindleg raising in sexual females of aphids (Homoptera: Aphidinea). *Entomologia Generalis* 15, 33–49.

Strong, F.E. (1967) Observation on aphid cornicle secretions. *Annals of the Entomological Society of America* 60, 668–673.

Stroyan, H.L.G. (1958) A contribution to the taxonomy of some British species of *Sappaphis* Matsumara 1918 (Homoptera:Aphididae). *Journal of the Linnean Society (Zool)* 43, 644–713.

Tamaki, G., Butt, B.A. and Landis, B.J. (1970) Arrest and aggregation of male *Myzus persicae* (Hemiptera: Aphididae). *Annals of the Entomological Society of America* 63, 955–960.

Thieme, T. and Dixon, A.F.G. (1996) Mate recognition in the *Aphis fabae* complex: daily rhythm of release and specificity of sex pheromones. *Entomologia Experimentalis et Applicata* 79, 85–89.

Uesato, S., Ikeda, H., Fujita, T., Inouye, H. and Zenk, M.H. (1987) Elucidation of iridodial formation mechanism – partial purification and characterisation of the novel monoterpene cyclase from *Rauwolfia serpentina* cell suspension cultures. *Tetrahedron Letters* 28, 4431–4434.

Ward, S.A. (1991) Reproduction and host selection by aphids: the importance of 'rendezvous' hosts. In: Bailey, W.J. and Ridsdill-Smith, J. (eds) *Reproductive Behaviour of Insects, Populations and Individuals*. Chapman & Hall, London, pp. 202–226.

Weber, H. (1935) Hemiptera III. *Biologie der Tiere Deutschlands* 38, 210–286.

Wientjens, W.H.J.M., Lawijk, A.C. and van der Marel, T. (1973) Alarm pheromone of grain aphids. *Experientia* 29, 658–660.

Wohlers, P. (1981) Effects of the alarm pheromone (*E*)-β-farnesene on dispersal behaviour of the pea aphid *Acyrthosiphon pisum*. *Entomologia Experimentalis et Applicata* 29, 117–124.

Wohlers, P. and Tjallingii, W.F. (1983) Electroantennogram responses of aphids to the alarm pheromone (*E*)-β-farnesene. *Entomologia Experimentalis et Applicata* 33, 79–82.

Wynn, G.G. and Boudreaux, H.B. (1972) Structure and function of aphid cornicles. *Annals of the Entomological Society of America* 65, 157–166.

10 Scale Insects

EZRA DUNKELBLUM

Institute of Plant Protection, The Volcani Center, Agricultural Research Organization, PO Box 6, Bet Dagan 50250, Israel

Introduction

Scale insects are widely distributed in the world and many of them are serious pests of plants, particularly perennial fruit trees, forest trees and ornamentals. They belong to the superfamily Coccoidea and reproduction is either parthenogenic or bisexual. The adult males of the biparental species are winged insects, small and fragile, and their life span is very short – not more than a few hours to a few days. Conversely, females live for weeks or even a few months; they are wingless and may be sedentary (Fig. 10.1).

Damage by scale insects may be caused directly by feeding, by injecting toxins that disrupt essential processes in the plant or by attracting plant pathogens. Early monitoring by visual inspection is often difficult because of their small size and their behaviour. Traps baited with sex pheromones might be convenient and useful for detecting scale insect populations, but only ten sex pheromones have been identified and synthesized so far. They belong to three families – Diaspididae, Pseudococcidae and Margarodidae – out of about 20 families of the Coccoidea (Ben-Dov, 1985). This limited number of known scale insect sex pheromones contrasts sharply with the large number of sex pheromones identified from other insects, particularly moths and beetles (Mayer and McLaughlin, 1991).

History

The existence of a female sex pheromone in a scale insect was demonstrated for the first time in the red pine scale, *Matsucoccus resinosae* Bean and Godwin (Doane, 1966). Soon after that, the production of a sex pheromone was detected in the California red scale, *Aonidiella aurantii* (Maskell) (Tashiro and Chambers, 1967). Capture of males in traps baited with virgin females or Petri dish bioassays of solutions prepared by extracts of whole virgin females were used in early attempts

(a)

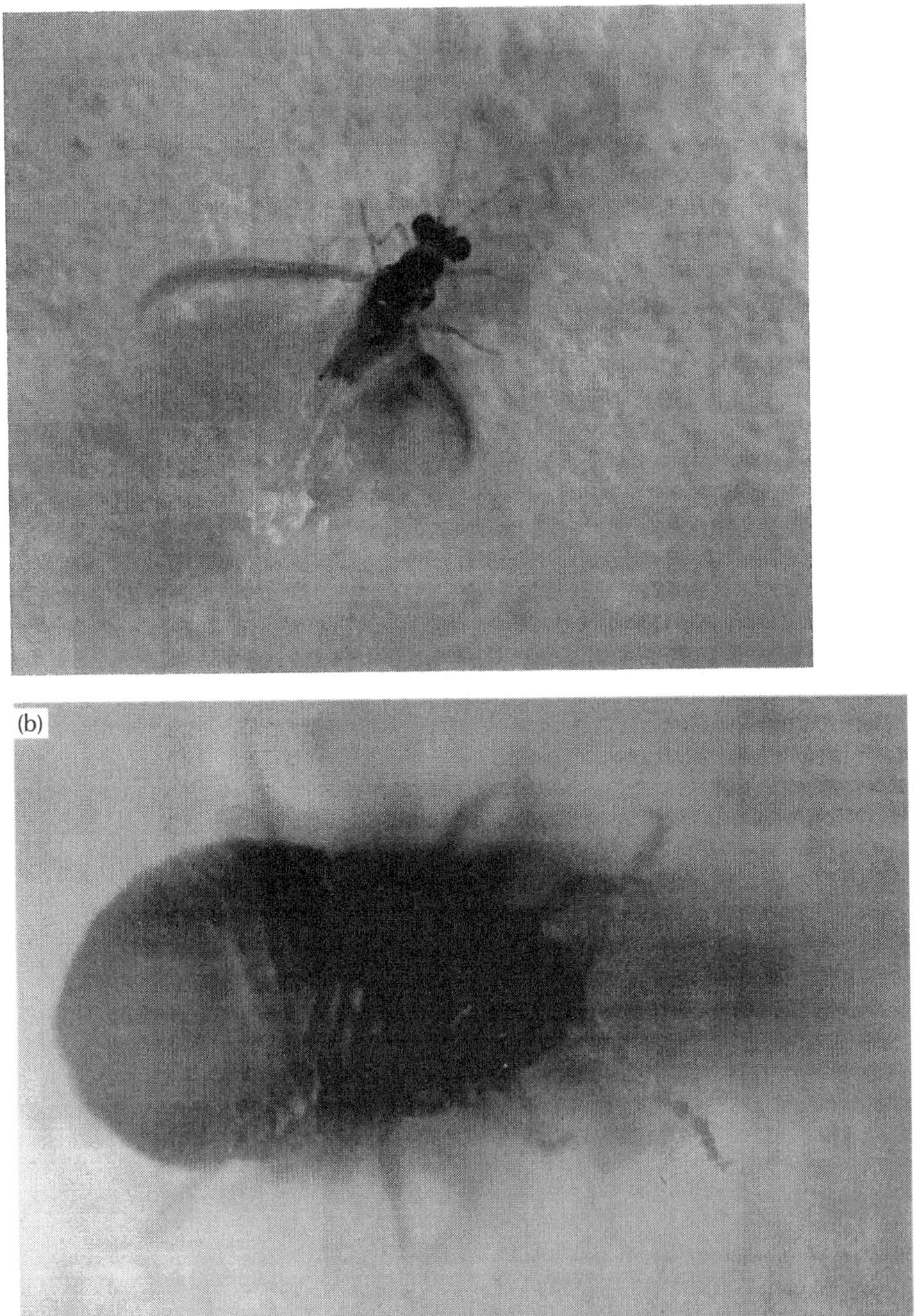

Fig. 10.1. Adult *Matsucoccus josephi* pine bast scales: (a) male; (b) female.

to prove the occurrence of sex pheromones in scale insects. Several diaspidid and pseudococcid species were investigated, and it became evident that adult females of biparental scale insects utilize sex pheromones to attract conspecific males. Subsequently, the presence of sex pheromones was demonstrated for the citrus mealybug, *Planococcus citri* (Risso) (Gravitz and Wilson, 1968); the yellow scale, *A. citrina* (Coquillett) (Moreno *et al.*, 1972a); the Comstock mealybug, *Pseudococcus comstocki* (Kuwana) (Moreno *et al.*, 1972b); the San José scale, *Quadraspidiotus perniciosus*, (Comstock) (Rice, 1974); the grape mealybug, *Planococcus ficus* (Signoret) (Rotundo and Tremblay, 1975); *Pseudococcus calceolariae* (Maskell) (Rotundo and Tremblay, 1976); the white peach scale, *Pseudaulacaspis pentagona* (Targioni Tozzetti) (Heath *et al.*, 1979); the Israeli pine bast scale, *Matsucoccus josephi* Bodenheimer & Harpaz (Dunkelblum and Sternlicht, unpublished results in Moore, 1981); and several other *Matsucoccus* spp. (Park *et al.*, 1986; Lanier *et al.*, 1989; Einhorn *et al.*, 1990). Finally the sex pheromone of the oleander scale insect, *Aspidiotis nerii* Bouché, has recently been identified (Einhorn *et al.*, 1998).

Early attempts to isolate and identify coccoid sex pheromones failed, the main reason being the difficulties involved in mass rearing, pheromone collection and the lack of analytical expertise and sensitive equipment. The first target was *A. aurantii*, which was studied in the USA by Tashiro and Chambers (1967) and Warthen *et al.* (1970). In the late 1970s, Italian researchers tried to isolate and identify the sex pheromones of *Ps. calceolariae* and *P. citri* but were not able to complete the chemical identification (Rotundo *et al.*, 1979; Rotundo and Tremblay, 1980). The sex pheromone of *A. aurantii* eventually became the first coccoid pheromone to be identified and to be synthesized (Roelofs *et al.*, 1978). This publication ought to be considered as a turning point. It demonstrated that coccoid sex pheromones are characteristic semiochemicals, emitted by females to attract conspecific males, and can be isolated and purified in sufficient amounts to enable the complete elucidation of their structure.

Isolation of Coccoid Sex Pheromones

Owing to the extremely small quantity of female-produced pheromone, continuous mass rearing is the first prerequisite in order to provide sufficient numbers of virgin females for chemical identification. At the same time, a continuous supply of males is required for the bioassays. Two methods of sexual separation were employed: application of juvenoids, which selectively prevent male maturation (Moreno *et al.*, 1976), or mechanical separation of the males. The first method has been applied to obtain virgin females of *A. aurantii* using dichlorovos or triprene (Roelofs *et al.*, 1978), of *A. citrina* using triprene (Gieselmann, 1990), of *P. citri* using triprene (Bierl-Leonhardt *et al.*, 1981) or RO 10–3108 (Rotundo and Tremblay, 1980), of *Ps. calceolariae* using RO 10–3108 (Rotundo *et al.*, 1979) and of *Ps. comstocki* using epofenonane (Bierl-Leonhardt *et al.*, 1980, 1982). Although the use of juvenoids is convenient, this method cannot always be used for the sexual separation of scale insects because it often also kills the females, as was found in the case of *Q. perniciosus* (Gieselmann *et al.*, 1979b). Mechanical separation was used for *Q. perniciosus* (Gieselmann *et al.*, 1979b), *P. pentagona* (Heath *et *al.*, 1979), *Ps. comstocki* (Negishi *et al.*, 1980) and for all *Matsucoccus* species studied: *M. resinosae* and synonym *M. matsumurae* (Kuwana) and *M. thunbergianae* Miller and Park (Park *et al.*, 1986; Lanier *et al.*, 1989), *M. feytaudi* Duc (Einhorn *et al.*, 1990) and *M. josephi* (Mendel *et al.*, 1990a). This method is based on the fact that females feed on the host for much longer than males and therefore males can be removed selectively from the host with a brush, as, for example, for *Q. perniciosus* (Gieselmann *et al.*, 1979b), or with a water spray in the case of *P. pentagona* (Heath *et al.*, 1979). In the case of *M. josephi*, males complete feeding at an early stage and develop into a mobile third instar which leaves the plant and descends to pupate at the bottom of the growing bags or containers. This observation enabled a simple separation of the two sexes (Mendel *et al.*, 1990a).

There is no visible external pheromone gland in scale insects and the discovery of sex pheromones in *A. aurantii* (Tashiro and Chambers, 1967) and in *A. citrina* (Moreno *et al.*, 1972a) prompted investigations concerning the pheromone-producing organ in these two species. Various parts of virgin females of *A. aurantii* and *A. citrina* were presented to males of the respective species and it was found that the pygidium of both females was active. Dissection of the internal pygidial glands showed that they were the site of pheromone production and a connection to the anus via a fragile duct was also found (Moreno, 1972; Moreno and Fargerlund, 1975). So the pheromone produced by the pygidial glands is apparently emitted via the rectum and anus. Additionally, the ultrastructure of epidermal glands in the *M. matsumurae* female has been studied by Young and Hong-Ren (1986) and it was suggested that one type was the sex pheromone production site.

Sex pheromones of some coccoids were obtained by extraction from whole virgin females in hexane (Park *et al.*, 1986). However, almost all coccoid sex pheromones were collected and identified from airborne material, mostly in flow systems, trapped on Porapak Q and Tenax absorbents or by cold-trapping. In one case, filter paper was used to absorb pheromone emitted from *Ps. comstocki* females (Negishi *et al.*, 1980). Initially, the crude pheromones were recovered from the absorbents by Soxhlet extraction, but later, all scale insect pheromones were obtained by eluting the absorbent tubes with pentane or hexane.

Most of the coccoid sex pheromones were isolated from scale insects that had been mass reared in the laboratory. Either natural host plants or other appropriate plants were used for growing scale insects. For pheromone collection, the best choice is the host that emits the least plant volatiles. That is why lemons were replaced with potato tubers as host material used in the isolation of the *A. aurantii* sex pheromone (Roelofs *et al.*, 1978). Potato tubers or potato sprouts were also the favoured hosts for several diaspidid and pseudococcid species, although in some cases banana squash (Gieselmann *et al.*, 1979b), pumpkins (Negishi *et al.*, 1980) or even lemons (Gieselmann *et al.*, 1979a) were used. Only one margarodid species, *M. josephi*, could be reared on its natural host, i.e. young *Pinus halepensis* saplings (Mendel *et al.*, 1990a); the others – *M. resinosae*, *M. matsumurae*, *M. thunbergianae* (Lanier *et al.*, 1989) and *M. feytaudi* (Einhorn *et al.*, 1990) – were collected from infested pine trees. These cases were exceptional in coccoid sex pheromone research, since it was necessary to ship infested pine branches from the forest to the laboratory or alternatively to collect crude pheromone in the vicinity of the infested pines and then to dispatch the Porapak Q or Tenax tubes to the chemical laboratory.

The standard unit of collected pheromone was defined as the equivalent of pheromone emitted per female per day (FDE). This is a practical unit and not an exact amount of pheromone: it may vary significantly among different scale insects, from day to day, and it depends on the efficiency of the collecting device. In some cases, one FDE was reported to be a few picograms; in others it was found to be as much as a few nanograms (Gieselmann, 1990). The number of FDEs collected from various scale insect species ranged from a few million to a few hundred million, and quantities isolated ranged from several micrograms of the sex pheromones of *P. pentagona* and *M. josephi* (Heath *et al.*, 1979; Dunkelblum *et al.*, 1993), to approximately 200 μg from *M. feytaudi* (Einhorn *et al.*, 1990) and to the unexpectedly large amount of 30 mg from *Ps. comstocki* (Bierl-Leonhardt *et al.*, 1982). The duration of pheromone collection varied among the different species. Females, as long as they do not mate, attract males for prolonged periods. It was shown, for example, that *A. aurantii* virgin females emit pheromone, although at lower rates, for as long as 84 days (Tashiro and Moffit, 1968). In most cases, adult virgin females were kept in the collecting devices between a few days and 2 weeks. It seems that the pheromone release is continuous, as demonstrated for *P. citri* and *Ps. calceolariae* (Rotundo and Tremblay, 1980) and for *M. josephi* (Mendel *et al.*,1990b).

Bioassay of Coccoid Sex Pheromones

The second prerequisite for successful isolation and identification of scale insect pheromones is the availability of a simple and reliable bioassay in the laboratory or the greenhouse. Limited field bioassays may also be performed, but these require larger amounts of pheromones which can be acquired only by synthesis. An additional complication is that male coccoids are usually active for a few hours only.

In the early research stages, bioassays were performed mainly in Petri dishes. Crude pheromone preparations from hexane extracts of whole females or collected airborne material were dispensed on filter paper disks, paraffin drops or stainless-steel planchets, and the number of males on or touching the pheromone source was recorded (Moreno *et al.*, 1972a; Heath *et al.*, 1979). In some cases, the typical response of males consisted of rapid sweeping abdominal movements and copulatory attempts (Rotundo and Tremblay, 1981). In the case of *M. josephi*, the response was characterized by a dance around and on the paper disk and by attempts to copulate with it (Mendel *et al.*, 1990b). Despite the simplicity of the method, the Petri dish bioassay gave reliable results and was successfully used in the identification of the sex pheromones of *P. pentagona* (Heath *et al.*, 1979), *M. resinosae* (Lanier *et al.*, 1989), *M. feytaudi* (Einhorn *et al.*, 1990) and *M. josephi* (Dunkelblum *et al.*, 1993). A modified Petri dish bioassay, consisting of tracing the path of the females in the dish and recording the males that were able to follow the trails, was used to follow scent trailing of *Ps. calceolariae* (Rotundo and Tremblay, 1981). For the cross-attraction study of two *Matsucoccus* species, a 'puffing' test was used to record male response to the sex pheromones. Samples of pheromone in hexane were placed in micropipettes and, after expelling the hexane, puffs of air containing pheromone were directed to the antennae of the males, whose trailing responses after the pheromone plume were recorded according to a specific protocol (Park *et al.*, 1986). A Y-olfactometer biaossay was used to compare the *M. josephi* male response to virgin and mated females (Mendel *et al.*, 1990b). A narrow-tube wind tunnel was designed for the study of behavioural patterns of *Ps. calceolariae* males under varying wind velocities. This bioassay demonstrated the ability of the males to reach the pheromone source upwind at a distance of approximately 1 m (Rotundo *et al.*, 1980).

A free-flight bioassay in a greenhouse was used in the identification of most of the sex pheromones of the diaspidid and pseudococcid species. Whenever available, it is preferred, as it resembles a natural field response of males to virgin females and it was developed initially for the bioassay of the *A. aurantii* sex pheromone (Tashiro *et al.*, 1969). Infested plants were placed in the greenhouse and various pheromone samples or fractions were located on the perimeter of a turntable, which rotated slowly in order to eliminate position bias. Males were able to fly freely within the room, and to respond to a relatively large number of samples. This method enabled a continuous bioassay to be performed without the necessity to collect males and to use them only during their short active period. It was successfully used in the identification of the sex pheromones of *A. aurantii* (Roelofs *et al.*, 1978), *A. citrina* (Gieselmann *et al.*, 1979a), *Q. perniciosus* (Gieselmann *et al.*, 1979b), *Ps. comstocki* (Negishi *et al.*, 1980; Bierl-Leonhardt *et al.*, 1982) and *P. citri* (Bierl-Leonhardt *et al.*, 1981).

The final confirmation of the identity of the various coccoid sex pheromones was obtained by field bioassays with synthetic pheromones. Such experiments were used to assess the activity of stereoisomers and pheromone analogues. The chirality of most of the coccoid pheromones was also deduced from these tests.

Sex Pheromone Identification

The general procedure to identify coccoid sex pheromones consisted of two main stages, as in most sex pheromone identification studies. The first stage comprised the isolation of the active component(s) and involved various chromatographic techniques, accompanied by a reliable bioassay. The second stage involved the actual chemical identification of the sex pheromone using microreactions and an array of spectroscopic techniques, adjusted to a microscale. As the amount of purified sex pheromone was very small in most cases, gas chromatography (GC) alone or coupled to mass spectrometry (GC–MS) in various modes (EI, CI, MS–MS) played a crucial role in the identification. In most studies, an effort was made to collect sufficient material for proton nuclear magnetic resonance (NMR) spectra.

Diaspidid sex pheromones

The common name of diaspidid species is armoured scale insects, derived from the dorsal scale that covers their body (Ben-Dov, 1985). Four of these insects – A.

aurantii, *A. citrina*, *Q. perniciosus* and *P. pentagona*, all agricultural pests – were the object of intensive research by Roelofs and co-workers and by Heath and co-workers in the late 1970s and early 1980s. The pheromone identification and chemistry of these four coccoid pheromones were described in a review by Gieselmann (1990) and by Fletcher and Bellas (1988).

The California red scale, *A. aurantii*, sex pheromone was the first diaspidid pheromone to be fully identified (Roelofs *et al.*, 1978). The purification and identification microtechniques developed for this pheromone were subsequently applied to the identification of other coccoid pheromones. Two active components were isolated by various column chromatography methods and further purification was achieved by means of non-polar and polar packed GC columns. The pheromone components were shown to be acetates by reduction to the corresponding alcohols and subsequent acetylation that restored their biological activity. Hydrogenation of the two components produced the same compound, indicating a similar carbon skeleton, one having two units of unsaturation and the second three such units. The reactions were monitored by GC–MS, which provided additional information on the functional group and carbon branching. Conjugated dienes were ruled out by UV measurements. Further microreactions, particularly ozonolysis, provided more information regarding branching. Accumulation of about 30 μg of each active component enabled measurement of NMR spectra in C_6D_6; these, combined with all the other data, provided sufficient information to postulate two isoprenoid structures (Fig. 10.2, **1** and **2**). Compound **1** has two chiral centres; compound **2** has one chiral centre and one trisubstituted double bond. The two compounds were isolated in an approximate ratio of 1:1 (Gieselmann, 1990).

Since the stereochemistry of the trisubstituted double bond and the configuration of the chiral centres in the two components could not be deduced from the accumulated information, it was necessary to synthesize all four stereoisomers of **1** and **2** (Roelofs *et al.*, 1978; Anderson *et al.*, 1980). Greenhouse and field bioassays of all synthetic stereoisomers (Gieselmann *et al.*, 1980) indicated that the active components **1** and **2** are (3*S*,6*R*)-3-methyl-6-isopropenyl-9-decenyl acetate ($C_{16}H_{28}O_2$) and (3*Z*,6*R*)-3-methyl-6-isopropenyl-3,9-decadienyl acetate ($C_{16}H_{26}O_2$), respectively. The common chiral centre in the two components has the same absolute configuration.

The scheme of chemical identification by microreactions and spectroscopic means, resulting in the postulation of tentative structures which are ultimately confirmed by synthesis, is typical for the identification of insect sex pheromones and is particularly characteristic for coccoid sex pheromones, which have novel structures containing double bonds and chiral centres.

Almost concurrently with the above work on *A. aurantii*, Roelofs and co-workers identified the sex pheromones of the yellow scale, *A. citrina*, and the San José scale, *Q. perniciosus*. In the first case, a single active isoprenoid acetate compound was isolated (Gieselmann *et al.*, 1979a). The configuration of the trisubstituted double bond was confirmed by synthesis of a racemic *E*/*Z* mixture (Anderson and Henrick, 1979) and the configuration of the chiral centre was deduced by bioassaying the optically active synthetic *R*- and *S*- enantiomers (Mori and Kuwahara, 1982). The combined data established the structure of the *A. citrina* pheromone as (3*S*,5*E*)-3,9-dimethyl-6-isopropyl-5,8-decadienyl acetate ($C_{17}H_{30}O_2$) (Fig. 10.2, **3**).

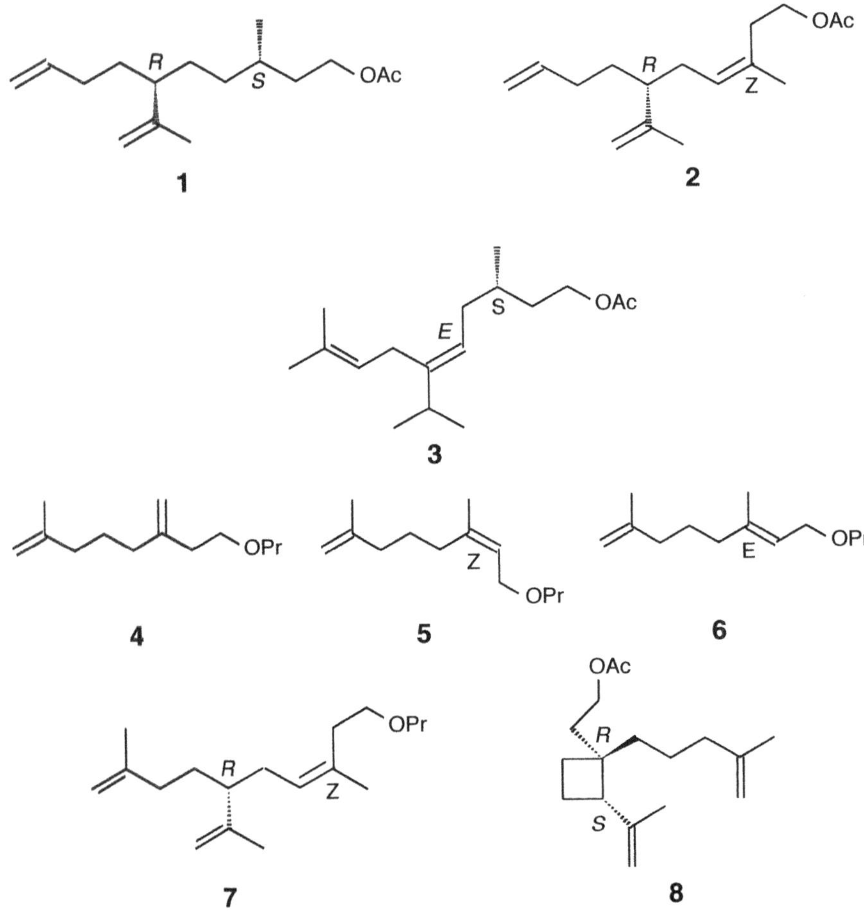

Fig. 10.2. Chemical structures of the five known diaspidid sex pheromones: compounds **1** and **2** of the California red scale, *Aonidiella aurantii*; compound **3** of the yellow scale, *Aonidiella citrina*; compounds **4**, **5** and **6** of the San José scale, *Quadraspidiotus perniciosus*; compound **7** of the white peach scale, *Pseudaulacaspis pentagona*; and compound **8** of the oleander scale, *Aspidiotus nerii*.

The sex pheromone of *Q. perniciosus* comprises three isomeric isoprenoid propanoates (Fig. 10.2, **4**, **5** and **6**) in a ratio of 45.8:46.7:4.8. The first two components, 7-methyl-3-methylene-7-octenyl propanoate ($C_{13}H_{22}O_2$) (**4**) and (Z)-3,7-dimethyl-2,7-octadienyl propanoate ($C_{13}H_{22}O_2$) (**5**), were initially identified in purified natural pheromone (Gieselmann *et al.*, 1979b). The minor component **6**, (*E*)-3,7-dimethyl-2,7-octadienyl propanoate ($C_{13}H_{22}O_2$) was identified only after it had been synthesized together with the Z-isomer **5**, and found to be biologically active (Anderson *et al.*, 1981).

The white peach scale, *Pseudaulacapsis pentagona*, sex pheromone was isolated and identified by means of similar microtechniques to those developed independently for the identification of the other diaspidid pheromones (Heath *et al.*, 1979). Some of these techniques were refined and, for the first time, Fourier-transform infrared (FT–IR) spectroscopy was used in the identification of a coccoid sex pheromone. The stereochemistry was deduced through synthesis of all four stereoisomers. The final structure was established as (3*Z*,6*R*)-3,9-dimethyl-6-isopropenyl-3,9-decadienyl propanoate ($C_{18}H_{30}O_2$) (Fig. 10.2, **7**).

The alcohol corresponding to **7** was identified as a second component, in large proportions, in a population of *P. pentagona* in France (J. Einhorn, France, 1998, personal communication). The alcohol, at high dosages, had an inhibitory effect in the field but enhanced the copulatory responses of males in a Petri dish bioassay (Einhorn *et al.*, 1983).

Very recently, the structure of the sex pheromone of the oleander scale, *Aspidiotus nerii* has been elucidated as (1*R*,2*S*)-*Z*-2-isopropenyl-1-(4′-methyl-4′ penten-1′-yl)-cyclobutaneethanol acetate ($C_{17}H_{28}O_2$) (Fig. 10.2, **8**). Synthesis of all four stereoisomers of **8**, followed by greenhouse and field bioessays confirmed the structure of the pheromone. This new scale insect sex pheromone contains an unusual trisubstituted cyclobutane structure, being the first cyclic diaspidid sex pheromone (Einhorn *et al.*, 1998).

Pseudococcid sex pheromones

Generally, members of this family have a mealy waxy secretion covering their body, which is why pseudococcid insects are commonly referred to as mealybugs (Ben-Dov, 1985). Only two sex pheromones of mealybugs have been identified so far, and the data are briefly described by Fletcher and Bellas (1988).

The sex pheromone of the Comstock mealybug, *Ps. comstocki* was identified by two independent groups (Bierl-Leonhardt *et al.*, 1980, 1982; Negishi *et al.*, 1980). The pheromone compound was identified as (*R*)-3-acetoxy-2,6-dimethyl-1,5-heptadiene ($C_{11}H_{18}O_2$) (Fig. 10.3, **9**). The corresponding alcohol was also detected, in a concentration approximately 0.5% of that of the acetate (Bierl-Leonhardt *et al.*, 1982). Reaction of the alcohol derived from the natural pheromone with Mosher's reagent and comparison by GC with the corresponding derivatives from the synthetic material indicated that the natural pheromone was 98% enantiomerically pure. The natural pheromone was found to be dextrorotatory (Bierl-Leonhardt *et al.*, 1982); the absolute configuration was established later by synthesis and bioassay of the two enantiomers (Mori and Ueda, 1981). The *Ps. comstocki* pheromone is the only example of a secondary acetate among all known coccoid sex pheromones.

The citrus mealybug, *P. citri* sex pheromone (Fig. 10.3, **10**) was identified by a combination of microreactions, derivatization and spectrometry as (1*R*,3*R*)-3-isopropenyl-2,2-dimethylcyclobutanemethyl acetate ($C_{12}H_{20}O_2$). The absolute configuration of the pheromone was established through synthesis of the two enantiomers from *Z*-(+)- and *Z*-(−)-verbenone, which were subsequently tested in a laboratory bioassay (Bierl-Leonhardt *et al.*, 1981). This was the first cyclobutane structure found among coccoid sex pheromones.

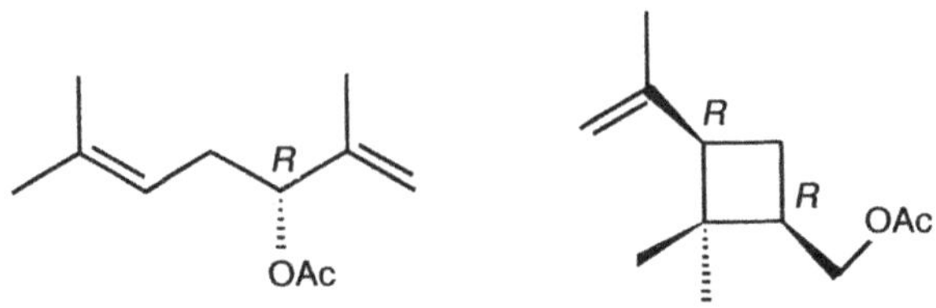

9 10

Fig. 10.3. Chemical structures of the two known preudococcid sex pheromones: compound **9** of the Comstock mealybug, *Pseudococcus comstocki*; compound **10** of the citrus mealybug, *Planococcus citri*.

Margarodid sex pheromones

All three known sex pheromones of margarodid scale insects belong to the species of the genus *Matsucoccus*. Members of this genus are referred to as pine bast scales, and they develop exclusively on *Pinus* spp. as host trees (Park *et al.*, 1986; Mendel *et al.*, 1991).

The first *Matsucoccus* sex pheromone (Fig. 10.4, **11**) to be identified was that of M. *resinosae* (red pine bast scale), M. *matsumurae* and M. *thunbergianae* (black pine bast scale). It was suggested by McLure (1983) and supported by Young *et al.* (1984) that M. *resinosae* might be a junior synonym of M. *matsumurae*. The other species, M. *thunbergianae*, has been discovered in Korea and it is morphologically distinct from, although closely resembling, M. *matsumurae* (Park *et al.*, 1986). A combined effort by scientists from the USA, China and Korea succeeded in identifying the pheromone (Lanier *et al.*, 1989), except for the configuration of the two chiral centres, as (2*E*,4*E*,6*R*,10*R*)-4,6,10,12-tetramethyl-2,4-tridecadiene-7-one ($C_{17}H_{30}O$), named 'matsuone'. The identification of **11** involved extensive use of MS, proton NMR (including decoupling experiments) and GC–FT-IR. The accumulated evidence led to four tentative isomeric structures, eventually synthesized, which confirmed the structure of the pheromone. The 6*R*,10*R* chirality was later established through synthesis of stereoisomers and field bioassays (Cywin *et al.*, 1991; Mori and Harashima, 1993a). The structure of **11** differed significantly from those of the known coccoid pheromones, being the first ketone identified among these pheromones.

The sex pheromone of the maritime pine bast scale, M. *feytaudi*, was identified in France. Extensive use of advanced spectroscopic techniques, such as GC–MS–MS, GC–FT-IR and NMR decoupling and two-dimensional experiments coupled with microreactions, particularly a Diels–Alder cycloaddition with tetracyanoethylene, led to the identification of two isomeric pheromone components (Fig. 10.4, **12** and **13**) in the ratio 97:3 as (3*S*,7*R*,8*E*,10*E*)-3,7,9-trimethyl-8,10-dodecadien-6-one ($C_{15}H_{26}O$) and (8*Z*,10*E*)-3,7,9-trimethyl-8,10-dodecadien-6-one ($C_{15}H_{26}O$), respectively (Einhorn *et al.*, 1990). The absolute configuration of the two chiral centres in the main component **12** was established later by synthesis and field bioassays (Cywin and Kallmerten, 1991; Mori and Harashima, 1993b; Jactel *et al.*, 1994).

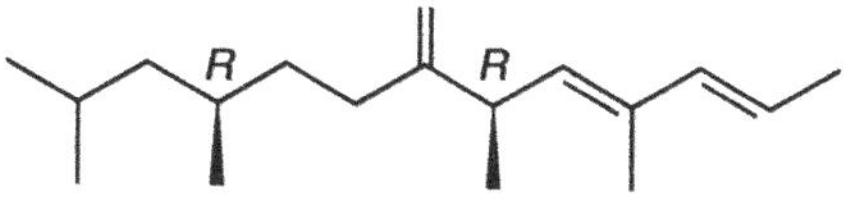

11

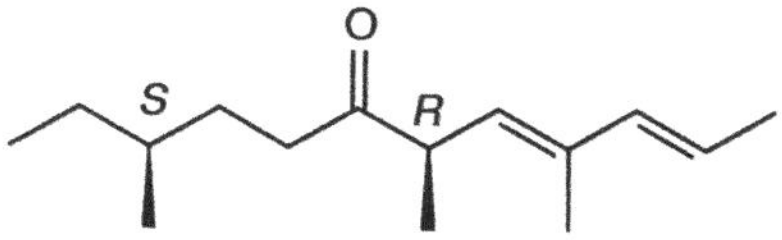

12

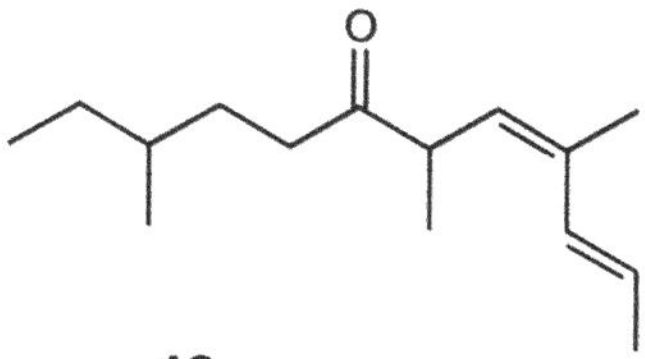

13

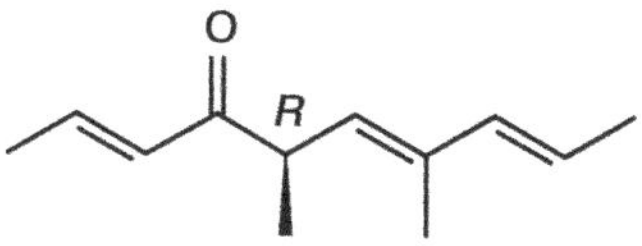

14

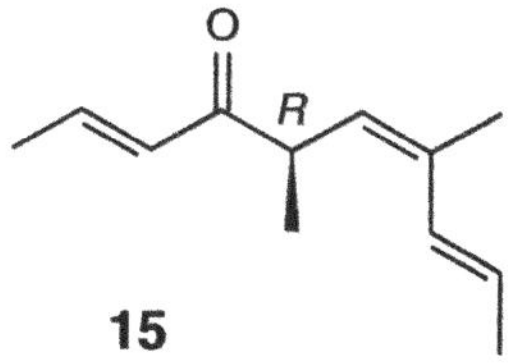

15

Fig. 10.4. Chemical structures of the three known margarodid sex pheromones: compound **11** of the red pine bast scale, *Matsucoccus resinosae (M. matsumurae, M. thunbergianae)*; compounds **12** and **13** of the maritime pine bast scale, *Matsucoccus feytaudi*; compounds **14** and **15** of the Israeli pine bast scale, *Matsucoccus josephi*.

The sex pheromone of the Israeli pine bast scale, M. *josephi*, was identified, by a joint effort, in Israel and France. Extensive use of MS techniques, including GC–MS–MS, and comparison with fragmentation of specially designed standard compounds, GC–FT-IR coupled with microreactions, particularly a Diels–Alder cycloaddition with tetracyanoethylene, enabled the identification of two isomeric components (Fig. 10.4, **14** and **15**) in a ratio of 75:25 as (2*E*,5*R*,6*E*,8*E*)-5,7-dimethyl-2,6,8-decatrien-4-one ($C_{12}H_{18}O$) and (2*E*,5*R*,6*Z*,8*E*)-5,7-dimethyl-2,6,8-decatrien-4-one ($C_{12}H_{18}O$), respectively (Dunkelblum *et al.*, 1993, 1995). The gross structure of the two components was confirmed by synthesis of racemic **14** and **15** (Zegelman *et al.*, 1993). The absolute configuration of the chiral centres in the major component **14**, and in the minor one **15**, were directly determined by chiral capillary GC and comparison with chiral synthetic *R*- and *S*-enantiomeric standards of **14** which also contained small amounts of the corresponding enantiomers of **15** (Mori and Amaike, 1994; Dunkelblum *et al.*, 1995). This is the only case of a chiral coccoid sex pheromone for which the stereoisomeric identity could be established directly without any derivatization (Fig. 10.5). The chirality of the natural pheromone was also determined by a field bioassay and, for the first time, by GC and electroantennography (GC–EAD) (Dunkelblum *et al.*, 1995).

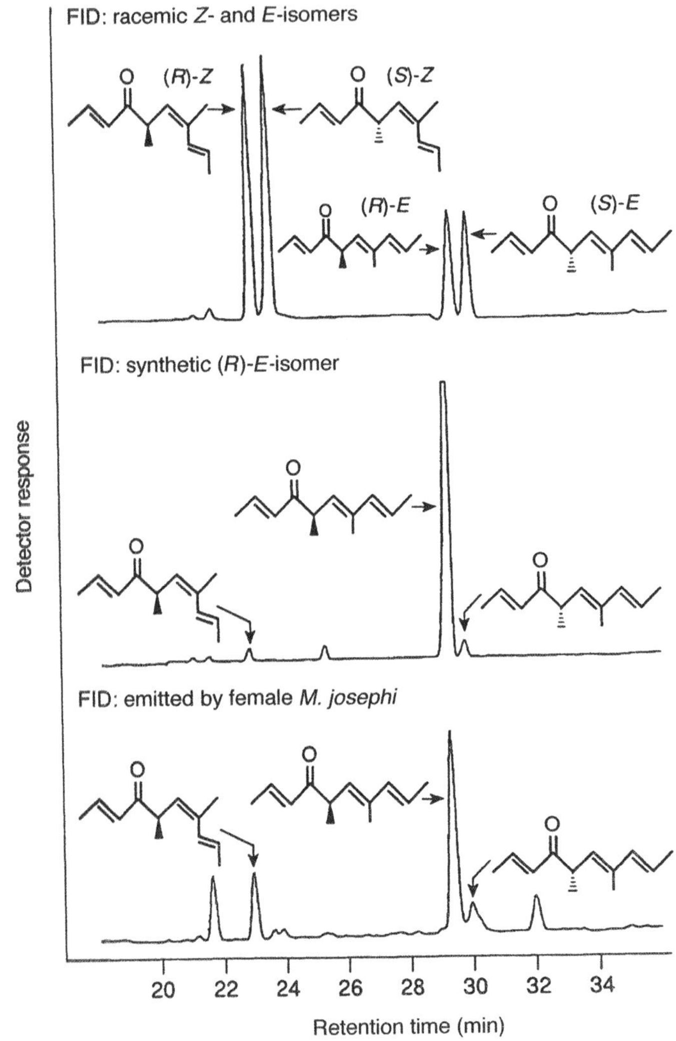

Fig. 10.5. Chiral separation of synthetic and natural *Matsucoccus josephi* sex pheromone stereoisomers on a Cyclodex-B column, isothermal at 120°C.

Synthesis

Synthesis is the very last step in the identification procedure and the first step in biological research and further field experiments. Many syntheses of coccoid sex pheromones have been published, but a comprehensive review would be beyond the scope of this chapter. Fortunately, Mori (1992) includes most of the syntheses of diaspidid and pseudococcid sex pheromones. This was published before the identification of the *Matsucoccus* sex pheromones (Fig. 10.4), so a brief description of their syntheses is included here.

M. matsumurae *sex pheromone* **11** *(matsuone)*

The first synthesis was accomplished by Cywin *et al.* (1991) using as key steps a [2,3]-Wittig rearrangement and, subsequently, a copper-assisted coupling of an organomagnesium reagent, derived from an optically active bromide, with a chiral epoxide. This synthesis enabled the establishment of the *syn* relationship between the two chiral centres of the pheromone **11** but did not determine their absolute configuration. The preparation of the two enantiomers of **11**, (6*R*,10*R*) and (6*S*,10*S*), was carried out by Mori and Harashima (1993a). The absolute configuration of the two stereoisomers was derived from the synthetic route, starting from (*R*)- and (*S*)-citronellol and employing the Sharpless asymmetric epoxidation and a palladium-catalysed reductive cleavage of the epoxy-ring.

Another stereoselective synthesis of **11** was developed by Mori *et al.* (1995), using a chiral epoxy synthon which was derived from a lipase-hydrolysed epoxy-*meso*-diaceate. The synthesis of all four possible stereoisomers of **11** was described by Lin and Xu (1993), who started from (*R*)- and (*S*)-citronellal and used asymmetric aldol reactions to construct the second chain of the molecules. The same group (Lin and Xu, 1996) developed a new stereoselective synthesis of **11**, based on a lipase-catalysed trans-esterification of *meso*-2,4-dimethyl-1,5-propanediol. This enzymatic reaction provided a chiral synthon which was used to control both chiral centres in **11**. A general scheme, based on a diastereoselective [2,3]-Wittig rearrangement of a tertiary bisallylic ether, was developed to prepare all three *Matsucoccus* pheromones **11**, **12** and **14** (Shi *et al.*, 1995a). This scheme, combined with the use of (*S*)-(+)- and (*R*)-(−)-camphanyl chlorides, applied to resolve a racemic key intermediate, provided a stereoselective route for all four possible stereoisomers of **11** (Shi *et al.*, 1995b). A new concise synthesis of racemic and diastereomeric mixtures of all three *Matsucoccus* sex pheromones, **11**, **12** and **14**, was published recently (Watanabe *et al.*, 1997). The key reaction is a Lewis acid-mediated cyanation to provide a common intermediate for the preparation of the three pheromones.

M. feytaudi *sex pheromone* **12**

The first synthesis of this pheromone was performed by Cywin and Kallmerten (1991), according to the route used for the preparation of the M. *matsumurae* pheromone, **11** (Cywin *et al.*, 1991). The relative stereochemistry was found to be *syn*, the same as in **11**. The synthesis of all four possible stereoisomers of **12** was developed by Mori and Harashima (1993b), starting from (*R*)- and (*S*)-citronellol along the same lines as for the preparation of **11** (Mori and Harashima, 1993a). The M. *feytaudi* pheromone was also prepared according to two recently developed general schemes for all *Matsucoccus* pheromones (Shi *et al.*, 1995a; Watanabe *et al.*, 1997).

M. josephi *sex pheromone* **14** *and* **15**

A racemic *E*/*Z* mixture (approximately 1:1) of the pheromone was prepared using the Reformatzki and Wittig reactions as key steps (Zegelman *et al.*, 1993). The chiral pheromone, **14**, and its *S*-enantiomer were prepared by a stereoselective synthesis (Mori and Amaike, 1994), with an enzymatic asymmetric hydrolysis of *meso*-epoxy-diacetate as a key step. The pheromone has also been prepared by two general syntheses (Shi *et al.*, 1995a; Watanabe *et al.*, 1997).

Chemical Structure

All identified diaspidid and pseudococcid sex pheromones are isoprenoid-type compounds, some lacking one methyl group. However, only the *Q. perniciosus* pheromone components have a common terpenoid skeleton (monoterpene type), and this is the only known coccoid pheromone that is not chiral. The other diaspidid pheromones have similar carbon skeletons with two stereogenic moieties, i.e. either two chiral centres or one *E*/*Z* double bond and one chiral centre. All are esters, acetates or propionates of primary alcohols, with the exception of the *Ps. comstocki* pheromone. The structures of the two pseudococcid sex pheromones differ significantly from each other, the *Ps. comstocki* pheromone is an aliphatic secondary acetate, while the *P. citri* pheromone is a cyclobutane compound. The three known margarodid sex pheromones are unsaturated ketones with one identical side chain as displayed in Fig. 10.6 (**16**). This chiral keto-diene moiety, with the same absolute configuration, is present in all three pheromones. The second side chain, R, has either a second chiral centre as in **11** and **12** or an *E* double bond as in **14**.

The number of identified coccoid sex pheromones is too small to draw any conclusions concerning structure relationships in families and genera. At present, the chemical diversity of the pseudococcid sex pheromones appears larger than that of other coccoid pheromones. However, this may change when more pseudococcid pheromones are identified.

Analogues

Sex pheromone analogues are usually prepared to replace field-sensitive pheromones (e.g. formates instead of aldehydes) or to probe structure–activity relationships. In some cases, analogues have been prepared during the course of chemical identification of new pheromones before the final structure was confirmed (Lanier *et al.*, 1989). There are only a few reports of coccoid sex pheromone analogues. Field tests showed that, in all cases, they were significantly less active than the corresponding natural pheromones.

The most studied examples are the *Ps. comstocki*, *P. citri*, *M. matsumurae* and *M. josephi* sex pheromone analogues. A series of analogues of the *Ps. comstocki* pheromone (Fig. 10.3) was prepared by shifting the acetoxy group, one methyl

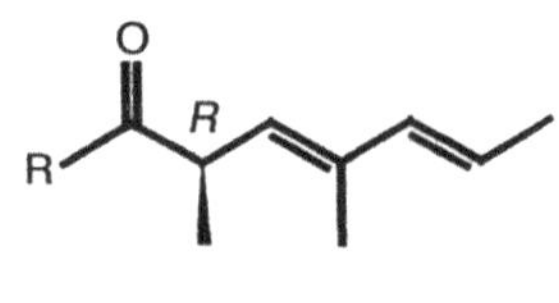

16

Fig. 10.6. The unsaturated ketone structure with one identical side chain (**16**) as observed in the three known margarodid sex pheromones.

group or a double bond in compound **9** (Uchida *et al.*, 1981; Bierl-Leonhardt *et al.*, 1982). Some of them exhibited considerable activity, but much less than **9**. A diverse group of analogues of the *P. citri* pheromone **10** was synthesized through modifications of the acetoxy group, the double bond and the cyclobutane ring (Dunkelblum *et al.*, 1987). The results indicated that all functional groups were essential for full activity. Lengthening of the cyclobutanemethanol moiety in **10** to cyclobutaneethanol produced the most active analogue, which displayed about half the attractivity of the pheromone. The three *M. matsumurae* (Hibbard *et al.*, 1991) and two *M. josephi* (Dunkelblum *et al.*, 1996) analogues had negligible activity. Interestingly, one of the *M. josephi* analogues had significant kairomonal activity (see also Chapter 16, this volume).

Biological Activity and Specificity

Almost all scale insect sex pheromones are chiral, and some comprise more than one component. The important question concerns the biological activities of the individual components and of the various stereoisomers, as compared with those of the natural stereoisomer. This issue has already been raised for the first identified coccoid sex pheromone, that of *A. aurantii*. All four possible stereoisomers of each component (Fig. 10.2, **1** and **2**) were prepared and bioassayed (Roelofs *et al.*, 1978; Gieselmann *et al.*, 1980). The results showed that there was enantiomeric and geometric specificity: only one stereoisomer of each component was biologically active and each had the same individual activity. Neither synergistic nor inhibitory effects of any unnatural stereoisomer could be detected. The origin of some activity of the 3*S*,6*S* stereoisomer of **1** probably lies in the presence of approximately 7% of the natural 3*S*,6*R* stereoisomer in the sample.

The potential synergistic and inhibitory effects of all stereoisomers of the *P. pentagona* sex pheromone (Fig. 10.2, **7**) were carefully studied by means of a laboratory bioassay of all four individual stereoisomers and of binary mixtures, at three dosages (Heath *et al.*, 1979). The results presented in Table 10.1 show clearly that the *R*,*Z* stereoisomer is the most active compound. The enantiomeric purity of the various compounds was about 95% and, therefore, the low activity of the *S*,*Z* stereoisomer and of some mixtures can be attributed to the presence of a small percentage of the active isomer. Again, no synergistic or inhibitory effects could be observed.

The sex pheromone of *Q. perniciosus* (Fig. 10.2) consists of the three isomeric components, **4**, **5** and **6**, in a ratio of 45.8:46.7:4.8. All three components were found to be individually active, exhibiting similar attractivity to males in a greenhouse bioassay. The three components and their mixtures were also tested in the field and again no synergistic effect was detected (Anderson *et al.*, 1981). The activity of the minor component, **6**, was slightly higher than that of compounds **4** or **5** (Hoyt *et al.*, 1983; Angerilli and Logan, 1986).

The *M. josephi* sex pheromone (Fig. 10.4, **14**) (*R*-*E*), its enantiomer (*S*-*E*) and a racemic mixture of **14** + **15** in an approximate ratio of 1:1 (Dunkelblum *et al.*, 1995) have recently been tested in the field. The results (Table 10.2) indicated that the antipode of the natural pheromone stereoisomer showed very slight activity,

Table 10.1. Mean (± SE) percentage response of male *Pseudaulacaspis pentagona* to stereoisomers of compound **7** in Fig. 10.2 and their mixtures.

Compound	10 pg	1 pg	100 fg
Test 1			
R,Z	75 ± 6	48 ± 4	36 ± 6
R,E	3 ± 3	0 ± 0	0 ± 0
S,Z	30 ± 12	12 ± 7	0 ± 0
S,E	3 ± 3	0 ± 0	7 ± 4
Test 2			
R,Z	84 ± 3	50 ± 7	38 ± 8
R,Z + *R,E*	84 ± 4	56 ± 3	40 ± 7
R,Z + *S,Z*	84 ± 11	56 ± 3	40 ± 7
R,Z + *S,E*	84 ± 5	52 ± 3	37 ± 9
S,Z + *S,E*	46 ± 0	31 ± 3	3 ± 3
S,E + *R,E*	15 ± 4	19 ± 7	2 ± 2

while the activity of a racemic mixture containing an equivalent amount of **14** was the same as that of the natural pheromone. Racemic component **15** showed negligible activity (Zegelman *et al.*, 1993). The activity of the pheromone **14** was not affected by any stereoisomer of **14** or **15**.

An interesting observation has been made regarding the activity of the four stereoisomers of the M. *feytaudi* pheromone (Fig. 10.4, **12**) and the M. *matsumurae* pheromone, **11**. Both have two chiral centres and, in each case, one pair of diastereoisomers has the *R*-configuration α to the ketone (Fig. 10.6, **16**). Field and laboratory bioassays indicated that both diastereoisomers, in each case, had comparable attractancy to males (Jactel *et al.*, 1994; Shi *et al.*, 1995b), despite the fact that only one is the natural pheromone. The significant activity of the unnatural diastereoisomer of **11** and **12**, which has the *R*- configuration α to the ketone, indicates that this chiral centre dominates the pheromonal activity.

The fact that the unnatural stereoisomers of coccoid sex pheromones do not adversely affect trap catch of males enables the use of racemic scale insect sex pheromones. This observation has practical implications, because racemic pheromones are usually cheaper and easier to prepare than chiral ones.

Another issue of interest relates to the specificity of individual scale insect sex pheromones. Female-emitted coccoid sex pheromones are expected to (and

Table 10.2. Captures of *Matsucoccus josephi* males (mean number trap^{-1} day^{-1}) in traps baited with pheromone stereoisomers (see Fig. 10.4). Different letters show a significant difference between catches.

Compounds	Amount (μg)	Trap catch
Natural **14** (*R*)	50	68.3a
Antipode of **14** (*S*)	50	6.9b
Racemic mixture of **14** + **15**	200	56.6a

normally do) attract only conspecific males. There are, however, a few exceptions: males of *M. matsumurae* and *M. thunbergianae* exhibit strong cross-attraction to crude hexane extracts of virgin females of the other species (Park *et al.*, 1986) and to synthetic matsuone, **11** (Hibbard *et al.*, 1991). The females of these species seem to produce the same sex pheromone (Lanier *et al.*, 1989), but it must be kept in mind that the taxonomic identity of these *Matsucoccus* species is still under discussion.

An interesting observation has been made recently regarding the sex pheromone of the San José scale, *Q. perniciosus*. This scale insect is a quarantine pest in Switzerland, where it is monitored with pheromone traps. Males have been caught in locations where *Q. perniciosus* was not expected, and this suggested attraction of another species. Eventually, *Q. zonatus* males were identified as a second *Quadraspidiotus* species that is attracted by the *Q. perniciosus* sex pheromone. The final identification of the closely related species was based on a DNA (RAPD–PCR) probe. The analysis was performed on males. The individuals were from a laboratory rearing, field collected and recovered from sticky traps. The method enabled the establishment of a molecular identification key for six European *Quadraspidiotus* species (Frey and Frey, 1995). The molecular identification of closely related scale insect species may be of great assistance in taxonomy studies of Coccoidea, and this method has been used recently to examine the genetic similarity among different populations of *M. josephi* (Mendel *et al.*, 1994b).

Kairomonal Activity

Some coccoid sex pheromones display kairomonal activity towards parasitoids and predators (see also Chapter 16, this volume). Prior to the identification of any coccoid sex pheromones, parasitic wasps had been observed to be attracted to sticky traps baited with virgin females and it had been suggested that the female pheromone exhibited kairomonal activity. However, these observations could not be investigated properly until synthetic scale insect sex pheromones had become available. The parasitoids *Aphytis africanus* Quednau and *A. melinus* DeBach responded to the synthetic *A. aurantii* sex pheromone (Grout and Richards, 1991), but they were equally attracted to unbaited yellow traps (Samways, 1988).

The aphelinid endoparasite *Prospaltella perniciosi* Tower (currently named *Encarsia perniciosi*) is a common natural enemy of the San José scale, *Q. perniciosus*. The two main individual pheromone components of the scale were found to attract significant numbers of the endoparasite, suggesting that the *Q. perniciosus* pheromone is a kairomone of *P. perniciosi* (Rice and Jones, 1982). Another study investigated the influence of a combination of trap colour and *Q. perniciosus* sex pheromone on trap catches of ten aphelinid parasitoids (McLain *et al.*, 1990). Only *E. perniciosi* displayed a significantly stronger response to traps baited with the pheromone than to blank traps.

The kairomonal activity of the *M. josephi* sex pheromone (Fig. 10.4, **14**) has been studied recently. Preliminary results (Zegelman *et al.*, 1993) indicated that the pheromone attracted the predator, *Elatophilus hebraicus* Pericart. A thorough investigation, using GC–EAD and field tests, indicated that **14** is a potent kairomone and that it attracted both males and females of the predator. Interestingly, the *M. matsumurae* and *M. feytaudi* sex pheromones (Fig. 10.4, **11** and **12**) exhibited kairomonal activity comparable with that of **14** despite the fact that *E. hebraicus* is associated only with *M. josephi* (Dunkelblum *et al.*, 1996). Traps baited with these pheromones (kairomones) exposed in the area of *Pinus brutia* pines infested with *M. josephi* (Near East and Crete) attracted only one predator, namely *E. hebraicus* (Mendel *et al.*, 1994a, 1997). The same traps installed in several countries in western Europe, China and Japan, which are outside the range of *M. josephi*, did not attract any predators except in Portugal and the Bordeaux area, where both *Elatophilus* spp. and *Hemerobius stigma* Stephens were caught in significant numbers (Mendel *et al.*, 1997; H. Jactel, France, 1997, personal communication).

Application

Many scale insect pests are small, and their early detection by visual inspection is difficult and tedious. Proper application of pheromone traps, followed by correct interpretation of the trap catch, could prove to be convenient and reliable for monitoring these pests. After the first coccoid sex pheromones had been identified and synthesized, male trap catch experiments were initiated in several countries, in order to assess the suitability of pheromone traps for monitoring.

The general technique of monitoring involved the use of sticky traps of various designs (e.g. Hoyt *et al.*, 1983) baited with rubber septa impregnated with pheromone. All coccoid pheromones are sufficiently stable, and commonly used dosages ranged from 50 to 1000 μg. Most of the work was carried out with pheromones of the California red scale, *A. aurantii*, the San José scale, *Q. perniciosus*, and the citrus mealybug, *P. citri*. Recently, *Matsucoccus* pheromones became available for the detection of pine bast scales. The main problem with the use of scale insect pheromone traps for monitoring purposes was the establishment of a relationship between trap catch and the infestation level in the field.

The flight phenology of *A. aurantii* males was initially monitored by means of traps baited with virgin females (Rice and Moreno, 1969, 1970) and subsequently, from the late 1970s onwards (Roelofs *et al.*, 1978), with synthetic pheromone lures. In California, four generations were detected and a predictive model was developed, based on the male catch in the first, second or fourth flight (Moreno, 1983; Moreno and Kennett, 1985). It was concluded that the pheromone traps could be incorporated as a valuable tool for managing of the pest. A cost–benefit analysis was also carried out (Gardner *et al.*, 1983) to assess and justify the investment in the *A. aurantii* pheromone project in the USA. A positive conclusion was reached in view of the importance and assistance of the programme for the citrus growers. At present, large numbers of pheromone traps are utilized in California to provide a basis for pest management decision making (P.A. Kirsch, Oregon, 1998, personal communication). A small-scale programme for the prediction of *A. aurantii* infestation was carried out in Israel. An attempt was made to determine thresholds for spraying, based on the male catch (Sternlicht *et al.*, 1981). Extensive work was conducted in South Africa, where pheromone traps were exposed in climatically differing locations (Wessels and Meyer, 1982; Samways, 1988; Grout and Richards, 1991) and the flight phenology was studied. It was concluded that pheromone traps might be used in some locations for red scale monitoring (Samways, 1988). However, in areas where the activity of *Aphytis* parasitoids was high, correlation between trapped males and fruit infestation was not possible (Grout and Richards, 1991). Prediction of fruit infestation was also difficult because of the two- and sometimes three-cohort structure of the scale population per intergeneration period (Grout *et al.*,1989). Some limited attempts have been made to use the pheromone for control – for example, in disrupting mating of *A. aurantii* in Israel (Hefetz *et al.*, 1988). The use of pheromone traps for monitoring, in combination with biological control, has been reported from Sicily (Raciti *et al.*, 1995).

Various experiments have been carried out with the *Q. perniciosus* pheromone in order to optimize the use of pheromone traps for early detection of the scale. Attempts were made to correlate trap catch with degree-day models (Hoyt *et al.*, 1983; Angerilli and Logan, 1986). Most of the early work was done in California, Oregon and Washington (USA) and in British Columbia (Canada) (Jorgenson *et al.*, 1981). The flight phenology of the male scale was also studied in New York State (Mague and Reissig, 1983). It was concluded that pheromone traps could be useful in monitoring spring populations, in the assessment of previously applied control measures and in tracing the spread of scale populations into new locations. Pheromone traps have also been used in Europe for quarantine purposes (Smetnik *et al.*, 1986; Frey and Frey, 1995). Moderate success in mating disruption of *Q. perniciosus* was observed in 1977 in California, using paraffin emulsions as the carrier for the pheromone. However, the high price of the pheromone (c. US\$325 g^{-1}) made the method economically prohibitive (R.E. Rice, California, 1977, personal communication).

Pheromone traps have been used for early detection of *P. citri* mainly in citrus orchards. Extensive field tests in Sardinia established the necessary parameters for an efficient population survey (Ortu and Delrio, 1982). The programme has been expanded and is now combined with the application of natural enemies (Fronteddu *et al.*, 1996). Seasonal population fluctuations and trends of *P. citri* have been studied in Israel (Hefetz and Tauber, 1990), where traps are being used in conjunction with biological control (Z. Mendel, Israel, 1998, personal communication).

Monitoring of *Matsucoccus* spp. with pheromone traps has been initiated recently. A survey of *M. josephi* in parts of the Near East indicated that this pest is not limited to Israel but is present also in Lebanon and Jordan (Mendel *et al.*, 1994a). Population trends of *M. josephi* and its predator *E. hebraicus* have been studied in various locations in Israel (Mendel *et al.*, 1997). The use of *M. feytaudi* pheromone traps has enabled the early detection of this pest in the forests of Corsica (Jactel *et al.*, 1996).

This record on the application of scale insect pheromones is not complete; there are sketchy reports of surveys and monitoring in various countries, published in local journals or summarized in conference abstracts. Despite the economic importance of scale insect pests, early detection and monitoring with pheromone traps has not progressed as much as was earlier anticipated. The commercial sources of coccoid sex pheromones are very limited and good quality pheromones are expensive. Some pheromone lures are available from various companies, such as Isagro in Italy, IPM Technologies in Oregon and other specialist suppliers.

More extensive and careful field work with coccoid pheromones is required for the proper assessment of their suitability for use in early detection and monitoring. It is important to establish correlations between male trap catches and infestation. Prediction of infestation levels may still remain difficult, but early detection and survey of scale insects, using pheromone traps, is certainly possible and has been successful in several cases. Reduction of pesticide usage and widespread application of biocontrol agents will be very difficult without the use of pheromone traps. Another open question is the feasibility of using these pheromones for control of scale insect pests by mass trapping or mating disruption. So far, no significant research has been carried out to evaluate these possibilities.

Conclusions

Ten coccoid sex pheromones are known so far and many more await identification and synthesis. The practical use of these pheromones is still limited, although they seem to be a convenient tool for early detection of these insect pests. Cheaper commercial pheromones and extensive field research are urgently required to promote the use of these pheromones, and to make them a key element in appropriate integrated pest management programmes.

Acknowledgements

The studies on *Matsucoccus josephi* from our group described here were supported by the Forest Department of the Jewish National Fund (KKL). I would like to thank Dr Yair Ben-Dov, The Volcani Center, for reviewing the manuscript and making valuable comments.

References

Anderson, R.J. and Henrick, C.A. (1979) Synthesis of (E)-3,9-dimethyl-6-isopropyl-5,8-decadien-1-yl acetate, the sex pheromone of the yellow scale. *Journal of Chemical Ecology* 5, 773–779.

Anderson, R.J., Adams, K.G., Chinn, H.R. and Henrick, C.A. (1980) Synthesis of the optical isomers of 3-methyl-6-isopropenyl-9-decen-1-yl acetate, a component of the California red scale pheromone. *Journal of Organic Chemistry* 45, 2229–2236.

Anderson, R.J., Gieselmann, M.J., Chinn, H.R., Adams, K.G., Henrick, C.A., Rice, R.E. and Roelofs, W.L. (1981) Synthesis and identification of a third component of the San José scale sex pheromone. *Journal of Chemical Ecology* 7, 695–706.

Angerilli, N.P.D. and Logan, D.M. (1986) The use of pheromone and barrier to monitor San José scale (Homoptera: Diaspididae) phenology in the Okanagan valley of British Columbia. *Canadian Entomologist* 118, 767–774.

Ben-Dov, Y. (1985) Coccoidea. In: Scholtz, C.H. and Holm, E. (eds) *Insects of Southern Africa.* Butterworths, Durban, pp. 168–175.

Bierl-Leonhardt, B.A., Moreno, D.S., Schwarz, M., Forster, H.S., Plimmer, J.R. and DeVilbiss, E.D. (1980) Identification of the pheromone of the Comstock mealybug. *Life Sciences* 27, 399–402.

Bierl-Leonhardt, B.A., Moreno, D.S., Schwarz, M., Fargerlund, J. and Plimmer, J.R. (1981) Isolation, identification and synthesis of the sex pheromone of the citrus mealybug, *Planococcus citri* (Risso). *Tetrahedron Letters* 22, 389–392.

Bierl-Leonhardt, B.A., Moreno, D.S., Schwarz, M., Forster, H.S., Plimmer, J.R. and DeVilbiss, E.D. (1982) Isolation, identification, synthesis and bioassay of the pheromone of the Comstock mealybug and some analogs. *Journal of Chemical Ecology* 8, 689–699.

Cywin, C.L. and Kallmerten, J. (1991) Enantioselective synthesis of a sex attractant pheromone of the pine scale *Matsucoccus feytaudi*. *Journal of Natural Products* 54, 1664–1667.

Cywin, C.L., Webster, F.X. and Kallmarten, J. (1991) Synthesis of (−)-(6*R*,10*R*)-matsuone. Assignment of relative stereochemistry to a pheromone of *Matsucoccus* pine bast scale. *Journal of Organic Chemistry* 56, 2953–2955.

Doane, C.C. (1966) Evidence for a sex attractant in females of the red pine scale. *Journal of Economic Entomology* 59, 1539–1540.

Dunkelblum, E., Ben-Dov, Y., Goldschmidt, Z., Wolk, J. and Somekh, L. (1987) Synthesis and field bioassay of some analogs of sex pheromone of citrus mealybug, *Planococcus citri* (Risso). *Journal of Chemical Ecology* 13, 863–871.

Dunkelblum, E., Mendel, Z., Assael, F., Harel, M., Kerhoas, L. and Einhorn, J. (1993) Identification of the female sex pheromone of the Israeli pine bast scale, *Matsucoccus josephi*. *Tetrahedron Letters* 34, 2805–2808.

Dunkelblum, E., Gries, R., Gries, G., Mori, K. and Mendel, Z. (1995) Chirality of Israeli pine bast scale, *Matsucoccus josephi* (Homoptera: Matsucoccidae) sex pheromone. *Journal of Chemical Ecology* 21, 849–858.

Dunkelblum, E., Mendel, Z., Gries, G., Gries, R., Zegelman, L., Hassner, A. and Mori, K. (1996) Antennal response and field attraction of the predator *Elatophilus hebraicus* (Hemiptera: Anthocoridae) to sex pheromones and analogues of the three *Matsucoccus* spp. (Homoptera: Matsucoccidae). *Bioorganic and Medicinal Chemistry* 4, 489–494.

Einhorn, J., Bianch, H. and Benassy, C. (1983) Effets combinés de constituants phéromonaux sur le compartement sexuel des mâles d'une Cochenille diaspine, *Pseudaulacaspis pentagona* Targ (Homoptera, Diaspididae). *Comptes Rendus de l'Académie des Sciences*, Série III, 296, 861–863.

Einhorn, J., Menassieu, P., Malosse, C. and Ducrot, P.H. (1990) Identification of the sex pheromone of the maritime pine scale, *Matsucoccus feytaudi*. *Tetrahedron Letters* 31, 6633–6636.

Einhorn, J., Guerrero, A., Ducrot, P.H., Boyer, F.D., Gieselmann, M. and Roelofs, W.L. (1998) Sex pheromone of the oleander scale, *Aspidiotus nerii*: structural characterization and absolute configuration of an unusual functionalized cyclobutane. *Proceedings of the National Academy of Sciences USA* 95, 9867–9872.

Fletcher, B.S. and Bellas, T.E. (1988) Pheromones of Hemiptera, Blattoidea, Orthoptera, Mecoptera, other insects and Acari. In: *Handbook of Natural Pesticides*, Vol. IV *Pheromones*, *Part B*. CRC Press, Boca Raton, Florida, pp. 207–271.

Frey, J.E. and Frey, B. (1995) Molecular identification of six species of scale insects (*Quadraspidiotus* spp.) by RAPD–PCR: assessing the field-specificity of pheromone traps. *Molecular Ecology* 4, 777–780.

Fronteddu, F., Canu, D., D'Amico, R., Delpiano, N., Fancello, F. and Nanni, G. (1996) Applications of integrated control methods in citrus cultivation: biotechnical control against *Ceratitis capitata* and biological control of *Planococcus citri*. *Informatore Fitapatologico* 46, 34–39.

Gardner, P.D., Ervin, R.T., Moreno, D.S. and Baritelle, J.L. (1983) California red scale (Homoptera: Diaspididae): cost analysis of a pheromone monitoring program. *Journal of Economic Entomology* 76, 601–604.

Gieselmann, M.J. (1990) Pheromone identification and chemistry. In: Rosen, D. (ed.) *Armored Scale Insects: their Biology, Natural Enemies and Control. World Crop Pests*, Vol. 4A. Elsevier, Amsterdam, pp. 225–232.

Gieselmann, M.J., Moreno, D.S., Fargerlund, J., Tashiro, H. and Roelofs, W.L. (1979a) Identification of the sex pheromone of the yellow scale. *Journal of Chemical Ecology* 5, 27–33.

Gieselmann, M.J., Rice, R.E., Jones, R.A. and Roelofs, W.L. (1979b) Sex pheromone of the San José scale. *Journal of Chemical Ecology* 5, 891–900.

Gieselmann, M.J., Henrick, C.A., Anderson, R.J., Moreno, D.S. and Roelofs, W.L. (1980) Responses of male California red scale to sex pheromone isomers. *Journal of Insect Physiology* 26, 179–182.

Gravitz, N. and Wilson, C. (1968) A sex pheromone from the citrus mealybug. *Journal of Economic Entomology* 61, 1458–1459.

Grout, T.G. and Richards, G.I. (1991) Value of pheromone traps for predicting infestations of red scale, *Aonidiella aurantii* (Maskell) (Hom., Diaspididae), limited by natural enemy activity and insecticides used to control citrus thrips, *Scirtothris aurantii* Faure (Thys., Thripidae). *Journal of Applied Entomology* 111, 20–27.

Grout, T.G., Du Toit, W.J., Hofmeyr, J.H. and Richards, G.I. (1989) California red scale (Homoptera: Diaspididae) phenology on citrus in South Africa. *Journal of Economic Entomology* 82, 793–798.

Heath, R.R., McLaughlin, J.R., Tumlinson, J.H., Ashley, T.R. and Doolittle, R.E. (1979) Identification of the white peach scale sex pheromone. An illustration of micro-techniques. *Journal of Chemical Ecology* 5, 941–953.

Hefetz, A. and Tauber, O. (1990) Male response to the synthetic sex pheromone of *Planococcus citri* (Risso) (Hom., Pseudococcidae) and its application for population monitoring. *Journal of Applied Entomology* 109, 502–506.

Hefetz, A., Kronenberg, S., Peleg, B.A. and Bar-Zakay, I. (1988) Mating disruption of the California Red Scale, *Aonidiella aurantii* (Homoptera: Diaspididae). In: Goren, R. and Mendel, K. (eds) *Proceedings of the Sixth International Citrus Congress*. Tel-Aviv, pp. 1121–1127.

Hibbard, B.E., Lanier, G.N., Park, S.C., Qi, Y.T., Webster, F.X. and Silverstein, R.M. (1991) Laboratory and field tests with the synthetic sex pheromone of three *Matsucoccus* pine bast scales. *Journal of Chemical Ecology* 17, 89–102.

Hoyt, S.C., Westigard, P.H. and Rice, R.E. (1983) Development of pheromone trapping techniques for male San José scale (Homoptera: Diaspididae). *Environmental Entomology* 12, 371–375.

Jactel, H., Menassieu, P., Lettere, M., Mori, K. and Einhorn, J. (1994) Field response of maritime pine scale, *Matsucoccus feytaudi* Duc. (Homoptera: Margarodidae), to synthetic sex pheromone stereoisomers. *Journal of Chemical Ecology* 20, 2159–2170.

Jactel, H., Menassieu, P. and Burban, C. (1996) Découverte en Corse de *Matsucoccus feytaudi* Duc. (Homoptera: Margarodidae), cochenille du pin maritime. *Annales des Sciences Forestières* 53, 145–152.

Jorgensen, C.D., Rice, R.E., Hoyt, S.C. and Westigard, P.H. (1981) Phenology of the San José scale (Homoptera: Diaspididae). *Canadian Entomologist* 113, 149–159.

Lanier, G.N., Qi, Y.T., West, J.R., Park, S.C., Webster, F.X. and Silverstein, R.M. (1989) Identification of the sex pheromone of three *Matsucoccus* pine bast scales. *Journal of Chemical Ecology* 15, 1645–1659.

Lin, G.Q. and Xu, W.C. (1993) Enantioselective synthesis of all four stereoisomers of (2*E*,4*E*)-4,6,10,12-tetramethyl-2,4-tridecadien-7-one, the sex pheromone of *Matsucoccus* pine bast scale. *Tetrahedron Letters* 34, 5931–5934.

Lin, G.Q. and Xu, W.C. (1996) Enzymatic synthesis of (2*E*,4*E*)-4,6,10,12-tetramethyl-2,4-tridecadien-7-one, the sex pheromone of *Matsucoccus matsumurae* Japanese pine bast scale. *Bioorganic and Medicinal Chemistry* 4, 375–380.

Mague, D.L. and Reissig, W.H. (1983) Phenology of the San José scale (Homoptera: Diaspididae) in New York State apple orchards. *Canadian Entomologist* 115, 717–722.

Mayer, M.S. and McLaughlin, J.R. (1991) *Handbook of Insect Pheromones and Sex Attractants*. CRC Press, Boca Raton, Florida, 1083 pp.

McLain, D.C., Rock, G.C. and Woolley, J.B. (1990) Influence of trap color and San José scale (Homoptera: Diaspidididae) pheromone on sticky trap catches of 10 aphelenid parasitoids (Hymenoptera). *Environmental Entomology* 19, 926–931.

McLure, M.S. (1983) Temperature and host availability affect the distribution of *Matsucoccus matsumurae* (Kuwana) (Homoptera: Margarodidae) in Asia and North America. *Annals of the Entomological Society of America* 76, 761–765.

Mendel, Z., Saphir, N. and Robison, D. (1990a) Mass rearing of the Israeli pine bast scale, *Matsucoccus josephi*, with notes on its biology and mating behaviour. *Annals of the Entomological Society of America* 83, 532–537.

Mendel, Z., Dunkelblum, E. and Robison, D. (1990b) Sexual behavior of *Matsucoccus josephi* (Homoptera: Margarodidae): asynchronous adult emergence and release of female sex pheromone. *Journal of Chemical Ecology* 16, 2341–2352.

Mendel, Z., Carmi, E. and Podoler, H. (1991) Relations between the genera *Matsucoccus* (Homoptera: Margarodidae) and *Elatophilus* (Hemiptera: Anthocoridae) and their significance. *Annals of the Entomological Society of America* 84, 502–507.

Mendel, Z., Assael, F., Saphir, N., Zehavi, A. and Kafisheh, W. (1994a) New distribution records of *Matsucoccus josephi* and *Pineus pini* (Homoptera) on pine trees in parts of the Near East. *Phytoparasitica* 22, 9–18.

Mendel, Z., Nestel, D. and Gafni, R. (1994b) Examination of the origin of the Israeli population of *Matsucoccus josephi* (Homoptera: Matsucoccidae) using random amplified polymorphic DNA–polymerase chain reaction method. *Annals of the Entomological Society of America* 87, 165–169.

Mendel, Z., Adar, K., Nestel, D. and Dunkelblum, E. (1997) Sex pheromone traps as a tool for the study of population trends of the predator of a scale insect and for the identification of potential predators for biological control. *IOBC/WPRS Bulletin* 20(1), 231–240.

Moore, I. (1981) Biological amplification for increasing electroantennogram discrimination between two female sex pheromones of *Spodoptera littoralis* (Lepidoptera: Noctuidae). *Journal of Chemical Ecology* 7, 791–798.

Moreno, D.S. (1972) Location of the site of production of the sex pheromone in the yellow scale and the California red scale. *Annals of the Entomological Society of America* 65, 1283–1286.

Moreno, D.S. (1983) Efficiency of pheromone traps in citrus pest detection. *Citrograph* 68, 77–79.

Moreno, D.S. and Fargerlund, J. (1975) Histology of the sex pheromone glands in the yellow scale and the California red scale. *Annals of the Entomological Society of America* 68, 425–428.

Moreno, D.S. and Kennett, C.E. (1985) Predictive year-end California red scale (Homoptera: Diaspididae) orange fruit infestation based on catches of males in the San Joaquin Valley. *Journal of Economic Entomology* 78, 1–9.

Moreno, D.S., Carman, G.E., Rice, R.E., Shaw, J.G. and Bain, N.S. (1972a) Demonstration of a sex pheromone of the yellow scale, *Aonidiella citrina*. *Annals of the Entomological Society of America* 65, 443–436.

Moreno, D.S., Reed, D.K., Shaw, J.G. and Newel, I.M. (1972b) Sex lure trap for Comstock mealybug. *Citrograph* 58, 43–44.

Moreno, D.S., Fargerlund, J. and Shaw, J.G. (1976) California red scale: development of a method for testing juvenoids. *Journal of Economic Entomology* 69, 292–297.

Mori, K. (1992) The synthesis of insect pheromones, 1979–1989. In: ApSimon, J. (ed.) *The Total Synthesis of Natural Products*, Vol. 9. John Wiley & Sons, New York, pp. 280–332.

Mori, K. and Amaike, M. (1994) Synthesis of (2*E*,5*R*,6*E*,8*E*)-5,7-dimethyl-2,6,8-decatrien-4-one, the major component of the sex pheromone of the Israeli pine bast scale and its antipode. *Journal of the Chemical Society Perkin Transaction I*, 2727–2733.

Mori, K. and Harashima, S. (1993a) Synthesis of (2*E*,4*E*,6*R*,10*R*)-4,6,10,12-tetramethyl-2,4-tridecadien-7-one (matsuone) – the primary component of the sex pheromone of three *Matsucoccus* pine bast scales – and its antipode. *Liebigs Annalen der Chemie*, 993–1001.

Mori, K. and Harashima, S. (1993b) Synthesis of (2*E*,4*E*,6*R*,10*S*)-4,6,10-trimethyl-2,4-dodecadiene-7-one – the major component of the sex pheromone of the maritime pine bast scale (*Matsucoccus feytaudi*) – and its three stereoisomers. *Liebigs Annalen der Chemie*, 391–401.

Mori, K. and Kuwahara, S. (1982) Synthesis of optically active forms of (*E*)-6-isopropyl-3,9-dimethyl-5,8-decadienyl acetate, the pheromone of the yellow scale. *Tetrahedron* 38, 521–525.

Mori, K. and Ueda, H. (1981) Synthesis of the optically active forms of 2,6-dimethyl-heptadien-3-ol acetate, the pheromone of the Comstock mealybug. *Tetrahedron* 37, 2581–2583.

Mori, K., Furuuchi, T. and Matsuyama, K. (1995) A new synthesis of (2*E*,4*E*)-4,6,10,12-tetramethyl-2,4-tridecadien-7-one (Matsuone), the primary component of the sex pheromone of three *Matsucoccus* pine bast scales. *Liebigs Annalen der Chemie*, 2093–2099.

Negishi, T., Uchida, M., Tamaki, Y., Mori, K., Ishiwatari, T., Asano, S. and Nakagawa, K. (1980) Sex pheromone of the Comstock mealybug, *Pseudococcus comstocki* Kuwana: isolation and identification. *Applied Entomology and Zoology* 15, 328–333.

Ortu, S and Delrio, G. (1982) Osservazioni sull'impiego in campo del feromone sessuali di sintesi di *Planococcus citri* (Risso) (Homoptera, Coccoidae). *Estratto da Redia* 65, 341–353.

Park, S.C., West, J.R., Abrahamson, L.P., Lanier, G.N. and Silverstein, R.M. (1986) Cross-attraction between two species of *Matsucoccus*. Extraction, bioassay, and isolation of the sex pheromone. *Journal of Chemical Ecology* 12, 609–617.

Raciti, E., Tuminnelli, R., Campo, G. and Cutuli, G. (1995) Strategies of integrated defence in citrus crops. *Informatore Agrario* 51, 73–76.

Rice, R.E. (1974) San José scale: field studies with a sex pheromone. *Journal of Economic Entomology* 67, 561–562.

Rice, R.E. and Jones, R.A. (1982) Collections of *Prospaltella perniciosi* Tower (Hymenoptera: Aphelinidae) in San José scale (Homoptera: Diaspididae) pheromone traps. *Environmental Entomology* 11, 876–880.

Rice, R.E. and Moreno, D.S. (1969) Marking and recapture of California red scale for field studies. *Annals of the Entomological Society of America* 62, 558–560.

Rice, R.E. and Moreno, D.S. (1970) Flight of male California red scale. *Annals of the Entomological Society of America* 63, 91–96.

Roelofs, W.L., Gieselmann, M.J., Cardé, A.M., Tashiro, H., Moreno, D.S., Henrick, C.A. and Anderson, R.J. (1978) Identification of the California red scale sex pheromone. *Journal of Chemical Ecology* 4, 211–224.

Rotundo, G. and Tremblay, E. (1975) Studi su un feromone sessuale del *Planococcus citri* (Risso) (Homoptera, Coccoidea) III. Specificita. *Bolletino del Laboratorio di Entomologia Agraria 'Filippo Silvestri' di Portici* 32, 125–130.

Rotundo, G. and Tremblay, E. (1976) Osservazioni sull'attivita divolo dei maschi di *Pseudococcus calceolariae* (Maskell). *Bolletino del Laboratorio di Entomologia Agraria 'Filippo Silvestri' di Portici* 33, 108–112.

Rotundo, G. and Tremblay, E. (1980) Evaluation of the daily rate of sex pheromone release by the females of two mealybug species (Homoptera Coccoidea Pseudococcidae). *Bolletino del Laboratorio di Entomologia Agraria 'Filippo Silvestri' di Portici* 37, 167–170.

Rotundo, G. and Tremblay, E. (1981) Scent trailing by virgin females of *Pseudococcus calceolariae*. *Journal of Chemical Ecology* 7, 85–88.

Rotundo, G., Gaston, L.K. and Shorey, H.H. (1979) Collection and purification of the female sex pheromone of *Pseudococcus calceolariae* (Homoptera: Coccoidea). *Bolletino del Laboratorio di Entomologia Agraria 'Filippo Silvestri' di Portici* 36, 160–171.

Rotundo, G., Tremblay, E. and Papa, P. (1980) Short range orientation of *Pseudococcus calceolariae* (Mask.) males in a wind tunnel. *Bolletino del Laboratorio di Entomologia Agraria 'Filippo Silvestri' di Portici* 37, 31–37.

Samways, M.J. (1988) Comparative monitoring of red scale *Aonidiella aurantii* (Mask.) (Hom., Diaspididae) and its *Aphytis* spp. (Hym., Aphelinidae) parasitoids. *Journal of Applied Entomology* 105, 483–489.

Shi, X., Webster, X.F., Kallmerten, J. and Meinwald, J. (1995a) Diastereoselective [2,3]-Wittig rearrangement of a tertiary bisallylic ether. *Tetrahedron Letters* 36, 7197–7200.

Shi, X., Webster, X.F., Kallmerten, J. and Meinwald, J. (1995b) Synthesis and determination of the absolute configuration of matsuone, sex pheromone of female *Matsucoccus* pine scales. *Tetrahedron Letters* 36, 7201–7204.

Smetnik, A.I., Konstaninova, G.M., Maksimova, V.I., Rozinskaya, E.M. and Shelukhin, V.I. (1986) Application of coccid sex pheromones in the practice of the USSR quarantine services. In: *Proceedings of the Fifth International Symposium of Scale Insect Studies*, Portici, Italy, pp. 209–213.

Sternlicht, M., Bar Zacay, I., Szivos, Y. and Dunkelblum, E. (1981) Prediction of infestation by and control of the California red scale, *Aonidiella aurantii* (Maskell), through use of its pheromone. In: *Proceedings of the International Society of Citriculture* 2, pp. 695–700.

Tashiro, H. and Chambers, D.L. (1967) Reproduction in the California red scale, *Aonidiella aurantii* (Homoptera: Diaspididae). I. Discovery and extraction of a female sex pheromone. *Annals of the Entomological Society of America* 60, 1166–1170.

Tashiro, H. and Moffit, C. (1968) Reproduction in the California red scale, *Aonidiella aurantii* II. Mating behavior and postinsemination female changes. *Annals of the Entomological Society of America* 61, 1014–1020.

Tashiro, H., Chambers, D.L., Moreno, D. and Beavers, J. (1969) Reproduction in the California red scale, *Aonidiella aurantii* III. Development of an olfactometer for bioassay of the female sex pheromone. *Annals of the Entomological Society of America* 62, 935–940.

Uchida, M., Nakagawa, K., Negishi, T., Asano, S. and Mori, K. (1981) Synthesis of 2,6-dimethyl-1,5-heptadien-3-ol acetate, the pheromone of the Comstock mealybug *Pseudococcus comstocki* Kuwana, and its analogs. *Agricultural Biological Chemistry* 45, 369–372.

Warthen, J.D. Jr, Rudrum, M., Moreno, D.S. and Jacobson, M. (1970) *Aonidiella aurantii* sex pheromone isolation. *Journal of Insect Physiology* 16, 2207–2209.

Watanabe, H., Watanabe, T., Kitahara, T. and Mori, K. (1997) Concise synthesis of a racemic and diastereomeric mixture of the sex pheromones of *Matsucoccus* pine scales. *Bioscience, Biotechnology and Biochemistry* 61, 127–130.

Wessels, A.B. and Meyer, S.R. (1982) Pheromone traps in citrus orchards; their use in monitoring red scale. *Citrus and Subtropical Fruit Journal* 578, 7–9.

Young, B. and Hong-Ren, Y. (1986) Studies on ultrastructure and function of epidermal glands on female *Matsucoccus matsumurae* (Kuwana) (Coccoidea: Margarodidae). *Bolletino del Laboratorio di Entomologia Agraria 'Filippo Silvestri' di Portici* 43, 79–82.

Young, B., Miller, D.R. and McLure, M.S. (1984) Attractivity of the female sex pheromone of Chinese *Matsucoccus matsumurae* (Kuwana) to males of M. *matsumurae* in Japan and to males of M. *resinosae* in the United States (Margarodidae, Coccoidea, Homoptera). *Contributions Shanghai Institute of Entomology* 4, 1–20.

Zegelman, L., Hassner, A., Mendel, Z. and Dunkelblum, E. (1993) Synthesis and field bioassay of the Israeli pine bast scale, *Matsucoccus josephi* female sex pheromone. *Tetrahedron Letters* 34, 5641–5644.

11 Phytophagous Bugs

HEATHER L. MCBRIEN AND JOCELYN G. MILLAR

Department of Entomology, University of California, Riverside, CA 92521, USA

Introduction

The true bugs (order Hemiptera, suborder Heteroptera) are characterized by piercing and sucking mouthparts, and about 50,000 species are known (Arnett, 1993). Most are phytophagous, and true bugs have been ranked fourth among the economically important insect orders (Arnett, 1993). They are found in all types of crops, from forage and fibre crops, through vegetable and row crops, to perennial tree crops. Both immatures and adults damage plants by injection of saliva which liquefies tissues. The dissolved materials are then sucked out, leaving a necrotic lesion. Bug damage results in premature abortion or deformation of fruits, leaf wilt and disease transmission (Hoffmann *et al.*, 1987; Panizzi, 1997).

Many bug species are polyphagous, and the adults are winged and highly mobile, exacerbating monitoring and control problems. Some crops (e.g. cotton, lucerne (alfalfa), legumes) have chronic infestations, whereas damage to other crops is sporadic, occurring only when bugs migrate into a crop from the surrounding habitat. Migrations occur in response to senescence of native hosts, or mowing or harvesting of nearby crops harbouring bug populations (Toscano and Stern, 1976). Effective bug control hinges on rapid detection of these invasions.

Bug sampling methods usually consist of sweep-net or beating-tray sampling, or visual inspections for damage, requiring multiple replications to generate accurate assessments. Unlike the situation with many other pest insect species, the development of semiochemically based sampling methods for bugs is in its infancy. This chapter reviews progress towards identifying and using pheromones of phytophagous bugs in integrated pest management (IPM) programmes. The overall semiochemistry of true bugs, including sex, aggregation and alarm pheromones, and defensive secretions, has been reviewed (Aldrich, 1988a,b, 1995, 1996; see also Chapter 14, this volume).

The need for better bug sampling methods has acquired new urgency because of recent fundamental changes in pest control for major crops. In the past, bugs were often controlled by pesticides applied against other insects. However, reduced pesticide usage in transgenic crops and in crops treated with species-specific pheromone-based mating disruption has resulted in a resurgence of bugs and other secondary pests (e.g. Turnipseed and Greene, 1996; Layton *et al.*, 1997). These examples dramatically illustrate unintended consequences that may result if all parts of a crop system are not given adequate consideration when implementing new pest control strategies.

Complications in Studying Bug Pheromones

The identification of bug pheromones is complicated by several factors. Firstly, bugs produce voluminous quantities of defensive secretions that may mask pheromone components, and these defensive compounds at lower concentrations may also have alternative roles as pheromone components (e.g. Millar *et al.*, 1997). Secondly, the sexes producing and responding to attractant pheromones differ between families and even genera. Extracts made from males and females of some species contain sex-specific compounds (e.g. *Phytocoris* spp.: Millar *et al.*, 1997; Millar and Rice, 1998), whereas those of other species do not (e.g. *Lygus* bugs: Blumenthal, 1978; Gueldner and Parrott, 1978; Aldrich *et al.*, 1988; the alydid bug *Leptocorisa chinensis* (Dallas): Leal *et al.*, 1996). For mirid bugs, females are the attractive sex and attract only adult males, but the compounds used may either be female-specific or produced by both sexes. In the known examples from pentatomids, coreids and scutellerids, males produce the attractants. The situation becomes complex in other families. For example, males of the alydid bug *Riptortus clavatus* (Thunberg) attract adults of both sexes and nymphs (Leal *et al.*, 1995); but with another alydid bug, *L. chinensis*, the pheromone compounds are produced by both sexes but only males are attracted (Leal *et al.*, 1996). Thus, careful determination of the (producing) attractive and responding sex(es) is crucial.

This confusing picture of pheromone producers and responders can be clarified by considering the role of a pheromone and the context in which it is produced. Mirid bug attractants are clearly sex pheromones used to attract mates. In contrast, some species (e.g. *Leptoglossus occidentalis* Heidemann: Blatt and Borden, 1996) produce aggregation pheromones to assemble overwintering individuals. In other species, the attractants have dual roles as both sex and aggregation pheromones (e.g. *R. clavatus*: Leal *et al.*, 1995). However, frequently too little is known about the biology of various bug species to explain their usage of pheromones adequately.

The third factor is that bugs must be in an appropriate physiological state to produce or respond to pheromones. In long-lived species, adults may take a number of days to mature sexually and begin producing and responding to pheromone. For species that overwinter as adults, bugs enter reproductive diapause in response to seasonal conditions and become unresponsive to pheromones (e.g. Strong *et al.*, 1970; Aldrich *et al.*, 1991). Aggregation pheromones may follow a similar pattern of seasonal use (e.g. Blatt and Borden, 1996).

Fourthly, bugs have a plethora of glands producing compounds for communication and defence. In some species, pheromone compounds are produced in discrete, multicellular glands; in others, they are produced in patches of unicellular glands

(Evans *et al.*, 1990). Unlike with many other insects, the pheromone glands are not necessarily located in characteristic sites where they can be readily found and dissected for study of their contents.

Finally, some species (e.g. *Chlorochroa* and *Thyanta* spp.) are difficult to maintain in laboratory culture. Amongst other factors, crowding may encourage chronic release of defensive secretions in laboratory cultures, so that cages should be well ventilated (L. Ehler, University of California, Davis, 1997, personal communication). Some species also appear to be sensitive to inbreeding depression.

The overall picture has been complicated by the premature, inappropriate or incorrect designation of bug semiochemicals as pheromones. Many compounds have been identified from bugs, but the number of studies in which bug pheromone blends have been completely identified by systematic, bioassay-driven fractionation and recombination of components is small. Such methodical studies, although tedious and expensive, are critical to the further development of our understanding of the use of pheromones by bugs.

Methodology

Collection and analysis of pheromones

Pheromone extracts from bugs have been prepared in several ways (Evans *et al.*, 1990). They have been prepared by soaking whole insects or body sections in solvent (e.g. *Nezara viridula* (L.): Pavis and Malosse, 1986; *Campylomma verbasci* (Meyer): Smith *et al.*, 1991). In a related method, whole insects have been thermally desorbed (van Beek *et al.*, 1998). The value of these techniques is debatable because most of the compounds obtained may be unrelated to pheromone components.

A second method is to prepare extracts by soaking glands or by collecting gland contents with a glass capillary (e.g. Aldrich *et al.*, 1993a). A general procedure has been developed by J.R. Aldrich (1995, Tsukuba, personal communication). The legs and wing covers are removed from the anaesthetized bug, and the bug is pinned ventral side down. The outer margins of the abdominal cuticle and the cuticle at the junction between the thorax and abdomen are cut and the dorsal abdominal cuticle is folded back with its attached glands. The metathoracic glands are then dissected out. It is crucial to identify and dissect all glands, reservoirs and associated structures. However, this method has very serious limitations. Although preparing gland extracts can be straightforward, reconstituting active pheromone blends from their contents may be difficult or impossible because it is not yet clear which glands are the sources of pheromones in different subfamilies and genera. Furthermore, some bugs (e.g. *N. viridula*) do not appear to have discrete macroscopic pheromone glands that can be dissected.

Another possibility is to collect volatiles produced by live bugs on a suitable adsorbent, and then to recover them by solvent extraction. This method is probably the most useful because the extracts obtained are representative of what bugs actually release, aerations can be continued for days or even weeks, changing the collectors as required, and the bugs can be aerated on food, which enhances both longevity and pheromone production (e.g. Numata *et al.*, 1990; Leal *et al.*, 1995). Dead bugs must be removed as soon as possible because they 'leak' defensive compounds, which contaminate the extracts. However, aeration extracts are not without limitations. Aerations are time averages, and if pheromone is produced only in short bursts, even aeration extracts may be misleading (e.g. *L. chinensis*: Leal *et al.*, 1996).

All extract preparation methods have additional, potential weaknesses. For example, the methods described above recover polar, water-soluble compounds poorly, if at all; very volatile compounds may be obscured by solvent peaks during analysis; or compounds may degrade during extraction or analysis (e.g. Millar, 1997).

Laboratory bioassay methods

As with any insect, bugs must be in the appropriate physiological state for pheromone studies. For studies of sex pheromones, bugs must be sexually mature and reproductively active. Diel periodicity must also be taken into account (Blumenthal, 1978; Harris and Todd, 1980). It is best to work with virgins, but for those species that mate several times, bugs may be reused after a refractory period. Finally, some species only produce pheromones in the presence of food (Leal *et al.*, 1995). Analogous considerations apply for studies of aggregation pheromones (e.g. *L. occidentalis*, Blatt and Borden, 1996).

Development of reliable laboratory bioassays for bug pheromones has proved problematic, possibly because bugs are sensitive to the unnatural conditions. Bugs may also be distracted by movement and noise (H.L. McBrien and J.G. Millar, 1998, unpublished results). Often, the typical bioassays used with volatile stimuli (e.g. olfactometers, wind tunnels) have met with limited success; consequently, many workers have moved directly to field trials. Nevertheless, laboratory bioassays have been used in a number of studies. Two-choice moving air bioassays were used with *Lygus* spp. (Graham, 1987, 1988) and *Euschistus obscurus* (Palisot de Beauvois) (Borges and Aldrich, 1994), and Y-tubes have been used with lygaeids (Aldrich *et al.*, 1997) and pentatomids (Mitchell and Mau, 1971; Leal *et al.*, 1998). In our laboratory, a vertical Y-tube with an overhead light to induce phototaxis has been used in studies with five pentatomid species (H.L. McBrien and J.G. Millar, 1998, unpublished results). Wind tunnels have been used with mirid (Groot *et al.*, 1998b) or pentatomid bugs (Leal *et al.*, 1998). Bioassays based on the elicitation of specific behaviours associated with courtship or mating also have proved useful (Groot *et al.*, 1998a,b; Leal *et al.*, 1998).

Coupled gas chromatography–electroantennogram detection (GC–EAD) has been used (Leal *et al.*, 1995, 1996, 1998; Aldrich *et al.*, 1997), but has limitations because bug antennae may respond to defensive compounds. Conversely, behaviourally active pheromone components may not elicit detectable EAD signals (Leal *et al.*, 1998). Thus, GC–EAD may be helpful, but results need to be thoroughly checked with behavioural bioassays.

Bioassay strategies merit careful consideration because bug extracts contain a number of compounds, and the number of possible combinations to test rapidly becomes large. Furthermore, defensive compounds or excessive doses may inhibit responses. Several possible testing strategies may be tried:

1. Reconstruction of the complete blend, followed by elimination of components until the minimum set is identified. This strategy was successful with two *Phytocoris* spp. (Millar *et al.*, 1997; Millar and Rice, 1998), but would not work for species whose extracts contain inhibitory compounds (e.g. *L. chinensis*: Leal *et al.*, 1996).
2. Additive bioassays, in which the attractive blend is built up from a single active component with other compounds added subsequently to see if the activity increases. This strategy has been used with several stink bug species (see below).
3. Bioassay of sex-specific compounds. This strategy worked for the mirid *C. verbasci*, in which the pheromone consisted of two female-specific compounds, but it will not work with species in which the pheromone blend consists of combinations of female-specific compounds with compound(s) produced by both sexes.
4. GC–EAD guided testing. The limitations of GC–EAD have been discussed above.

Identified Bug Pheromones

Structures of known and suspected pheromone components for phytophagous bug species are shown in Fig. 11.1. The following sections review the identification and testing of these pheromones by bug family.

The plant bugs (Miridae)

The family Miridae contains about 10,000 species worldwide, making it the largest heteropteran family (Aldrich, 1996). Despite their importance to agriculture, sex pheromones have been identified for only three species, *Campylomma verbasci* (Smith *et al.*, 1991), *Phytocoris relativus* Knight (Millar *et al.*, 1997) and *P. californicus* Knight (Millar and Rice, 1998). Attraction of males to females has been demonstrated for a number of other species, including several *Lygus* species (Scales, 1968; Strong *et al.*, 1970; Graham 1988), the cocoa capsid bug, *Distantiella theobroma* (Dist.) (King, 1973), the cocoa mirid, *Helopeltis clavifer* (Walker) (Smith, 1977), the green apple bug, *Lygocoris communis* (Knight) (Boivin and Stewart, 1982), the apple brown bug, *Atractotomus mali* (Meyer) (Smith *et al.*, 1994), the green capsid bug, *Lygocoris pabulinus* (L.) (Blommers *et al.*, 1988), the tea mosquito bug, *Helopeltis antonii* (Sign.) (Sundararaju *et al.*, 1994), *Calocoris norvegicus* (Gmelin) (Welter, 1991), *Neurocolpus longirostris* Knight (Rice and Jones, 1989) and the rice leaf bug, *Trigonotylus caelestialium* (Kirkaldy) (Kakizaki and Sugie, 1997).

The mullein bug, Campylomma verbasci

The first mirid sex pheromone was identified from *C. verbasci* (Fig. 11.2), a pest of apples (Thistlewood *et al.*, 1989a; Smith, 1991; Beers, 1992; Stigter, 1996). Males

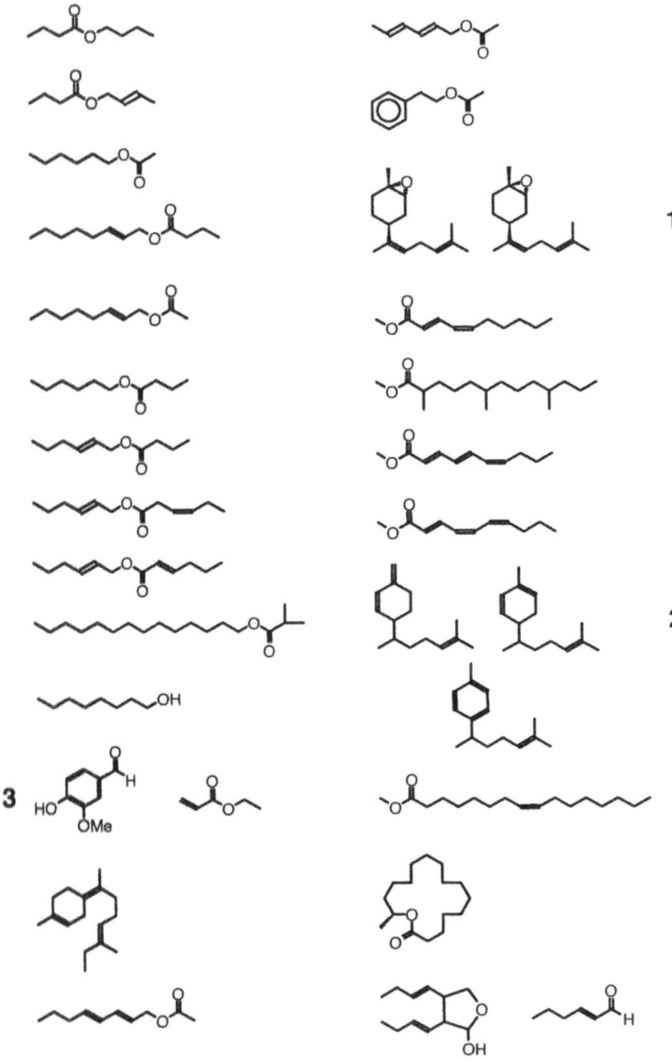

Fig. 11.1. Structures of pheromone components for phytophagous bug species.

were attracted to live females, extracts of crushed females, and volatiles from females feeding on mullein, *Verbascum thapsis* L. (Thistlewood *et al.*, 1989b). Extracts from the head and thorax of females were most attractive to males, suggesting that the metathoracic glands may be the source of pheromone. The pheromone blend consisted of two female-specific components, butyl butyrate and E2-butenyl butyrate (**1** and **2**, Fig. 11.1), in a ratio of *c.* 94:6 (Smith *et al.*, 1991; McBrien *et al.*, 1994b). Neither component was active alone, whereas the two-component blend was as attractive as five live virgin females when released at rates of 91 and 183 μg day^{-1}.

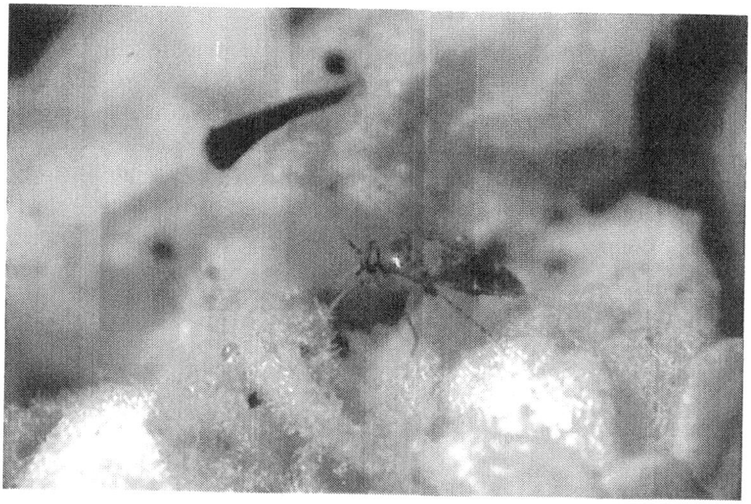

Fig. 11.2. The mullein bug, *Campylomma verbasci*, on a mullein plant. (Courtesy of Mark Gardiner.)

Phytocoris relativus *and* P. californicus

Phytocoris relativus and *P. californicus* are pests of pistachio (Rice *et al.*, 1988). Identification of the sex pheromone blends of *P. relativus* and *P. californicus* was straightforward, with females of both species producing sex-specific compounds. Laboratory bioassays (Y-tubes, pitfall and arena bioassays) produced contradictory results, but field bioassays using sticky traps baited with rubber septum lures gave consistent results. Complete reconstructions of the blends of female volatiles attracted males, and several iterative bioassay cycles eliminated inactive compounds. The blends were species-, sex- and stage-specific, and, in contrast with *C. verbasci*, the attractants of *P. relativus* and *P. californicus* consisted of 2:1 blends of a compound produced by both sexes (hexyl acetate, **3** in Fig. 11.1), with the female-specific compounds *E*2-octenyl butyrate and *E*2-octenyl acetate, respectively (**4** and **5**, Fig. 11.1) (Millar *et al.*, 1997; Millar and Rice, 1998). *E*2-Octenyl butyrate inhibited attraction of *P. californicus* males, but *E*2-octenyl acetate did not inhibit *P. relativus* (Millar and Rice, 1998). The *P. relativus* pheromone components were produced in the thorax. Subsequent work with other species has revealed that the complete reconstruction strategy is not universally effective, probably due to defensive or other inhibitory compounds in crude extracts (Millar *et al.*, 1997).

Lygus *species*

Lygus spp. are notorious pests of numerous crops (Strong *et al.*, 1970), and pheromones for use in IPM programmes would be immensely valuable (Aldrich *et al.*, 1988). However, efforts spanning several decades have failed to identify pheromone components for any *Lygus* species, because of several factors. Firstly, there are minimal differences in the chemical profiles of males and females. Secondly, no research group has been able to produce consistently attractive extracts. It is still unknown whether the activity of female extracts is being masked by inhibitors, or the extract preparation methods are failing to collect or are destroying the active compounds. Thirdly, reproducible and reliable laboratory bioassays have not yet been developed for *Lygus* spp.

Numerous investigators have established that female *Lygus* bugs produce a sex pheromone which attracts males (e.g. *L. lineolaris* (Palisot de Beauvois): Scales, 1968; Blumenthal, 1978; Slaymaker and Tugwell, 1984; *L. hesperus* Knight: Strong *et al.*, 1970). *L. lineolaris* females were also weakly attractive to males of the mirid bugs *Adelphocoris lineolatus* (Goeze) and *Stenotus binotatus* (Fabricius) (Slaymaker and Tugwell, 1984). Graham (1987) demonstrated interspecific attraction between *L. desertinus* Knight, *L. elisus* Van Duzee and *L. lineolaris*, but female *L. hesperus* attracted conspecifics only. A follow-up study in which body sections of live *L. hesperus* females were sealed with impermeable nail varnish suggested that the pheromone was associated with the abdomen tip (Graham, 1988).

Gueldner and Parrott (1978) found that hexyl butyrate and *E*2-hexenyl butyrate (**6** and **7**, Fig. 11.1) comprised 60–90% of steam distillates from male and female *L. lineolaris*, with *E*2-hexenyl butyrate being more prevalent in females. Hexyl butyrate and *E*2-hexenyl butyrate also comprised 70–80% of the total volatiles from male and female *L. lineolaris*, *L. hesperus* and *L. elisus* (Aldrich *et al.*, 1988). In bioassays, single compounds and blends of several *L. lineolaris* esters were no more attractive to males than controls (Gueldner and Parrott, 1978; Hedin *et al.*, 1985). Furthermore, hexyl butyrate decreased the attractiveness of five caged females (Blumenthal, 1978). Communications by the authors with investigators who have studied *Lygus* spp. indicate that a large number of unsuccessful bioassays have been conducted with various compounds from *Lygus* extracts.

Electroantennogram (EAG) studies have not shed further light on *Lygus* pheromones. EAGs recorded from *L. lineolaris* showed that hexyl butyrate and *E*2-hexenyl butyrate elicited greater responses from male than female antennae (Chinta *et al.*, 1994; Dickens *et al.*, 1995) but, as described above, these compounds have consistently failed to attract male bugs. Thus, even with abundant evidence that females use a volatile sex attractant, *Lygus* spp. pheromones have resisted all attempts at identification.

The green capsid bug, Lygocoris pabulinus

Females of *L. pabulinus* (Fig. 11.3), a pest of apples and pears (Blommers *et al.*, 1997), attract males with a sex pheromone (Blommers *et al.*, 1988; Groot *et al.*, 1998b). However, no consistent sex-specific differences have been found in extracts (A.T. Groot, Wageningen, 1998, personal communication), and female *L. pabulinus* showed no behaviour that could be associated with courtship or calling (Groot *et al.*, 1998b). However, males vibrate the abdomen before copulation and this behaviour has been used to bioassay potential pheromone compounds (Groot *et al.*, 1998a,b). A selection of insect- and plant-produced chemicals that elicited strong EAG responses from male antennae were not attractive to males (Groot *et al.*, 1998a; A.T. Groot, Wageningen, 1998, personal communication), although *E*2-octenyl butyrate (**4**, Fig. 11.1) elicited abdominal vibration by males.

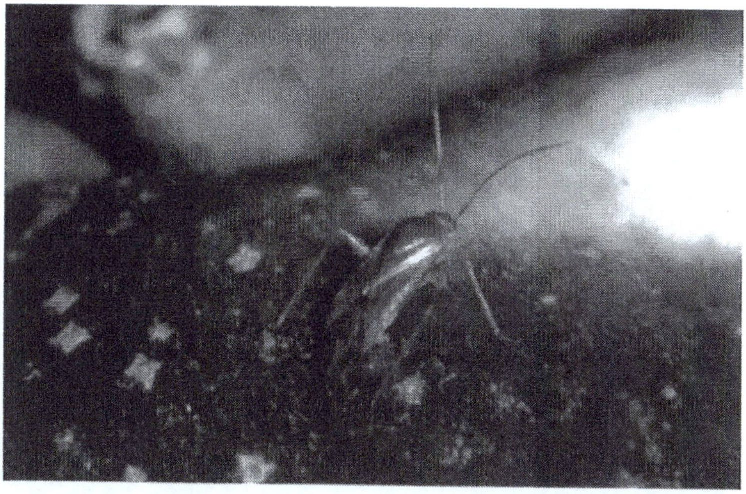

Fig. 11.3. The green capsid bug, *Lygocoris pabulinus,* on apple. (Courtesy of Astrid Groot.)

Calocoris norvegicus

The univoltine C. *norvegicus* is a pest of potatoes, carrots, strawberries, pistachios and tree fruit (Wheeler and Henry, 1992). Virgin females attract males (Welter, 1991) and several pheromone components have been identified from aerations (Millar *et al.*, 1998). The pheromone consists of at least three components, two of which are female-specific, but identification of the pheromone blend has been hindered by highly variable trap catches and random catches of females (J.G. Millar and R.E. Rice, 1998, unpublished results). The latter problem was ameliorated but not eliminated by switching to triangular Jackson traps with smaller openings (Rice, 1998).

The buckeye bug, Neurocolpus longirostris

Rice and Jones (1989) demonstrated that females attract males, and Millar *et al.* (1998) identified a number of compounds from aeration extracts. Trap catches have been extremely variable, but suggest that the pheromone blend consists of at least four components of widely different volatility, two of which are female-specific (J.G. Millar and R.E. Rice, 1998, unpublished results). Males have also been observed trying to copulate with the rubber septum lures.

The broad-headed bugs (Alydidae)

Attractant pheromones have been identified for two alydid species. Males of the bean bug *R. clavatus* (Fig. 11.4), a pest of soybeans, attracted non-diapausing adults of both sexes (Numata *et al.*, 1990). GC–EAD analysis of aerations and whole-body extracts of males on food revealed four antennal-stimulatory peaks (Leal *et al.*,

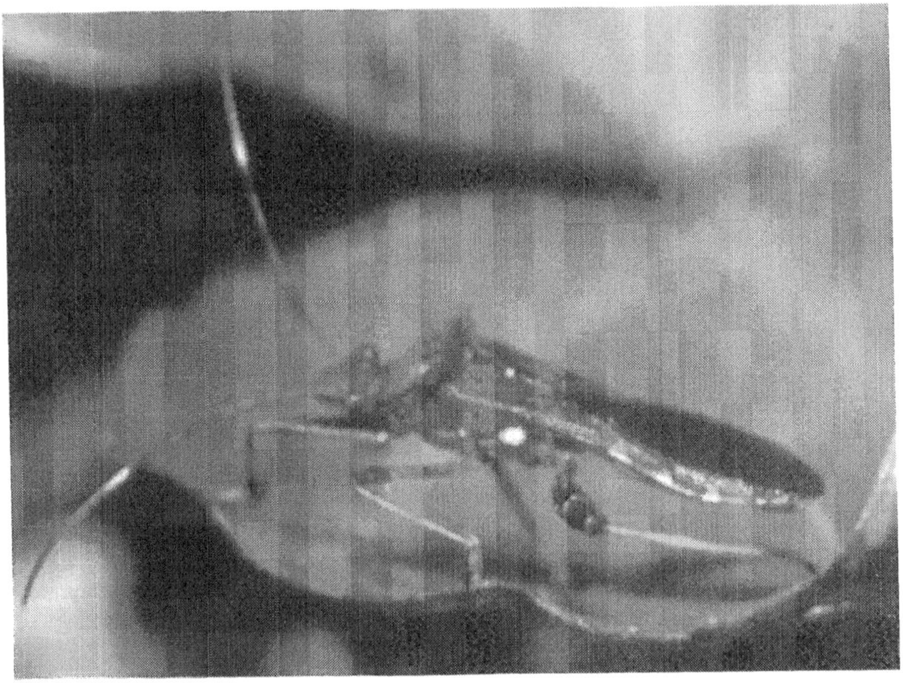

Fig. 11.4. The bean bug, *Riptortus clavatus.* (Courtesy of Walter Soares Leal.)

1995), one of which (*E*2-hexenyl hexanoate) was an alarm pheromone (Leal and Kadosawa, 1992). A 1:5:1 blend of the three remaining male-specific compounds – *E*2-hexenyl *Z*3-hexenoate, *E*2-hexenyl *E*2-hexenoate and myristyl isobutyrate (MI) (**8**, **9** and **10** in Fig. 11.1) – attracted adults and second-instar nymphs to water-pan traps baited with plastic pellet lures (*c.* 100 mg doses) (Leal *et al.*, 1995). A further study found that 50 mg of MI was attractive by itself. This high dose of MI was just as attractive as 50 mg of the three-component blend (containing 7 mg MI), while neither of the other two components was attractive alone (Mizutani *et al.*, 1997).

Analyses of aeration and metathoracic gland extracts of the congeneric *Riptortus serripes* Fabricius found a number of male-specific unsaturated esters and acids, including *E*2-hexenyl *E*2-octenoate and *E*2-hexenyl *Z*3-hexenoate, whereas both males and females produced *E*2-hexenal and *E*2-hexenyl hexanoate (Aldrich *et al.*, 1993b). The roles of these compounds as pheromone components remains unknown.

Identification of the pheromone of the rice bug *L. chinensis* illustrates several interesting points. Volatiles from adults of both sexes showed no qualitative sex-specific differences, but eight compounds elicited strong GC–EAD responses from antennae of both sexes (Leal *et al.*, 1996). Four of the EAD-active chemicals were not tested as attractants because they were probably defensive compounds, but the mixture of the remaining four compounds, or any mixture of three of the four compounds, was not attractive. Only a specific blend of two components, *E*2-octenyl acetate and octanol (**5** and **11**, Fig. 11.1), attracted adult males. The fact that the sex pheromone consists of this specific subset of compounds from the indistinguishable and unattractive extracts from males and females is remarkable. Other components of the time-averaged trapped volatiles must inhibit response to the pheromone, and so the insects must be able to regulate the blend of volatiles they produce.

The squash bugs and leaf-footed bugs (Coreidae)

There is evidence for male-produced aggregation pheromones for two coreid species. Male *Leptoglossus australis* Fabricius attracted conspecific adults of both sexes and nymphs (Yasuda, 1990; Yasuda and Tsurumachi, 1994, 1995) while male western conifer seed bugs, *L. occidentalis*, produced an aggregation pheromone in autumn which attracted adults of both sexes (Blatt, 1994; Blatt and Borden, 1996). Pheromone components for these two species have not been identified.

There is limited evidence that male *Amblypelta lutescens lutescens* Distant produce an aggregation pheromone (Aldrich *et al.*, 1993b). Male-specific compounds included *E*,*E*-α-farnesene, (3*R*)-nerolidol, β-ocimene, linalool, *E*-β-farnesene and two unidentified components. In preliminary field tests, a few adult bugs were observed within 30 cm of baited traps.

Male-specific compounds in glands and extracts of volatiles from several other coreid species have been proposed as attractants, but as yet there is little supporting behavioural or bioassay data (*Leptoglossus phyllopus* (L.): Aldrich *et al.*, 1993b; *Leptoglossus clypealis* (Heidemann): Millar and Wang, 1994; Aldrich 1988b, 1995).

The shield bugs (Scutelleridae)

Males of the sun bug, *Eurygaster integriceps* Puton, a pest of wheat and other cereals, produce a sex attractant (Zdárek and Kontev, 1975). Several male-specific compounds have been identified, including vanillin and ethyl acrylate (**12** and **13** in Fig. 11.1) (Ubik *et al.*, 1975; Vrkoč *et al.*, 1977), and homo-γ-bisabolene (**14**, Fig. 11.1) (Staddon *et al.*, 1994), but the roles of these compounds await clarification. Male-specific secretions produced by *Tectoris diophthalmus* (Thunberg), *Hotea gambiae* (Westwood) and *Sphaerocoris annulus* (Fabricius) have been summarized (Aldrich 1988a,b, 1995) but there are no data on their biological function.

The seed bugs (Lygaeidae)

Attractant pheromones have been identified for two species: *Tropidothorax cruciger* (Motschulsky) and *Neacoryphus bicrucis* Say (Aldrich *et al.*, 1997). In the former, two components (*E*2-octenyl acetate and *E*2,*E*4-octadienyl acetate: **5** and **15** in Fig. 11.1) found in extracts from both sexes elicited strong GC–EAD responses from antennae of both sexes. Males and females were attracted to 500 ng of a 10:1

blend in Y-tube bioassays, and to rubber septa dosed with 10 μl of the neat blend in field tests, but whether both components were required for attraction was not reported.

The second species, *N. bicrucis*, produces male-specific pheromone components (*E*2, *E*4-hexadienyl acetate and phenethyl acetate: **16** and **17** in Fig. 11.1) in the sexually dimorphic tubular glands and secondary reservoir of the metathoracic scent gland complex (Aldrich *et al.*, 1997). Both sexes were attracted to rubber septa loaded with 5 μl of a 92:8 blend of the two components.

The stink bugs (Pentatomidae)

Nezara *and* Acrosternum *species*

Nezara viridula (Fig. 11.5) is one of the most important phytophagous pentatomids (Todd 1989) for predaceous pentatomids, (see Chapter 14). Several groups have reported that virgin or mated males attract females (Mitchell and Mau, 1971; Brennan *et al.*, 1977; Aldrich *et al.*, 1987) or both sexes plus nymphs (Harris and Todd, 1980). Pavis and Malosse (1986) prepared active extracts from portions of the ventral abdominal cuticle, and developed a laboratory olfactometer bioassay. Borges *et al.* (1987) made the crucial observation that extracts of males elicited only long-range orientation from females. The major component of the pheromone, produced in unicellular secretory structures associated with the ventral abdominal cuticle (Pavis and Malosse, 1986; Lucchi, 1994; Borges, 1995), was finally identified as a bisabolene epoxide (BAE), (*Z*)-(1′*S*,3′*R*,4′*S*)-(−)-2-(3′4′-epoxy-4′methylcyclohexyl)-6-methylhepta-2,5-diene (*trans*-*Z*-BAE: **18** in Fig. 11.1). The *trans* designation refers to the relationship between the pendant group on the ring and the epoxide group (Baker *et al.*, 1987). Only the (−)-enantiomer was attractive in bioassays, and 12 ng doses on filter paper were as attractive as live males. Bioassays were not conducted with blends of *trans*-*Z*-BAE and any other male-specific compounds, such as *cis*-*Z*-BAE (**19**, Fig. 11.1) or *Z*-α-bisabolene.

Almost simultaneously with the Borges *et al.* (1987) report, Aldrich *et al.* (1987) confirmed the presence of *trans*- and *cis*-*Z*-BAE (ratio 3:1, chirality unknown), *Z*-α-bisabolene, *E*-nerolidol and nonadecane in male volatiles. The role of each compound as a pheromone component was not clear. In field tests in the USA and Brazil, blends of *trans*- and *cis*-*Z*-BAE attracted low numbers of bugs, but few experimental details were given (Aldrich *et al.*, 1993a).

The pheromone chemistry of a French *N. viridula* strain has been studied in detail using a walking bioassay (Brézot *et al.*, 1993). Adult females were attracted to mature males, but nymphs, mature males, or sexually immature adults of both sexes were not, in contrast to results obtained previously in field tests (see above). Females were attracted to 100 ng to 5 μg doses of a 10:1 blend of bug-produced *trans*:*cis*-*Z*-BAEs on filter paper, but not to a 1:12 *trans*:*cis* blend, nor to 100 ng doses of the minor male-specific compounds *E*-nerolidol or nonadecane. Blends of *trans*- and *cis*-BAEs with these minor components were not tested. Females of a Guadeloupe population of *N. viridula*, whose males produced an 84:16 blend of *trans*:*cis*-*Z*-BAE, were attracted to both racemic *cis*- and *trans*-*Z*-BAE (each contaminated with < 3% of the *trans* and *cis* isomers, respectively), with the

Fig. 11.5. A mating pair of the southern green stink bug, *Nezara viridula*. (Courtesy of J. Steven McElfresh.)

response being strongest to the *trans* isomer (Brézot *et al.*, 1994). The presence of the unnatural enantiomers in the synthetic blends caused neither synergism nor inhibition. Furthermore, 3:1 to 1:3 blends of (−)-*trans*:(−)-*cis*-Z-BAEs (500 ng) were more attractive than either compound alone, with the 3:1 blend, closest to the bugs' natural ratio, attracting the largest percentage of females. This provided the first evidence that the *N. viridula* pheromone is a blend.

The relative roles of the *trans*- and *cis*-Z-BAEs as *N. viridula* pheromone components remain controversial. Bugs from different populations apparently produce different *trans*:*cis* ratios (Baker *et al.*, 1987; Aldrich *et al.*, 1987, 1993a; Brézot *et al.*, 1994), but variability within populations is also high (e.g. Guadeloupe, 42–82% *trans*: Brézot *et al.*, 1994; Australia, 33–75% *trans*: Ryan *et al.*, 1995). The importance of the *trans*:*cis* ratio remains unclear, particularly as Brézot *et al.* (1994) showed that females responded to a wide range of ratios.

The profile of male-specific volatiles from the congeneric *N. antennata* Scott is similar to that of *N. viridula* males, containing Z-α-bisabolene, *trans*-Z-BAE, *cis*-Z-BAE and *E*-nerolidol, all of unknown chirality (Aldrich *et al.*, 1989, 1993a). Males of several species in the sister genus *Acrosternum* also produce Z-α-bisabolene, and *cis*- and *trans*-BAE of unknown chirality. In three of these species (*A. aseadum* Rolston, *A. hilare* (Say), *A. marginatum* Palisot de Beauvois), the *trans*:*cis* ratio is heavily skewed towards *cis*-BAE, whereas in *A. pennsylvanicum* (DeGeer), the ratio is about 1:1. However, for all species except *A. hilare*, the role of these compounds remains unknown.

In current studies with *A. hilare*, a vertical Y-tube bioassay has been used (H.L. McBrien, J.G. Millar, J.S. McElfresh and L. Gottlieb, 1998, unpublished results). Mature males attract mature females but not males, and females are unattractive to either sex. Males produce (−)-*cis*-Z-BAE and (−)-*trans*-Z-BAE (**19** and **18** in Fig. 11.1; ratio *c.* 19:1). Attractant activity was confined to the BAE-containing fraction, and so other compounds in the extracts are probably not pheromone components. Finally, the male-produced blend of the two BAEs was more attractive to females than other blends, and the individual components were not attractive, indicating that the blend is important. Field trials are awaiting synthesis of gram quantities of the two stereoisomers.

In summary, many questions remain to be resolved with *Nezara* and *Acrosternum* pheromones. *N. viridula* is the only species in this group for which any bioassay data with synthetic pheromone blends have been published, and these only for laboratory bioassays. Even with *N. viridula*, the roles of minor components, optimal ratios and doses, and mechanisms for maintaining reproductive isolation between sympatric *Nezara* and *Acrosternum* species await clarification.

Finally, production of multigram quantities of the pure BAE stereoisomers for field tests is not trivial. Several syntheses have been reported (Baker *et al.*, 1987; Marron and Nicolau, 1989, in which structures are mislabelled; Baptistella and Alexio, 1994; Kuwahara *et al.*, 1998), but most are only amenable to production of small quantities. A simpler synthesis of a mixture of all BAE stereoisomers has been reported (Tomioka and Mori, 1992), but the biological activity of this mixture may be questionable.

Euschistus *species*

Progress has been made in the identification of sex pheromone compounds for two groups of species in the genus *Euschistus*, a number of which are pests of fruit and nut crops (Aldrich *et al.*, 1991). Males of the first group of species comprising *E. conspersus* Uhler (Fig. 11.6), *E. tristigmus* (Say), *E. politus* Uhler, *E. servus* (Say) and *E. ictericus* (L.) produce methyl E2,Z4-decadienoate (E2,Z4–10:COOMe: **20** in Fig. 11.1) as a major component (Aldrich *et al.*, 1991), with E2,E4–10:COOMe being a minor component in all species except *E. tristigmus*. Other minor male-specific compounds included geranylacetone (*E. conspersus*, *E. tristigmus* and *E. servus*), decanoic acid (*E. tristigmus*) and Z2,E4–10:COOMe. In field trials in Maryland, E2,Z4–10:COOMe attracted a few *E. conspersus*, *E. tristigmus*, *E. politus* and *E. servus* immatures and adults of both sexes (Aldrich *et al.*, 1991). Addition of minor male-specific components (decanoic acid for *E. tristigmus* and *E. politus*, geranylacetone for *E. servus*) did not improve attraction, nor did mixing E2,Z4–10:COOMe with male dorsal abdominal gland (DAG) compounds (Aldrich *et al.*, 1995). The DAG blends did not contain sex-specific compounds but did differ between species. Significant proportions of the bugs attracted (up to c. 50%) were collected within 1 m of the trap, suggesting that the pheromone attracts bugs only to the vicinity of a male. Based on low catches of adult and nymphal *E. conspersus* in field trials using traps baited with E2,Z4–10:COOMe, Wilson *et al.* (1988) and Wilson (1990) suggested that the pheromone may act over a relatively short range. Clear plastic cylinder traps with inverted cone entrances performed better than several other trap designs (Zalom, 1992). Overall, it is clear that E2,Z4–10:COOME is a male-produced pheromone component for several *Euschistus* spp., but the reasons for the low trap catches remain elusive.

In the second group of *Euschistus* spp., comprising *E. heros* Fabricius and *E. obscurus* (Aldrich *et al.*, 1991, 1994), male-specific volatiles of *E. heros* were dominated by methyl 2,6,10-trimethyltridecanoate (2,6,10–13:COOMe: **21** in Fig.

Fig. 11.6. The consperse stink bug, *Euschistus conspersus*. (Courtesy of J. Steven McElfresh.)

11.1), with 2,6,10–12:COOMe and *E*2,*Z*4–10:COOMe being minor components. In *E. obscurus*, both 2,6,10–13:COOMe and *E*2,*Z*4–10:COOMe were major components, with traces of 2,6,10–12:COOMe and *E*2,*E*4–10:COOMe. The 12- and 13-carbon methyl esters have eight possible stereoisomers, complicating their identification and synthesis, and 2,6,10–13:COOMe has been synthesized as a stereoisomeric mixture (Mori and Murata, 1994a), as a mixture of the 2*R*/*S*, 6*S*,10*S* stereoisomers (Ferreira and Zarbin, 1996), and as the individual stereoisomers (Mori and Murata, 1994b).

In laboratory bioassays, female *E. obscurus* were attracted to extracts from live males, but fractions of the extracts gave equivocal results (Borges and Aldrich, 1994). Borges *et al.* (1998a) demonstrated that female *E. heros* were attracted to 10 μg of a synthetic mixture of all eight stereoisomers of 2,6,10–13:COOMe, with attraction being strongest during the latter half of the day. In follow-up trials with the stereoisomeric blend or with a blend of the 2*R*/*S*, 6*S*,10*S*-13:COOMe isomers, a few *E. heros* were trapped (Borges *et al.*, 1998b), along with larger numbers of another pentatomid, *Piezodorus guildinii* (Westwood).

Other stink bug species

Adults of both sexes of an orchard pest, the brown-winged green bug *Plautia stali* Scott, were attracted to live males (Moriya and Shiga, 1984) but showed no inclination to copulate, suggesting that the pheromone had an aggregative rather than a sexual function. The pheromone was identified by bioassay-driven fractionation as a single component, methyl *E*2,*E*4,*Z*6-decatrienoate (*E*2,*E*4,*Z*6–10:COOMe: **22** in Fig. 11.1) (Sugie *et al.*, 1996), a more unsaturated analogue of *E*2,*Z*4–10:COOMe, the *Euschistus* pheromone component. Water-pan traps baited with 30 μg of pheromone dispensed from cotton wicks caught 3.2 bugs $trap^{-1}$ day^{-1} (56% females).

Mature *Thyanta pallidovirens* Stål females were attracted to live males and aeration extracts from live males in a vertical Y-tube bioassay (H.L. McBrien and J.G. Millar, 1998, unpublished results). The pheromone blend consisted of thermally unstable *E*2,*Z*4,*Z*6–10:COOMe (**23**, Fig. 11.1) (Millar, 1997) plus one or more of the sesquiterpenes β-sesquiphellandrene, zingiberene, and α-curcumene (**24**, **25** and **26** in Fig. 11.1). The individual components were not attractive. Determination of the chiralities and syntheses of the sesquiterpenes are in progress. The ester compound, *E*2,*Z*4,*Z*6–10:COOMe, is an isomer of the pheromone of *P. stali* (Sugie *et al.*, 1996).

Male-produced pheromone components attractive to both sexes have also been identified from aerations of *Piezodorus hybneri* (Gmelin) (Fig. 11.7) using GC–EAD (Leal *et al.*, 1998). The bioactive components also elicited diagnostic behaviours, consisting of wing-fanning by females and exposure of the pygophor by males. The pheromone blend consisted of methyl *Z*8-hexadecenoate (*Z*8–16:COOMe), β-sesquiphellandrene (suggested to be the (−)-enantiomer), and (*R*)-(−)-15-hexadecanolide (**24**, **27** and **28** in Fig. 11.1) in a ratio of 1:10:4. All three compounds elicited male display, but only the first two were attractive alone. Optimal attraction was obtained with the blend of three components. Interestingly, β-sesquiphellandrene failed to elicit antennal responses in GC–EAD assays. Field tests have not yet been carried out.

The sex pheromone of a second *Piezodorus* species, *P. guildinii*, caught serendipitously in moderate numbers during field tests of *E. heros* pheromone (2,6,10–13:COOMe), is under investigation (Borges *et al.*, 1998b).

Male spined citrus bugs, *Biprorulus bibax* Breddin (Fig. 11.8), produce aggregation pheromone in enlarged sexually dimorphic DAGs, as occurs in some predacious pentatomids (Aldrich, 1995; Chapter 14, this volume), instead of from secretory cells in the ventral abdomen. Male DAGs contained a hemiacetal, (3*R*,4*S*,1′*E*)-3,4-bis(1′-butenyl)tetrahydro-2-furanol (**29**, Fig. 11.1) and linalool, with lesser amounts of two farnesol isomers and nerolidol (Oliver *et al.*, 1992; Mori *et al.*, 1993; James *et al.*, 1994). The much smaller female DAGs contained only linalool. In a field cage, females were attracted to the racemic synthetic hemiacetal

Fig. 11.7. The stink bug, *Piezodorus hybneri*. (Courtesy of Walter Soares Leal.)

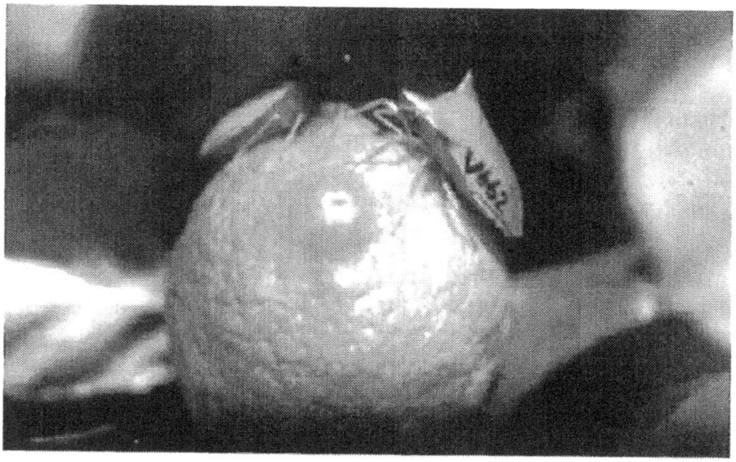

Fig. 11.8. The spined citrus bug, *Biprorulus bibax*. (Courtesy of David James.)

alone, and to a blend (using racemic components) roughly mimicking the male DAG extract (James *et al.*, 1994). Diapausing females were unresponsive. In field trials, bugs would not enter traps, but citrus trees containing traps and adjacent trees became heavily infested with bugs of both sexes and all reproductive stages. Further tests determined that *E*2-hexenal (**30**, Fig. 11.1), a common bug chemical, was also attractive to reproductively active bugs.

Applications

Pheromone traps for monitoring: practical considerations

Pheromone-baited traps would be of tremendous value in providing rapid, sensitive and selective methods for monitoring adult bug populations. However, several practical factors need to be considered before and during development of applications.

Synthesis of pheromones

Bug pheromones vary widely in structural complexity, from simple readily available compounds (e.g. alcohols and esters in mirid pheromones) to complex compounds with several possible stereoisomers (e.g. *B. bibax*, *T. pallidovirens*, *N. viridula*) which may be difficult to produce in bulk. For the latter species, limited or no data are available on the biological activity of the unnatural stereoisomers, and it is unclear whether stereochemically pure materials will be required.

Pheromone dispensers and formulations

A wide variety of dispensers has been used for trap baits, including pheromone-impregnated rubber septa (e.g. Aldrich *et al.*, 1991; Millar *et al.*, 1997) or plastic pellets (Leal *et al.*, 1995, 1996), polymer slices (Aldrich *et al.*, 1997; McBrien *et al.*,

1997), capillary tubes (Smith *et al.*, 1991), vials with small holes in the lids (Smith *et al.*, 1991; James *et al.*, 1996), bubble caps or plastic centrifuge tubes (McBrien *et al.*, 1994b), and cotton ropes (Sugie *et al.*, 1996). Because some pheromone components are highly volatile, rubber septum dispensers may have short field lifetimes (Millar *et al.*, 1997), while plastic tubes and polymer or wax formulations may last longer (McBrien *et al.*, 1994b; Rice, 1997, 1998). A more complex problem is the large difference in volatility between pheromone components. When formulated in a single dispenser, the more volatile compounds are released quickly, causing changes in blend ratios and short lure lifetimes. The only solution seems to be dispensers with two or more parts, each optimized for a different component(s).

Optimum doses and release rates remain largely untested. The three mirid bug species for which pheromones are known all respond to large release rates (several milligrams per day: McBrien *et al.*, 1994b; Millar *et al.*, 1997; Millar and Rice, 1998), but methodical dosage trials remain to be done with most other bug species.

Effective range of pheromone lures

Limited information is available on the distance over which bug pheromone lures act. For some mirids (*C. verbasci* and *P. relativus*), large trap catches suggest that these bugs respond over at least several metres. However, highly variable catches of other mirids (e.g. *C. norvegicus*) that may be in clumped distributions centred on patches of host plants or other patchy resources suggest that lure ranges may be quite short, with traps close to clumps catching many bugs, and traps more distant from clumps catching few or no bugs. Trap catches of phytophagous pentatomids have generally been low, suggesting that baits have a short range, but this may be a result of other factors. In contrast, adult *R. clavatus* were attracted to traps on the roof of a three-storey building (Numata *et al.*, 1990).

Trap design: small bugs

Mirids are small enough to be captured effectively on trap surfaces coated with sticky materials (e.g. Stickem® and Tanglefoot®), such as commercial or homemade sticky traps (Hedin *et al.*, 1985; Graham, 1987; Blommers *et al.*, 1988; Smith *et al.*, 1994; McBrien *et al.*, 1994b, 1996; Millar *et al.*, 1997; Millar and Rice, 1998; Rice, 1998). Penetration of light into the trap interior does not seem to be critical, and mirids enter traps hung from branches or mounted on stakes, indicating that the bugs fly into traps. The size of the trap opening may influence the shape of the pheromone plume and the entry of insects into the trap (Lewis and Macaulay, 1976; McBrien *et al.*, 1994b, 1996). In contrast, use of smaller, more enclosed Jackson traps rather than Pherocon 1C wing traps greatly reduced the numbers of female *C. norvegicus* and *N. longirostris* randomly trapped, which had complicated testing of pheromone formulations (Rice, 1997, 1998).

A few studies have reported the use of traps other than sticky traps. Kakizaki and Sugie (1997) used water-pan traps to catch rice leaf bugs, *T. caelestialium*, while King (1973) used Fluon®-coated screen traps with the cocoa capsid, *D. theobroma*. Female cocoa mirids, *H. clavifer*, in mesh cages attracted males to trees containing cages, rather than directly to the cages (Smith, 1977), indicating attraction to the vicinity of a pheromone source rather than right up to a source. This type of behaviour has been reported for pentatomids (see below), but not for other mirids.

The importance of trap height and placement was shown with *C. verbasci* trapped in apple orchards: traps placed higher in the canopy captured more bugs than traps at lower elevations (Smith, 1991; McBrien *et al.*, 1996).

Trap design: large bugs

Sticky traps were used to capture the alydid, *R. clavatus* (Mizutani *et al.*, 1997), the coreid, *L. chinensis* (Leal *et al.*, 1996), and the lygaeid, *Tropidothorax cruciger* (Aldrich *et al.*, 1997). However, sticky traps may be ineffective for large bugs such as pentatomids if they are strong enough to pull themselves free. Furthermore, pentatomids may not enter traps readily, particularly those whose inside areas are dark or enclosed (Mizell and Tedders, 1995; James *et al.*, 1996). Pentatomid females also appear to walk rather than fly towards a pheromone source and so traps must be placed accordingly – for example, in contact with foliage or in tree crotches. Alternatively, it has been suggested that tall traps placed on the ground will intercept bugs flying close to the ground (Mizell and Tedders, 1995). Another critical factor to consider is the complex reproductive behaviour of pentatomids, and the 'strength' and range of their pheromones is unclear. Females orient to the vicinity of males producing pheromones, but then become quiescent, while the male initiates a complex courtship involving visual, tactile and acoustic cues (e.g. Borges *et al.*, 1987; Wang and Millar, 1997). Trap design and trapping strategies must take all these factors into account (Wilson *et al.*, 1988; Wilson, 1990; Zalom, 1992).

Aldrich *et al.* (1984) developed a trap consisting of a clear plastic container with two inverted funnels for insect entry ('jug' trap). This trap has been used in field trials for *N. viridula* (Aldrich *et al.*, 1987), several *Euschistus* spp. (Aldrich *et al.*, 1991, 1995) and the lygaeid, *N. bicrucis* (Aldrich *et al.*, 1997). In field trials with *Euschistus*, a significant proportion (28–61%) of the bugs attracted were actually collected from the outside of the trap or within 1 m of the trap (Aldrich *et al.*, 1991), indicating that some bugs were attracted to the vicinity of a pheromone source rather than right up to the source.

Pheromone-baited water-pan traps were used for *P. stali* (Sugie *et al.*, 1996) and the alydid, *R. clavatus* (Numata *et al.*,1990), but only about 50% of the bugs attracted were actually caught, with the rest perched on or near the trap. Water-pan traps were also used with the coreid, *L. australis* (Yasuda and Tsurumachi, 1995).

The pyramidal Tedders trap (0.6–1.2 m high) may solve some of the problems with trapping pentatomids (Mizell and Tedders, 1995; Mizell *et al.*, 1997; Yonce and Mizell, 1997). The base of the trap is painted yellow, which is attractive to the bugs, and the trap's shape directs the bugs into the screened collection cage on the top. Tedders traps baited with *Euschistus* pheromone captured 3600 hemipterans (93% *Euschistus* spp.) in a pecan orchard over 5 months (Yonce and Mizell, 1997).

Other types of traps have been used for large bugs with varying degrees of success, including the Lindgren (1983) multiple funnel trap, which was effective with the coreid, *L. occidentalis* (Blatt and Borden, 1996).

Seasonality

The biological context in which a pheromone is used must be considered when developing applications. Growers want sampling systems when their crops are most susceptible, but this may not coincide with times when a bug species is using pheromones. Knowledge of bug life cycles is critically important when developing applications. For example, bugs that overwinter as adults enter reproductive diapause, during which they do not produce or respond to pheromone (Strong *et al.*, 1970; Brennan *et al.*, 1977; Aldrich *et al.*, 1991; James *et al.*, 1994, 1996). The response of bugs to aggregation pheromones may also be seasonal. The coreid *L. occidentalis* responded to aggregation pheromone only in autumn, when the bugs normally form large overwintering aggregations (Blatt and Borden, 1996). Interestingly, they did not respond to alarm pheromone at this time, presumably to prevent disruption of the aggregations (Blatt *et al.*, 1998).

Monitoring and determination of economic thresholds

Monitoring bug populations with pheromone traps has three main objectives. The first is to track the phenology of species which chronically infest a particular crop (e.g. *Phytocoris* spp. in pistachios: Millar *et al.*, 1997; Rice 1997, 1998; Millar and Rice, 1998). The second objective is to detect the unpredictable immigration of bugs into crops in response to conditions in the surrounding habitat (Toscano and Stern, 1976; Slaymaker and Tugwell, 1984). To date this objective has not been achieved for any of the polyphagous bug species, primarily due to problems with trap design, uncertainties as to the exact composition of pheromone blends, and incomplete understanding of the insects' biology.

The third objective is to relate trap catches of bugs to population levels and to economic damage thresholds. This objective is closest to achievement for *C. verbasci* on apple. McBrien *et al.* (1994a) demonstrated a correlation between catches of males in the autumn and the density of nymphs the following spring in conventional apple orchards. However, organic orchards did not show this relationship, for reasons which were unclear. Although no *Lygus* sex pheromones have been identified, Slaymaker and Tugwell (1984) described an inexpensive female-baited trap for monitoring *Lygus* bugs. Adult female *L. lineolaris* remained attractive for 2 weeks. Similar results may be obtained with other species.

Control: mating disruption

In addition to the normal technical problems associated with pheromone-based mating disruption (e.g. stable long-term release rates, point sources per hectare, pheromone stability, etc.), mating disruption of bugs must take into account that many species are highly mobile and polyphagous. Consequently, treatment of a crop with pheromones for mating disruption may have limited impact on polyphagous species which are probably not in the crop for most of the treatment period, or which enter the crop after mating. Mating disruption may be feasible only for those species whose populations are predictable in time and space, and which are closely tied to particular host plants.

With *C. verbasci* in apple orchards, the two-component pheromone blend demonstrated potential for mating disruption, while the individual components did not (McBrien *et al.*, 1996). Overwintering populations of *C. verbasci* were suppressed by treating orchards in the autumn, and first-generation nymphs were reduced by 71–81% the following spring (McBrien *et al.*, 1997). Experiments in orchards treated with the pheromone blend or each of the single components probed mating disruption mechanisms (Judd *et al.*, 1995). Exposure of male *C. verbasci* to the two-component blend or each compound alone caused males to respond to skewed blends. Individual components appeared to disrupt communication by sensory imbalance and modified interpretation of blends, while disruption by the complete blend apparently involved several mechanisms, including false trail following and camouflage (Bartell, 1982; Cardé, 1990).

Conclusions

In this chapter we have seen that the development of bug pheromones for use in IPM has lagged far behind the applications developed for other insect orders. Pheromones of a number of bug species have been identified, and their potential has been demonstrated for several species, with high trap catches in field trials. However, for other species, particularly the pentatomids, difficult and unexpected problems have surfaced. Nevertheless, the economic losses caused by direct and indirect bug damage warrant redoubled efforts to circumvent these problems. Particular attention must be directed towards developing a better understanding of the role of pheromones in bug biology. For example, it is clear that female *Lygus* bugs use a sex attractant pheromone, but no research group has been able to prepare consistently active extracts. Detailed re-examination of the insects' mating behaviour may allow us to determine exactly when and under what conditions the pheromone is produced, and how to collect it for analysis. For other bugs, such as stink bugs, field studies must be done to examine the details of their pheromone-mediated behaviour. We need to determine whether stink bug pheromones are inherently short-range or weak attractants, or whether poorly understood aspects of the bugs' biologies are the cause of disappointing field results to date.

It has also become clear that trap design is critically important for larger bugs. Some progress has been made, but further advances are a necessity for the development of robust and sensitive trapping systems. In fact, low trap catches have even hindered determination of correct blend ratios and doses. Some blend optimization can be done in laboratory bioassays, but both dose and blend experiments require field corroboration.

Finally, we must make every effort to define bug pheromone blends clearly. In the past, the term 'pheromone' has often been applied loosely, and frequently incorrectly, to compounds isolated from bugs. This has created confusion in the literature and hindered the development of schemes showing similarities and differences between the blends produced by related species, and the mechanisms by which reproductive isolation may be maintained by sympatric congeners. Such schemes would be helpful in identifying pheromone blends for other species, as well as providing valuable data to corroborate taxonomic relationships.

References

Aldrich, J.R. (1988a) Chemistry and biological activity of pentatomid sex pheromones. In: Cutler, H.G. (ed.) *Biologically Active Natural Products: Potential Use in Agriculture*. ACS Symposium Series No. 380, Washington, DC, pp. 417–431.

Aldrich, J.R. (1988b) Chemical ecology of the Heteroptera. *Annual Review of Entomology* 33, 211–238.

Aldrich, J.R. (1995) Chemical communication in the true bugs and parasitoid exploitation. In: Cardé, R.T. and Bell, W.J. (eds) *Chemical Ecology of Insects II*. Chapman & Hall, New York, pp. 318–363.

Aldrich, J.R. (1996) Sex pheromones in Homoptera and Heteroptera. In: Schaefer, C.W. (ed.) *Studies on Hemipteran Phylogeny*. Entomological Society of America, Lanham, Maryland, pp. 199–233.

Aldrich, J.R., Kochansky, J.P. and Abrams, C.B. (1984) Attractant for a beneficial insect and its parasitoids: pheromone of the predatory spined soldier bug, *Podisus maculiventris* (Hemiptera: Pentatomidae). *Environmental Entomology* 13, 1031–1036.

Aldrich, J.R., Oliver, J.E., Lusby, W.R., Kochansky, J.P. and Lockwood, J.A. (1987) Pheromone strains of the cosmopolitan pest, *Nezara viridula* (Heteroptera: Pentatomidae). *Journal of Experimental Zoology* 244, 171–175.

Aldrich, J.R., Lusby, W.R., Kochansky, J.P., Hoffmann, M.P., Wilson, L.T. and Zalom, F.G. (1988) *Lygus* bug pheromones vis-à-vis stink bugs. In: Brown, J.M. and Richter, D.A. (eds) *Proceedings of the Beltwide Cotton Production Conferences*. National Cotton Council of America, Memphis, Tennessee, pp. 213–216.

Aldrich, J.R., Lusby, W.R., Marron, B.E., Nicolaou, K.C., Hoffmann, M.P. and Wilson, L.T. (1989) Pheromone blends of green stink bugs and possible parasitoid selection. *Naturwissenschaften* 76, 173–175.

Aldrich, J.R., Hoffmann, M.P., Kochansky, J.P., Lusby, W.R., Eger, J.E. and Payne, J.A. (1991) Identification and attractiveness of a major pheromone component for Nearctic *Euschistus* spp. stink bugs (Heteroptera: Pentatomidae). *Environmental Entomology* 20, 477–483.

Aldrich, J.R., Numata, H., Borges, M., Bin, F., Waite, G.K. and Lusby, W.R. (1993a) Artifacts and pheromone blends from *Nezara* spp. and other stink bugs (Heteroptera: Pentatomidae). *Zeitschrift für Naturforschung* 48C, 73–79.

Aldrich, J.R., Waite, G.K., Moore, C., Payne, J.A., Lusby, W.R. and Kochansky, J.P. (1993b) Male-specific volatiles from Nearctic and Australasian true bugs (Heteroptera: Coreidae and Alydidae). *Journal of Chemical Ecology* 19, 2767–2781.

Aldrich, J.R., Oliver, J.E., Lusby, W.R., Kochansky, J.P. and Borges, M. (1994) Identification of male-specific volatiles from Nearctic and Neotropical stink bugs (Heteroptera: Pentatomidae). *Journal of Chemical Ecology* 20, 1103–1111.

Aldrich, J.R., Rosi, M.C. and Bin, F. (1995) Behavioral correlates for minor volatile compounds from stink bugs (Heteroptera: Pentatomidae). *Journal of Chemical Ecology* 21, 1907–1920.

Aldrich, J.R., Leal, W.S., Nishida, R., Khrimian, A.P., Lee, C.-J. and Sakuratani, Y. (1997) Semiochemistry of aposematic seed bugs. *Entomologia Experimentalis et Applicata* 84, 127–135.

Arnett, R.H. Jr (1993) *American Insects. A Handbook of the Insects of North America North of Mexico*. The Sandhill Crane Press, Gainesville, Florida, 850 pp.

Baker, R., Borges, M., Cooke, N.G. and Herbert, R.H. (1987) Identification and synthesis of (Z)-(1′S,3′R,4′S)(−)-2-(3′,4′-epoxy-4′-methylcyclohexyl-6-methylhepta-2,5-diene, the sex pheromone of the southern green stink bug, *Nezara viridula* (L.). *Journal of the Chemical Society, Chemical Communications*, 414–416.

Baptistella, L.H.B., and Aleixo, A.M. (1994) Stereocontrolled synthesis of the sex pheromone of *Nezara viridula* (L.). *Liebigs Annalen der Chemie*, 785–789.

Bartell, R.J. (1982) Mechanisms of communication disruption by pheromone in the control of Lepidoptera: a review. *Physiological Entomology* 7, 353–364.

Beek, T.A. van, Drijfhout, F.P. and Boezem, J.J.N. van den (1998) On-line collection and GC analysis of pheromones from intact insects. *Abstracts of the 2nd International Symposium on Insect Pheromones*, Wageningen, the Netherlands, March 30–April 3, 1998, pp. 15–17.

Beers, E.H. (1992) *Campylomma* still a challenge for Northwest growers. *Good Fruit Grower* 43, 32–35.

Blatt, S.E. (1994) An unusually large aggregation of the western conifer seed bug, *Leptoglossus occidentalis* (Hemiptera: Coreidae), in a man-made structure. *Journal of the Entomological Society of British Columbia* 91, 71–72.

Blatt, S.E., and Borden, J.H. (1996) Evidence for a male-produced aggregation pheromone in the western conifer seed bug, *Leptoglossus occidentalis* Heidemann (Hemiptera: Coreidae). *Canadian Entomologist* 128, 777–778.

Blatt, S.E., Borden, J.H., Pierce, H.D. Jr, Gries, R. and Gries, G. (1998) Alarm pheromone system of the western conifer seed bug, *Leptoglossus occidentalis*. *Journal of Chemical Ecology* 24, 1013–1031.

Blommers, L., Bus, V., Jongh, E. de and Lentjes, G. (1988) Attraction of males by virgin females of the green capsid bug *Lygocoris pabulinus* (Heteroptera: Miridae). *Entomologische Berichten, Amsterdam* 48, 175–179.

Blommers, L.H.M., Vaal, F.W.N.M. and Helsen, H.H.M. (1997) Life history, seasonal adaptations and monitoring of common green capsid *Lygocoris pabulinus* (L.) (Hem., Miridae). *Journal of Applied Entomology* 121, 389–398.

Blumenthal, M.A. (1978) The metathoracic gland system of *Lygus lineolaris* (Heteroptera: Miridae). MSc thesis, Cornell University, Ithaca, New York.

Boivin, G. and Stewart, R.K. (1982) Attraction of male green apple bugs, *Lygocoris communis* (Hemiptera: Miridae) to caged females. *Canadian Entomologist* 114, 765–766.

Borges, M. (1995) Attractant compounds of the southern green stink bug, *Nezara viridula* (L.) (Heteroptera: Pentatomidae). *Anais da Sociedade Entomologica do Brasil* 24, 215–225.

Borges, M. and Aldrich, J.R. (1994) Attractant pheromone for nearctic stink bug, *Euschistus obscurus* (Heteroptera: Pentatomidae): insight into a neotropical relative. *Journal of Chemical Ecology* 20, 1095–1102.

Borges, M., Jepson, P.C. and Howse, P.E. (1987) Long-range mate location and close-range courtship behaviour of the green stink bug, *Nezara viridula* and its mediation by sex pheromones. *Entomologia Experimentalis et Applicata* 44, 205–212.

Borges, M., Mori, K., Costa, M.L.M. and Sujii, E.R. (1998a) Behavioural evidence of methyl 2,6,10-trimethyltridecanoate as a sex pheromone of *Euschistus heros* (Het., Pentatomidae). *Journal of Applied Entomology* 122, 335–338.

Borges, M., Schmidt, F.G.V., Sujii, E.R., Medeiros, M.A., Mori, K., Zarbin, P.H.G. and Ferreira, J.T.B. (1998b) Field responses of stink bugs to the natural and synthetic pheromone of the Neotropical brown stink bug, *Euschistus heros* (Heteroptera: Pentatomidae). *Physiological Entomology* 23, 202–207.

Brennan, B.M., Chang, F. and Mitchell, W.C. (1977) Physiological effects on sex pheromone communication in the southern green stink bug, *Nezara viridula*. *Environmental Entomology* 6, 169–173.

Brézot, P., Malosse, C. and Renou, M. (1993) Study of the attractivity of the male pheromone in *Nezara viridula* L. (Heteroptera: Pentatomidae). *Comptes Rendus de l'Académie des Sciences Série III – Sciences de la Vie* 316, 671–675.

Brézot, P., Malosse, C., Mori, K. and Renou, M. (1994) Bisabolene epoxides in sex pheromone in *Nezara viridula* (L.) (Heteroptera: Pentatomidae): role of *cis* isomer and relation to specificity of pheromone. *Journal of Chemical Ecology* 20, 3133–3147.

Cardé, R.T. (1990) Principles of mating disruption. In: Ridgway, R.L., Silverstein, R.M. and Inscoe, M.N. (eds) *Behavior-Modifying Chemicals for Insect Management: Applications of Pheromones and Other Attractants*. Marcel Dekker, New York, pp. 47–72.

Chinta, S., Dickens, J.C. and Aldrich, J.R. (1994) Olfactory reception of potential pheromones and plant odors by tarnished plant bug, *Lygus lineolaris* (Hemiptera: Miridae). *Journal of Chemical Ecology* 20, 3251–3267.

Dickens, J.C., Callahan, F.E., Wergin, W.P. and Erbe, E.E. (1995) Olfaction in a hemimetabolous insect: antennal-specific protein in adult *Lygus lineolaris* (Heteroptera: Miridae). *Journal of Insect Physiology* 41, 857–867.

Evans, C.E., Staddon, B.W. and Games, D.E. (1990) Analysis of gland secretions of Pentatomoidea (Heteroptera) by gas chromatography-mass spectrometry techniques. In: McCaffery, A.R. and Wilson, I.D. (eds) *Chromatography and Isolation of Insect Hormones and Pheromones*. Plenum Press, New York, pp. 321–328.

Ferreira, J.T.B. and Zarbin, P.H.G. (1996) Pheromone syntheses: a tropical approach. Enantioselective synthesis of the (2*R*,6*S*,10*S*) and (2*S*,6*S*,10*S*) isomers of methyl 2,6,10-trimethyldodecanoate. *Bioorganic and Medicinal Chemistry* 4, 381–388.

Graham, H.M. (1987) Attraction of *Lygus* spp. males by conspecific and congeneric females. *The Southwestern Entomologist* 12, 147–155.

Graham, H.M. (1988) Sexual attraction of *Lygus hesperus* Knight. *Southwestern Entomologist* 13, 31–37.

Groot, A.T., Visser, J.H., Geerts, R., Drijfhout, F.P. and Beek, T.A. van (1998a) Behavioural response of *Lygocoris pabulinus* (Heteroptera: Miridae) to potential sex pheromone compounds. *Proceedings of Experimental and Applied Entomology* 9, NEV, Amsterdam, pp. 191–195.

Groot, A.T., Wal, E. van der, Schuurman, A., Visser, J.H., Blommers, L.H.M. and Beek, T.A. van (1998b) Copulation behaviour of *Lygocoris pabulinus* (L.) under laboratory conditions. *Entomologia Experimentalis et Applicata* 88, 219–228.

Gueldner, R.C. and Parrott, W.L. (1978) Volatile constituents of the tarnished plant bug. *Insect Biochemistry* 8, 389–391.

Harris, V.E. and Todd, J.W. (1980) Male-mediated aggregation of male, female and 5th-instar southern green stink bugs and concomitant attraction of a tachinid parasite, *Trichopoda pennipes*. *Entomologia Experimentalis et Applicata* 27, 117–126.

Hedin, P.A., Parrott, W.L., Tedders, W.L. and Reed, D.K (1985) Field responses of the tarnished plant bug to its own volatile constituents. *Journal of the Mississippi Academy of Sciences*, 30, 63–66.

Hoffmann, M.P., Wilson, L.T. and Zalom, F.G. (1987) Control of stink bugs in tomatoes. *California Agriculture* (May–June), 4–6.

James, D.G., Mori, K., Aldrich, J.R. and Oliver, J.E. (1994) Flight-mediated attraction of *Biprorulus bibax* Breddin (Hemiptera: Pentatomidae) to natural and synthetic aggregation pheromone. *Journal of Chemical Ecology* 20, 71–80.

James, D.G., Heffer, R. and Amaike, M. (1996) Field attraction of *Biprorulus bibax* Breddin (Hemiptera: Pentatomidae) to synthetic aggregation pheromone and (*E*)-2-hexenal, a pentatomid defense chemical. *Journal of Chemical Ecology* 22, 1697–1708.

Judd, G.J.R., McBrien, H.L. and Borden, J.H. (1995) Modification of responses by *Campylomma verbasci* (Heteroptera: Miridae) to pheromone blends in atmospheres permeated with synthetic sex pheromone or individual components. *Journal of Chemical Ecology* 21, 1991–2002.

Kakizaki, M. and Sugie, H. (1997) Attraction of males to females in the rice leaf bug, *Trigonotylus caelestialium* (Kirkaldy) (Heteroptera: Miridae). *Applied Entomology and Zoology* 32, 648–651.

King, A.B.S. (1973) Studies of sex attraction in the cocoa capsid, *Distantiella theobroma* (Heteroptera: Miridae). *Entomologia Experimentalis et Applicata* 16, 243–254.

Kuwahara, S., Itoh, D., Leal, W.S. and Kodama, O. (1998) A convenient synthesis of (2*S*,3*R*,6*S*,7*Z*)- and (2*R*,3*S*,6*S*,7*Z*)-2,3-epoxy-7,10-bisaboladiene, the sex pheromones of the southern green stink bug (*Nezara viridula*). *Tetrahedron* 54, 11421–11430.

Layton, M.B., Williams, M.R. and Stewart, S. (1997) BT-cotton in Mississippi: the first year. *Proceedings of the 1997 Beltwide Cotton Conferences*. National Cotton Council of America, Memphis, Tennessee, pp. 861–863.

Leal, W.S. and Kadosawa, T. (1992) (*E*)-2-Hexenyl hexanoate, the alarm pheromone of the bean bug *Riptortus clavatus* (Heteroptera: Alydidae). *Bioscience, Biotechnology and Biochemistry* 56, 1004–1005.

Leal, W.S., Higuchi, H., Mizutani, N., Nakamori, H., Kadosawa, T. and Ono, M. (1995) Multifunctional communication in *Riptortus clavatus* (Heteroptera: Alydidae): conspecific nymphs and egg parasitoid *Ooencyrtus nezarae* use the same adult attractant pheromone as chemical cue. *Journal of Chemical Ecology* 21, 973–985.

Leal, W.S., Ueda, Y. and Ono, M. (1996) Attractant pheromone for male rice bug, *Leptocorisa chinensis*: semiochemicals produced by both male and female. *Journal of Chemical Ecology* 22, 1429–1437.

Leal, W.S., Kuwahara, S., Shi, X., Higuchi, H., Marino, C.E.B., Ono, M. and Meinwald, J. (1998) Male-released sex pheromone of the stink bug *Piezodorus hybneri*. *Journal of Chemical Ecology* 24, 1817–1829.

Lewis, T. and Macaulay, E.D.M. (1976) Design and elevation of sex attractant traps for the pea moth, *Cydia nigricana* (Steph.), and effect of plume shape on catches. *Ecological Entomology* 1, 175–187.

Lindgren, B.S. (1983) A multiple funnel trap for scolytid beetles (Coleoptera). *Canadian Entomologist* 115, 299–302.

Lucchi, A. (1994) The scent gland system of *Nezara viridula* (L.) (Heteroptera Pentatomidae). III. Male ventral abdominal glands. *Redia* 77, 1–10.

Marron, B.E. and Nicolau, K.C. (1989) Stereocontrolled synthesis of the sex pheromones of the green stink bug. *Synthesis*, 537–539.

McBrien, H.L., Judd, G.J.R. and Borden, J.H. (1994a) *Campylomma verbasci* (Heteroptera: Miridae): pheromone-based seasonal flight patterns and prediction of nymphal densities in apple orchards. *Journal of Economic Entomology* 87, 1224–1229.

McBrien, H.L., Judd, G.J.R., Borden, J.H. and Smith, R.F. (1994b) Development of sex pheromone-baited traps for monitoring *Campylomma verbasci* (Heteroptera: Miridae). *Environmental Entomology* 23, 442–446.

McBrien, H.L., Judd, G.J.R. and Borden, J.H. (1996) Potential for pheromone-based mating disruption of the mullein bug, *Campylomma verbasci* (Meyer) (Heteroptera: Miridae). *Canadian Entomologist* 128, 1057–1064.

McBrien, H.L., Judd, G.J.R. and Borden, J.H. (1997) Population suppression of *Campylomma verbasci* (Heteroptera: Miridae) by atmospheric permeation with synthetic sex pheromone. *Journal of Economic Entomology* 90, 801–808.

Millar, J.G. (1997) Methyl (2E,4Z,6Z)-deca-2,4,6-trienoate, a thermally unstable, sex-specific compound from the stink bug *Thyanta pallidovirens*. *Tetrahedron Letters* 38, 7971–7972.

Millar, J.G. and Rice, R.E. (1998) Sex pheromone of the plant bug *Phytocoris californicus* (Heteroptera: Miridae). *Journal of Economic Entomology* 91, 132–137.

Millar, J.G. and Wang, Q. (1994) Pheromones of stinkbugs and leaffooted bugs infesting pistachios. In: *California Pistachio Industry Annual Report, Crop Year 1993–1994*. California Pistachio Commission, Fresno, California, pp. 83–86.

Millar, J.G., Rice, R.E. and Wang, Q. (1997) Sex pheromone of the mirid bug *Phytocoris relativus*. *Journal of Chemical Ecology* 23, 1743–1754.

Millar, J.G., McBrien, H.L. and Gottlieb, L. (1998) Identification of pheromones of bug species infesting pistachios. In: *California Pistachio Industry Annual Report – Crop Year 1997–1998*. California Pistachio Commission, Fresno, California, pp. 115–119.

Mitchell, W.C. and Mau, R.F.L. (1971) Response of the female southern green stink bug and its parasite, *Trichopoda pennipes*, to male stink bug pheromones. *Journal of Economic Entomology* 64, 856–859.

Mizell, R.F. and Tedders, W.L. (1995) A new monitoring method for detection of the stinkbug complex in pecan orchards. *Proceedings of the Southeastern Pecan Growers Association* 95, 36–40.

Mizell, R.F., Tedders, W.L., Yonce, C.E. and Aldrich, J.R. (1997) Stink bug monitoring – an update. *Proceedings of the Southeastern Pecan Growers Association* 97, 50–52.

Mizutani, N., Wada, T., Higuchi, H., Ono, M. and Leal, W.S. (1997) A component of a synthetic aggregation pheromone of *Riptortus clavatus* (Thunberg) (Heteroptera: Alydidae), that attracts an egg parasitoid, *Ooencyrtus nezarae* Ishii (Hymenoptera: Encyrtidae). *Applied Entomology and Zoology* 32, 504–507.

Mori, K. and Murata, N. (1994a) Pheromone synthesis. 162. Synthesis of methyl 2,6,10-trimethyltridecanoate, the male-produced pheromone of the stink bugs *Euschistus heros* and *E. obscurus*, as a stereoisomeric mixture. *Liebigs Annalen der Chemie*, 637–639.

Mori, K. and Murata, N. (1994b) Synthesis of all of the eight stereoisomers of methyl 2,6,10-trimethyltridecanoate, the male-produced pheromone of the stink bugs, *Euschistus heros* and *E. obscurus*. *Liebigs Annalen der Chemie*, 1153–1160.

Mori, K., Amaike, M. and Watanabe, H. (1993) New synthesis and revision of the absolute configuration of the hemiacetal pheromone of the spined citrus bug *Biprorulus bibax*. *Liebigs Annalen der Chemie*, 1287–1294.

Moriya, S. and Shiga, M. (1984) Attraction of the male brown-winged green bug, *Plautia stali* Scott (Heteroptera: Pentatomidae) for males and females of the same species. *Applied Entomology and Zoology* 19, 317–322.

Numata, H., Kon, M. and Hidaka, T. (1990) Male adults attract conspecific adults in the bean bug, *Riptortus clavatus* Thunberg (Heteroptera: Alydidae). *Applied Entomology and Zoology* 25, 144–145.

Oliver, J.E., Aldrich, J.R., Lusby, W.R., Waters, R.M. and James, D.G. (1992) A male-produced pheromone of the spined citrus bug. *Tetrahedron Letters* 33, 891–894.

Panizzi, A.R. (1997) Wild hosts of pentatomids: ecological significance and role in their pest status on crops. *Annual Review of Entomology* 42, 99–122.

Pavis, C. and Malosse, C. (1986) A sex pheromone produced by mature males in the southern green stink bug, *Nezara viridula* (L.) (Heteroptera: Pentatomidae). *Comptes Rendus de l'Académie des Sciences Série III – Sciences de la Vie* 303, 273–276.

Rice, R.E. (1997) Field bioassays of synthetic sex pheromones for sucking bugs (Hemiptera). In: *California Pistachio Industry Annual Report, Crop Year 1996–1997*. California Pistachio Commission, Fresno, California, pp. 95–100.

Rice, R.E. (1998) Field bioassays of synthetic sex pheromones for sucking bugs (Hemiptera). In: *California Pistachio Industry Annual Report, Crop Year 1997–1998*. California Pistachio Commission, Fresno, California, pp. 104–111.

Rice, R.E. and Jones, R. (1989) Biology and control of insects and mites on pistachios. In: *California Pistachio Industry Annual Report, Crop Year 1988–1989*. California Pistachio Commission, Fresno, California, pp. 65–68.

Rice, R.E., Bentley, W.J. and Beede, R.H. (1988) Insect and mite pests of pistachios in California. *University of California, Division of Agriculture and Natural Resources Publication 21452*. Oakland, California.

Ryan, M.A., Moore, C.J. and Walter, G.H. (1995) Individual variation in pheromone composition in *Nezara viridula* (Heteroptera: Pentatomidae): how valid is the basis for designating 'pheromone strains'? *Comparative Biochemistry and Physiology B* 111, 189–193.

Scales, A.L. (1968) Female tarnished plant bugs attract males. *Journal of Economic Entomology* 61, 1466–1467.

Slaymaker, P.H. and Tugwell, N.P. (1984) Inexpensive female-baited trap for the tarnished plant bug (Hemiptera: Miridae). *Journal of Economic Entomology* 77, 1062–1063.

Smith, E.S.C. (1977) Presence of a sex attractant pheromone in *Helopeltis clavifer* (Walker) (Heteroptera: Miridae). *Journal of the Australian Entomological Society* 16, 113–116.

Smith, R.F. (1991) The mullein bug, *Campylomma verbasci*. In: Williams, K. (ed.) *New Directions in Tree Fruit Pest Management*. Good Fruit Grower, Washington State Tree Fruit Commission, Yakima, Washington, pp. 199–214.

Smith, R.F., Pierce, H.D. Jr and Borden, J.H. (1991) Sex pheromone of the mullein bug, *Campylomma verbasci* (Meyer) (Heteroptera: Miridae). *Journal of Chemical Ecology* 17, 1437–1447.

Smith, R.F., Gaul, S.O., Borden, J.H. and Pierce, H.D. Jr (1994) Evidence for a sex pheromone in the apple brown bug, *Atractotomus mali* (Meyer) (Heteroptera: Miridae). *Canadian Entomologist* 126, 445–446.

Staddon, B.W., Abdollahi, A., Parry, J., Rossiter, M. and Knight, D.W. (1994) Major component in male sex pheromone of cereal pest *Eurygaster integriceps* Puton (Heteroptera: Scutelleridae) identified as a homosesquiterpenoid. *Journal of Chemical Ecology* 20, 2721–2731.

Stigter, H. (1996) *Campylomma verbasci*, a new pest on apple in the Netherlands (Heteroptera: Miridae). In: Polesny, F., Müller, W. and Olszak, R.W. (eds) *Proceedings of an International Conference on Integrated Fruit Production, 28 August–2 September 1995, Cedzyna, Poland. IOBC/WPRS Bulletin* 19(4), 140–144.

Strong, F.E., Sheldahl, J.A., Hughes, P.R. and Hussein, E.M.K. (1970) Reproductive biology of *Lygus hesperus* Knight. *Hilgardia* 40, 105–147.

Sugie, H., Yoshida, M., Kawasaki, K., Noguchi, H., Moriya, S., Takagi, K., Fukuda, H., Fujiie, A., Yamanaka, M., Ohira, Y., Tsutsumi, T., Tsuda, K., Fukumoto, K., Yamashita, M. and Suzuki, H. (1996) Identification of the aggregation pheromone of the brown-winged green bug, *Plautia stali* Scott (Heteroptera: Pentatomidae). *Applied Entomology and Zoology* 31, 427–431.

Sundararaju, D., Bakthavatsalam, N., John, J.N. and Vidyasagar, P.S.P.V. (1994) Presence of female sex pheromone in tea mosquito bug, *Helopeltis antonii* Sign. (Heteroptera: Miridae). *Entomon* 19, 47–51.

Thistlewood, H.M.A., McMullen, R.D. and Borden, J.H. (1989a) Damage and economic injury levels of the mullein bug, *Campylomma verbasci* (Meyer) (Heteroptera: Miridae), on apple in the Okanagan Valley. *Canadian Entomologist* 121, 1–9.

Thistlewood, H.M.A., Borden, J.H., Smith, R.F., Pierce, H.D. Jr and McMullen, R.D. (1989b) Evidence for a sex pheromone in the mullein bug, *Campylomma verbasci* (Meyer) (Heteroptera: Miridae). *Canadian Entomologist* 121, 737–744.

Todd, J.W. (1989) Ecology and behavior of *Nezara viridula*. *Annual Review of Entomology* 34, 273–292.

Tomioka, H. and Mori, K. (1992) Simple synthesis of a diastereomeric mixture of 2-(3′,4′-epoxy-4′-methylcyclohexyl)-6-methyl-2,5-heptadiene, the pheromone of the green stink bug. *Bioscience, Biotechnology and Biochemistry* 56, 1001.

Toscano, N.C. and Stern, V.M. (1976) Dispersal of *Euschistus conspersus* from alfalfa grown for seed to adjacent crops. *Journal of Economic Entomology* 69, 96–98.

Turnipseed, S.G. and Greene, J.K. (1996) Strategies for managing stink bugs in transgenic B.T. cotton. *Proceedings of the 1996 Beltwide Cotton Conferences*. National Cotton Council of America, Memphis, Tennessee, pp. 935–936.

Ubik, K., Vrkoč, J., Žďárek, J. and Kontev, C. (1975) Vanillin: possible sex pheromone of an insect. *Naturwissenschaften* 62, 348.

Vrkoč, J., Ubik, K., Žďárek, J. and Kontev, C. (1977) Ethyl acrylate and vanillin as components of the male sex pheromone complex in *Eurygaster integriceps* (Heteroptera, Scutelleridae). *Acta Entomologica Bohemoslovaca* 74, 205–206.

Wang, Q. and Millar, J.G. (1997) Reproductive behavior of *Thyanta pallidovirens* (Heteroptera: Pentatomidae). *Annals of the Entomological Society of America* 90, 380–388.

Welter, S.C. (1991) Selective management of *Calocoris* and stink bugs in pistachio orchards. In: *California Pistachio Industry Annual Report – Crop Year 1990–1991*. California Pistachio Commission, Fresno, California, pp. 127–131.

Wheeler, A.G. Jr and Henry, T.J. (1992) *A Synthesis of the Holartic Miridae (Heteroptera): Distribution, Biology, and Origin, with Emphasis on North America*. The Thomas Say Foundation, Vol. 15, Entomological Society of America, Lanham, Maryland, 282 pp.

Wilson, L.T. (1990) Biological control and monitoring of southern green stink bug in fresh market tomatoes. In: *1989–1990 Annual Report California Fresh Market Tomato Advisory Board*, pp. 67–75.

Wilson, L.T., Zalom, F.G. and Hoffman, M.P. (1988) Pest management of stink bug on fresh market tomatoes. In: *1987–1988 Annual Report California Fresh Market Tomato Advisory Board*, pp. 31–32.

Yasuda, K. (1990) Ecology of the leaf footed plant bug, *Leptoglossus australis* Fabricius (Heteroptera: Coreidae), in the sub-tropical region of Japan. *Tropical Agriculture Research Series* 23, 229–238.

Yasuda, K. and Tsurumachi, M. (1994) Immigration of leaf-footed plant bug, *Leptoglossus australis* (Fabricius) (Heteroptera: Coreidae), into cucurbit fields on Ishigaki Island, Japan, and role of male pheromone. *Japanese Journal of Applied Entomology and Zoology* 38, 161–167.

Yasuda, K. and Tsurumachi, M. (1995) Influence of male adults of the leaf-footed plant bug, *Leptoglossus australis* (Fabricius) (Heteroptera: Coreidae), on host-searching of the egg parasitoid, *Gryon pennsylvanicum* (Ashmead) (Hymenoptera: Scelionidae). *Applied Entomology and Zoology* 30, 139–144.

Yonce, C. and Mizell, R. (1997) Stink bug trapping with a pheromone. *Proceedings of the Southeastern Pecan Growers Association* 97, 54–56.

Zalom, F.G. (1992) Biological control of southern green stink bug and development of pheromone traps for consperse stink bug. In: *1991–1992 Annual Report, California Tomato Board*, pp. 1–10.

Žďárek, J. and Kontev, C. (1975) Some ethological aspects of reproduction in *Eurygaster integriceps* (Heteroptera, Scutelleridae). *Acta Entomologica Bohemoslovaca* 72, 239–248.

12 Grasshoppers and Locusts

Ahmed Hassanali and Baldwyn Torto

Chemical Ecology Department, International Centre of Insect Physiology and Ecology (ICIPE), PO Box 30772, Nairobi, Kenya

Introduction

Chemical communication in the family Acrididae (grasshoppers and locusts) has not received much attention to date. Paradoxically, one of the earliest examples of pheromonal activity reported for any insect was the maturation effect of the gregarious-phase male of the desert locust *Schistocerca gregaria* (Forskål) (Norris, 1954). This relative neglect may have been in part due to the assumption that acridoids rely generally on non-chemical signals (visual, acoustic, tactile) in their communication systems (Uvarov, 1977; Whitman, 1990). However, interest in elucidating the factors underlying physiological and behavioural processes of some economically important acridoid pests and, in particular, the phenomenon of polymorphism, has demonstrated the mediation of a series of primer and releaser semiochemicals. Much of the earlier work implicating pheromonal effects and interspecific chemical signals in grasshoppers and locusts, and attempts to identify the mediating semiochemicals, has been reviewed by Loher (1990), Whitman (1990) and Byer (1991). Byer (1991) also summarized speculations on the ecological role and adaptive value of these semiochemicals. While providing adequate background to the earlier work, this chapter will focus largely on the results of recent studies on different acridoid processes such as gregarization, aggregation, maturation, oviposition and mating (Fig. 12.1 gives a summary of pheromones associated with the life cycle of *S. gregaria*). The implications of these findings for the development of new strategies for the control of the species studied, and the opportunities they have engendered for elucidating the semiochemical bases of similar processes in other grasshoppers and locusts, will also be highlighted.

Gregarization

Manifestations of gregarious behaviour are found in many species of acridoids. These vary from loose assemblages mediated by acoustic and/or visual cues, through

 Pheromones of Non-Lepidopteran Insects Associated with Agricultural Plants (eds J. Hardie and A.K. Minks)

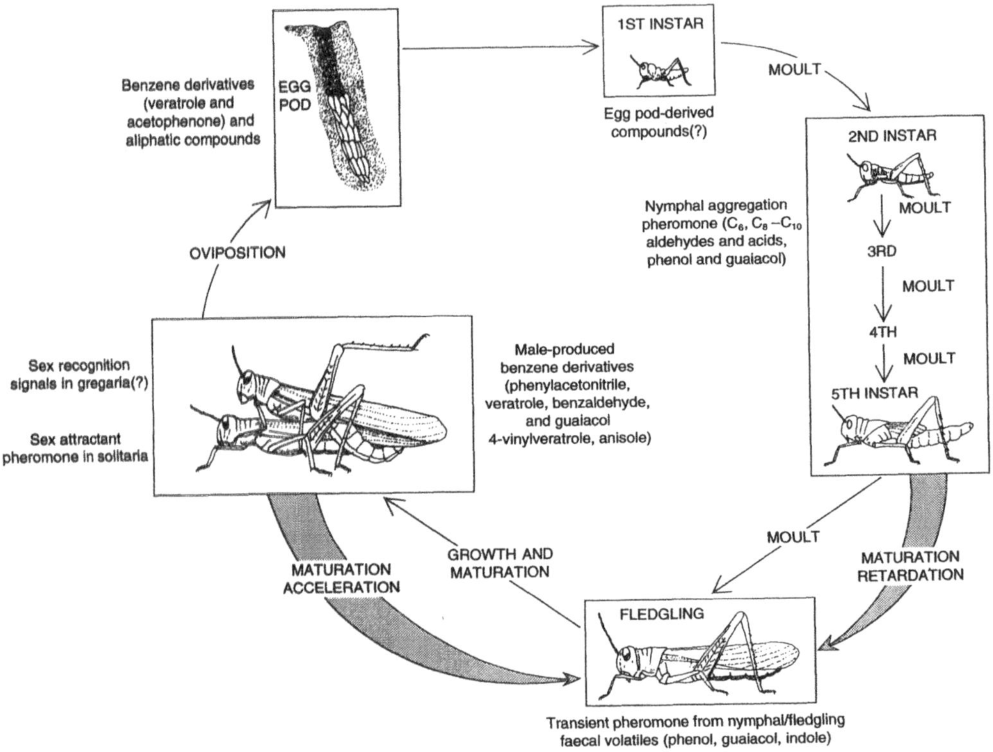

Fig. 12.1. Life cycle of the desert locust, *Schistocerca gregaria*, showing times at which pheromones operate.

temporary gregariousness associated with certain ecological situations and periods in the life cycles of some grasshoppers, to the deep-seated gregarious character manifested by polymorphic species (Uvarov, 1977). The last group is a characteristic feature of locusts which can transform reversibly in a graded manner between two extreme phases, the solitaria and gregaria, that differ in physiology, behaviour, pigmentation and morphology (Uvarov, 1921, 1966; Steedman, 1990). Gregarization is predicated on acridoid density, which may be promoted by a variety of environmental and biotic factors, including improved rains and vegetation, habitat topologies, favourable microclimates, convergent winds, and patchy distribution of food plants and oviposition sites (Kennedy, 1939; Rainey, 1962; Roffey and Popov, 1968; Uvarov, 1977; Bouaichi *et al.*, 1996). Several mediating factors have been implicated including visual (Ellis and Pearce, 1962), tactile (Chauvin, 1941; Ellis, 1959) and pheromonal (Nolte, 1963; Gillett, 1968) stimuli. Dietary factors (Jackson *et al.*, 1978) and previous phase history of the insect (Michel, 1980) have also been shown to influence the extent of gregarization. Two sets of pheromonal effects may be distinguished: releaser effects, resulting in behavioural aggregation, and primer effects, which may contribute to the physiological transformation (and associated behavioural, biochemical and morphological changes) of solitarious insects to the gregarious phase. The signals mediating the former are designated as aggregation (also referred to as 'cohesion') pheromones and the latter as gregarization pheromones.

Releaser effects

The widespread occurrence of grouping behaviour in acridoids suggests that this might in many cases be mediated by olfactory signals, acting with or without other cues. Unfortunately, much of the earlier behavioural work on *S. gregaria* and *Locusta migratoria migratorioides* (Reiche & Fairmaire) (Fig. 12.2) failed to establish conclusively the involvement of such signals (Haskell *et al.*, 1962; Gillett *et al.*, 1976). In a more recent study, Fuzeau-Braesch *et al.* (1988) sought to identify volatile compounds associated with crowd-reared populations of *S. gregaria* and *L. migratoria* and to investigate their effects on the aggregation behaviour of the locusts. Phenol, guaiacol and veratrole were identified which, although unattractive to the locusts, elicited significant clumping behaviour in both species and were considered to function as a 'cohesion' pheromone. Despite criticisms that may be levelled against the methodologies employed in some of these studies (Byer, 1991; Obeng-Ofori *et al.*, 1993), they provided a strong indication that locust aggregation might not be modulated by anemotactic responses but by a retentive mechanism that holds the insects within its perceptible range.

In more recent studies, the retentive responses of different stages and sexes of *S. gregaria* to potential sources of aggregating signals were studied in an enclosed arena, one side of which was permeated with clean air and the other with air enriched with volatiles associated with gregarious desert locusts of different stages, or their faeces (Obeng-Ofori *et al.*, 1993, 1994a,b; Torto *et al.*, 1994, 1996). Test locusts were released into the olfactometer in groups as well as individually. The results showed that volatiles emitted by locusts elicit grouping from receptive individuals by retaining them within the atmosphere enriched with the appropriate volatiles. The retentivity of the pheromones was independent of whether the test insects were released alone or in groups, indicating that non-olfactory stimuli are not important mediating components of the grouping response (Obeng-Ofori *et al.*, 1993). The studies also revealed some interesting differentiation between phases, sexes and the different stages of the insect in the production and chemistry of the

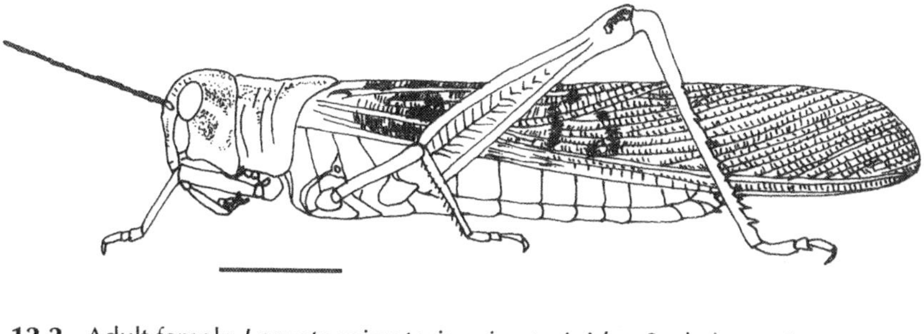

Fig. 12.2. Adult female *Locusta migratoria migratorioides*. Scale bar = 1 cm.

pheromones, as well as in their aggregative responses to different pheromonal emissions, as detailed below. The implications of these findings for other locusts and grasshoppers are also highlighted.

Aggregation pheromones of gregarious S. gregaria *nymphs*

The second to fifth instar nymphal stages share a common pheromone which has no effect on the adult (Obeng-Ofori *et al.*, 1993, 1994a; Torto *et al.*, 1996). Both sexes produce the pheromone. Analysis (GC–EAD and GC–MS) of the volatile emissions from live fifth instars led to the identification of C_6 and C_8–C_{10} straight-chain aldehydes and their corresponding fatty acids as the major electrophysiologically active compounds (Fig. 12.3, **1–8**) (Torto *et al.*, 1996). Smaller amounts of guaiacol and phenol were also present (Fig. 12.3, **9** and **10**). These phenols are the main constituents of nymphal faeces, which also elicited aggregation responses from the nymphs (Obeng-Ofori *et al.*, 1994b). In laboratory assays, synthetic blends of the eight aliphatic compounds, together with the two phenols, promoted nymphal aggregations to a level comparable to that induced by the natural volatile blend from live nymphs (Torto *et al.*, 1996).

The first gregarious instar nymphs are not aggregated by the pheromone of the higher instars and rely on a different set of signals to aggregate. However, its

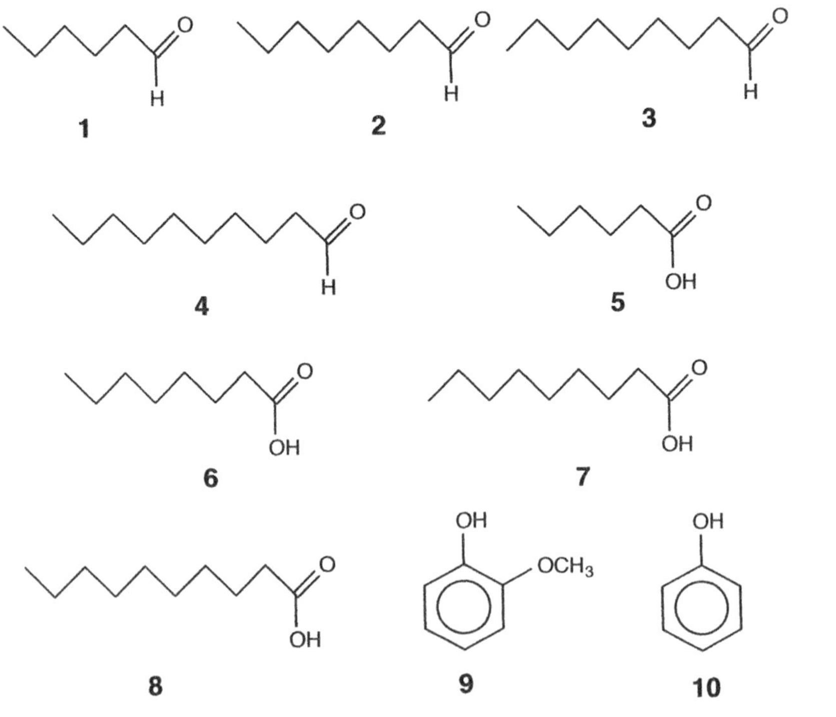

Fig. 12.3. Constituents of gregarious *S. gregaria* nymphal (second to fifth instar) aggregation pheromone: **1**, hexanal; **2**, octanal; **3**, nonanal; **4**, decanal; **5**, hexanoic acid; **6**, octanoic acid; **7**, nonanoic acid; **8**, decanoic acid; **9**, guaiacol; **10**, phenol.

pheromone is not yet fully characterized (B. Torto, 1998, unpublished results). The biological rationale for this pheromonal differentiation between the first instar and other nymphal stages is not clear and further studies on the behavioural ecology and physiology of these stages are needed to shed some light on this question.

Aggregation pheromones of gregarious S. gregaria *adults*

The adult stages are indifferent to the nymphal pheromones (Obeng-Ofori *et al.*, 1993; Torto *et al.*, 1994). On the other hand, males and females of all ages respond equally to a pheromone produced by older males. The existence of different pheromone systems for nymphal and adult stages may relate to the different behavioural ecologies of the two stages (Uvarov, 1966) and the need to separate them into different groupings. It also reflects the different roles played by the two blends in synchronizing maturation of adults (see below).

Analysis of airborne volatiles of older males showed the presence of six electrophysiologically active benzene derivatives comprising anisole, veratrole, benzaldehyde, guaiacol, phenylacetonitrile and phenol (Fig. 12.4) (Torto *et al.*, 1994). The first two did not elicit significant behavioural effects. Of the rest, phenylacetonitrile, which is the dominant component of the pheromone emission (*c.* 80%), was the most active. The neuronal activity generated by the six compounds in *S. gregaria* has been studied by single-cell recordings from antennal olfactory receptor neurones (Hansson *et al.*, 1996) and antennal lobe interneurons (Anton and Hansson, 1996). These studies demonstrated the presence of groups of neurones that were specifically excited by the six compounds, as well as others that were identified as 'blend specialists' which responded only to specific mixtures of the pheromone components. The roles of the two groups of neurones remain to be further investigated.

Since adults are indifferent to the nymphal pheromones and there is a delay in the production of the adult pheromone, how do early fledgers in a gregarious population maintain their cohesiveness? Field observations have shown a close association between hoppers and fledgling adults (Uvarov, 1977) and, associated with them, the presence of large amounts of faecal droppings (Obeng-Ofori *et al.*, 1994b). Young adults aggregate strongly in response to volatiles of their own faeces as well as to those of nymphs and older adults. GC–EAD and GC–MS analyses showed the presence of guaiacol, phenol and indole in the airborne volatiles of nymphal faeces and in those of the young adult. Faeces of the older adults contained benzaldehyde and phenylacetonitrile in addition to the two phenols. Thus, the blend of the two phenols and indole that make up nymphal and fledgling faecal volatiles acts as a transient aggregation pheromone for the young adults until these start to produce the adult pheromone (Obeng-Ofori *et al.*, 1994b). An important consequence of this is that early fledgers come into close contact with the nymphal pheromone, which can thus influence their maturation rate (see below).

Responses of solitarious S. gregaria *to the gregarious-phase pheromone*

Solitarious stages do not produce the pheromones of the corresponding gregarious counterparts in any significant amounts (Njagi *et al.*, 1996; Torto *et al.*, 1996). However, studies on the EAG responses of solitarious adults show a pattern of response, including threshold ranges, comparable to those of the gregarious

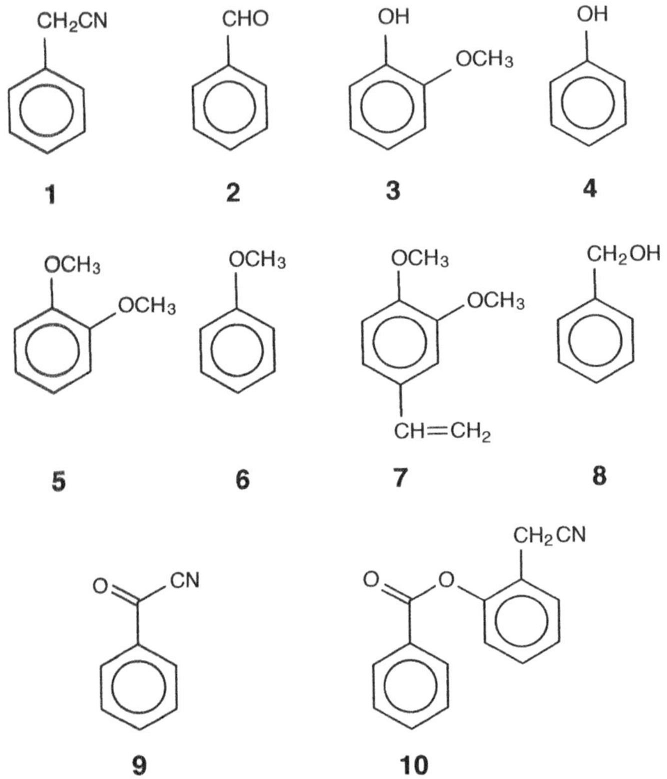

Fig. 12.4. Constituents of gregarious *S. gregaria* mature adult male volatile emissions: **1**, phenylacetonitrile; **2**, benzaldehyde; **3**, guaiacol; **4**, phenol; **5**, veratrole; **6**, anisole; **7**, 4-vinylveratrole; **8**, benzyl alcohol; **9**, benzoyl nitrile; **10**, 2-benzoyloxyphenylacetonitrile; **1–4**, adult aggregation pheromone; **3**, **4**, **8**, **9**, **10**, each present at < 1%.

counterparts with respect to phenylacetonitrile, benzaldehyde, guaiacol and phenol, although some phase-related differences were apparent at higher doses. Moreover, in behavioural assays, the responses of solitarious adult individuals were comparable to those of gregarious counterparts to each of the pheromone components, to different combinations of these compounds, or to crude volatiles derived from adult gregarious males. The implication of these findings is that in a gregarizing population of the desert locust, solitarious individuals that encounter foci of pheromone-emitting groups of gregarious insects would be arrested by the pheromone and become recruited into the group. If new solitarious recruits of either sex mate with the gregarious individuals they will produce offspring with gregarious characteristics (Dale and Tobe, 1990; Islam *et al.*, 1994a,b) and if their contact with gregarious groups is sufficiently prolonged, the new recruits will rapidly acquire gregarious characteristics (Ellis, 1963; Roessingh *et al.*, 1993; Bouaichi *et al.*, 1995). It has been proposed that this process might constitute an important mechanism of spread of the gregarious character across the population (Njagi *et al.*, 1996).

Aggregation pheromones of other locusts and grasshoppers

There may be a close parallel between pheromone-mediated behaviour of other locust species and that of *S. gregaria*. A recent study on the cross-responses of *S. gregaria* and *L. m. migratorioides* indicates that there is an overlap in the pheromone compositions of the two species (Niassy, 1997). However, there are also some significant differences. Compared with *S. gregaria*, *L. m. migratorioides* shows a lesser pattern of stage and sex differentiation in its aggregation pheromone biology. Thus, although fifth instar nymphs do not respond to the adult pheromone, adults respond significantly to that of nymphs. Production of the adult aggregation pheromone is not confined to males since both sexes induce aggregation from conspecifics. Detailed analysis of the pheromones of this species remains to be undertaken.

The aggregation behaviour of some grasshopper species appear to be reminiscent of that of *S. gregaria* and other locust species. Two aggregative stages have been observed for *Zonocerus variegatus* (L.) (Fig. 12.5) that involve early instar nymphs and mature adults (McCaffery *et al.*, 1978; Chapman *et al.*, 1986). Interestingly, aggregation in adults is initiated by a male pheromone that is attractive to both males and females (McCaffery and Page, 1982). Several other species are known to be gregarious as hoppers but not as adults. The hopper stages of *Z. elegans* (Thunberg) resemble those of locusts (Uvarov, 1977). Others are cohesive in early instars but the aggregation tendency becomes weaker in late instars. Conversely, some grasshoppers – e.g. *Patanga succincta* (L.), *Anacridium melanorhodon* (Walker), *Faureia milanjica* (Karsch), *Hieroglyphus nigrorepletus* (Bolivar) – show gregarious behaviour in the adult stages only, and may even swarm and undertake long flights (Sjöstedt, 1923; Ghouri and Ahmad, 1960; Rao, 1962; Roffey, 1965; Uvarov, 1977). Others exhibit some features of gregarious behaviour in situations of exceptional abundance (Uvarov, 1977). The factors mediating aggregation in these insects and the role of pheromone communication need to be established.

Primer effects

Two opposite primer effects have been documented in the literature: one which has implicated gregarization pheromones; and the other a solitarization factor in *S. gregaria*.

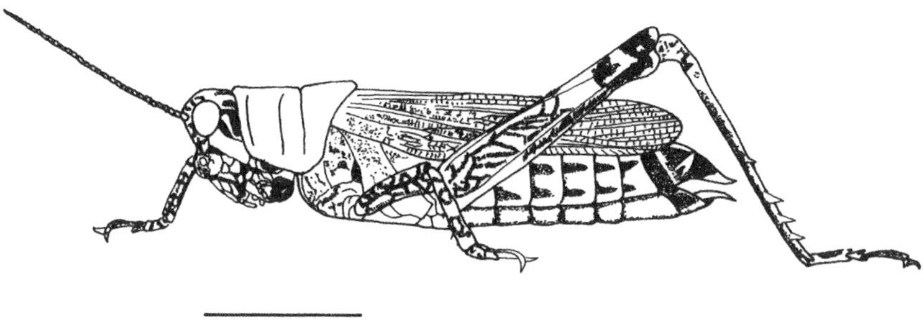

Fig. 12.5. Adult female *Zonocerus variegatus*. Scale bar = 1 cm.

Gregarization pheromone

Crowding of solitarious locusts has been known to lead to rapid development of some gregarious characters including cohesive behaviour (Ellis, 1963; Roffey and Popov, 1968), although the manifestation of all gregarious characters takes several generations (Pener, 1991). Recently, the effect of crowding and isolation of *S. gregaria* was studied by quantitative behavioural techniques (Roessingh and Simpson, 1994; Bouaichi *et al.*, 1995; Heifetz *et al.*, 1996). Crowding at nymphal or adult stage led to the acquisition of gregarious behavioural characters within 1–4 h. On the other hand, isolation of crowd-reared insects led to rapid loss of gregarious behaviour. In another study, the effects of shifting *S. gregaria* from solitary to crowded rearing conditions, and the reverse, were monitored by emission of the aggregation pheromone by adult males (Deng *et al.*, 1996). Crowding of solitaria resulted in the production of the pheromone within days. Even pairing of two solitarious adults led to detectable production of the pheromone, showing that two insects confined together are able to provide each other with the essential stimuli for inducing the onset of phase change. Conversely, adult males resulting from shifting gregarious insects to solitary conditions rapidly lost their ability to produce the pheromone.

The potential role of pheromones in phase gregarization was first recognized by Nolte (1963), and later confirmed by Gillett (1983), following observations that isolated hoppers of *S. gregaria, L. migratoria* and *Locustana pardalina* (Walker) kept in the same room with crowded hoppers continued to retain the pigmentary, morphometric and grouping traits of the gregaria. It was proposed that a volatile gregarizing pheromone from gregarious hoppers inhibited the solitarization of isolated hoppers. The source of the pheromone was traced to hopper faeces of both phases, but not those of adults (Nolte *et al.*, 1973). Examination of the constituents of the steam distillate from the faeces of crowded *L. migratoria* hoppers led to the identification of 5-ethylguaiacol (referred to as 'locustol') as the principal component of the gregarization pheromone. However, recent analyses in several different laboratories failed to demonstrate the presence of the compound in any of the airborne volatiles of locusts or their faeces (Fuzeau-Braesch *et al.*, 1988; Francke and Schmidt, 1994; Obeng-Ofori *et al.*, 1994b; Torto *et al.*, 1994, 1996).

Since solitarious stages do not emit the aggregation pheromones of their gregarious counterparts, it seems improbable that these would be involved in priming phase transformation of solitaria to the gregaria. Indeed, when solitarious nymphs or fledglings were exposed to volatiles from crowd-reared nymphs without visual and physical contact, no evidence of phase change was observed although some melanization occurred (Deng, 1995). Similar results were obtained when solitarious nymphs were exposed to airborne volatiles of faecal extracts (Heifetz *et al.*, 1996). However, it is possible that, once gregarious, the aggregation pheromones might provide sufficient primer stimuli to delay solitarization. This would account for Nolte's (1963) and Gillett's (1983) observations referred to above.

How important then are chemical cues in phase transformation of solitaria to the gregaria? In a recent study by Heifetz *et al.* (1996), gregarious nymphs were isolated from the second instar in individual chambers partitioned by either aluminium, glass or wire screening. In addition, the effects of faecal volatiles, faecal extracts and cuticular extracts of crowd-reared nymphs on gregarization were compared. Phase change was monitored using two behavioural indices based on nymphal activity and association. Visual stimuli alone were ineffective and so were faecal components. On the other hand, cuticular extracts significantly affected behavioural phase transition. Interestingly, antennectomy of crowd-reared nymphs, but not hind-tarsal amputation, induced solitarization. The results implicate contact or short-range signals perceived by antennal receptors as the primer pheromones.

Solitarization factors

Faeces of crowd-reared adult desert locusts were previously shown to have solitarizing effects on gregarious nymphs (Gillett *et al.*, 1976). Although the existence of a solitarization pheromone was hypothesized, the effect is accounted for by pheromonal differentiation between nymphal and adult stages (Obeng-Ofori *et al.*, 1993, 1994b; Torto *et al.*, 1994). Components specific to the adult pheromone blend leads to dispersal in nymphs, and vice versa. The faeces of older and mature male *S. gregaria* contain adult pheromone components, in addition to guaiacol and phenol (see above). Treatment of hopper bands in the field with the adult pheromone leads to disorientation and gradual dispersal of the insects (M. Bashir, Nairobi, 1998, personal communication), and a recent study has shown that phenylacetonitrile, the major component of the adult pheromone, inhibits perception by nymphs of their own aggregation pheromone (Ochieng, 1997).

Maturation Synchrony

Gregarization in polymorphic acridoids involves not only the physiological transformation of individual insects and associated behavioural and morphological changes, but also synchronization of the maturation and development of individuals in a population. Maturation synchrony is critical to the gregarious-phase locusts, and ensures that the succeeding generation of hoppers remains spatially and temporally cohesive. In the desert locust, two principal mechanisms have been implicated: (i) interspecific signals associated with desert shrubs that may affect, primarily, reproductively inactive solitarious adults; and (ii) intraspecific signals emanating from gregarious nymphs and mature males which inhibit and accelerate, respectively, the maturation of young adults.

Maturation synchrony in solitaria

During periods of drought, young solitarious adult locusts feeding largely on senescent plants remain sexually immature (Ellis *et al.*, 1965). Carlisle *et al.* (1965) have reported extensive observations that showed that maturation occurs more or less simultaneously at sites hundreds of kilometres apart a week or so before the actual arrival of rains and several weeks before the appearance of annual vegetation. Of special interest is that locust maturation appeared to coincide with the bud burst of certain desert shrubs, including species of *Boswelia* and *Commiphora*.

Follow-up laboratory studies showed that exposure to aromatic emissions of the floral parts and articles of commerce derived from the shrubs caused accelerated maturation of crowded locusts. A recent study compared the composition and maturation effects of the essential oil of the desert shrub C. *quadricincta* before and after winter rains (Assad *et al.*, 1997a). The oil collected before the rains was richer in terpenoids, particularly the volatile components, and immature males and females of crowd-reared S. *gregaria* exposed to its vapour matured faster compared with those exposed to the oil obtained after the rains. These studies have yet to be repeated with reproductively inactive solitarious S. *gregaria* adults under natural field conditions. Nevertheless, they do suggest that widely scattered solitarious insects might time their reproductive activity to the changing physiology of desert shrubs, including perhaps food and shelter plants, and thereby attain some measure of developmental synchrony (Carlisle *et al.*, 1965; Assad *et al.*, 1997a).

No comparable studies have been reported on other acridoids. However, it is conceivable that other insects, particularly those that are adapted to semi-arid and arid habitats, might rely on flora adapted to such habitats for dependable environmental cues to synchronize their reproductive activities with the on-coming rains.

Maturation synchrony in gregaria

Studies undertaken by Norris (1954, 1964) and her co-workers (Norris and Pener, 1965; Richards and El Mangoury, 1968) on the sexual maturation of S. *gregaria* and L. *migratoria* suggested that pheromones played an important role in the timing of maturation of these insects. The presence of mature males, but not females, accelerated the maturation of young male and female adults. On the other hand, S. *gregaria* nymphs of either sex retarded the maturation of adult conspecifics. It has been suggested that in nature the retarding influence of hoppers tends to delay the maturation of early fledgers until the majority of nymphs have moulted into adults (Norris, 1964; Richards and El Mangoury, 1968). In time, an increasing number of male adults will start to produce the maturation pheromone, which will accelerate the maturation of other adults, particularly the late fledgers, and result in maturation synchrony of the entire population (Richards and El Mangoury, 1968; Mahamat *et al.*, 1993).

Maturation accelerating signal

Loher (1960) found that solvent extracts of mature males of S. *gregaria* accelerated the maturation of young adult insects. This could be induced without actual contact with the stimulus, suggesting that a volatile pheromone was involved. Associated with the onset of maturation in the male desert locust is integumental yellowing of abdominal tergites, which rapidly cover much of the body as the insect matures (Norris, 1954; Loher, 1960). More recently, the effects of solvent extracts and airborne volatiles trapped from mature males on immature males were compared with those emanating from live mature males (Mahamat *et al.*, 1993). The signal sources (extracts, trapped volatiles or mature males) were kept in the upper chamber of two-storey aluminium cages which allowed exposure of the recipient immature insects placed in the lower chamber to the volatiles diffusing from the sources without visual or tactile contact. Both hexane extracts of body washes of mature males and trapped volatiles were as effective as mature males in accelerating maturation. Indeed, the level of acceleration attained with the mature male volatiles (in the absence of visual and tactile cues) was comparable with that obtained when immature insects were left in direct contact with mature males, showing that pheromonal effects constitute a dominant factor in accelerated maturation. In this study, yellowing as described by Norris (1954) coincided with the onset of copulatory activity. Interestingly, a relatively short contact of 1–4 days did not elicit accelerated maturation. Immature adults needed to be exposed to the adult stimulus for over a week before significant acceleration occurred (Mahamat *et al.*, 1993). In females, maturation acceleration was also reflected in early increase in haemolymph vitellogenin titres (Mahamat *et al.*, 1997).

Two attempts to characterize the maturation-accelerating constituents of the desert locust have been described (Blight *et al.*, 1969; Mahamat *et al.*, 1993). The earlier study led to the identification of acetone, methyl acetate and ethyl acetate in airborne collections of mature males and females, and 1-pyrroline in that of the mature males. None of these were reported to be active. In the recent analysis of volatile emissions of mature males, however, anisole, benzaldehyde, veratrole, phenylacetonitrile and 4-vinylveratrole were identified in the ratio 4.7:7.0:3.3:79.8:5.0 (Fig. 12.4) (Mahamat *et al.*, 1993). In addition, guaiacol, phenol, benzyl alcohol, 2-benzoyloxyphenylacetonitrile and benzoyl nitrile were present in trace amounts (< 1%). Blends of the five major components in the natural ratio, released at rates comparable with those from mature male *S. gregaria*, were shown to simulate the natural volatile emission from mature males (H. Mahamat, Nairobi, 1998, personal communication). The presence of the major component (phenylacetonitrile) was critical for full potency, although some of the other components also showed varying degrees of accelerating effects. Thus, the adult pheromonal emission contains components that have dual functions, as releaser signals (adult aggregants) and as primer signals (maturation accelerants) (Mahamat *et al.*, 1993).

Maturation retarding signal

Until recently, no follow-up studies were reported on the maturation inhibitory effect associated with nymphs of *S. gregaria*. In a recent reinvestigation, the effects of the presence of male or female fifth instar *S. gregaria* on the maturation of immature adults were compared with exposure of the latter to volatiles from fifth instars, their faeces, and synthetic blends of the nymphal aggregation pheromone comprising C_6, C_8–C_{10} aliphatic aldehydes and acids, guaiacol and phenol (Fig. 12.3) (Assad *et al.*, 1997b). Maturation was monitored by integumental yellowing, copulatory activity, phenylacetonitrile titres, ovulation (oocyte length in females) and oviposition. The results demonstrated that maturation-retarding effects of male and female nymphs are due to their aggregation pheromone. Interestingly, neither the aldehydes nor the acids nor the nymphal faecal volatiles alone had any significant effect on maturation (Assad *et al.*, 1997b). Thus, as in the case of adult *S. gregaria*, the nymphal pheromone functions parsimoniously as a nymphal aggregant (a releaser effect) and as an adult maturation retardant (a primer effect).

Primer and releaser effects in synchrony development

The dual roles of nymphal and adult pheromones, and the aggregative function of faecal volatiles, provide insights into the subtleties of the synchronization process in a gregarizing desert locust population. Firstly, adults (immature and mature) are behaviourally indifferent to the nymphal pheromone (Obeng-Ofori *et al.*, 1993). So how does the pheromone exert its primer effect on young adults? It does so because of varying periods of delay in the production of the adult pheromone (Mahamat *et al.*, 1993) and reliance by young adults on volatiles derived from their own faeces and those of nymphs for cohesiveness (Obeng-Ofori *et al.*, 1994a). As mentioned earlier, faecal volatiles from these sources, unlike those of the older and mature males, have the same composition (Obeng-Ofori *et al.*, 1994a). This ensures a close association of nymphs and immature adults in the field when the former are fledging (Uvarov, 1977; Assad *et al.*, 1997b). Once most nymphs have moulted into adults, however, there is no more inhibitory influence because, on their own, faecal constituents do not affect maturation. Secondly, as the older males in the fledgling population start to produce the adult pheromone, there is a gradual switch from a purely aggregative influence of faecal constituents to one that exerts both a releaser and a primer effect. Since maturation acceleration is associated with earlier production of the adult pheromone on recipients, increasing numbers of males would contribute to the pheromone pool earlier and speed up the maturation of the late fledgers. Conversely, maturation inhibition is associated with delayed production of adult pheromone. In the early stages of gregarization, fledglings are likely to co-occur not only with the final nymphal instar in different stages of development but also with earlier instars. In this situation, maturation of early fledgers would be slowed down further by prolonged exposure to nymphs and also by the delayed production of the adult pheromone. However, as the population develops greater synchrony, duration of contact with nymphs would be reduced and the adult pheromone would be produced earlier. The function of the two pheromone systems would then increasingly be to fine-tune the synchronization process (Assad *et al.*, 1997b).

Apart from *S. gregaria*, very little has been done to elucidate similar effects on other polymorphic acridoids. Maturation-accelerating effects implicating pheromones released by adult males have also been observed in *L. migratoria* and *S. cancellata* (Serville) (Amerasinghe, 1978; Ali, 1987; Hawkes *et al.*, 1987). However, the mediating signals have yet to be identified.

Oviposition-associated Signals

Two types of semiochemical effects have been described in acridoids: those due to releaser signals that elicit group oviposition behaviour (Fig. 12.6), and those due to primer signals that affect either the insect's gregarious characters or its reproductive rate (eggs per pod, oviposition time and frequency).

Releaser signals

Group oviposition is an important trait of acridoids that exhibit the gregarious character. Female *S. gregaria* have been observed to deposit their egg pods at sites

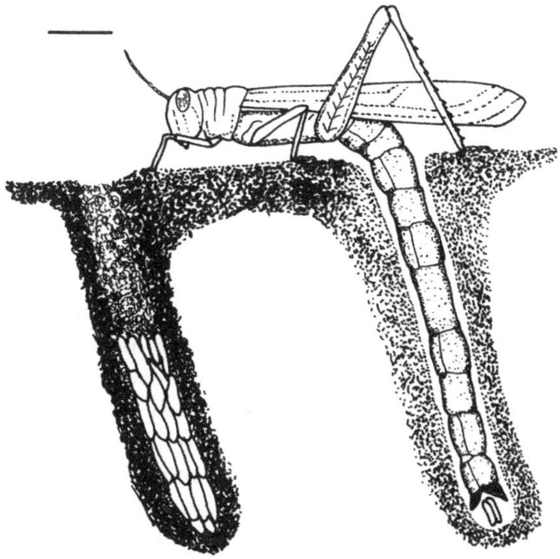

Fig. 12.6. Adult female desert locust, *Schistocerca gregaria,* ovipositing in sand (right) and newly laid eggs with covering and plug of foam (left). Scale bar = 1 cm.

where other females are ovipositing even though environmentally suitable alternatives may be present (Popov, 1958; Stower *et al.*, 1958; Uvarov, 1977). Sand in which females had laid previously was also attractive to gravid females (Saini *et al.*, 1995). Similar observations were made earlier with egg-laying females of *L. migratoria,* which were attracted over a distance of 0.5 m to sand used for previous egg-laying (Lauga and Hatté, 1977a,b, 1978), and in *Z. variegatus,* which moved upwind to oviposition sites (McCaffery and Page, 1982).

Attraction in crowd-reared *S. gregaria* was associated with the froth of egg pods (Fig. 12.6), the eggs on their own being unattractive (Saini *et al.*, 1995). Extracts of the froth obtained by sequential extraction with hexane, ethyl acetate and methanol evoked different levels of egg-laying. The hexane extract was most potent, followed by the methanolic extract, suggesting the mediation of less polar volatile and polar less volatile components. Indeed, volatiles trapped from the froth evoked strong oviposition, and ovipositing females were often seen touching sand contaminated with froth materials with their antennae and palpi, suggesting a search for short-range or contact signals (Saini *et al.*, 1995). A similar behaviour is shown by *L. migratoria,* which may even ingest contaminated sand prior to ovipositing (Lauga and Hatté, 1977a). Interestingly, solitarious *S. gregaria* females prefer to lay near host plants rather than near their own egg pods. However, egg pods originating from gregarious females are more attractive than host plants (M. Bashir, Nairobi, 1998, personal communication). Similarly, solitarious *L. migratoria* females are as responsive as their gregarious counterparts to the oviposition aggregative signal associated with egg pods of gregarious insects (Lauga and Hatté, 1977). Remarkably, eggs from both species of solitarious females laid in sand contaminated with egg pods of gregarious counterparts hatched into more gregarious nymphs (Lauga and Hatté, 1978; McCaffery *et al.*, 1998). Thus, both releaser and primer effects are associated with constituents of the egg pods of the two locust species (see below).

Rai *et al.* (1997) examined volatiles trapped from the egg froth of crowd-reared *S. gregaria* by GC–EAD and GC–MS techniques. Acetophenone and veratrole were identified (Fig. 12.7), each of which was shown to induce oviposition in gravid females. Examination of sand exposed to egg pods has revealed the presence of additional components that also induce oviposition (Torto *et al.*, 1999). Veratrole is also a constituent of the volatiles emanating from older gregarious males (Mahamat *et al.*, 1993; Torto *et al.*, 1994). Norris (1963, 1970) observed that ovipositing *S. gregaria* females were aggregated by other individuals at different developmental stages, alive or dead, as well as to locust extracts and shed skin. Extracts of older males were particularly aggregative, which may implicate veratrole as the active compound. However, communal ovipositing sites of acridoids are generally mixed-sex aggregations, since males and females remain paired following copulation (Popov, 1958; McCaffery and Page, 1982). It is possible then that oviposition behaviour in some of these insects is modulated by both egg froth-associated oviposition signals and the adult aggregation–maturation pheromone blend (Saini *et al.*, 1995).

Primer signals

An interesting signal is that associated with the egg pods of gregarious females which predisposes the offspring to emerge more gregarious, thus effecting transgenerational transfer of phase characters (Lauga and Hatté, 1978; Bouaichi *et al.*, 1995). In a recent study, eggs as well as foam plugs of egg pods of crowd-reared gregarious females were shown to be the source of the primer gregarizing factor in *S. gregaria* (McCaffery *et al.*, 1998). Preliminary experiments indicate that the factor is a small (< 3 kDa) hydrophilic substance which is believed to diffuse around and into the eggs in the pod and modify their embryonic development (McCaffery *et al.*, 1998).

Closely related to the maturation effects outlined above are the effects of volatile semiochemicals on egg production in females. In *Schistocerca vaga* (Scudder), virgin females exposed to male odours (in the absence of visual and physical contact) laid more eggs per pod than control females, but comparable to those laid by mated

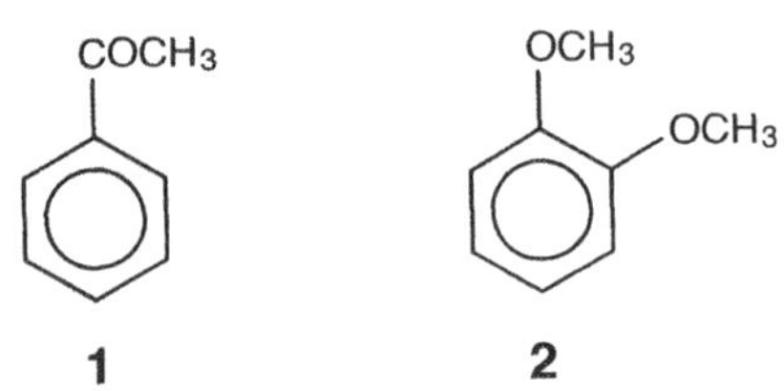

Fig. 12.7. Constituents of volatile emissions of froth of egg pods of gregarious *S. gregaria* mature adult females inducing oviposition in conspecifics: **1**, acetophenone; **2**, veratrole, also a constituent of mature adult male volatile emissions (see Fig. 12.4).

females (Okelo, 1979). Similar exposure to male volatiles in oedipodine *Aiolopus thalassinus* (Fabricius) led to earlier oviposition and also more eggs and egg pods (Schmidt and Osman, 1988). Oviposition rates in insects thus exposed were even higher than those in mated females. Identification of pheromonal constituents associated with the males of these acridoids may help to throw some light on their relationship with the maturation-accelerating compounds described earlier.

An interesting interspecific effect has also been reported in *S. gregaria* (Assad *et al.*, 1997a). Females exposed to volatiles of the desert plant *Commiphora quadricincta* collected before the winter rains produced larger numbers of eggs per pod compared with control insects. Those that were exposed to volatiles collected after the rains showed a drop (although not significant) in the number of eggs relative to controls. These findings may reflect the insect's reliance on the changing physiology of desert plants not only in timing its reproductive activity but also in controlling its reproductive potential in response to the food resources likely to be available (Assad *et al.*, 1997a).

A different type of primer effect has been reported in several different migratory acridoids comprising *S. gregaria, L. migratoria* and *Melanoplus sanguinipes* (Fabricius) (Norris, 1954; Pickford *et al.*, 1969; Lange and Loughton, 1985). In these insects, oviposition is stimulated by copulation. The effect has been traced to a water-soluble, proteinaceous pheromone, produced by the opalescent gland (Odhiambo, 1969). This is transferred via the spermatophore to the female during mating (Leahy, 1973a,b; Friedel and Gillot, 1976; Lange and Loughton, 1985). The pheromone has not been characterized, although the enzyme prostaglandin synthetase has been implicated (Whitman, 1990). In crickets, this enzyme and arachidonic acid are transferred to the female during mating, where they react to produce prostaglandins, which then stimulate oviposition (Loher *et al.*, 1981; Lange and Loughton, 1984; Loher, 1984).

Sex Attraction and Recognition

Although the use of visual and acoustic cues in the mating behaviour of acridoids has been well described (Faber, 1953; Jacobs, 1953; Otte, 1970), very little research has been conducted on the chemical cues that may be associated with this behaviour. Yet there are anecdotal bases for believing that the mediation of pheromones may be important in acridoids (Whitman, 1990). Moreover, pheromone communication has been demonstrated in the mating behaviour of many orthopteroid groups (Whitman, 1982), and Otte (1970) speculated that the ancestral stock of orthopteroids may have relied upon chemical cues for mate recognition. However, to date, the mediation of sex pheromones has been demonstrated in only a few grasshopper and one locust species.

Sex pheromones in grasshoppers

In the grasshopper *Hieroglyphus nigrorepletus* a sex pheromone of medium volatility is produced in the digestive tract and discharged with the excreta. The pheromone is attractive to the male conspecifics over a short distance (Siddiqi and Khan, 1981). In *Taeniopoda eques* (Butmeister), a non-volatile contact pheromone associated with the female defensive exudate elicits sexual behaviour in the males (Whitman, 1982). Similarly, behavioural observations have implicated sex pheromones in the neotropical grasshoppers *Elaeochlora trilineata* (Serville) and *Chromacris speciosa* (Thunberg) (Riede, 1987). None of these sex pheromones has been characterized.

Sex pheromones in polymorphic acridoids

Polymorphic acridoids present contrasting situations. In the gregarious phase, high population densities ($> 15 \times 10^3$ ha^{-1} in desert locusts: El Bashir *et al.*, 1993), maintained cohesive by the aggregation pheromones, make mate-finding relatively easy (Byer, 1991). However, recognition of conspecific mates may be mediated by short-range or contact signals. Females of crowd-reared *S. gregaria* are passive and display no calling behaviour (Norris, 1964; Strong and Amerasinghe, 1977; Amerasinghe, 1978). Males, on the other hand, are aggressive and appear to recognize females (Norris, 1962; Inayatullah *et al.*, 1994). In *L. migratoria*, sex-specific differences in cuticular protein components have been observed (Cassier *et al.*, 1980; Cassier and Papillon, 1983), which could allow discrimination between males and females during copulatory attempts.

Francke and Schmidt (1994) investigated high boiling constituents of pentane body washings of crowd-reared *S. gregaria* adult males. Although the washings were dominated by hydrocarbons, two interesting compounds with 3-methyl-2-one substructure were located in GC–MS studies. One was identified as 3,7-dimethyl-heptacosan-2-one, which bears a striking resemblance to the sex pheromone of the German cockroach, *Blatella germanica* (Fig. 12.8) (Francke and Schmidt, 1994). This finding points towards a need for more detailed investigations on constituents of low volatility of locusts (and grasshoppers) and their roles in mate recognition and mating.

In the solitarious phase, on the other hand, population densities can be extremely low (< 5 ha^{-1} in the desert locust: Uvarov, 1977), and although patchy distribution of host plants may contribute to bringing the insects together, successful mate finding may be dependent on long-range pheromone signals (Byer, 1991; Hassanali and Mahamat, 1991). A detailed behavioural study of solitary-reared *S. gregaria* confirmed that mature males are indeed attracted upwind to hidden solitarious mature females in wind tunnels (Inayatullah *et al.*, 1994). Use of females with glued wings did not affect male responses, demonstrating the mediation of a volatile signal. Efforts to characterize the pheromone have implicated long-range attractive (H. Mahamat, Nairobi, 1998, personal communication) and short-range arrestant constituents (B. Torto, 1998, unpublished results). However, the precise roles of these in mate-finding behaviour remain to be established.

Prospective Applications

As seen in this chapter, no semiochemicals of acridoids other than those of *S. gregaria* have been adequately characterized to allow elucidation of their roles and

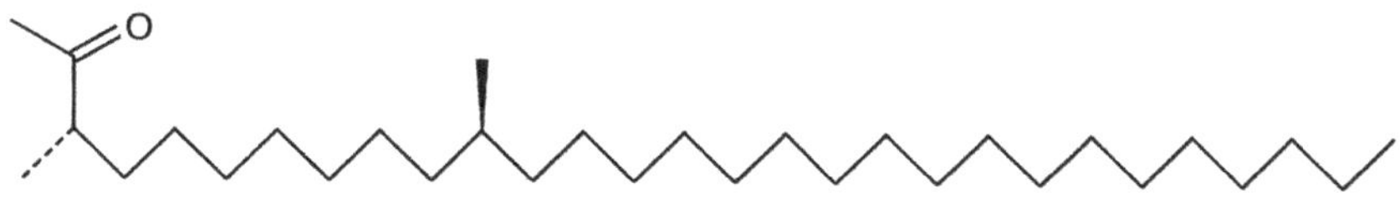

1

2

Fig. 12.8. 1, Sex pheromone of the German cockroach *Blatella germanica*; **2**, 3,7-dimethylheptacosan-2-one, detected in pentane body washings of adult males of *S. gregaria*, showing structural similarities of the two compounds.

exploration of their potential application. Thus, any speculation on the use of semiochemicals of Acridoidea must inevitably be biased by insights obtained on the chemical ecology of *S. gregaria*, although this may constitute a model for others.

Three categories of applications are proposed: (i) as phase markers to elucidate the factors underlying gregarization–solitarization dynamics; (ii) in the development of monitoring systems for detection and measurement of population build up; and (iii) in direct control.

Acridoid pheromones as phase markers

Gregarization–solitarization dynamics are central to the biology and pest status of locusts. Although the processes associated with gregarization have been described (Roffey and Popov, 1968), the underlying mechanisms are not well understood. Earlier field observations and recent studies have implicated several factors that may be important in the process. These include: (i) environmental heterogeneity and patchy distribution of vegetation, which may force solitarious insects to come together (Roffey and Popov, 1968; Bouachi *et al.*, 1996; Steedman, 1990); (ii) rapid production of aggregation pheromone resulting from forced crowding of solitaria (Deng *et al.*, 1996); and (iii) aggregative responses of solitaria to the gregarious-phase pheromone (Njagi *et al.*, 1996). Locust aggregation pheromones augmented with phase behavioural indices (Roessingh *et al.*, 1993; Heifetz *et al.*, 1996) provide powerful tools for studying the pace and extent of gregarization at different locust population levels in different natural and artificial habitats with varying host plant densities and compositions. Such studies may provide a much better characterization of habitats conducive to gregarization and a basis for refined early-warning systems.

Acridoid semiochemicals in monitoring

Early intervention in a preventive strategy has been a primary motive of much of the recent research on the desert locust. The development of a reliable monitoring system that can guide such interventions is critical. The richness of acridoid chemical ecology provides several options. Firstly, there is the changing composition of volatiles of flora associated with acridoid breeding areas (Assad *et al.*, 1997a), which could allow detection of the onset of breeding before the arrival of rains. Flora volatiles may be detected by electrophysiological or chemical techniques or, perhaps more conveniently, by remote sensing. Remote detection of moisture levels by microwave sensing technology is now well established. Could this be adapted to plant terpenoid constituents in the environment? Secondly, long-range sex pheromones, where they occur, would be very useful in the development of pheromone traps for early detection of reproductively active insects and monitoring their densities. Likewise, oviposition semiochemicals may be useful in monitoring gravid females. Lastly, it has been proposed that the absorption of the nitrile group of phenylacetonitrile in the infra-red (IR) spectrum may be used as a means of detecting aggregations of adults. IR-sensors might be installed in an aircraft used for surveillance of gregarious populations of *S. gregaria* adults (Francke and Schmidt, 1994).

Acridoid semiochemicals in control

Speculation on the use of pheromones in the control of polymorphic acridoids has focused on the possibility of disrupting gregarization in the early stages of population build up (Byer, 1991). Recent studies on desert locust pheromones suggest several ways to achieve this. Perhaps the most intriguing arises from cross-stage effects of nymphal and adult pheromones. The adult pheromone has been shown to be an effective solitarizing agent of hopper bands in the field (M. Bashir, Nairobi, 1998, personal communication). Oviposition pheromones may also be used to effect dispersal. Spreading the pheromone over a wide area has been proposed as a possible strategy (Byer, 1991; Saini *et al.*, 1995).

Since it is not yet known how dispersed, degregarized populations of nymphal or adult locusts would behave and develop, the concomitant deployment of entomopathogens or insecticides may be necessary. In the laboratory, desert locust hoppers treated with the adult pheromones have been found to be more susceptible to entomopathogens and synthetic insecticides (M. Bashir, Nairobi, 1998, personal communication). The use of semiochemicals to lure locusts or grasshoppers into pathogen traps for control or as a means of disseminating pathogens is another possibility.

Conclusions

This chapter has shown that we have just begun to unravel the rich and complex systems of pheromonal and related communication in grasshoppers and locusts.

Recent studies on the desert locust may provide a useful model in the process. However, even in this acridoid, many questions remain unanswered: the role and identity of primer gregarization pheromones; the role of aggregation pheromones during marching by hopper bands and swarming by adults; the signals involved in mating and mate recognition in the two phases; and the role of environmental signals, particularly those derived from host and other desert plants on the insect, to mention some. The chapter has also shown that the results obtained to date on the desert locust have provided useful new semiochemical tools for probing into these questions and also novel ways of manipulating the insect's behaviour and physiology. Some of these will undoubtedly lead to practical applications, and that represents the ultimate justification for continued research on the semiochemicals of acridoids. In this regard, it is pertinent to quote what Sir Boris Uvarov said more than two decades ago (Uvarov, 1977):

> ... it must be admitted that progress toward long-term ecological control of injurious acridoids lags behind the undoubted but short-term successes of direct control ... The successes achieved in control, which are only temporary and of only local significance, lead to the assertions, even by some acridologists, that the problem of acridoids as pests has been solved and that further research on it is pointless ... While recognizing that the direct control methods provide undoubted temporary protection during the upsurges and in the outbreak areas, it is essential to aim under their cover at a more radical solution of the problem.

Acridoid chemical ecology can make important contributions to that goal.

References

Ali, Y. (1987) Pheromone cells in epidermis of *Locusta migratoria*. *Pakistan Journal of Zoology* 19, 127–132.

Amerasinghe, F.P. (1978) Pheromonal effects on sexual maturation, yellowing and the vibration reaction in immature male desert locusts (*Schistocerca gregaria*). *Journal of Insect Physiology* 24, 309–314.

Anton, S. and Hansson, B.S. (1996) Antennal lobe interneurons in the desert locust *Schistocerca gregaria* (Forskål): processing of aggregation pheromones in adult males and females. *Journal of Comparative Neurology* 370, 85–96.

Assad, Y.O.H., Torto, B., Hassanali, A., Njagi, P.G.N., Bashir, N.H.H. and Mahamat, H. (1997a) Seasonal variation in the essential oil composition of *Commiphora quadricincta* and its effect on the maturation of immature adults of the desert locust, *Schistocerca gregaria*. *Phytochemistry* 44, 833–842.

Assad, Y.O.H., Hassanali, A., Torto, B., Mahamat, H., Bashir, N.H.H. and Bashir, S.E. (1997b) Effects of fifth instar nymphs on maturation of immature adults of the desert locust *Schistocerca gregaria* (Forskål) (Orthoptera: Acrididae). *Journal of Chemical Ecology* 23, 1373–1388.

Blight, M.M., Crove, J.F. and McCormick, A. (1969) Volatile nitrogenous bases emanating from laboratory-reared colonies of the desert locust *Schistocerca gregaria*. *Journal of Insect Physiology* 15, 259–272.

Bouaichi, A., Roessingh, P. and Simpson, S.J. (1995) Analysis of the behavioural effects of crowding and re-isolation on solitary-reared adult desert locusts (*Schistocerca gregaria*) and their offspring. *Physiological Entomology* 20, 199–208.

Bouaichi, A., Simpson, S.J. and Roessingh, P. (1996) The influence of environmental microstructure on the behavioural phase state and distribution of the desert locust *Schistocerca gregaria*. *Physiological Entomology* 21, 247–256.

Byer, J.A. (1991) Pheromones and chemical ecology of locusts. *Biological Reviews* 66, 347–378.

Carlisle, D.B., Ellis, P.E. and Betts, E. (1965) The influence of aromatic shrubs on sexual maturation in the desert locust *Schistocerca gregaria*. *Journal of Insect Physiology* 11, 1541–1558.

Cassier, P. and Papillon, M. (1983) Le dimorphisme sexuel des protéines cuticulaires du criquet migrateur, *Locusta migratoria migratorioides* (Reiche & Farmaire). *Canadian Journal of Zoology* 61, 2976–2986.

Cassier, P., Porcheron, P., Papillon, M. and Lensky, Y. (1980) Study of the cuticular proteins of the African migratory locust, *Locusta migratoria migratorioides* (R. and F.). Quantitative data. *Annales de Sciences Naturelles Zoologie* 2, 51–65.

Chapman, R.F., Page, W.W. and McCaffery A.R. (1986) Bionomics of the variegated grasshopper (*Zonocerus variegatus*) in West and Central Africa. *Annual Review of Entomology* 31, 479–505.

Chauvin, R. (1941) Contribution á l'étude physiologique du criquet pèlerin et du déterminisme des phénomènes grégaires. *Annales de la Société Entomologique de France* 1, 1–173.

Dale, J.F. and Tobe, S.S. (1990) The endocrine basis of locust phase polymorphism. In: Chapman, R.F. and Joern, A. (eds) *Biology of Grasshoppers*. John Wiley & Sons, New York, pp. 393–414.

Deng, A.L. (1995) Studies of some factors that influence phase dynamics of the desert locust, *Schistocerca gregaria* (Forskål) (Orthoptera: Acrididae). PhD thesis, University of Khartoum, Sudan.

Deng A. L., Torto, B. Hassanali, A. and Ali, E.E. (1996) Effects of shifting to crowded or solitary conditions on pheromone release and morphometrics of the desert locust, *Schistocerca gregaria* (Forskål) (Orthoptera:Acrididae). *Journal of Insect Physiology* 42, 771–776.

El-Bashir, S., Inayatullah, C., Ahmed, A.O. and Abdel Rahman, H.E. (1993) Models for estimation of population density of late instar nymphs and fledglings of the desert locust, *Schistocerca gregaria* (Forsk.). *International Journal of Pest Management* 39, 467–470.

Ellis, P.E. (1959) Some factors influencing phase characters in the nymphs of the locust, *Locusta migratoria migratorioides* (R.& F.). *Insectes Sociaux* 6, 21–39.

Ellis, P.E. (1963) Changes in the social aggregation of locust hoppers with changes in rearing conditions. *Animal Behaviour* 11, 152–160.

Ellis, P.E. and Pearce, A. (1962) Innate and learned behaviour patterns that lead to group formation in locust hoppers. *Animal Behaviour* 10, 305–318.

Ellis, P.E., Carlisle, D.B. and Osborne, O.P. (1965) Desert locust: sexual maturation delayed by feeding on senescent vegetation. *Science* 149, 546–547.

Faber, A. (1953) Laut und Gebärdensprache bei Insekten: Orthoptera (Geradflügler). I. *Mitteilungen der Staatliches Museum für Naturkunde, Stuttgart* 287, 1–198.

Francke, W. and Schmidt, G.H. (1994) Potential of semiochemicals for locust control. In: Rembold, H., Benson, J.A., Franzen, H., Weickel, B. and Schulz, F.A. (eds) *New Strategies for Locust Control*. ASTAF, Bonn, pp. 48–52.

Friedel, T. and Gillot, C. (1976) Male accessory gland substance of *Melanoplus sanguinipes*: an oviposition stimulant under control of the corpus allatum. *Journal of Insect Physiology* 22, 489–495.

Fuzeau-Braesch, S., Genin, E., Jullien, R., Knowles, E. and Papin, C. (1988) Composition and role of volatile substances in atmosphere surrounding two gregarious locusts, *Locusta migratoria* and *Schistocerca gregaria*. *Journal of Chemical Ecology* 14, 1023–1032.

Ghouri, A.S.K. and Ahmad, H. (1960) Swarming of *Hieroglyphus nigrorepletus*. *Plant Protection Bulletin FAO* 8, 135–136.
Gillett, S.D. (1968) Airborne factor affecting grouping behaviour in locusts. *Nature* 218, 782–783.
Gillett, S.D. (1983) Primer pheromones and polymorphism in the desert locust. *Animal Behaviour* 31, 221–230.
Gillett, S.D., Packham, J.M. and Papworth, S.J. (1976) Possible pheromonal effects on aggregation and dispersion in the desert locust, *Schistocerca gregaria* (Forsk.). *Acrida* 5, 287–297.
Hansson, B.S., Ochieng, S.A., Grosmaitre, X., Anton, S. and Njagi, P.G.N. (1996) Physiological responses and central nervous projections of antennal olfactory receptor neurons in the adult desert locust, *Schistocerca gregaria* (Orthoptera: Acrididae). *Journal of Comparative Physiology* A 179, 157–67.
Haskell, P.T., Paskin, M.W.J. and Moorhouse, J.E. (1962) Laboratory observations on factors affecting the movements of hoppers of the desert locust. *Journal of Insect Physiology* 8, 53–78.
Hassanali, A. and Mahamat, H. (1991) Semiochemical research on the desert locust *Schistocerca gregaria* (Forsk.) at ICIPE: rationale and scope. In: Njagi, P.G.N. and Chaudhury, M.F.B. (eds) *Proceedings – Workshop on Effective Networking of Research and Development on Environmentally Sustainable Locust Control Methods among Locust Affected Countries*. The International Centre of Insect Physiology and Ecology, ICIPE Science Press, Nairobi, pp. 5–12.
Hawkes, F., Rzepka, J. and Gontrand, G. (1987) The scent glands of male South American locust *Schistocerca cancellata*, a electron microscope study. *Tissue and Cell* 19, 687–703.
Heifetz, Y., Voet, H. and Applebaum, S.W. (1996) Factors affecting behavioural phase transition in the desert locust, *Schistocerca gregaria* (Forskål) (Orthoptera: Acrididae). *Journal of Chemical Ecology* 22, 1717–1734.
Inayatullah, C., El Bashir, S. and Hassanali, A. (1994) Sexual communication in the desert locust (Orthoptera: Acrididae): an evidence of a sex pheromone. *Environmental Entomology* 23, 1544–1551.
Islam, M.S., Roessingh, P., Simpson, S.J. and McCaffery, A.R. (1994a) Parental effects on the behaviour and colouration of nymphs of the desert locust *Schistocerca gregaria*. *Journal of Insect Physiology* 40, 173–181.
Islam, M.S., Roessingh, P., Simpson, S.J. and McCaffery, A.R. (1994b) Effects of population density experienced by parents during mating and oviposition on the phase of hatching desert locusts, *Schistocerca gregaria*. *Proceedings of the Royal Society of London* B 257, 93–98.
Jackson, G.T., Popov, G.B., Ibrahim, A.O., Alghamedi, S.A. and Khan, A.M. (1978) Effect of natural food plants on the development, maturation, fecundity and phase of the desert locust, *Schistocerca gregaria* (Forskål). Centre for Overseas Pest Research, London, and LRS/ARS (Jeddah), pp. 1–17.
Jacobs, W. (1953) Verhaltensbiologische Studien an Feldhausschrecken. *Zeitschrift für Tierpsychologie* 1, 1–228.
Kennedy, J.S. (1939) The behaviour of the desert locust, *Schistocerca gregaria* (Forskål) (Orthoptera) in an outbreak centre. *Transactions of Royal Entomological Society of London* 89, 385–452.
Lange, A.B. and Loughton, B.G. (1984) An analysis of the secretions of the male accessory reproductive gland of the African migratory locust. *International Journal of Invertebrate Reproduction and Development* 7, 73–81.
Lange, A.B. and Loughton, B.G. (1985) An oviposition-stimulating factor in the male accessory reproductive gland of the locust, *Locusta migratoria*. *Journal of Comparative Endocrinology* 57, 208–215.

Lauga, J. and Hatté, M. (1977a) Propriétés grégarisantes acquises par le sable dans lequel ont pondu à de nombreuses reprises des femelles grégaires de *Locusta migratoria migratorioides*, R. & F. (Orthoptère, Acrididae). *Acrida* 6, 307–311.

Lauga, J. and Hatté, M. (1977b) Mise en évidence d'une activité de type phéromone dans le milieu de ponte du criquet migrateur *Locusta migratoria* L. *Comptes-Rendus Hebdomadaire des Séances Académie Sciences, Série D* 285, 57–59.

Lauga, J. and Hatté, M. (1978) L'activité grégarisante du sable de ponte chez *Locusta migratoria* L.: action sur le comportement et la réproduction des individus. *Annales des Sciences Naturelles Zoologie Biologie Animale (12E)* 20, 37–52.

Leahy, M.G. (1973a) Oviposition of virgin *Schistocerca gregaria* (Forskål) (Orthoptera: Acrididae) after implant of the male accessory gland complex. *Journal of Entomology A: General Entomology* 48, 69–78.

Leahy, M.G. (1973b) Oviposition of *Schistocerca gregaria* (Forskål) (Orthoptera: Acrididae) mated with males unable to transfer spermatophores. *Journal of Entomology A: General Entomology* 48, 79–84.

Loher, W. (1960) The chemical acceleration of the maturation process and its hormonal control in the male of the desert locust. *Proceedings of the Royal Society B* 153, 381–397.

Loher, W. (1984) Behavioural and physiological changes in cricket-females after mating. *Advances in Invertebrate Reproduction* 3, 189–201.

Loher, W. (1990) Pheromones and phase transformation in locusts. In: Chapman, R.F. and Joern, A. (eds) *Biology of Grasshoppers*. John Wiley & Sons, New York, pp. 337–355.

Loher, W., Ganjian, I., Kubo, I., Stanley-Samuelson, D. and Tobe, S.S. (1981) Prostaglandins: their role in egg-laying of the cricket, *Teleogryllus commodus*. *Proceedings of the National Academy of Sciences USA* 78, 7835–7838.

Mahamat, H., Hassanali, A., Odongo, H., Torto, B. and El Bashir, S. (1993) Studies on the maturation-accelerating pheromone of the desert locust *Schistocerca gregaria* (Orthoptera: Acrididae). *Chemoecology* 4, 159–164.

Mahamat, H., Hassanali, A. and Ferenz, H.J. (1997) Vitellogenin titres in normal and accelerated maturation of gregarious-phase *Schistocerca gregaria*. *Comparative Biochemistry and Physiology* 116, 447–451.

McCaffery, A.R. and Page, W.W. (1982) Oviposition behaviour of the grasshopper *Zonocerus variegatus*. *Ecological Entomology* 7, 85–90.

McCaffery, A.R., Cook, A.G., Page, W.W. and Perfect, T.J. (1978) Utilisation of food by *Zonocerus variegatus* (L) (Orthoptera: Acridoidea). *Bulletin of Entomological Research* 68, 589–606.

McCaffery, A.R., Simpson, S.S., Islam, M.S. and Roessingh, P. (1998) A gregarising factor present in the egg pod foam of the desert locust *Schistocerca gregaria*. *Journal of Experimental Biology* 201, 347–363.

Michel, R. (1980) Development of flight behaviour of successive generations of desert locust, *Schistocerca gregaria* (Forskål) raised in isolation then in groups. *Animal Behaviour* 28, 1288–1289.

Niassy, A. (1997) Interaction between *Schistocerca gregaria* (Forskål) and *Locusta migratoria migratorioides* (Reich & Farmaire) in relation to phase polymorphism. PhD thesis, University of Ghana, Legon, Ghana.

Njagi, P.G.N., Torto, B., Obeng-Ofori, D. and Hassanali, A. (1996) Phase-independent responses to phase-specific aggregation pheromone in adult desert locust, *Schistocerca gregaria* (Forskål) (Orthoptera: Acrididae). *Physiological Entomology* 21, 131–137.

Nolte, D.J. (1963) A pheromone for melanization of locusts. *Nature* 200, 660–661.

Nolte, D.J., Eggers, S.H. and May, I.R. (1973) A locust pheromone: locustol. *Journal of Insect Physiology* 19, 1547–1554.

Norris, M.J. (1954) Sexual maturation in the desert locust (*Schistocerca gregaria* Forskål) with special reference to the effects of grouping. *Anti-Locust Bulletin* 18, 1–44.
Norris, M.J. (1962) Group effects on the activity and behaviour of adult males of the desert locust (*Schistocerca gregaria* Forsk.) in relation to sexual maturation. *Animal Behaviour* 10, 275–291.
Norris, M.J. (1963) Laboratory experiments on gregarious behaviour in ovipositing females of the desert locust (*Schistocerca gregaria* [Forskål]). *Entomologia Experimentalis et Applicata* 6, 279–303.
Norris, M.J. (1964) Accelerating and inhibiting effects of crowding on sexual maturation in two species of locusts. *Nature* 203, 784–785.
Norris, M.J. (1970) Aggregation response in ovipositing females of the desert locust, with special reference to the chemical factor. *Journal of Insect Physiology* 16, 1493–1515.
Norris, M.J. and Pener, M.P. (1965) An inhibitory effect of allatectomized males and females on the sexual maturation of young male adults of *Schistocerca gregaria* (Forsk.) (Orthoptera: Acrididae). *Nature* 208, 1122.
Obeng-Ofori, D., Torto, B. and Hassanali, A. (1993) Evidence for mediation of two releaser pheromones in aggregation behaviour of the gregarious desert locust *Schistocerca gregaria* (Orthoptera: Acrididae). *Journal of Chemical Ecology* 19, 1665–1676.
Obeng-Ofori, D., Njagi, P.G.N., Torto, B., Hassanali, A. and Amiani, H. (1994a) Sex differentiation studies relating to releaser pheromones of the desert locust, *Schistocerca gregaria* (Forskål). *Entomologia Experimentalis et Applicata* 73, 85–91.
Obeng-Ofori, D., Torto, B., Njagi, P.G.N., Hassanali, A. and Amiani, H. (1994b) Fecal volatiles as part of the aggregation pheromone complex of the desert locust, *Schistocerca gregaria* (Forskål) (Orthoptera: Acrididae). *Journal of Chemical Ecology* 20, 2077–2087.
Ochieng, S. (1997) Odour detection in the desert locust *S. gregaria*: antennal structure and function. PhD thesis, Lund University, Sweden.
Odhiambo, T.R. (1969) The architecture of the accessory reproductive glands of the male desert locust 1: types of glands and their secretions. *Tissue and Cell* 1, 155–182.
Okelo, O. (1979) Influence of male presence on clutch size in *Schistocerca vaga* Scudder (Orthoptera: Acrididae). *International Journal of Invertebrate Reproduction* 1, 317–321.
Otte, D. (1970) A comparative study of communicative behaviour in grasshoppers. *Miscellaneous Publication Museum Zoology, University of Michigan* 141, 1–168.
Pener, M.P. (1991) Locust phase polymorphism and its endocrine relations. *Advances in Insect Physiology* 23, 1–79.
Pickford, R., Ewen, A.B. and Gillott, C. (1969) Male accessory gland substance: an egg-laying stimulant in *Melanoplus sanguinipes* (F.) (Orthoptera: Acrididae). *Canadian Journal of Zoology* 47, 1199–1203.
Popov, G.B. (1958) Ecological studies on oviposition by swarms of the desert locust (*Schistocerca gregaria* Forskål) in eastern Africa. *Anti-Locust Bulletin* 31, 1–70.
Rai, M.M., Hassanali, A., Saini, R.K., Odongo, H. and Kahoro, H. (1997) Identification of components of the oviposition aggregation pheromone of the gregarious desert locust *Schistocerca gregaria*. *Journal of Insect Physiology* 43, 83–97.
Rainey, R.C. (1962) Some effects of environmental factors on movements and phase-change of locust populations in the field. In: *Physiologie Comportement et Écologie des Acridiens en Rapport avec la Phase*. CNRS, Paris, pp. 175–199.
Rao, R.Y. (1962) Fluctuations in infestation of the Bombay Locust *Patanga succincta* L. in India, correlated with weather data. *Proceedings of the 1st All-India Congress of Zoology, 1959 Jabalpur* 4, 25–28.
Richards, M.J. and El Mangoury, M.A. (1968) Further experiments on the effects of social factors on the rate of sexual maturation in the desert locust. *Nature* 219, 865–866.

Riede, K. (1987) A comparative study of mating behaviour in some neotropical grasshoppers (Acridoidea). *Ethology* 76, 265–296.

Roessingh, P. and Simpson, S.J. (1994) The time-course of behavioural phase change in nymphs of the desert locust, *Schistocerca gregaria. Physiological Entomology* 19, 191–197.

Roessingh P., Simpson, S.J. and James, S. (1993) Analysis of phase related changes in behaviour of desert locust nymphs. *Proceedings of the Royal Society London* B 252, 43–49.

Roffey, J. (1965) *Report to the Government of Thailand on Locust and Grasshopper Control.* Report No. 2109, FAO, Rome.

Roffey, J. and Popov, G. (1968) Environmental and behavioural processes in a desert locust outbreak. *Nature* 219, 446–450.

Saini, R.K., Rai, M.M., Hassanali, A., Wawiye, J. and Odongo, H. (1995) Semiochemicals from froth of egg pods attract ovipositing female *Schistocerca gregaria. Journal of Insect Physiology* 41, 711–716.

Schmidt, G.H. and Osman, K.S.A. (1988) Male pheromones and egg production in Acrididae. In: Sehnal, F., Zabza, A. and Denlinger, D.L. (eds) *Endocrinological Frontiers in Physiological Insect Ecology*. Wroclaw Technical University Press, Wroclaw, pp. 701–706.

Siddiqi, J.I. and Khan, M.A. (1981) The secretion and perception of a sex pheromone in the grasshopper *Hieroglyphys nigroreplatus* Bolivar (Orthoptera: Acrididae). *Acrida* 10, 233–242.

Sjöstedt, Y. (1923) Zoological results of the Swedish expedition to Central Africa 1921. Insecta I. Acridoidea. *Ark Zoology* 15, 39 pp.

Steedman, A. (1990) *Locust Handbook*, 3rd edn. Natural Resources Institute, Chatham, UK, 204 pp.

Stower, W.J., Popov, G.B. and Greathead, D.J. (1958) Oviposition behaviour and egg mortality of the desert locust (*Schistocerca gregaria* Forskål) on the coast of Eritrea. *Anti-Locust Bulletin* 30, 1–331.

Strong, L. and Amerasinghe, F.P. (1977) Allatectomy and sexual receptivity in females of *Schistocerca gregaria. Journal of Insect Physiology* 23, 131–135.

Torto, B., Obeng-Ofori, D., Njagi, P.G.N., Hassanali, A. and Amiani, H. (1994) Aggregation pheromone system of adult gregarious desert locust, *Schistocerca gregaria* (Forskål) (Orthoptera: Acrididae). *Journal of Chemical Ecology* 20, 1749–1762.

Torto, B., Njagi, P.G.N., Hassanali, A. and Amiani, H. (1996) Aggregation pheromone system of nymphal gregarious desert locust, *Schistocerca gregaria* (Forskål) (Orthoptera: Acrididae). *Journal of Chemical Ecology* 22, 2273–2281.

Torto, B., Assad, Y.O.H., Njagi, P.G.N. and Hassanali, A. (1999) Evidence for additional pheromonal components mediating oviposition aggregation in *Schistocerca gregaria* (Forskål). *Journal of Chemical Ecology* 25, 835–845.

Uvarov, B.P. (1921) A revision of the genus *Locusta* L. (Pacytylus L. Fieb.), with a new theory as to the periodicity and migrations of locusts. *Bulletin of Entomological Research* 12, 135–163.

Uvarov, B.P. (1966) *Grasshoppers and Locusts*, Vol. 1. Cambridge University Press, Cambridge, 481 pp.

Uvarov, B.P. (1977) *Grasshoppers and Locusts*, Vol. 2. Centre for Overseas Pest Research, London, 613 pp.

Whitman, D.W. (1982) Grasshopper sexual pheromone: a component of the defensive secretion in *Taenipoda eques. Physiological Entomology* 7, 111–115.

Whitman, D.W. (1990) Grasshopper chemical communication. In: Chapman, R.F. and Joern, A. (eds) *Biology of Grasshoppers*. John Wiley & Sons, New York, pp. 357–391.

13 Termites

MANFRED KAIB

Department of Animal Physiology, University of Bayreuth, D–95440 Bayreuth, Germany

Introduction

Social insects occur in most terrestrial environments with their success and dominance reliant on the evolution and maintenance of sociality. In some species this can lead to insect colonies with more than one million nestmates (Wilson, 1990). The highest degree of sociality can be found in the order Isoptera (termites) and in Hymenoptera, ants and in some species of bees and wasps. Termites are eusocial insects with overlapping generations occurring together in a colony but, in contrast with hymenopterous species, the only adults present in termite colonies are the king and queen. The social organization in such colonies depends on the control of the proportion of different castes, and on an efficient recognition and communication system. Like all social insects, termites primarily rely on chemicals during caste regulation and communication, and in particular they employ pheromones produced by various glands. Both primer and releaser pheromones can be distinguished. *Primer* pheromones cause long-term physiological changes in nestmates by controlling their endocrine and/or reproductive systems (Wilson, 1971). This type of pheromone is usually dispersed by only one or a few individuals, e.g. the queen(s), and may regulate sexuality and caste expression. In contrast, chemical signals that cause immediate behavioural changes in conspecifics are defined as *releaser* pheromones and are produced by numerous nestmates.

Primer Pheromones and Caste Differentiation

Castes, in the traditional sense, are groups of individuals that are morphologically differentiated to perform specific tasks such as reproduction, foraging for food, or colony defence (see also Peeters, 1990). However, in recent years, mainly for Hymenoptera, castes have also been defined according to their behavioural role. In this chapter the term 'caste' is used in the traditional, morphological sense.

In termites, the one adult caste is initially winged (alate) and develops from the nymphs (Fig. 13.1a). These alates, of both sexes, fly from the nest to establish new colonies and after swarming lose their wings (Fig. 13.1b) to form tandems, to mate and become the reproductives of a new colony. In some species, mainly in the 'lower' termites, these so-called primary reproductives may later be replaced in the colony by secondary or even tertiary reproductives, which develop from nymphal stages (nymphoid) or from workers (ergatoid). Termite queens typically become physogastric, due to an enormous growth of the fat body and the ovaries (Fig. 13.2), while the males remain smaller.

The second true caste in termites comprises the soldiers. These are always non-reproductive and are more sclerotized and more heavily pigmented than the other sterile individuals (Fig. 13.2). Worker termites may be more or less differentiated, depending on the evolutionary status of the species. In primitive termites, social tasks are accomplished by unspecialized larvae or nymphs. In the higher developed Termitidae, and some other termite groups, workers constitute a true caste, specialized in morphology and behaviour and permanently excluded from the nymphal developmental pathway. Though controversial, the term 'worker caste' is well established in the literature and will also be used in this chapter.

Evolution of soldier castes

Caste differentiation and the proportion of individuals belonging to different castes is regulated by extrinsic and intrinsic factors. Extrinsic factors may serve as a major cause in the evolution of the soldier caste where morphological adaptations, such as highly sclerotized and powerful mandibles, make them suitable for colony defence. Ants are major predators of termite nests and predator–prey interactions

Fig. 13.1. (a) After swarming, female alates raise their abdomen, expose the sternal and tergal glands, flutter their wings, and attract males by their pheromones. (b) When the female has been successful in calling, female and male shed their wings and form a tandem, during which the male closely follows the female.

Fig. 13.2. In termites, queens become physogastric when laying eggs. The species presented here, *Macrotermes michaelseni,* has evolved two soldier castes: major (MS) and minor (mS). There are also two types of workers, major and minor workers, which differ in head capsule width (difficult to see on this photograph). Some larvae (white individuals in the foreground) can also be seen.

between termites and ants occur continuously, unlike the occasional or seasonal predatory activities of vertebrates (Kaib, 1985, 1987). Consequently, the literature is dominated by descriptions of defence interactions with these equal-sized predators; and the constant predation by ants, as well as the life habits of termite species, are the driving force for adaptive evolution of a soldier defence strategy. Drywood termites and dampwood termites tend to have lower proportions of soldiers than subterranean termites (Haverty, 1977). In termite species relying on chemical defence (Deligne *et al.*, 1981) and in species whose soldiers have additional tasks, such as initiation for food foraging (Traniello, 1981; Kaib, 1990), the proportion of soldiers may reach up to one third of the total population of steriles in a colony. In addition, the proportion of soldiers may be modulated by the production of alates (Lenz, 1994) or by environmental factors such as a temporarily elevated predation by ants (Abe and Darlington, 1985).

Control of sexuality

There is evidence that juvenile hormones (JHs), released by the corpora allata, act as primer pheromones to maintain and regulate caste expression or caste suppression in termites (Lüscher and Springhetti, 1960; Lüscher, 1960, 1976). Acting in association with other endocrine secretions, JHs were first shown to control metamorphosis in insects. JHs or JH analogues fed or topically applied to termites also have pheromonal effects, suggesting that JH could circulate between individuals and act as primer pheromones (Noirot and Bordereau, 1991).

The reproductive pair (king and queen) in a colony appear to exert a strong influence, inhibiting the development of further reproductives. It can be shown that removal of the king or queen from a termite colony causes workers or nymphs to develop as reproductives to replace them (Lüscher, 1961; Springhetti, 1985). Similar results could be achieved by using JH analogues (Wanyonyi, 1974; Yin and Gillot, 1975; Doki *et al.*, 1984). Under natural conditions, when termite colonies become very large, the control of the sexual pair appears to wane in distant sections of the colony. Workers or nymphs may then attain reproductive status (McMahan, 1966; Springhetti and Pinamonti, 1977; Grassé, 1982). Such secondary reproductives have frequently been described in termites (Thorne, 1982a, 1985, 1997).

Caste regulation

Because of their specialized mandibles, termite soldiers cannot feed themselves and have to be fed by workers. Thus caste proportions have to be optimized to meet the prevailing environmental demands (Lenz, 1976; Oster and Wilson, 1978). The rate at which castes differentiate appears to become self-regulated once a caste-specific proportion has been reached in a colony (Henderson, 1998). Soldier termites, for example, regulate their own numbers by inhibiting the larval development of other soldiers. In *Nasutitermes lujae* (Wasmann) the differentiation of new soldiers is apparently suppressed or delayed by a primer pheromone from the soldiers' frontal gland (Lefeuve and Bordereau, 1984). However, this result cannot be generalized for all termites, as soldier inhibition also exists in species devoid of a frontal gland (e.g. Castle, 1934; Grassé, 1982). Soldier production is controlled by JH or, influenced experimentally, by its analogues (Castle, 1934; Hrdy and Křeček, 1972). Lenz (1976) noted that species-specific proportions of soldiers may be a result of variable thresholds for JH. On the other hand, by increasing the overall JH levels, a colony may increase the number of soldiers at a given time, e.g. at high rates of predation. For termite species faced with this high predation due to their life habits, Henderson (1998) suggested that, over time, a reduced JH threshold for worker-to-soldier moults might have evolved.

Self-regulation

The non-volatile JHs appear to be transmitted to other nestmates by allogrooming or by trophallaxis. Egg-laying queens, for example, produce together with the eggs a secretion rich in JH, which is readily picked up by termite workers and presumably spread over the entire colony. Self-regulation has also been reported for alates (e.g. Miller, 1969; Grassé, 1982; Goldberg and Tobe, 1985; Springhetti, 1985; Henderson and Delaplane, 1994). Conversely, one caste can regulate differentiation in another caste. Springhetti (1985) and Bordereau (1985), for example, have shown that the production of nymphs may stimulate the production of soldiers and that soldiers may stimulate the production of secondary reproductives. Seasonal fluctuations in soldier levels, e.g. prior to swarming, are often correlated with the number of alates (Bodot, 1969; Sen-Sarma and Mishra, 1972; Howard and Haverty, 1981).

Although numerous studies on caste differentiation and caste regulation have been published in the last four decades, the role of primer pheromones in termites is not yet completely understood (for more details see Noirot and Bordereau, 1991; Henderson, 1998).

Releaser Pheromones and Communication

Chemical communication is fairly well understood in social Hymenoptera (see Chapter 17, this volume). However, comparatively little is known about the chemical communication in termites and the pheromone systems involved. Reviews of early research on termite pheromones is summarized in Stuart (1969, 1970), Moore (1974) and Prestwich (1984). More recently, Pasteels and Bordereau (1998) focused their review on pheromones involved in foraging, sexual behaviour and defence. In contrast to the Hymenoptera, for which some 63 different exocrine glands are known (Billen and Morgan, 1998), only a few exocrine glands have been described in termites. The chemical signals may either be released as airborne volatiles or may be deposited as low-volatile or non-volatile substances on solid substrates such as soil or food. The low number and low variability of the exocrine glands in termites are unusual in social insects and may be explained by the facts that: (i) termites are phylogenetically an old clade with a hemimetabolous development and a conservative morphology; (ii) unlike ants, termites are never predatory but feed on wood or plant litter which is available over long periods of time. In consequence, there were no ecological constraints to evolve complex and diverse communication strategies to maintain territories or to rapidly monopolize limited food resources.

Exocrine glands, the sources of pheromones

The exocrine glands discussed here focus on those cases for which detailed morphological and functional knowledge is available (Table 13.1).

Labial or salivary gland

The termite labial or salivary gland is located in the meso- and metathorax and has the same general structure as that observed in Orthoptera and Dictyoptera (Noirot,

Table 13.1. Exocrine glands in termites and the behavioural patterns released by their secretions.

Exocrine gland	Released behaviour
Labial gland	Colony defence
	Nest or gallery building
	Aggregation at feeding sites
Sternal gland	Trail-following
	Sexual communication
Tergal gland	Sexual communication
Frontal gland	Alarm
Cuticle?	Recognition

1969). Labial glands have been investigated in workers from various termite families and their morphology appears to be uniform. The paired gland consists of numerous spherical acini which are composed of different types of secretory cells and of canalicular cells that collect the secretion from the secretory cells and form the duct leaving the acini (Billen *et al.*, 1989; Costa-Leonardo and Cruz-Landim, 1991; Kaib and Ziesmann, 1992; Costa-Leonardo, 1997). The glandular ducts, in the head, join with those coming from the water sacs and open into the oral cavity from where the chemical signal is released, together with the saliva, on to the substrate (Kaib and Ziesmann, 1992; Reinhard *et al.*, 1997). Although labial glands are also found in termite soldiers, where the secretion may serve for colony defence (Maschwitz and Tho, 1974; Moore, 1974; Wood *et al.*, 1975), data on their function focus mainly on the labial glands of termite workers. Their secretion serves various functions during nest construction (Noirot, 1969; Bruinsma, 1979). Apparently, in some species the labial gland of termite workers has a fungistatic or fungicidal function for non-termitophilic fungi (Grassé, 1982; Brian, 1983) or plays a role in the breakdown and digestion of cellulose (Martin and Martin, 1978; Hogan *et al.*, 1988; Veivers *et al.*, 1991). The pheromonal function of the labial gland secretion is aimed at the aggregation of termite workers at gnawing sites (see below).

Sternal gland

The sternal gland is an unpaired gland, first described by Grassi and Sandias (1893). It is found in workers, soldiers and alates and varies between termite families. In Mastotermitidae, the sternal gland is found in a chain of three parts located in the middle of the third, fourth and fifth sternite. In all other taxa the sternal gland is a single organ; in Hodotermitidae it is located on the fourth sternite; in Termopsidae, on either the fourth or the fifth sternite; and in the other termite taxa, in the front of the fifth sternite (Noirot, 1969, 1995). At least in the phylogenetically more advanced termite taxa, the sternal gland includes two types of secretory cells and in some species the anterior part of the gland differs from the posterior part (Noirot, 1969). This difference gave rise to the hypothesis that different types of secretion might be produced in the sternal gland; however, as yet there is no unequivocal evidence for this assumption. The sternal gland is the source of the trail pheromone and, in some species, plays a role in sex communication (see below).

Tergal glands

The tergal glands appear to exist only in alates and are located in the last three to five abdominal tergites. The unpaired tergal glands not only vary morphologically between species but are apparently not developed at all in Hodotermitidae and Rhinotermitidae and in some species of Termitidae (Noirot, 1969). In summary, our knowledge of tergal glands is rather fragmentary, in part controversial, and further investigation is needed for a better understanding of their phylogenetic and functional role.

Frontal gland

The frontal gland is the most studied exocrine gland in termites, in both morphological and chemical respects. It is present only in Rhinotermitidae and Termitidae and differs considerably between species in structure, size and function. This unpaired cephalic gland is formed by a deep invagination of the cuticle that opens to the exterior by a frontal pore. The glandular epithelium is composed of different types of secretory cells and is covered, at least partly, by a modified cuticle which forms the reservoir for the secretion (Deligne *et al.*, 1981). In most termite taxa, the frontal gland is restricted to the head, but within the Rhinotermitidae it often extends deeply into the abdomen (Renoux, 1976; Deligne *et al.*, 1981; Kaib, 1985, 1990). The frontal gland is well developed in termite soldiers, but in workers it is rudimentary and consists only of a thickening of the cephalic integument without a frontal pore. Male and female alates of *Schedorhinotermes lamanianus* (Sjöstedt) have a well developed frontal gland and its secretion is composed of the same chemical constituents as that of the minor soldiers.

In different termite species, the frontal gland secretions cover a wide range of chemical classes – terpenoids, quinones, lactones, hydrocarbons, nitroalkanes, ketones and ketoaldehydes as well as proteins and mucopolysaccharides (Prestwich, 1984). They function as irritants, glues or contact poisons in colony defence. Additional functions in the regulation of caste determination or as fungistatic substances are likely. In some termite species volatiles from the frontal gland serve as alarm pheromones (see below).

Cuticle

The termite cuticle is (as in other social insects) probably the largest exocrine gland. Complex mixtures of chemicals, mainly hydrocarbons, are believed to be cues for nestmate recognition (Howard, 1993; Clément and Bagnères, 1998). However, this belief is mainly based on the fact that the mixtures of hydrocarbons differ between conspecific colonies (e.g. Howard, 1993) or that the level of such differences in mixtures correlate well with the level of intercolonial aggression (Jmhasly *et al.*, 1998). Additionally, in ants, clear evidence has been found that an exocrine gland in the head, the postpharyngeal gland, produces these hydrocarbon mixtures, which are then transferred to nestmates by allogrooming and form a colony-specific recognition label (Hefetz *et al.*, 1996; Soroker *et al.*, 1998). Comparable studies in termites are currently lacking.

Some other exocrine glands have been described in termites but their possible function in pheromonal communication is poorly understood (Noirot, 1969; Bacchus, 1979).

Releaser Pheromones

Trail pheromones

Trail-following by termites was first reported by Dudley and Beaumont (1889). Approximately 70 years later, Lüscher and Müller (1960) and Stuart (1961) independently found that the sternal gland was the source of trail pheromones. To date no other source for trail pheromones is known. Hence, the source of trail pheromones in termites is homogeneous – very different from ants, where about ten distinct sources for a variety of trail pheromones have been reported (Hölldobler, 1984).

Chemical identification of trail pheromones was hindered by the fact that many termite species follow artificial trails produced by a wide diversity of chemical components such as glycol derivatives, phthalates, fatty acids and saturated (mostly primary) alcohols (Verron and Barbier, 1962; Becker, 1966; Becker and Mannesmann, 1968; Hummel and Karlson, 1968; Karlson *et al.*, 1968; Klochov and Zhuzhikov, 1990). Furthermore, extracts from the wood on which termites feed or from wood infested by the fungus *Gloeophyllum trabeum* (= *Lenzites trabea Rers.* ex Fr.) induce trail-following (Matsumura *et al.*, 1969; Ritter *et al.*, 1973, 1977). Consequently Pasteels and Bordereau (1998) stressed that compounds inducing trail-following on artificial trails can only be claimed to be a trail pheromone with certain confidence when, firstly, the compound is present in the sternal gland and, secondly, the termites respond to a physiologically realistic concentration of the compound.

Only two different trail pheromones have been chemically identified. In *Reticulitermes flavipes* (Kollar), (*Z*,*Z*,*E*)-(3,6,8)-dodecatrienol (Fig. 13.3a) is the trail pheromone (Matsumura *et al.*, 1968, 1969; Tai *et al.*, 1969) and its trail-following threshold is 0.01 pg cm^{-1} trail. *R. flavipes* shows remarkable specificity to the pheromone as stereoisomers and derivatives of the pheromone are less effective by up to seven orders of magnitude (Tai *et al.*, 1971). In contrast with this high chemical specificity, this dodecatrienol is not species-specific. The same dodecatrienol isomer has been found in *Coptotermes formosanus* (Shiraki) (Tokoro *et al.*, 1989), *R. speratus* (Kolbe) (Tokoro *et al.*, 1990, 1991), *R. santonensis* (Shiraki) (Laduguie *et al.*, 1994a) and *R. hesperus* (Banks) (Grace *et al.*, 1995). In addition, species from different and not closely related termite subfamilies show trail-following at more or less the same sensitivity when the dodecatrienol is offered (Table 13.2). Only one other trail pheromone has been identified in addition to the dodecatrienol, also from the sternal gland: (*E*)-6-cembrene (Fig. 13.3b), which was formally called neocembrene (Moore, 1966; Birch *et al.*, 1972; Kodama *et al.*, 1975). This compound was found in the Australian *Nasutitermes exitiosus* (Hill) and later in the African *Trinervitermes bettonianus* (Sjöstedt) (McDowell and Oloo, 1984). Only a few quantitative behavioural data are available for cembrene and it appears that, at least for *T. bettonianus*, it is needed in quantities about four orders of magnitude larger than dodecatrienol to induce trail-following (McDowell and Oloo, 1984). Interestingly, Kaib *et al.* (1982) could demonstrate that *T. bettonianus* responds with trail-following to sternal gland extracts from other termite species at the same sensitivity as those species were responding to their own extracts. These species included a number of those listed in Table 13.2, which are very sensitive to (*Z*,*Z*,*E*)-(3,6,8)-dodecatrienol. (*E*)-6-Cembrene could not be identified in other Nasutitermitinae (Laduguie *et al.*, 1994b).

With only two trail pheromone compounds identified so far and the fact that numerous species from distantly related termite taxa responded more or less equally to the same trail pheromone (the dodecatrienol, Table 13.2), the assumption arose that this pheromone system had remained evolutionarily conservative (Pasteels and Bordereau, 1998). This assumption is further supported by the fact that, unlike in ants, only the sternal gland has been identified as the production site in termites.

(a) (b)

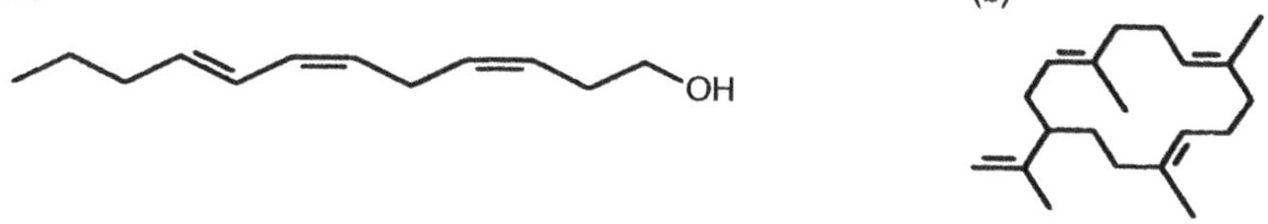

Fig. 13.3. Two trail pheromone compounds have been chemically identified in termites: (a) (*Z*,*Z*,*E*)-(3,6,8)-dodecatrienol and (b) (*E*)-6-cembrene (formally named neocembrene).

Table 13.2. Termite species (workers) responding to (*Z*,*Z*,*E*)-(3,6,8)-dodecatrienol with trail-following (NR = no response in the range 10–0.01 pg cm^{-1} trail) (modified from Pasteels and Bordereau, 1998).

Termite species	Threshold concentration (pg cm^{-1} trail)	Amount (pg per worker)
Kalotermitidae		
Kalotermes flavicollis	NR	
Rhinotermitidae		
Heterotermes sp.	1	
Schedorhinotermes lamanianus	1	
Reticulitermes flavipes	< 0.01	
Reticulitermes virginicus	0.01	
Reticulitermes speratus	0.02	3
Reticulitermes santonensis	0.1	1–10
Coptotermes formosanus	≈ 0.02	50
Termitidae		
Nasutitermitinae		
Nasutitermes lujae	0.01	
Nasutitermes diabolus	0.01	
Apicotermitinae		
Astalotermes sp.	10	
Termitinae		
Amitermes evuncifer	10	
Basidentitermes sp.	1	
Cephalotermes rectangularis	1	
Cubitermes fungifaber	1	
Cubitermes severus	0.1	
Microcerotermes sp.	0.1	
Noditermes sp.	0.1	
Ophiotermes sp.	0.1	
Thoracotermes sp.	10	
Macrotermitinae		
Pseudacanthotermes spiniger	0.1	≈ 10
Ancistrotermes cavithorax	1	
Macrotermes bellicosus	NR	
Macrotermes subhyalinus	NR	
Microtermes sp.	NR	

This evolutionary conservatism does not imply that the sternal gland secretion might not be species-specific. Behavioural tests with both artificial and natural trails showed species-specific responses, particularly in choice tests. Thus, a multi-component pheromonal system has been suggested (Matsumura *et al.*, 1972; Howard *et al.*, 1976; Kaib *et al.*, 1982; Runcie, 1987) in which (Z,Z,E)-(3,6,8)-dodecatrienol serves as a general orientation signal which is supplemented by other, species-specific cues.

Trail communication during exploration is a collective process. The exploration of a new food source, for example, is followed by a recruitment phase which, once a large number of nestmates has been guided to the food source, will be followed by the phase of food exploitation. There is convincing evidence that during all three phases, trail pheromone plays a key role. Species-specific differences concern the relative importance of soldiers and workers or of groups of individuals during exploration, gallery building and recruitment. In *N. lujae* the older workers, which have well developed sternal glands, initiate trail laying (Pasteels, 1965). In *Macrotermes bellicosus* (Smeatman) trail recruitment is induced by major workers (Lys and Leuthold, 1991). In contrast, in *Nasutitermes costalis* (Holmgren) (Traniello and Busher, 1985) and in *S. lamanianus* (Schedel and Kaib, 1987; Kaib, 1990), exploration is initiated by termite soldiers. Differences between soldiers and workers or between groups of workers of different age or morph are assumed to be due to different response thresholds to the trail pheromone or quantitative differences of the same trail pheromone.

In *S. lamanianus*, minor soldiers explore promising feeding sites within the colony's territory. On their excursions they lay pheromone trails by means of the volatile secretion from their well developed sternal glands. During the initial phase of foraging, pheromone concentration on trails is low and trails are not followed by termite workers, which remain close to the ends of the foraging galleries (Fig. 13.4a). Termite workers will only follow the trail after frequent commuting by minor soldiers and reinforcement of the chemical trails. On established trails in the open, minor soldiers deviate more laterally from the trail where the concentration of the pheromone is lower (Fig. 13.4b). Thus, termite workers operating outside the galleries are always guided and flanked by minor soldiers. The caste-specific polyethism during the search for food can be most easily explained by differences in pheromone sensitivity. As shown in Fig. 13.5, workers are less sensitive to the trail pheromone than minor soldiers (Schedel and Kaib, 1987; Kaib, 1990).

Once a new food source has been detected by *S. lamanianus*, workers lay a recruitment trail on their route back to the nest, which instantly leads to mass recruitment of nestmates. It has been postulated that the change of the function of the trail from orientation to recruitment is not due to qualitative differences in the trail pheromone, but is a quantitative effect (Schedel and Kaib, 1987). However, Lys and Leuthold (1991) suggest the opposite for *M. bellicosus*.

Trail pheromone does not control collective foraging in all termite species. Leuthold *et al.* (1976) have shown that *Hodotermes mossambicus* (Hagen) uses visual orientation during foraging. This species collects grass as food up to 3 m away from the foraging exits of their underground tunnel network. During daytime, exploration and foraging for food are based on light menotaxis. When sun (or full moon) is not available as an orientation cue, scout workers during their search for food lay pheromone

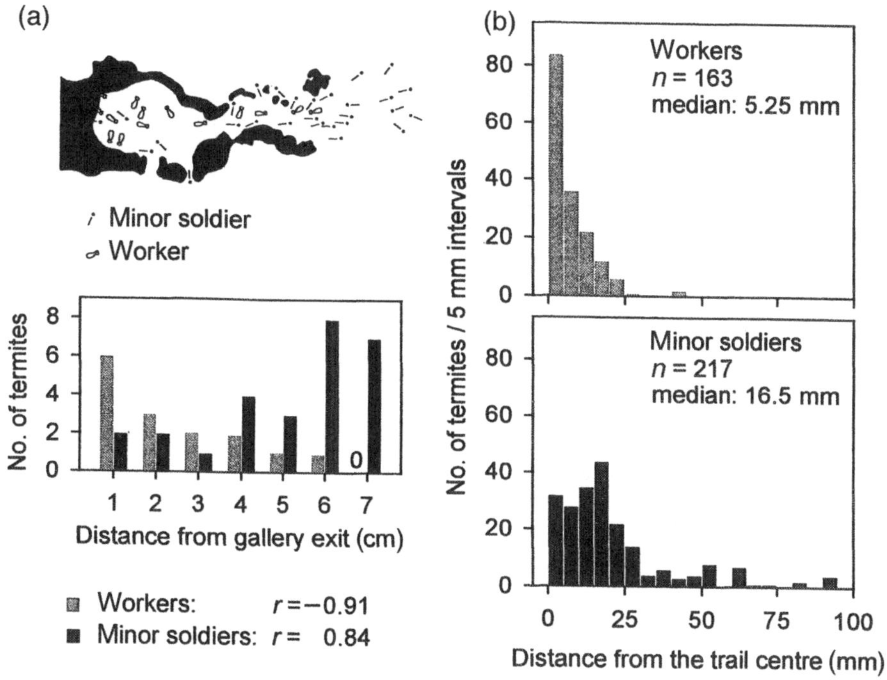

Fig. 13.4. (a) At the exit of foraging galleries, minor soldiers of *Schedorhinotermes lamanianus* initiate foraging and move out of the galleries over larger distances than the workers. (b) In addition, minor soldiers deviate more laterally from the pheromone trail than workers do. Therefore, outside the shelter of the galleries, workers are always guided and secured by minor soldiers.

trails which guide them back to the foraging exits (Heidecker and Leuthold, 1983). As scout workers leave these exits and spread in all directions, they progressively build up a radial trail pattern. Close to the foraging exits this pattern causes an overall pheromone gradient which rises in concentration towards the exits. Termites then orient osmoklinotactically towards the higher pheromone concentration and thus find the entrances of the underground tunnel network (Leuthold, 1975).

Trail pheromone is also used as cue for building foraging galleries. Preliminary studies in *S. lamanianus* have shown that workers build galleries only along established pheromone trails. The width of such galleries depends on the concentration of the trail pheromone. As newly built galleries, both in the field and in the laboratory, have a minimum width of 9 mm, one may postulate that the trail pheromone has to reach a concentration well above the threshold for trail-following before gallery building behaviour in termite workers is induced.

Sex pheromones

After swarming, male and female termites perform postflight behaviour. Females raise their abdomen, expose the sternal gland, and flutter their wings (Fig. 13.1a).

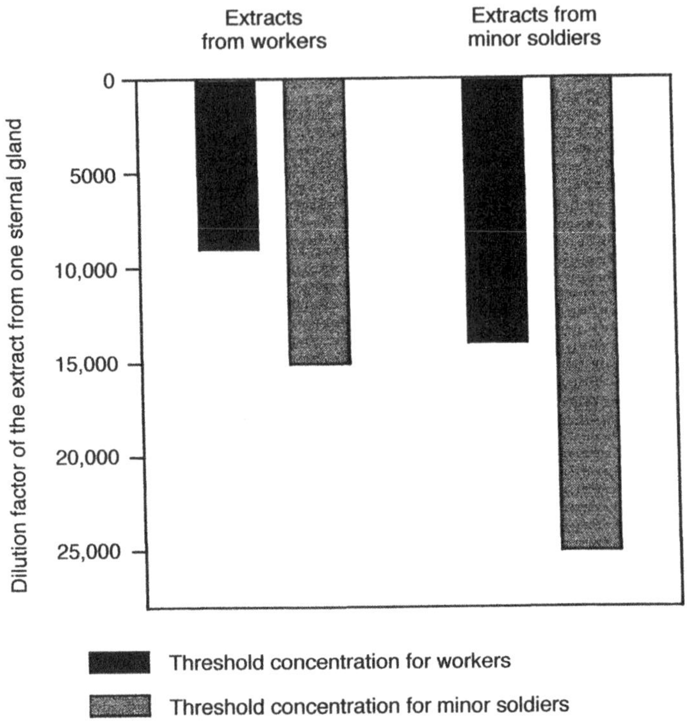

Fig. 13.5. In *Schedorhinotermes lamanianus,* the sternal gland of minor soldiers carries more trail pheromone than that of the workers. Minor soldiers are also more sensitive to the trail pheromone. Hexane extracts of dissected sternal glands were diluted until the threshold concentration for trail-following was reached. The threshold concentration was defined as 1 trail unit (TU).

By this mechanism they disperse sex pheromone and attract males, with which they form 'tandems' (Fig. 13.1b). This postflight behaviour seems to be characteristic for most termite species and was first described by Fuller (1915). For several termite species the sternal gland secretion has been identified as a source of the sex pheromone (Pasteels and Bordereau, 1998), so it was not surprising to find that the sex attractant substances are identical to the trail pheromone, but are secreted in much larger quantities (Bordereau *et al.*, 1991; Laduguie *et al.*, 1994a). *Pseudacanthotermes spiniger* (Sjöstedt) female alates, for example, have 15–20 ng pheromone in the sternal gland, but males only 1–3 ng and workers 0.01 ng. This pheromone is also used in 'tandem running', during which a calling female is closely followed by one or several males. The (Z,Z,*E*)-(3,6,8)-dodecatrienol acts at a short range in *P. spiniger*, which is in agreement with the observation by Leuthold (1975) that the female sternal gland secretion of *Trinervitermes bettonianus* is an airborne attractant being active only at a short range of about 4 cm. Calling seems to be an attribute of female alates as well as dealates, but this might be not a general rule. In *Hodotermes mossambicus*, Leuthold and Bruinsma (1977) reported that only males call, whereas Hewitt and Nel (1969) found calling behaviour in both females and males. Caution is needed in the interpretation of such observations.

As with the trail pheromone, there is strong evidence that the sex pheromones are multi-component systems and that the constituents seem to act synergistically (Laduguie *et al.*, 1994a). This view is supported by the fact that in *Reticulitermes* species, female sternal gland extracts are species-specific (Clément, 1982; Clément *et al.*, 1986). Furthermore, sex-specific pheromones have been demonstrated for *Zootermopsis angusticollis* (Hagen) (Pasteels, 1972).

In addition to the sternal gland secretion, the secretion of the tergal glands helps unpaired alates to encounter a partner (Fig. 13.1), although no description of a pheromone has been published. Females of *T. bettonianus* expose their tergal glands during tandem running and apparently attract males over a distance of approximately 10 cm (Leuthold, 1975).

Knowledge of the function and chemistry of sex pheromones in termites is still limited and is in part anecdotal. Detailed work on these subjects is badly needed.

Alarm pheromones and colony defence

Colony defence is a collective process, which is coordinated by a variety of signals. Vibration produced by soldiers banging with their heads on the ground was described by Stuart (1969) and was investigated in more detail by Kirchner *et al.* (1994). This type of mechanical signal is quickly transmitted via the substrate throughout the termite colony.

In addition to vibration, chemical signals play an important role in alarm and colony defence. Chemical defence is usually aimed at the antagonist and is thus an individual rather than a collective behaviour. In their reviews, Pasteels *et al.* (1983) and Prestwich (1984) reported that research was mainly focused on the role of the frontal gland secretion from the termite soldiers. In many species, soldiers explore the environment and are usually the first individuals of a colony encountering an antagonist. Chemical defence is also reported for termite workers, which smear an odoriferous slime or defecation on the antagonist during self-defence (e.g. Williams, 1959; Sands, 1982; Thorne, 1982b).

Alarm responses in termites were first demonstrated in soldiers of the genus *Nasutitermes* (Ernst, 1959). Soldiers spray frontal gland secretion over considerable distances on to attacking ants. From this sticky secretion, volatiles evaporate which alert soldiers to the site of combat. In the genus *Nasutitermes* these volatiles are mostly monoterpenes, sesquiterpenes and diterpenes, as was shown by numerous authors (see Prestwich, 1984; Pasteels and Bordereau, 1998). Although a variety of substances from the frontal gland have been chemically identified, very few of them function as an alarm pheromone. Detailed studies within several *Nasutitermes* species demonstrated that workers and soldiers respond differently to the monoterpene alarm pheromones and these castes often have a complementary role during defence (Eisner *et al.*, 1976; Kriston *et al.*, 1977; Vrkoč *et al.*, 1978; Stuart, 1981; Traniello, 1981; Traniello and Busher, 1985; Roisin *et al.*, 1990). Some of these monoterpenes are produced with remarkably high enantiomeric purity

(Everaerts *et al.*, 1990; Lindström *et al.*, 1990; Valterová *et al.*, 1992, 1993), but the level of enantiomeric purity may vary considerably between allopatric populations of the same species (Valterová *et al.*, 1993). Again for the genus *Nasutitermes*, it has been shown that the alarm reaction can be enantiomer-specific. In *N. principes* (Desneux), the (+)-enantiomer of α-pinene is more active in eliciting alarm behaviour than the (−)-enantiomer (Roisin *et al.*, 1990).

Caste-specific responses to an alarm pheromone could also be demonstrated in *S. lamanianus* (Kaib and Kriston-Meissl, 1979; Kaib, 1990). This termite produces a blend of ketones, vinylketones and dienones in its frontal gland which cause disorientation and cessation of recruiting in ants (Kaib, 1982). These ketones are present in the same ratio in major and minor soldiers. Minor soldiers also produce β-ketoaldehydes (Kaib, 1990). Several behavioural responses by workers are shown towards both types of secretion, but the secretion from minor soldiers always releases higher response levels. Alarm and attack behaviour in workers, however, is only induced by the secretion of minor soldiers (Kaib and Kriston-Meissl, 1979). This difference in response to the two types of frontal gland secretion becomes meaningful when considering the biology of *S. lamanianus*. Major soldiers remain in the nest centres, where reproductives and brood are located, and protect the nest against invading ants. In contrast, the much more agile minor soldiers guide and accompany workers to the foraging sites, where interactions with antagonists may occur and alarm communication helps colony defence (Kaib, 1985).

Frontal gland secretion and associated alarm pheromones may interact with other pheromonal systems. In *S. lamanianus* the search for food is initiated by minor soldiers and workers which are less sensitive to the trail pheromone. When predatory ants intrude into the termite foraging parties outside the galleries, they are first confronted by minor soldiers which release frontal gland secretion during combat. At low concentrations of trail pheromone, as found on foraging trails outside galleries, frontal gland secretion inhibits dose-dependent trail-following behaviour in termite workers (Fig. 13.6). As a consequence, during combat the unarmed termite workers no longer leave the foraging galleries which provide them with shelter. Foraging for food is then immediately interrupted (Kaib, 1990).

Aggregation pheromone and feeding stimulant

The role of pheromones in the process of exploitation of a food source has gained attention only in recent years. During gnawing, termites release saliva which makes workers aggregate at the feeding sites and thus leads to a highly contiguous pattern of clusters of gnawing workers. For *S. lamanianus* (Kaib and Ziesmann, 1992) and *Reticulitermes santonensis* (Reinhard *et al.*, 1997) it has been demonstrated that the aggregation pheromone is produced in the paired labial gland and released via the oral cavity on to the food. No other body compartment, including the water sacs, carries this signal. However, the content of the water sacs is assumed to be the solvent or carrier for the labial gland secretion. This induces aggregation of workers even when the signal is deposited without food. Bioassays were started after a drop of labial gland secretion had dried up on a glass plate. Workers aggregated at the

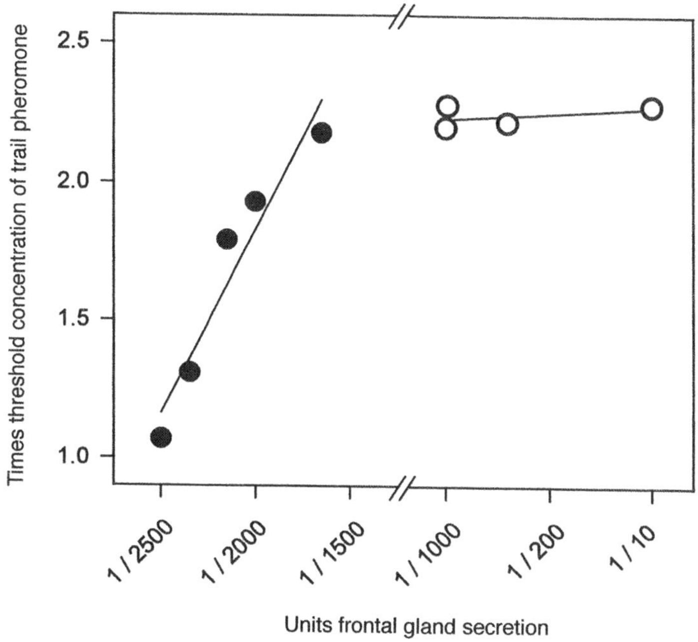

Fig. 13.6. In *Schedorhinotermes lamanianus,* frontal gland secretion inhibits trail-following behaviour of termite workers at pheromone concentrations just above threshold concentration (= 1.0). The plot shows that low doses of pheromone inhibit trail-following in a dose-dependent way (●); whereas at high doses (> 1/1000), as present on well established trails or in galleries, trail-following is not affected by the frontal gland secretion (○). One unit of frontal gland secretion is the average amount released by a minor soldier during combat.

spot with the labial gland secretion and tried to gnaw at the glass (Reinhard *et al.*, 1997). The labial gland secretion thus acts alone and not in association with food.

At the feeding sites, termite workers start gnawing and form food pellets, which are mostly transferred to other individuals before being transported back to the nest. Transfer of food pellets as well as their transport to the nest primarily occurs in commuting zones, whereas gnawing is concentrated at special gnawing zones. The development of two distinct zones during food exploitation suggests the interaction between different pheromone systems – that from the labial gland and that from the sternal gland.

Like the pheromone trails, the commuting zones on gnawing sites are marked by the trail pheromone from the sternal gland. In a detailed study it was demonstrated that *S. lamanianus* workers avoided areas for gnawing in which trail pheromone was present (Reinhard and Kaib, 1995). It not only inhibited, in a dose-dependent manner, the development of gnawing aggregations, but also deterred workers from sites already marked by labial gland secretion. Furthermore, it dispersed even well established aggregations of gnawing workers. When the sternal gland secretion was removed, workers re-aggregated (Fig. 13.7).

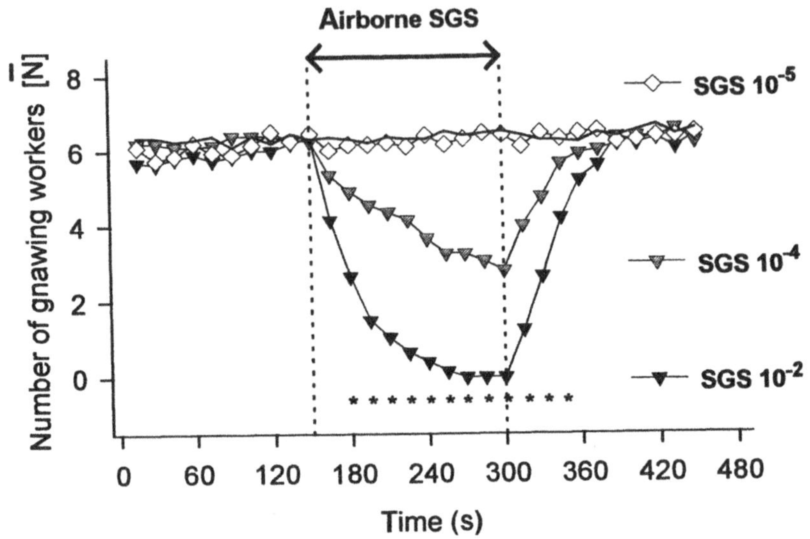

Fig. 13.7. In *Schedorhinotermes lamanianus*, the trail pheromone from the sternal gland secretion (SGS), introduced as an airborne signal from above (⟵⟶), inhibits gnawing aggregations in a dose-dependent way where labial gland secretion has been applied by termite workers at high concentrations. Gnawing aggregations may even become dispersed. After removal of the inhibitory signal the gnawing aggregations return to initial levels. *Significant difference between the SGS and control at $P \leq 0.001$. (Modified from Reinhard and Kaib, 1995.)

The division of the feeding site into zones for feeding and for commuting is regulated by the balance between the two pheromone systems. In commuting zones the level of airborne trail pheromone from the sternal gland is high, due to frequent reinforcement of the trail pheromone. Here, gnawing is totally inhibited, even if workers should deposit labial gland secretion. In a network of trails with different pheromone concentrations, termites tend to choose the stronger trails. Trails with lower quantities of trail pheromone are abandoned and are not reinforced again. Because of the relatively high volatility of the trail pheromone, its concentration falls below the threshold for gnawing inhibition and workers start releasing labial gland secretion. Consequently, areas of low trail pheromone concentration will gradually be turned from commuting zones into gnawing zones (Reinhard and Kaib, 1995).

The balance between gnawing and commuting is governed not only by the frequencies of workers performing the two behavioural sequences, but also by the significant difference in the volatility of the two pheromones involved. In contrast to the volatile trail pheromone from the sternal gland, the aggregation pheromone from the labial gland is non-volatile and persists over long periods of time. This difference in the evaporation rate again leads to a constant shift between the two zones of activity. Individual gnawing aggregations, initially separated by zones for commuting, will finally merge and the food source will be completely used up. In the termite *S. lamanianus*, foraging for food and exploitation of a food source is a dynamic process governed by different pheromone systems, the frontal gland secretion of the minor soldiers that inhibits trail-following in foraging areas outside the galleries, the sternal gland secretion of both workers and soldiers, and the labial gland secretion of the workers.

The role of termites in agroecosystems and their control

Termites play an important role in agroecosystems throughout the tropics and subtropics where they are also considered as key species in natural ecosystems, particularly in African rangelands. Termites are likely to play a major beneficial role by promoting essential ecological processes such as the decomposition of organic matter, thus facilitating the carbon and nitrogen fluxes in the soil, recycling plant particles at various soil depths and making nutrients available to the vegetation (Black and Okwakol, 1997). Two examples derived from African savannah ecosystems, which are well known for their richness in livestock, may help to elucidate the ecological role of termites. Firstly, grass-feeding termites of the genus *Macrotermes* reach population densities of about 50 kg dry weight per hectare, which equals that of the biomass of large herbivores. These termites, together with the harvester termite genus *Hodotermes*, consume about twice as much plant litter as the herbivorous livestock. Termites may thus achieve pest status, especially in drought periods or in situations with too much livestock. The level of losses in crops in terms of ecosystem-level energy transfer may well be significant but remains to be quantitatively investigated (e.g. Lepage, 1981; Collins, 1982). Secondly, Macrotermitidae, while building their conspicuous mounds and maintaining extended tunnel systems for food foraging, recycle approximately 2000 kg soil ha^{-1} $year^{-1}$.

Relatively few of the more than 2500 described termite species are agricultural pests but their damage to cultivations and pastures can be significant. Termite populations and their activity increase with mulching or green manure decomposition (e.g. Tian *et al.*, 1993; Lehmann *et al.*, 1995). In a long-term study, Wood *et al.* (1977) demonstrated that, after wood clearing, the termite population had increased in maize and yam cultivations but remained low in sweet potatoes, pasture and groundnut. In maize, a crop particularly susceptible to termite attack, the decline in species richness was associated with large increases in the abundance of *Amitermes* and *Microtermes* pest species (Black and Okwakol, 1997). Significant damage in postharvest products such as wooden structures, particularly in human settlements, is attributed to termites and can exceed US$3 billion annually (Su and Scheffrahn, 1990).

A number of ecological, social, political and economic factors provide strong motivation for developing control schemes for termites. However, with such cryptic insects, detection prior to the occurrence of significant damage remains particularly problematic. To date, control of termites is predominantly based on chemical methods. Control of subterranean termites of the genus *Reticulitermes* and *Coptotermes* with baits appears to be fairly successful, but there is considerable disagreement among researchers who have developed these baits on whether elimination and suppression are realistic goals in termite control (Lewis, 1997).

Therefore, prospects for the development of alternative termite control methods appear good. Current research focuses on baits containing small amounts of toxins offered in a portable matrix that termites can pass along to nest mates. The original belief was that the selection of suitable bait toxins was the key issue in the success of bait systems in controlling termite infestations. However, it has become increasingly clear that the attraction of termites to the bait stations, thus ensuring a significant build-up in termite numbers at the stations, is of critical importance to the viability of any commercial bait system. Improving current bait systems by the addition of either the trail pheromones or the gnawing aggregation pheromone may stimulate high numbers of termites to come to and feed on the baits; and when the toxin is applied in a portable matrix, these first termites can contaminate foraging and feeding nest mates and even cause the death of a whole termite colony. For the development of such alternative control programmes it even might be of advantage that both trail pheromones and gnawing aggregation pheromones appear not to be species-specific.

Conclusions

Despite the significant ecological and economic role of termites, the present knowledge of their pheromone systems remains well behind that of other groups of social insects and of insects as a whole. The cryptic life style of termites and the extremely large number of individuals in a colony make detailed behavioural studies difficult to perform. Furthermore, when termites are isolated from their colony for a bioassay, they often tend to display abnormal behaviour. Clearly, it is very difficult to develop effective and reliable bioassays for termite studies. This may be the reason that earlier work on termite pheromones focused primarily on trail communication and alarm and other defensive semiochemicals. The only real possibility is to perform behavioural tests with small experimental colonies. To date, no firm ethological and chemical evidence is available for sexual behaviour mediated by pheromones.

The pest status of termites has been well established and there is renewed interest in novel approaches to pest management, such as the development of alternative control strategies based on the use of termite pheromones. Many questions remain to be answered. Primer pheromones are basic to colony development; they function in all social insect groups and hold promise for the development of specific bioregulators to control termite populations and infestations. Releaser pheromones, such as sex pheromones, may be used for trapping swarming alates, or trail pheromone and gnawing aggregation pheromones may help to attract termites to baits with toxins. Due to the great economic significance of termites in several parts in the world and with better instrumentation and new techniques being developed, important progress in termite pheromone research can be expected in the near future.

Acknowledgements

I thank B. Eisermann for help during reviewing the termite literature and for preparing the figures. M. Lenz, S. Mikus, J. Pasteels and Y. Roisin gave valuable comments on an earlier version of the manuscript. The work was supported by the German Science Foundation (DFG).

References

Abe, T. and Darlington, J.P.E.C. (1985) Distribution and abundance of a mound-building termite, *Macrotermes michaelseni*, with special reference to its subterranean colonies and ant predators. *Physiology and Ecology* 22, 59–74.

Bacchus, S. (1979) New exocrine gland on the legs of some Rhinotermitidae (Isoptera). *International Journal of Insect Morphology and Embryology* 8, 135–142.

Becker, G. (1966) Spurfolge-Reaktion von Termiten auf Glycol-Verbindung. *Zeitschrift für Angewandte Zoologie* 53, 495–498.

Becker, G. and Mannesmann, R. (1968) Untersuchungen über das Verhalten von Termiten gegenüber einigen spurbildenden Stoffen. *Zeitschrift für Angewandte Entomologie* 62, 399–436.

Billen, J. and Morgan, E.D. (1998) Pheromone communication in social insects: sources and secretion. In: Vander Meer, R.K., Breed, M.D., Winston, M.L. and Espalie, K.E. (eds) *Pheromone Communication in Social Insects*. Westview Press, Oxford, pp. 3–33.

Billen, J., Joye, L. and Leuthold, R.H. (1989) Fine structure of the labial gland in *Macrotermes bellicosus* (Isoptera, Termitidae). *Acta Zoologica* 70, 37–45.

Birch, A.J., Brown, W.V., Corrie, J.E.T. and Moore, B.P. (1972) Neocembrene A, a termite trail pheromone. *Journal of Chemical Society Perkin Transactions* 1, 2653–2658.

Black, H.I.J. and Okwakol, M.J.N. (1997) Agricultural intensification, soil biodiversity and agroecosystem function in the tropics: the role of termites. *Applied Soil Ecology* 6, 37–53.

Bodot, P. (1969) Composition des colonies de termites: ses fluctuations au cours du temps. *Insectes Sociaux* 16, 39–53.

Bordereau, C. (1985) The role of pheromones in termite caste differentiation. In: Watson, J.A.L., Okot-Kotber, B.M. and Noirot, C. (eds) *Caste Differentiation in Social Insects*. Pergamon Press, New York, pp. 221–226.

Bordereau, C., Robert, A., Bonnard, O. and Le Quéré, J.L. (1991) (3Z,6Z,E8)-Dodecatrien-1-ol: sex pheromone in a higher fungus-growing termite, *Pseudacanthotermes spiniger* (Isoptera, Macrotermitidae). *Journal of Chemical Ecology* 17, 2177–2191.

Brian, M.V. (1983) *Social Insects, Ecology and Behavioral Biology*. Chapman & Hall, London, New York, 377 pp.

Bruinsma, O.H. (1979) An analysis of building behaviour of the termite *Macrotermes subhyalinus*. PhD thesis, Wageningen Agricultural University, the Netherlands.

Castle, G.B. (1934) The damp-wood termites in western United States, genus *Zootermopsis* (formerly *Termopsis*). In: Kofoid, C.A., Light, S.F., Horner A.C., Randell, M., Herms, W.B. and Bowe, E.E. (eds) *Termite and Termite Control*. University of California Press, Berkeley, pp. 273–310.

Clément, J.L. (1982) Phéromones d'attraction sexuelle des termites européens du genre *Reticulitermes* (Rhinotermitidae). Mécanismes comportementaux et isolements spécifiques. *Biology of Behaviour* 7, 55–68.

Clément, J.L. and Bagnères, A.-G. (1998) Nestmate recognition in termites. In: VanderMeer, R.K., Breed, M.D., Winston, M.L. and Espalie, K.E. (eds) *Pheromone Communication in Social Insects*. Westview Press, Oxford, pp. 126–155.

Clément, J.L., Howard, R., Blum, M.S. and Lloyd, H. (1986) L'isolement spécifique des termites du genre *Reticulitermes* du Sud-Est des Etats-Unis. Mise en évidence grâce à la chimie et au comportement d'une espèce jumelle de *R. virginicus* = *R. malletei* sp. nov. et d'une semi-species de *R. falvipes*. *Centre Royal Academy of Sciences Paris* 302, 67–70.

Collins, N.M. (1982) The interaction and impact of domestic stock and termites in a Kenyan rangeland. In: Breed, M.D., Michener, C.D. and Evans, E.E. (eds) *The Biology of Social Insects*. Westview Press, Boulder, Colorado, pp. 80–84.

Costa-Leonardo, A.M. (1997) Secretion of salivary glands of the Brazilian termite *Serritermes serrifer* Hagen & Bates (Isoptera: Serritermitidae). *Annales de la Société Entomologique de France* 33, 29–37.

Costa-Leonardo, A.M. and Cruz-Landim, C. (1991) Morphology of the salivary gland acini in *Grigiotermes bequaerti* (Isoptera: Termitidae: Apicotermitidae). *Entomologia Generalis* 16, 13–21.

Deligne, J., Quennedey, A. and Blum, M.S. (1981) The enemies and defense mechanisms of termites. In: Hermann, H.R. (ed.) *Social Insects*, Vol. II. Academic Press, New York, pp. 1–76.

Doki, H.K., Tsunoda, K. and Nishimoto, K. (1984) Effect of juvenile hormone analogues on caste-differentiation in the termite, *Reticulitermes speratus* (Kolbe) (Isoptera: Rhinotermitidae). *Material und Organismen* 19, 175–187.

Dudley, P.H. and Beaumont, J. (1889) Observations on the termites of white ants of the isthmus of Panama. *Transactions of the New York Academy of Sciences* 8, 85–114.

Eisner, T., Kriston, I. and Aneshansley, D.J. (1976) Defensive behavior of a termite (*Nasutitermes exitiosus*). *Behavioral Ecology and Sociobiology* 1, 83–125.

Ernst, E. (1959) Beobachtungen beim Spritzakt der *Nasutitermes*-Soldaten. *Revue Suisse de Zoologie* 66, 289–295.

Everaerts, C., Bonnard, O., Pasteels, J.M., Roisin, Y. and König, W.A. (1990) (+)-α-Pinene in the defensive secretion of *Nasutitermes princeps* (Isoptera, Termitidae). *Experientia* 46, 227–230.

Fuller, C. (1915) Observations on some South African termites. *Annals of the Natal Museum* 3, 329–505.

Goldberg, S. and Tobe, S.S. (1985) Adaptation of a radiochemical assay for juvenile hormone biosynthesis to study caste differentiation in a primitive termite. *Journal of Insect Physiology* 31, 347–352.

Grace, J.K., Wood, D.L., Kobo, I. and Kim, M. (1995) Behavioural and chemical investigation of trail pheromone from the termite *Reticulitermes hesperus* Banks (Isoptera: Rhinotermitidae). *Journal of Applied Entomology* 119, 501–505.

Grassé, P.-P. (1982) *Termitologia*, Vol. 1. Masson, Paris, 676 pp.

Grassi, B. and Sandias, A. (1893) Costituzione e sviluppo della società dei termitidi. *Atti Accademica Gioenia Scientia Naturale di Catania* 6, 1–75.

Haverty, M.I. (1977) The proportion of soldiers in termite colonies: a list and a bibliography (Isoptera). *Sociobiology* 2, 199–216.

Hefetz, A., Errard, C., Chambris, A. and Le Negrate, A. (1996) Postpharyngeal gland secretion as a modifier of aggressive behavior in myrmicine ant *Manica rubida*. *Journal of Insect Behavior* 9, 709–717.

Heidecker, J.L. and Leuthold, R.H. (1983) The organisation of collective foraging in the harvester termite *Hodotermes mossambicus* (Isoptera). *Behavioral Ecology and Sociobiology* 14, 195–202.

Henderson, G. (1998) Primer pheromones and possible soldier caste influence on the evolution of sociality in lower termites. In: Vander Meer, R.K., Breed, M.D., Winston, M.L. and Espalie, K.E. (eds) *Pheromone Communication in Social Insects*. Westview Press, Oxford, pp. 314–343.

Henderson, G. and Deleplane, K.S. (1994) Formosan subterranean termite swarming behaviour and alate sex-ratio (Isoptera: Rhinotermitidae). *Insectes Sociaux* 41, 19–28.

Hewitt, P.H. and Nel, J.J. (1969) The influence of group size on the sarcosomal activity and the behaviour of *Hodotermes mossambicus* alate termites. *Journal of Insect Physiology* 15, 2169–2177.

Hogan, M., Veivers, P.C., Slaytor, M. and Czolij, R.T. (1988) The site of cellulose breakdown in higher termites (*Nasutitermes walkeri* and *Nasutitermes exitiosus*). *Journal of Insect Physiology* 34, 891–899.

Hölldobler, B. (1984) Evolution of insect communication. In: Lewis, T. (ed.) *Insect Communication*. Academic Press, London, pp. 349–377.

Howard, R.W. (1993) Cuticular hydrocarbons and chemical communication. In: Stanley-Samuelson, D.W. and Nelson, D.R. (eds) *Insect Lipids: Chemistry, Biochemistry and Biology*. University of Nebraska Press, Lincoln and London, pp. 179–226.

Howard, R.W. and Haverty, M.L. (1981) Seasonal variation in caste proportions of field colonies of *Reticulitermes flavipes* (Kollar). *Environmental Entomology* 10, 546–549.

Howard, R.W., Matsumura, F. and Coppel, H.C. (1976) Trail-following pheromones of the Rhinotermitidae: approaches to their authentication and specificity. *Journal of Chemical Ecology* 2, 147–166.

Hrdy, I. and Krečeк, J. (1972) Development of superfluous soldiers induced by juvenile hormone in the termite, *Reticulitermes lucifugus santonensis*. *Insectes Sociaux* 19, 105–109.

Hummel, H. and Karlson, P. (1968) Hexansäure als Bestandteil des Spurpheromons der Termite *Zootermopsis nevadensis* Hagen. *Hoppe-Seyler's Zeitschrift für Physiologische Chemie* 349, 725–727.

Jmhasly, P., Brandl, R., Leuthold, R.H. and Kaib, M. (1998) Aggression and cuticular hydrocarbon profiles in a termite. *Zoology* 101 (Suppl. I), 11.

Kaib, M. (1982) Beeinflussung des Rekrutierverhaltens von Ameisen durch das Frontaldrüsensekret der Termite *Schedorhinotermes lamanianus*. *Verhandlungen der Deutschen Zoologischen Gesellschaft* 75, 315.

Kaib, M. (1985) Defense strategies of termites: a review exemplified by *Schedorhinotermes lamanianus*. *Mitteilungen der Deutschen Gesellschaft für Allgemeine und Angewandte Entomologie* 4, 302–306.

Kaib, M. (1987) Frontal gland secretion and colony defence in termites: evolutionary aspects. In: Eder, J. and Rembold, H. (eds) *Chemistry and Biology of Social Insects*. Peperny-Verlag, München, pp. 406–407.

Kaib, M. (1990) Intra- and interspecific chemical signals in the termite *Schedorhinotermes* – production sites, chemistry, and behaviour. In: Gribakin F.G., Wiese, K. and Popov, A.V. (eds) *Sensory Systems and Communication in Arthropods*. Birkhäuser Verlag, Basel and Berlin, pp. 26–32.

Kaib, M. and Kriston-Meissl, I. (1979) Frontaldrüsensekret von Termitensoldaten: Chemische Zusammensetzung und biologische Bedeutung. *Verhandlungen der Deutschen Zoologischen Gesellschaft* 72, 232.

Kaib, M. and Ziesmann, J. (1992) The labial gland in the termite *Schedorhinotermes lamanianus* (Isoptera: Rhinotermitidae): morphology and function during communal food exploitation. *Insectes Sociaux* 39, 373–384.

Kaib, M., Bruinsma, O. and Leuthold, R.H. (1982) Trail-following in termites. Evidence for a multicomponent system. *Journal of Chemical Ecology* 8, 1193–1205.

Karlson, P., Lüscher, M. and Hummel, H. (1968) Extraktion und biologische Auswertung des Spurpheromones der Termite *Zootermopsis nevadensis*. *Journal of Insect Physiology* 14, 1763–1771.

Kirchner, W.H., Bröcker, I. and Tautz, J. (1994) Vibrational alarm communication in the damp-wood termite *Zootermopsis nevadensis*. *Physiological Entomology* 19, 187–190.

Klochov, S.G. and Zhuzhikov, D.D. (1990) Termite trail pheromones: specificity and biosynthesis. In: Gribakin, F.G., Wiese, K. and Popov, A.V. (eds) *Sensory Systems and Communication in Arthropods*. Birkhäuser Verlag, Basel and Berlin, pp. 40–43.

Kodama, M., Matsuki, Y. and Itô, S. (1975) Syntheses of macrocyclic terpenoids by intramolecular cyclization I. (±)-Cembrene-A, a termite trail pheromone, and (±)-nephthenol. *Tetrahedron Letters* 35, 3065–3068.

Kriston, I., Watson, J.A.L. and Eisner, T. (1977) Non-combative behaviour of large soldiers of *Nasutitermes exitiosus* (Hill): an analytical study. *Insectes Sociaux* 24, 103–111.

Laduguie, N., Robert, A., Bonnard, O., Vieau, F., Le Quéré, J.L., Semon, E. and Bordereau, C. (1994a) Isolation and identification of (3Z,6Z,8E)-3,6,8-dodecatrien-1-ol in *Reticulitermes santonensis* Feytaud (Isoptera, Rhinotermitidae): roles in worker trail-following and in alate sex-attraction behaviour. *Journal of Insect Physiology* 40, 781–787.

Laduguie, N., Robert, A., Salini, N. and Bordereau, C. (1994b) Les termites et le dodecatrienol. *Actes des Colloques Insectes Sociaux* 9, 27–34.

Lefeuve, P. and Bordereau, C. (1984) Soldier formation regulated by a primer pheromone from the soldier frontal gland in a higher termite, *Nasutitermes lujae*. *Proceedings of the National Academy of Science USA* 81, 7665–7668.

Lehmann, J., Schroth, G. and Zech, W. (1995) Decomposition and nutrient release from leaves, twigs and roots of three alley-cropped tree legumes in central Togo. *Agroforestry Systems* 29, 21–36.

Lenz, M. (1976) The dependence of hormone effects in termite caste determination on external factors. In: Lüscher, M. (ed.) *Phase and Caste Determination in Insects: Endocrine Aspects*. Pergamon Press, Oxford, pp. 73–89.

Lenz, M. (1994) Food resources, colony growth and caste development in wood-feeding termites. In: Hunt, J.H. and Nalepa, C.A. (eds) *Nourishment and Evolution in Insect Societies*. Westview Press, Boulder, Colorado, pp. 159–210.

Lepage, M. (1981) L'impact des populations récoltantes de *Macrotermes michaelseni* (Sjöstedt) (Isoptera; Macrotermitinae) dans une écosystème sémi-arid (Kajiado, Kenya). 1. L'activité de récolte et son déterminisme. *Insectes Sociaux* 28, 207–308.

Leuthold, R.H. (1975) Orientation mediated by pheromones in social insects. In: Noirot, C., Howse, P.E. and Le Masne, G. (eds) *Pheromones and Defensive Secretions in Social Insects*. Université de Dijon, France, pp. 197–211.

Leuthold, R.H. and Bruinsma, O. (1977) Pairing behavior in *Hodotermes mossambicus* (Isoptera). *Psyche* 84, 109–119.

Leuthold, R.H., Bruinsma, O. and van Huis, A. (1976) Optical and pheromonal orientation and memory for homing distance in the harvester termite *Hodotermes mossambicus* (Hagen). *Behavioral Ecology and Sociobiology* 1, 127–139.

Lewis, V.R. (1997) Alternative control strategies for termites. *Journal of Agricultural Entomology* 14, 291–307.

Lindström, M., Norin, T., Valterová, I. and Vrkoč, J. (1990) Chirality of the monoterpene alarm pheromones of termites. *Naturwissenschaften* 77, 134–135.

Lüscher, M. (1960) Hormonal control of caste differentiation in termites. *Annals of the New York Academy of Science* 89, 549–563.

Lüscher, M. (1961) Social control of polymorphism in termites. In: Kennedy, J.S. (ed.) *Insect Polymorphism*. Royal Entomological Society, London, pp. 57–67.

Lüscher, M. (1976) Evidence for an endocrine control of caste determination in higher termites. In: Lüscher, M. (ed.) *Phase and Caste Determination in Insects: Endocrine Aspects*. Pergamon Press, Oxford, pp. 91–103.

Lüscher, M. and Müller, B. (1960) Ein spurbildendes Sekret bei Termiten. *Naturwissenschaften* 21, 503.

Lüscher, M. and Springhetti, A. (1960) Untersuchungen über die Bedeutung der Corpora allata für die Differenzierung der Kasten bei der Termite *Kalotermes flavicollis* F. *Journal of Insect Physiology* 5, 190–212.

Lys, J.A. and Leuthold, R.H. (1991) Task-specific distribution of the two worker castes in extranidal activities in *Macrotermes bellicosus* (Smeathman): observation of behaviour during food acquisition. *Insectes Sociaux* 38, 161–170.

Martin, M.M. and Martin, J.S. (1978) Cellulose digestion in the midgut of the fungus-growing termite *Macrotermes natalensis*: the role of acquired digestive enzymes. *Science* 119, 1453–1455.

Maschwitz, U. and Tho, Y.P. (1974) Chinone als Wehrsubstanzen bei einigen orientalischen Macrotermitinen. *Insectes Sociaux* 21, 231–234.

Matsumura, F., Coppel, H.C. and Tai, A. (1968) Isolation and identification of termite trail-following pheromone. *Nature* 219, 963–964.

Matsumura, F., Tai, A. and Coppel, H.C. (1969) Termite trail-following substance, isolation and purification from *Reticulitermes virginicus* and fungus infected wood. *Journal of Economic Entomology* 62, 599–603.

Matsumura, F., Jewett, D.W. and Coppel, H.C. (1972) Interspecific response to synthetic trail-following substances. *Journal of Economic Entomology* 65, 600–602.

McDowell, P.G. and Oloo, G.W. (1984) Isolation, identification and biological activity of trail-following pheromone of termite *Trinervitermes bettonianus* (Sjöstedt) (Termitidae: Nasutitermitinae). *Journal of Chemical Ecology* 10, 835–851.

McMahan, E.A. (1966) Food transmission with the *Cryptotermes brevis* colony (Isoptera: Kalotermitidae). *Annals of the Entomological Society of America* 59, 1131–1137.

Miller, E.M. (1969) Caste differentiation in the lower termites. In: Krishna, K. and Weesner, F.M. (eds) *Biology of Termites*, Vol. 1. Academic Press, New York, pp. 283–310.

Moore, B.P. (1966) Isolation of the scent-trail pheromone of an Australian termite. *Nature* 211, 746–747.

Moore, B.P. (1974) Pheromones in termite societies. In: Birch, M.C. (ed.) *Pheromones*. Elsevier, Amsterdam, pp. 250–266.

Noirot, C. (1969) Glands and secretions. In: Krishna, K. and Weesner, F.M. (eds) *Biology of Termites*, Vol. 1. Academic Press, New York, pp. 89–123.

Noirot, C. (1995) The sternal glands of termites: segmental pattern, phylogenetic implications. *Insectes Sociaux* 42, 321–323.

Noirot, C. and Bordereau, C. (1991) Termite polymorphism and morphogenetic hormones. In: Gupta, A.P. (ed.) *Morphogenetic Hormones of Arthropods*. Rutgers University Press, New Brunswick, New Jersey, pp. 293–324.

Oster, G.F. and Wilson, E.O. (1978) *Caste and Ecology in the Social Insects*. Princeton University Press, Princeton, New Jersey, 352 pp.

Pasteels, J.M. (1965) Polyéthisme chez les ouvrier de *Nasutitermes lujae* (Termitidae Isoptères). *Biologia Gabonica* 1, 191–205.

Pasteels, J.M. (1972) Sex-specific pheromones in a termite. *Experientia* 28, 105–106.

Pasteels, J.M. and Bordereau, C. (1998) Releaser pheromones in termites. In: VanderMeer, R.K., Breed, M.D., Winston, M.L. and Espelie, K.E. (eds) *Pheromone Communication in Social Insects*. Westview Press, Oxford, pp. 193–215.

Pasteels, J.M., Grégoire, J.-C. and Rowell-Rahier, M. (1983) The chemical ecology of defense in arthropods. *Annual Review of Entomology* 28, 263–289.

Peeters, C. (1990) The double meaning of the term caste. In: Veeresh, G.K., Mallik B. and Viraktamath, C.A. (eds) *Social Insects and the Environment. Proceedings of the XIth Congress of IUSSI*. IBH Publishing Co., New Delhi, p. 192.

Prestwich, G.D. (1984) Chemical systematics of termite exocrine secretions. *Annual Review of Ecology and Systematics* 14, 287–311.

Reinhard, J. and Kaib, M. (1995) Interaction of pheromones during food exploitation by the termite *Schedorhinotermes lamanianus*. *Physiological Entomology* 20, 266–272.

Reinhard, J., Hertel, H. and Kaib, M. (1997) Systematic search for food in the subterranean termite *Reticulitermes santonensis* De Feytaud (Isoptera: Rhinotermitidae). *Insectes Sociaux* 44, 147–158.

Renoux, J. (1976) Polymorphisme de *Schedorhinotermes lamanianus* (Sjöstedt) (Isoptera – Rhinotermitidae). Essai d'interprétation. *Insectes Sociaux* 23, 279–494.

Ritter, F.J., van Oosten, A.M. and Persoons, C.J. (1973) Attractants and trail-following compounds for the Saintonge termite: tricycloekasantalal and dihydroagarofuran from sandalwood oils and a sesquiterpenoid $C_{15}H_{24}O$ from pinewood infested with *Lenzites trabea*. In: *Proceedings of the VIIth Congress IUSSI*. Royal Entomological Society, London, pp. 330–331.

Ritter, F.J., Bruggeman, I.E.M., Persoons, C.J., Talman, E., van Oosten, A.M. and Verwiel, P.E.J. (1977) Evaluation of social insect pheromones in pest control, with special reference to subterranean termites and Pharaoh's ants. In: McFarlane, N.R. (ed.) *Crop Protection Agents: Their Biological Evaluation*. Academic Press, New York, pp. 210–216.

Roisin, Y., Everaerts, C., Pasteels, J.M. and Bonnard, O. (1990) Caste-dependent reaction to soldier defensive secretion and chiral alarm/recruitment pheromone in *Nasutitermes princeps*. *Journal of Chemical Ecology* 16, 2865–2875.

Runcie, C.D. (1987) Behavioral evidence for multicomponent trail pheromone in the termite, *Reticulitermes flavipes* (Kollar) (Isoptera: Rhinotermitidae). *Journal of Chemical Ecology* 13, 1967–1978.

Sands, W.A. (1982) Agonistic behavior of African soldierless Apicotermitinae (Isoptera: Termitidae). *Sociobiology* 7, 61–72.

Schedel, A. and Kaib, M. (1987) Polyethism during foraging in *Schedorhinotermes lamanianus* in unprotected areas: the role of exocrine glands. In: Eder, J. and Rembold, H. (eds) *Chemistry and Biology of Social Insects*. Peperny-Verlag, Munich, p. 416.

Sen-Sarma, P.K. and Mishra, S.C. (1972) Seasonal fluctuations of colony composition and population in *Neotermes bosei* Snyder (Insecta: Isoptera: Kalotermitidae). *Journal of the Indian Academy for Wood Science* 3, 43–48.

Soroker, V., Fresneau, D. and Hefetz, A. (1998) Formation of colony odor in Ponerine ant *Pachycondyla apicalis*. *Journal of Chemical Ecology* 24, 1077–1090.

Springhetti, A. (1985) The function of the royal pair in the society of *Kalotermes flavicollis* (Fagr.) (Isoptera: Kalotermitidae). In: Watson, J.A.L., Okot-Kotber, B.M. and Noirot, C. (eds) *Caste Differentiation in Social Insects*. Pergamon Press, New York, pp. 165–176.

Springhetti, A. and Pinamonti, S. (1977) Some effects of queen bee extracts (*Apis mellifera* L.) on *Kalotermes flavicollis* Fabr. (Isoptera). *Insectes Sociaux* 24, 61–70.

Stuart, A.M. (1961) Mechanism of trail-laying in two species of termites. *Nature* 189, 419.

Stuart, A.M. (1969) Social behavior and communication. In: Krishna, K. and Weesner, F.M. (eds) *Biology of Termites*, Vol. 1. Academic Press, New York, pp. 193–232.

Stuart, A.M. (1970) The role of chemicals in termite communication. In: Johnston, J.W., Moulton, D.G. and Turk, A. (eds) *Communication by Chemical Signals. Advances in Chemoreception*, Vol. 1. Appleton-Century-Crofts, New York, pp. 79–105.

Stuart, A.M. (1981) The role of pheromones in the initiation of foraging, recruitment and defence by the soldiers of a tropical termite, *Nasutitermes corniger* (Motschulsky). *Chemical Senses* 6, 409–420.

Su, N.-Y. and Scheffrahn, R.H. (1990) Economically important termites in the United States and their control. *Sociobiology* 17, 77–94.

Tai, A., Matsumura, F. and Coppel, H.C. (1969) Chemical identification of the trail-following pheromone for a southern subterranean termite. *Journal of Organic Chemistry* 34, 2180–2182.

Tai, A., Matsumura, F. and Coppel, H.C. (1971) Synthetic analogues of the termite trail-following pheromone, structure and biological activity. *Journal of Insect Physiology* 17, 181–188.

Thorne, B.L. (1982a) Polygyny in termites: Multiple primary queens in colonies of *Nasutitermes corniger* (Motschulsky) (Isoptera: Termitidae). *Insectes Sociaux* 29, 102–117.

Thorne, B.L. (1982b) Termite–termite interaction: workers as an agonistic caste. *Psyche* 89, 133–150.

Thorne, B.L. (1985) Termite polygyny: the ecological dynamics of queen mutualism. In: Hölldobler, B. and Lindauer, M. (eds) *Experimental Behavioral Ecology and Sociobiology*. Fischer Verlag, Stuttgart, pp. 325–341.

Thorne, B.L. (1997) Evolution of eusociality in termites. *Annual Review of Ecology and Systematics* 28, 27–54.

Tian, G., Brussaard, L. and Kang, B.T. (1993) Biological effects of plant residues with contrasting chemical compositions under humid tropical conditions: effects on soil fauna. *Soil Biology and Biochemistry* 25, 731–737.

Tokoro, M., Takahashi, M., Tsunoda, K. and Yamaoka, R. (1989) Isolation and primary structure of the trail pheromone of the termite, *Coptotermes formosanus* Shiraki (Isoptera: Rhinotermitidae). *Wood Research* 76, 29–38.

Tokoro, M., Yamaoka. R., Hayashiya, K., Takahashi, M. and Nishimoto, K. (1990) Evidence for trail pheromone precursor in the termite *Reticulitermes speratus* (Kolbe) (Rhinotermitidae: Isoptera). *Journal of Chemical Ecology* 16, 2549–2557.

Tokoro, M., Yamaoka. R., Hayashiya, K., Takahashi, M. and Tsunoda, K. (1991) Isolation and identification of the trail pheromone of the subterranean termite *Reticulitermes speratus* (Kolbe) (Rhinotermitidae: Isoptera). *Wood Research* 78, 1–14.

Traniello, J.F.A. (1981) Enemy deterrence in the recruitment strategy of a termite: soldier-organized foraging in *Nasutitermes costalis*. *Proceedings of the National Academy of Science USA* 78, 1976–1979.

Traniello, J.F.A. and Busher, C. (1985) Chemical regulation of polyethism during foraging in the neotropical termite *Nasutitermes costalis*. *Journal of Chemical Ecology* 11, 319–332.

Valterová, I., Vrkoč, J., Lindström, M. and Norin, T. (1992) On the natural occurrence of (−)-3-carene, a component of termite defense secretions. *Naturwissenschaften* 79, 416–417.

Valterová, I., Vrkoč, J. and Norin, T. (1993) The enantiomeric composition of monoterpene hydrocarbons in the defensive secretions of *Nasutitermes* termites (Isoptera): inter- and intraspecific variations. *Chemoecology* 4, 120–123.

Veivers, P.C., Mühlemann, R., Slaytor, M., Leuthold, R.H. and Bignell, D.E. (1991) Digestion, diet and polyethism in two fungus-growing termites: *Macrotermes subhyalinus* Rambur and *M. michaelseni* Sjöstedt. *Journal of Insect Physiology* 37, 675–682.

Verron, H. and Barbier, M. (1962) L'hexène-3-ol-1, substance attractive des termites *Calotermes flavicollis* et *Microcerotermes edentatus*. *Comptes-Rendus de l'Académie des Sciences, Paris* 254, 4089–4091.

Vrkoč, J., Krečeк, J. and Hrdy, I. (1978) Monoterpenic alarm pheromones in two *Nasutitermes* sp. *Acta Entomologica Bohemoslovaca* 75, 1–8.

Wanyonyi, K. (1974) The influence of juvenile hormone analogues on caste development in *Zootermopsis necadensis* (Hagen) (Isoptera). *Insectes Sociaux* 21, 35–44.

Williams, R.M.C. (1959) Colony development in *Cubitermes ugandensis* Fuller (Isoptera: Termitidae). *Insectes Sociaux* 6, 291–304.

Wilson, E.O. (1971) *The Insect Societies*. Belknap Press, Cambridge, Massachusetts, 548 pp.

Wilson, E.O. (1990) *Success and Dominance in Ecosystems: the Case of the Social Insects*. Ecological Institute Nordbünte, Oldendorf, Germany, 104 pp.

Wood, T.G., Johnson, R.A. and Ohiagu, C.E. (1977) Populations of termites (Isoptera) in natural and agricultural ecosystems in Southern Guinea savanna near Mokwa, Nigeria. *Geo-Eco-Tropics* 1, 139–148.

Wood, W.F., Truckenbrodt, W. and Meinwald, J. (1975) Chemistry of the defensive secretion from the African termite *Odontotermes badius*. *Annals of the American Entomological Society* 68, 359–360.

Yin, C.-M. and Gillot, C. (1975) Endocrine activity during caste differentiation in *Zootermopsis angusticollis* Hagen (Isoptera): a morphometric and autoradiographic study. *Canadian Journal of Zoology* 53, 1690–1700.

II Beneficials

14 Predators

JEFFREY R. ALDRICH

USDA–ARS, Insect Chemical Ecology Laboratory, Agricultural Research Center West, Building 007, Beltsville, MD 20705, USA

Introduction

Generalist predators function well as low-density population regulators to prevent pest outbreaks (Whitcomb and Godfrey, 1991; O'Neil *et al.*, 1996; Riechert and Lawrence, 1997). This review considers how pheromones could be used to understand and manipulate the behaviour of this important group of beneficials, especially for agriculturally significant predators. The search for pheromones of predators is still in its infancy, and semiochemicals are just beginning to be used to manipulate them for improved biological control (Aldrich, 1998). There are actually more examples of predators exploiting the semiochemicals of prey as host-finding and host-recognition kairomones than there are known pheromones of predators (Greany and Hagen, 1981; Mayer and McLaughlin, 1991). Kairomonal responses of predators will be included in this review primarily in cases involving host pheromones. Tritrophic examples of predators responding to herbivore-induced plant volatiles (synomones) will not be described since this phenomenon has been thoroughly reviewed elsewhere (e.g. Takabayashi and Dicke, 1996; Scutareanu *et al.*, 1997; Shimoda *et al.*, 1997; Dicke, 1998).

Mites (Acari)

Predaceous mites account for over half of the commercial predator market for predator augmentation (Coll and Ruberson, 1998) (Fig. 14.1a), being used principally against other phytophagous mites such as the two-spotted spider mite, *Tetranychus urticae* Koch. In the spider mite predators *Amblyseius fallacis* (Garman) and *Metaseiulus occidentalis* (Nesbitt) and the nematode predator *Macrocheles muscaedomesticae* (Scopoli), immature and adult females reportedly produce sex pheromones (Rock *et al.*, 1976; Hoy and Smilanick, 1979; Yasui, 1992). These pheromones function as arrestants or, at most, very short-range attractants, and none have been chemically identified.

Fig. 14.1. (a) *Phytoseiulus persimilis* Athias-Henriot adult with spider mite eggs. (b) Male *Xysticus ferox* (Hentz) crab spider. (c) Eastern yellowjacket worker, *Vespula maculifrons* (Buysson). (d) Adult goldeneye lacewing, *Chrysopa oculata* Say. (e) Adult syrphid fly. (f) *Lebia grandis* Hentz carabid adult feeding on Colorado potato beetle eggs.

Although the two-spotted spider mite produces a sex pheromone (Mayer and McLaughlin, 1991), there is no indication that predaceous mites use this or any other prey sex pheromone as a kairomone. However, the predaceous mites *A. fallacis* and *Phytoseiulus macropilis* (Banks) themselves deposit marking pheromones (yet to be identified) which reduce subsequent searching by conspecifics (Hislop and Prokopy, 1981). Another level of complexity is exhibited by *Phytoseiulus persimilis* Athias-Henriot, individuals of which release an odour that elicits the production of a second odour from prey spider mites (possibly an alarm pheromone

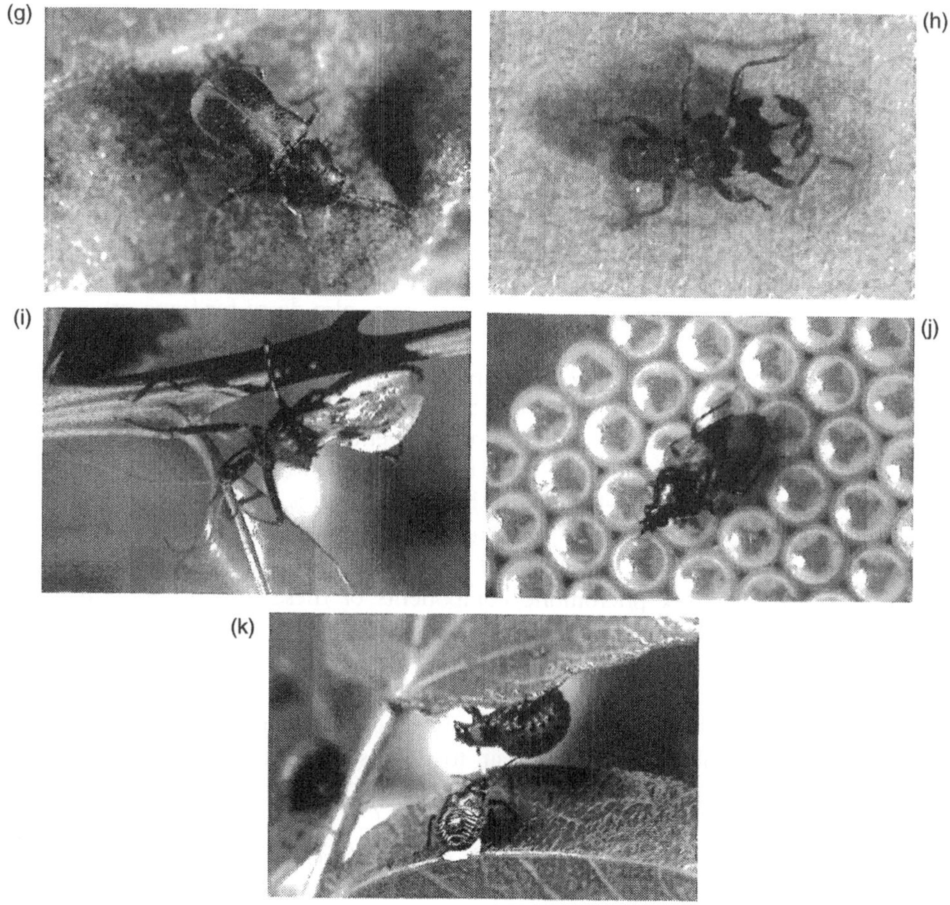

Fig. 14.1 (*Continued*). (g) *Collops vittatus* (Say) adult. (h) *Thanasimus dubius* (Fabricius) adult. (i) Australian assassin bug, *Pristhesancus plagipennis* Walker. (j) Adult *Orius insidiosus* (Say) pirate bug feeding on southern green stink bug eggs. (k) Fifth instar nymph of the spined soldier bug, *Podisus maculiventris* (Say), feeding on a Colorado potato beetle larva.

of the spider mite) causing conspecific predators to avoid occupied prey patches (Janssen *et al.*, 1997).

One kairomonal response by predaceous mites to thrips has been chemically characterized: *Amblyseius cucumeris* (Oudemans) is arrested by the alarm pheromone components, decyl and dodecyl acetates, of western flower thrips, *Frankliniella occidentalis* (Pergande) (Teerling *et al.*, 1993; Teerling, 1995).

Spiders (Araneae)

The first and only sex pheromone identified for a spider is for *Linyphia triangularis* (Clerck) (Schulz and Toft, 1993). Receptive females deposit copious amounts of (*R*,*R*)-3-(3-hydroxybutyryloxy)butyric acid on to their web which slowly deteriorates under natural acidic conditions to release the monomer, (*R*)-3-hydroxybutyric acid, plus (*E*)-2-butyric acid. Male *L. triangularis* are attracted to (*R*)-3-hydroxybutyric acid, roll the web into a ball (thereby minimizing further pheromone release) and then court the female. Compounds that evoke the web reduction behaviour of *L. triangularis* males also do so for some other linyphyiids; and males of another species, *Neriene montana* (Clerck), are attracted to (*E*)-2-butyric acid (Schulz, 1997). Sympatric linyphyiids probably recognize their species' web by characteristic web lipids (Schulz, 1997).

Curiously, adult males of ground-dwelling crab spiders (Thomisidae: *Xysticus* spp.) (Fig. 14.1b) are attracted to high concentrations of (*E*)-octenal and (*E*)-decenal, but not (*E*)-hexenal (Aldrich and Barros, 1995). All of these alk-2-enals are allomones of various 'true bugs' (Heteroptera), but crab spiders are sedentary predators so it was hypothesized that (*E*)-octenal and (*E*)-decenal are related or identical to sex pheromone components of these spiders (Aldrich and Barros, 1995). Alternatively, *Xysticus* males may use C_8 and C_{10} 2-alkenals, from heteropteran prey captured by females, in lieu of a sex pheromone.

Bolas spiders (Araneidae) provide a fascinating glimpse of chemical espionage (Yeargan, 1994). Instead of using a typical web to capture prey, adult females swing a sticky bolas on a thread at flying insects. Female *Mastophora* species lure unsuspecting male moths to within range of their bolas by emitting (*Z*)-9-tetradecenyl acetate, (*Z*)-9-tetradecenal, (*Z*)-11-hexadecenal and other moth pheromone components (Stowe *et al.*, 1987). *Mastophora* spiderlings do not produce a bolas; instead they sit on the edges of leaves at night and grab small male *Psychoda* flies that are attracted by pseudo-pheromones from the spiderlings (Yeargan and Quate, 1996). Aggressive chemical mimicry occurs in at least three other genera of araneids (Yeargan, 1994), and webs of some linyphiid spiders reportedly contain (*Z*)-11-octadecenyl acetate and (*Z*)-9-tricosene, pheromones of *Drosophila* spp. and *Musca domestica* L., respectively (Schulz, 1997). Also, the hunting spider *Habronestes bradleyi* Walckenaer uses the alarm pheromone of *Iridomyrmex purpureus* Smith (6-methyl-5-hepten-2-one) as a kairomone to home in on these ants (Allan *et al.*, 1996).

Despite the importance of spiders and predaceous mites in suppressing pests, their chemical ecology is just beginning to be elucidated and practical applications of this emerging knowledge have yet to be devised.

Ants and Social Wasps (Hymenoptera: Formicidae and Vespidae)

The existence of extrafloral nectaries in diverse plant species is evolutionary testimony to the effectiveness of ants as biocontrol agents (e.g. Oliveira, 1997) and, since antiquity, the Chinese have practised husbandry of nests of the citrus ant, *Oecophylla smaragdina* Fabricius, for pest control (Huang and Yang, 1987).

None the less, the possibility that ants and wasps respond to volatiles from extrafloral nectar seems not to have been investigated, nor have pheromones been used to manipulate these social insects for pest control (M.S. Blum, University of Georgia, Athens, and R.K. Vander Meer, US Department of Agriculture, Gainesville, Florida, 1997, personal communications).

Yellowjacket wasps (Vespinae) are beneficial predators of caterpillars and other pests but, of course, are dangerous because of their stings (Fig. 14.1c). Some progress has been made to use semiochemicals to mollify their presence. Heptyl butyrate is highly attractive to the western US yellowjacket, *Vespula pensylvanica* (Saussure) and, though it remains a mystery why heptyl butyrate is so attractive to this wasp, traps baited with the ester can depress yellowjacket populations (Akre *et al.*, 1981). However, eastern US yellowjacket species, including *V. germanica* (Fabricius) which is now common throughout much of the USA, are only slightly attracted to this or related esters. In New Zealand, *V. germanica* and *V. vulgaris* (L.) are not just a nuisance; they may be adversely affecting the native fauna by monopolizing the available honeydew supply. Weston *et al.* (1997) found that low concentrations of venom spiroacetals, *N*-isopentylacetamide and 2-heptanone were weakly attractive to *V. vulgaris* workers, but the same compounds were repellent at higher concentrations. Landolt (1998) reported the first potent chemical attractant for the German yellowjacket: a synergistic blend of acetic acid and isobutanol. This response is probably a food-finding behaviour, because both compounds were isolated from fermented molasses solutions.

Two cases have been documented in which foraging yellowjackets responded to the pheromones of other insects. Hendrichs *et al.* (1994, 1998) discovered that *V. germanica* workers home in on leks of pheromone-calling Mediterranean fruit fly males; and eastern yellowjackets, *V. maculifrons* (Buysson), are attracted to the pheromones of certain predaceous bugs (Pentatomidae: *Podisus* spp.; discussed below) (Aldrich *et al.*, 1985, 1986a). In the former case, the yellowjackets apparently learned to associate the presence of medfly males with their prey's pheromone (Hendrichs *et al.*, 1994, 1998). Recruitment of nestmates to carbohydrate food sources (but not protein sources) has since been demonstrated for the German yellowjacket (Overmyer and Jeanne, 1998). The reason that *V. maculifrons* workers are attracted to pheromones of predaceous bugs is not obvious. It is possible that the wasps seek to steal prey located by the bugs or that the wasps prey upon the tachinid fly parasitoids that themselves use the bugs' pheromones as host-finding kairomones (Aldrich 1995a,b). Alternatively, the chemicals involved, (*E*)-2-hexenal (known as leaf aldehyde) with either linalool or α-terpineol, may mimic a damaged-leaf odour used by yellowjackets to find defoliating larvae and, coincidentally, these compounds evolved as pheromones for the *Podisus* bugs.

Lacewings (Neuroptera: Chrysopidae and Hemerobiidae)

Common lacewings (Chrysopidae) and brown lacewings (Hemerobiidae) are entirely beneficial groups of insects. Some chrysopid adults are predaceous (e.g. *Chrysopa* s. str.), while in other genera (e.g. *Chrysoperla*) the adults feed exclusively on sweet juices and pollen (Principi and Canard, 1984) (Fig. 14.1d). Hemerobiids are generally found in wooded areas, and both the adults and larvae are predaceous. Adult chrysopids communicate over short distances by substrate-borne vibrations, and no pheromones have been identified for these important biological control agents. However, males of several *Chrysopa* and *Anisochrysa* spp. have exocrine glands that probably produce some kind of pheromone (Bitsch, 1984).

All lacewing larvae are predaceous, but their orientation to prey-associated allelochemicals is minimal and so the adult female must locate prey-infested habitats in which to oviposit (Greany and Hagen, 1981). Certain plant volatiles (e.g. caryophyllene and terpenyl acetate) are thought to be involved in habitat location by chrysopids (Flint *et al.*, 1979; Greany and Hagen, 1981; Pickett *et al.*, 1992). However, Hagen *et al.* (1976) made the most noteworthy discovery concerning chrysopid chemotaxis: many lacewing adults are attracted to honeydew containing tryptophan. Actually, a volatile breakdown product of tryptophan (indole acetaldehyde) is apparently the active attractant, but tryptophan remains a convenient slow-release source of indole acetaldehyde (van Emden and Hagen, 1976). Spraying artificial honeydew in crops can effectively attract chrysopids, but supplemental food sprays have not found widespread usage for augmentation because *Chrysoperla* spp. are now commercially available and attraction of native chrysopids depends on local population levels (Daane *et al.*, 1996). Attraction of chrysopid adults to tryptophan may yet find practical application in agroecosystems; for example, McEwen *et al.* (1994) suggested that tryptophan spray could conserve lacewings by concentrating adults in certain olive trees while other trees are treated with insecticides.

Efforts to augment chrysopids in agricultural fields, either by mass release of insectary-reared individuals or semiochemical attraction of wild adults, must take into account the discovery that larvae of at least some *Chrysopa* spp. produce potent oviposition-deterring pheromones (Ruzicka, 1994, 1996, 1997a,b). Larvae of *Chrysopa oculata* Say and *C. perla* (L.) chemically mark the substrate with a substance from the tip of the abdomen which deters females from ovipositing in the area. These repellent pheromones are most abundantly produced by first instar larvae, and persist for up to 3 weeks under laboratory conditions. Future studies should examine whether chrysopids whose adults feed on pollen, such as the commercially available common green lacewing *Chrysoperla carnea* (Stephens), respond to *Chrysopa* oviposition-deterring pheromones or themselves produce such pheromones (Ruzicka, 1996).

The chemical ecology of brown lacewings has not been studied, but a recent chance discovery that *Hemerobius stigma* Stephens is attracted to the sex pheromone of a pine bast scale insect (see Anthocoridae section) raises new possibilities for semiochemically manipulating hemerobiids (Mendel *et al.*, 1997).

Hoverflies (Diptera: Syrphidae)

Adult syrphids are not predaceous, and females require pollen in order to produce eggs (Stubbs and Falk, 1993) (Fig. 14.1e). Syrphid larvae, however, are important predators of aphids and related homopterans, while others specialize on eggs and larvae of leaf beetles (Chrysomelidae).

Köpf *et al.* (1997) showed that larvae of *Parasyrphus nigritarsis* Zett. exhibit a characteristic to and fro motion of the anterior portion of the body as they move toward salicylaldehyde or cyclopentanoid monoterpenes emanating from chrysomelid prey. At least some hoverflies specializing on homopterans use unidentified honeydew volatiles as ovipositional kairomones (Budenberg and Powell, 1992). It is possible that ammonia is involved in the attraction of adult hoverflies since aphids are unique among insects in excreting ammonia rather than uric acid (Sasaki *et al.*, 1990), and ammonia is attractive to various other flies (e.g. Mazor *et al.*, 1987). The rapid decline in the kairomonal activity of honeydew to certain parasitoids (Shaltiel and Ayal, 1998) is consistent with the evaporation of a volatile component of honeydew such as ammonia. There is also some evidence that normal non-volatile hydrocarbons in the range C_{22}–C_{28}, possibly emitted as part of the cornicle fluid secretion of aphids, stimulate syrphids to oviposit (Shonouda, 1996; Shonouda *et al.*, 1998a,b). On the other hand, females of aphidophagous syrphids often avoid ovipositing near large aphid aggregations in favour of small, young colonies more likely to ensure that the offspring will find suitable and sufficient food (e.g. Kan, 1989), but the mechanism of aphid colony assessment by syrphids is not precisely known.

Ground Beetles (Coleoptera: Carabidae)

Pheromones have yet to be identified for any ground beetle (Mayer and McLaughlin, 1991), although carabids are important predators in forest and agroecosystems (Lövei and Sunderland, 1996; Riddick *et al.*, 1998) (Fig. 14.1f). Kielty *et al.* (1996) found that some carabids are attracted to (*E*)-β-farnesene, suggesting that ground beetles utilize this alarm pheromone component to locate aphids that drop from their host plant.

Soft-winged Flower Beetles (Coleoptera: Melyridae)

Melyrids are poorly known generalist predators whose impact on agricultural pests is uncertain (Greany and Hagen, 1981) (Fig. 14.1g). Gut content immunoassays of the common North American species, *Collops vittatus* (Say), indicate that this beetle is, in fact, a significant predator of the sweetpotato whitefly and the pink bollworm (Hagler and Naranjo, 1996).

There is strong circumstantial evidence that *Collops* females produce a sex pheromone. Caryophyllene oxide is significantly attractive to *C. vittatus*, and caryophyllene alcohol is an order of magnitude more active and attracts only males (Flint *et al.*, 1981). Additionally, males of *C. hirtellus* were attracted to unidentified volatiles from a commercial wood preservative to the extent that 'beetles were swarming around houses so thickly that they were a nuisance' (White, 1985).

Ladybird Beetles (Coleoptera: Coccinellidae)

Ladybird beetles have played a prominent role in classical biological control (*c.* 180 species introduced in the USA). Generalist species such as *Coleomegilla maculata*

(De Geer) appear to be particularly compatible with transgenic crops where secondary pests or plant pollen are available (Riddick *et al.*, 1998). Despite the importance of ladybirds, the chemical identification of pheromones for these predators has been elusive (Hedin *et al.*, 1988; Veilleux *et al.*, 1993). Hemptinne *et al.* (1996) demonstrated that males of the two-spot ladybird beetle, *Adalia bipunctata* (L.), recognize females by a pheromone excreted from glands opening on to the outer surface of the elytra, but the pheromone is only perceived upon contact and has not been chemically identified. The best known aspect of coccinellid chemical ecology is their chemical defence: the conspicuous ladybird colours signal to predators that they contain toxic alkaloids, which can be externalized by reflex bleeding (King and Meinwald, 1996). In addition, volatile alkylmethoxypyrazines associated with the blood alkaloids act as olfactory alerting signals – the chemical equivalent of aposematic coloration (Moore *et al.*, 1990).

In the autumn, many ladybird beetles migrate from feeding sites to form overwintering aggregations (Majerus, 1994). Aggregating beetles often visually orient toward some prominent feature of the landscape, such as mountain outcroppings (e.g. *Hippodamia convergens* Guerin-Meneville) or tree trunks (e.g. *C. maculata*) (Roach and Thomas, 1991). However, aggregations of coccinellid beetles (Majerus, 1994), and those of a related species (Endodmychidae: *Stenotarsus rotundus* Arrow) (Denlinger, 1994), are frequently patchy or occur at the same site year after year, suggesting that pheromones may also be involved in the gregarious behaviour of these beetles. It now appears that alkylmethoxypyrazines serve a dual function in coccinellid chemical ecology, not only as warning odours to would-be predators but also as essential aggregation pheromone components (al Abassi *et al.*, 1998). 2-Isopropyl-3-methoxypyrazine was identified from the seven-spot ladybird, *Coccinella septempunctata* L., and shown in a laboratory bioassay to act as an attractant and arrestant for the adult beetles (al Abassi *et al.*, 1998). The persistent nature of alkylmethoxypyrazines and the co-occurrence of the same pyrazines in various coccinellid species (Moore *et al.*, 1990) are consistent with these compounds being involved in the formation of ladybird aggregations at the same place year after year and with the frequent coexistence of two or more coccinellid species in the same aggregation (Roach and Thomas, 1991; Majerus, 1994). Other compounds may also be involved in the gregarious behaviour of ladybird beetles because females of the multicoloured Asian lady beetle, *Harmonia axyridis* (Pallas), arrive at aggregation sites first (Nalepa *et al.*, 1996, 1998), and females produce sesquiterpenoids that are undetectable in spring-collected beetles (J.R. Aldrich, 1998, unpublished results). Semiochemicals mediating the formation of overwintering ladybird aggregations could prove useful in preserving populations of these predators (al Abassi *et al.*, 1998), for example, by luring beetles to specific sites where they are better protected against inclement weather and could be provided with food.

Whether or not ladybird adults or larvae use kairomones to locate prey has been a controversial issue (Majerus, 1994). Ladybird beetles reportedly orient toward volatiles emanating from prey (Obata, 1986; Sengonca and Liu, 1994), and there is even a commercially available Ladybug Lure™ (Consep, Inc., Bend, Oregon, USA) based on the aphid alarm pheromone component, (*E*)-β-farnesene. The product supposedly 'draws ladybugs and other aphid-eating bugs like a magnet' although the patent claims only a method to increase the number of syrphid larvae, making no mention of ladybird beetles (Banfield, 1994). At most, responses to these kairomones for ladybird beetles are weak (Nakamuta, 1991). On the other hand, honeydew from homopteran prey is a potent arrestant for various coccinellid adults and larvae (Greany and Hagen, 1981; Carter and Dixon, 1984; van den Meiracker *et al.*, 1990). Evans and Richards (1997) showed that ladybird densities increased two- to 13-fold within 48 h in the centre of plots sprayed with sucrose solutions, with a concomitant decline in beetle densities in the surrounding unsprayed portions of plots. These authors concluded that it may be feasible to use artificial honeydew application to concentrate, and thereby conserve, ladybird adults in the centre of a field when the surrounding crop area is treated with insecticide. In view of the responses of lacewings and hoverflies to honeydew (discussed above), it seems possible that inclusion of tryptophan or ammonia-generating salts might improve the attractiveness of sucrose solutions to ladybird beetles.

Natural ladybird populations have generally proved ineffective for the biological control of aphids, whereas many ladybird species are effective against coccid prey (Dixon *et al.*, 1997). The evolution of telescoping generations in aphids resulted in much shorter generation times for aphids than for coccids, making it more difficult for ladybird beetles to locate aphid populations. The discrepancy in the efficacy of aphidophagous versus coccidophagous ladybirds may also be due to semiochemically mediated reproductive behaviour of aphidophagous ladybirds. Hemptinne *et al.* (1992) found that female *A. bipunctata* avoided ovipositing in areas occupied by conspecific larvae. More recent research has confirmed the presence of a larval-produced oviposition-deterring pheromone in *A. bipunctata* and demonstrated the existence of a similar species-specific pheromone in *C. septempunctata* (Doumbia *et al.*, 1998). Interestingly, females of *C. septempunctata* were deterred from ovipositing at sites previously exposed to conspecific larvae as well as sites previously occupied by larvae of the goldeneye lacewing, *C. oculata* (Ruzicka, 1997b). Theoretically, adult ladybirds should leave an aphid aggregation as soon as a certain number of eggs are laid and/or larvae are present, and if ladybirds lay the optimal number of eggs their offspring have only a slight effect on the peak number of aphids (Kindlmann and Dixon, 1993). However, females of the coccidophagous ladybird *Cryptolaemus montrouzieri* Mulsant also avoid ovipositing in the presence of conspecific larvae (Merlin *et al.*, 1996).

Bark Beetle Predators (Coleoptera and Diptera)

A complex of predators uses the aggregation pheromones of bark beetles (Scolytidae) as kairomones, including flies (Dolichopodidae: *Medetera* spp.) and a multitude of beetles (Cleridae, Histeridae, Rhizophagidae, Staphylinidae, Trogostidae, Tenebrionidae and Colydiidae) (e.g. Weslien, 1992; Wigger, 1996). A particularly intriguing discovery is that scolytid predators in one geographical region are often more attracted to prey from distant regions (Raffa and Klepzig, 1989). For example, in Wisconsin, where the pine engraver, *Ips pini* (Say), is most attracted to a 25:75 blend of (*R*)-(−)-:(*S*)-(+)-ipsdienol (2-methyl-6-methylene-2,7-octadien-4-ol), the locally common predators *Thanasimus dubius* (Fabricius) (Fig. 14.1h) and *Platysoma* (= *Cylistix*) *cylindrica* (Paykull) are primarily attracted to the reverse blend of the enantiomers (Fig. 14.2a) (Raffa and Klepzig, 1989). A subsequent reciprocal experiment between California and Wisconsin showed that another abundant predator, *Enoclerus lecontei* (Wolcott), also clearly prefers 'exotic' prey pheromone blends (Fig. 14.2b) (Raffa and Dahlsten, 1995). In addition, biosynthesis patterns of, and response to, the *I. pini* minor synergist component lanierone (2-hydroxy-4,4,6-trimethyl-2,5-cyclohexadien-1-one) corroborate this pattern for some predators, including *E. lecontei* (Raffa and Dahlsten, 1995; Miller *et al.*, 1997). Thus, pressure from predators seems to have selected for shifts in bark beetle pheromone blends (Raffa and Klepzig, 1989), supporting the hypothesis of locally driven semiochemical adaptations and counter-adaptations among predators and prey (Raffa and Dahlsten, 1995).

Differential responses to semiochemicals have also been detected for *Dendroctonus* bark beetles and their predators. *Dendroctonus frontalis* Zimmermann females produce a mixture of 85% (1*S*,5*R*)-(−)- and 15% (1*R*,5*S*)-(+)-frontalin (1,5,-dimethyl-6,8-dioxabicyclo[3.2.1]octane) as an aggregation pheromone (Mayer and McLaughlin, 1991), but *T. dubius* (which prefers *D. frontalis* over *Ips* spp. as prey) responds electrophysiologically and behaviourally only to (−)-frontalin (Payne *et al.*, 1984). Payne *et al.* (1984) speculated that *D. frontalis* may be evolving as a result of predation pressure so that (+)-frontalin becomes a component of its aggregation pheromone to which the predator does not yet respond. Furthermore, *D. frontalis* males eventually produce verbenone and *endo*-brevicomin to repel other males (Mayer and McLaughlin, 1991), but these compounds have no influence on the arrival of *T. dubius* (Salom *et al.*, 1992).

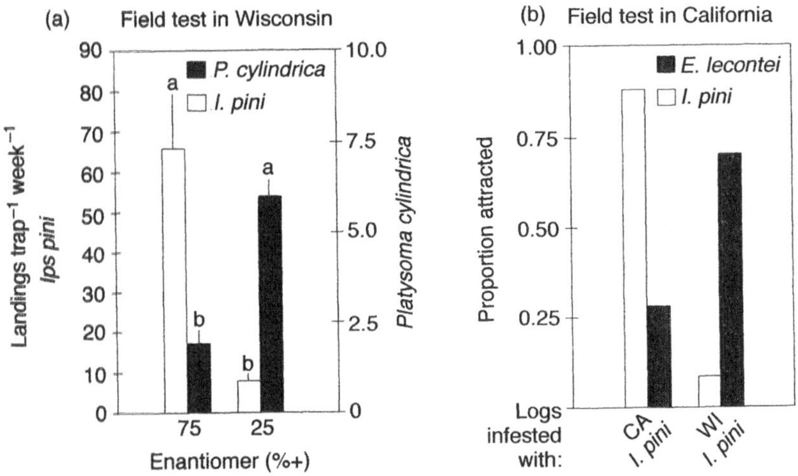

Fig. 14.2. (a) Responses of *Ips pini* (Say) and the predator, *Platysoma cylindrica* (Paykull), to chiral blends of ipsdienol (modified from Raffa and Klepzig, 1989). (b) Proportions of *I. pini* and the predator, *Enoclerus lecontei* (Wolcott), from California (CA) that were attracted to logs infested with *I. pini* from California versus logs infested with *I. pini* from Wisconsin (WI). (Modified from Raffa and Dahlsten, 1995.)

Similarly, attraction of Douglas-fir beetles, *Dendroctonus pseudotsugae* Hopkins, to aggregation pheromone is inhibited by an anti-aggregation pheromone, 3-methyl-cyclohex-2-en-1-one, while abundant coleopteran predators of *D. pseudotsugae* in the area are unaffected (Ross *et al.*, 1996). Finally, fungi associated with *D. frontalis* apparently induce the production of 4-allylanisole in loblolly pine which acts as a repellent to later-arriving bark beetles, but 4-allylanisole does not repel the major predator, *T. dubius* (Hayes *et al.*, 1994).

For the greater European spruce bark beetle, *Dendroctonus micans* (Kugelann), monitoring the levels of its specific predator, *Rhizophagus grandis* Gyllenhal, may actually be the best means to ascertain if the prey is established in an area. From a probable origin in Siberia, *D. micans* has and is continuing to spread westward into Europe, but the pest cannot reliably be detected at low population densities using pheromone-baited traps (Grégoire, 1988). On the other hand, *R. grandis* can be monitored at low densities using traps baited with synthetic kairomones mimicking monoterpenoids in the frass from *D. micans* (Grégoire *et al.*, 1992a,b; Wainhouse *et al.*, 1992), and the predator can only become established if *D. micans* is present. Thus, if the predators released one year before in an area thought to be colonized by the European spruce bark beetle are caught in kairomone-baited traps the next year, then *D. micans* is definitely established in the area (J.-C. Grégoire, Brussels, 1998, personal communication). In fact, semiochemicals have actually been used to facilitate bark beetle biological control of *D. micans* with *R. grandis*. Grégoire *et al.* (1992a) treated *Calliphora* larvae with a terpenoid kairomone to induce oviposition by the predator on these easily reared fly larvae to mass-produce predators for augmentative biological control.

Understanding the semiochemical idiosyncrasies of bark beetles and their predators should at least lead to improved population estimates and better guidelines for timing silvicultural practices (see also Chapter 6, this volume). Proactive control tactics might also be feasible, such as treating incipient scolytid infestations with pheromones tailored to local populations, thereby curtailing beetle attack while maximizing predator attraction (Miller *et al.*, 1997). Indeed, transplantation of indigenous natural enemies within the ranges of scolytid pests is a possible alternative to importing exotic natural enemies (Raffa and Dahlsten, 1995).

Damsel Bug and Bigeyed Bugs (Heteroptera: Nabidae and Lygaeidae: Geocorinae)

Species in these groups are among the most important predators in agroecosystems (Yeargan, 1998), but they are semiochemically unknown. Potent pheromones have recently been discovered for Lygaeinae (Aldrich *et al.*, 1997a), raising hopes that pheromones exist in *Geocoris* spp., and there is morphological evidence that nabids produce pheromones (Schuh and Slater, 1995).

Assassin Bugs (Heteroptera: Reduviidae)

Reduviids constitute the largest group of predaceous land bugs (Schuh and Slater, 1995), and the Triatominae include haematophagous vectors of American trypanosomiasis or Chagas' disease. *Triatoma infestans* Klug has recently been shown to possess specialized ammonia receptors used to locate hosts and to form resting aggregations (Taneja and Guerin, 1997), raising some hope that means can be devised to trap-out the vectors of Chagas' disease from infested domiciles.

The sole pheromone identified for a predaceous reduviid is for *Pristhesancus plagipennis* Walker, an important natural biocontrol agent of the Australian citrus pest *Biprorulus bibax* Breddin (James *et al.*, 1994) (Fig. 14.1i). Adults of *P. plagipennis* have three median dorsal abdominal glands, with the anterior and posterior glands being much larger in males than in females (Aldrich, 1998), a pattern that may be a general feature of harpactocorines because males of the North American 'bee killer', *Apiomerus crassipes* (Fabricius), have similarly enlarged dorsal abdominal glands (J.R. Aldrich, 1998, unpublished results). In laboratory olfactometer and outdoor flight-cage bioassays, female *P. plagipennis* were significantly attracted to calling males, male posterior dorsal abdominal gland extracts, and the (2*R*)-enantiomer of the major component of the secretion, (*Z*)-3-hexenyl 2-hydroxy-3-methylbutanoate (Moore *et al.*, 1993; James *et al.*, 1994). Males were also attracted to (*Z*)-3-hexenyl (*R*)-2-hydroxy-3-methylbutanoate (James *et al.*, 1994). Females evidently produce a pheromone from their anterior dorsal abdominal glands that attracts males at close range (D.G. James, Yanco, New South Wales, Australia, 1998, personal communication). In irrigated areas of New South Wales, *B. bibax* has become a major pest of citrus, largely because *P. plagipennis* does not naturally occur in this region. Therefore, a synthetic lure for *P. plagipennis* may prove useful in an unusual 'classical' biocontrol programme within the country to mass-trap predators in Queensland for establishment in the irrigated citrus groves of southern Australia.

Minute Pirate Bugs (Heteroptera: Anthocoridae)

Orius and *Anthocoris* spp. are important predators of aphids, thrips, mites and lepidopteran eggs in a wide variety of field and greenhouse cropping systems (Yeargan, 1998) (Fig. 14.1j). Some anthocorids (e.g. *Xylocoris* and *Lyctocoris* spp.) are important biocontrol agents of stored-product pests (Glenister, 1998). Linalool, α-terpineol, nerol and geraniol are biosynthesized in the metathoracic scent gland secretion of *Xylocoris flavipes* (Reuter) (Phillips *et al.*, 1995), but although some of these compounds are known components of other heteropteran pheromones (Aldrich, 1998), there is no evidence that they play a pheromonal role in *X. flavipes*. For *O. insidiosus* (Say), Y-tube olfactometer bioassays showed that males are attracted to volatiles from virgin females, and (2*E*,7)-octadienal, (2*E*,7)-octadienoic acid and (2*E*,9)-decadienal were identified as female-specific volatiles (J.R. Aldrich and P. Puapoomchareon, 1996, unpublished results). A blend of (2*E*,7)-octadienal and (*E*)-2-octenal was moderately, but significantly, attractive to *O. insidiosus* males in the field (C.L. Smith and J.R. Aldrich, 1998, unpublished results). The fact that only males are attracted to suspected anthocorid pheromones, as in the related Miridae which also include predaceous species (see also Chapter 11, this volume), limits the potential usefulness of pheromones for biocontrol programmes involving these bugs.

The first evidence of a predator using a sex pheromone to locate its homopteran prey has been reported for the anthocorid *Elatophilus hebraicus* Pericart, which specializes on the Israeli pine bast scale, *Matsucoccus josephi* Bodenheimer and Harpaz (Mendel *et al.*, 1995). Both sexes of *E. hebraicus* use (2*E*,5*R*,6*E*,8*E*)-5,7-dimethyl-2,6,8,-decatrien-4-one (Dunkelblum *et al.*, 1996), the sex pheromone of this serious pest of pine stands in Israel, as a kairomone to find prey (Chapter 9). The sex pheromones of two allopatric *Matsucoccus* spp. (*M. feytaudi* Ducasse and *M. matsumurae* (Kuwana)) also elicited strong electrophysiological responses in *E. hebraicus* antennae, and attracted significant numbers of *E. hebraicus* in Israel, even though this anthocorid does not occur in the eastern Mediterranean countries where *M. feytaudi* and *M. matsumurae* exist (Dunkelblum *et al.*, 1996). The synthetic pheromones of *M. josephi*, *M. feytaudi* and *M. matsumurae* have been tested in many countries of the Palaearctic region (Mendel *et al.*, 1997); the only country other than Israel where *Matsucoccus* pheromone-baited traps attracted significant numbers of predators was in Portugal, where two predators were caught: *E. crassicornis* Reuter and *Hemerobius stigma*. These new potential biological control agents of *M. josephi* are being evaluated for possible introduction into Israel, and *E. hebraicus* is now considered a candidate to improve the biological control of the eastern Mediterranean *Matsucoccus* spp. (Mendel *et al.*, 1997).

Orius spp. are more generalized predators than is *E. hebraicus*, and they seem to use relatively close-range prey- or food-associated cues to direct their foraging. For example, within its microhabitat, *O. tristicolor* (White) evidently finds western flower thrips (as do *Amblyseius* phytoseiid mites) via a kairomonal response to their alarm pheromone (decyl and dodecyl acetates) (Teerling *et al.*, 1993; Teerling, 1995).

At present, the cost of anthocorids restricts their augmentative use to controlled environments (greenhouses and grain storage) where semiochemicals may find use in stimulating their searching behaviour (Glenister, 1998); semiochemical manipulation of anthocorids in field crops is farther from being a reality (but see Scutareanu *et al.*, 1997).

Predaceous Stink Bugs (Heteroptera: Pentatomidae: Asopinae)

Stink bugs are phytophagous (Chapter 11) except for one monophyletic predaceous subfamily, the Asopinae (Fig. 14.1k) (Thomas, 1992). Males of many (but not all) asopines have discrete pheromone glands, in addition to the metathoracic scent gland responsible for production of their allomones. Two types of pheromone glands have been found: sternal (= androconial) glands, and dorsal abdominal glands (Aldrich, 1995a, 1998). Asopines possessing sexually dimorphic dorsal abdominal glands are usually common polyphagous predators, whereas those with male-specific sternal glands are more often specialized predators (Thomas, 1992). Some asopines have neither of these types of glands, but may still produce attractant pheromones from cells in the abdominal epidermis (Aldrich, 1995a).

Sternal gland secretions are chemically known for *Perillus*, *Stiretrus* and *Oplomus* spp. (Aldrich *et al.*, 1986b; Aldrich, 1995a, 1998). Many of these asopines are specialists on chrysomelid beetles; for example, *Perillus bioculatus* (Fabricius) preys on the Colorado potato beetle, *Leptinotarsa decemlineata* (Say) (Hough-Goldstein and Whalen, 1993). *Stiretrus anchorago* (Fabricius) males secrete the 6,10,13-trimethyltetradecanol, and adults and immatures are attracted in low but significant numbers to the synthetic racemate (Kochansky *et al.*, 1989); however, the bugs did not significantly discriminate between stereoisomers in the field (Mori and Wu, 1991; J.R. Aldrich, 1998, unpublished results). In general, asopines with sternal glands produce relatively non-volatile compounds, which probably evaporate continuously for days once the pheromone is excreted on to the sternum.

Asopine males with enlarged dorsal abdominal glands (e.g. *Podisus*, *Zicrona*, *Arma*, *Alcaeorrhynchus*, *Supputius*, *Tynacantha*, *Oechalia* and *Brontocoris*) are able to control pheromone emission behaviourally (Aldrich, 1995a; Aldrich *et al.*, 1996). (*E*)-2-Hexenal is a major component of most of these asopines' pheromones (*Podisus*, *Supputius*, *Brontocoris* spp.) or suspected pheromones (*Zicrona* and *Arma* spp.), with the essential major synergistic components differing between species (Aldrich, 1995a, 1998; Aldrich *et al.*, 1997b; J.C. Zanuncio and J.R. Aldrich, 1998, unpublished results; J.R. Aldrich and F. Bin, 1998, unpublished results). For example, α-terpineol is the essential synergist of (*E*)-2-hexenal in *Podisus maculiventris* (Say), whereas in the sympatric species, *P. neglectus* Westwood (= *P. fretus*), linalool replaces α-terpineol as the main synergist (Aldrich *et al.*, 1984; Aldrich, 1995a).

The best-known asopine pheromone system is that of the spined soldier bug, *P. maculiventris* (Aldrich, 1995a, 1998). Pheromone-calling males attract females, as well as males and nymphs. Immature spined soldier bugs presumably evolved responsiveness to the pheromone because its presence is associated with available prey (Sant'Ana *et al.*, 1997). A complex of parasitoids and commensals exploits the pheromone of *P. maculiventris* as a host-finding kairomone (Aldrich, 1995a). In fact, the capability to catch hundreds of a tachinid fly, *Euclytia flava* (Townsend), in traps baited with the pheromone of *P. maculiventris* provided an opportunity to test the premise of the 'new associations' biological control concept (Hokkanen and Pimentel, 1989). Wild *E. flava* females preferred to oviposit on exotic pentatomid species rather than on *P. maculiventris* and other indigenous host species (Aldrich, 1995b), a situation analogous to that described above for bark beetle predators.

The spined soldier bug overwinters in the adult stage, and a massive synchronized emergence occurs just prior to the bud burst of deciduous trees in the eastern USA (Fig. 14.3) (Aldrich, 1995a). New-generation adults emerge asynchronously and few of these predators will enter pheromone-baited traps, in part because of tachinid parasitoids perched on the traps (Aldrich *et al.*, 1984). Therefore, summertime trap catches of *P. maculiventris* are much lower than during the spring emergence. *Podisus* and *Supputius* spp. in the New World tropics behaved similarly, but emergences (possibly synchronized by changes in rainfall) were much less striking in terms of the numbers of predators and parasitoids attracted (Aldrich *et al.*, 1997b).

Augmentation of *P. maculiventris* can significantly suppress targeted insect pests (Hough-Goldstein and Whalen, 1993; Hough-Goldstein, 1998), but mass production of this predator is not yet economically or logistically practical (Mahr, 1997). However, an artificial diet for *P. maculiventris* is nearing commercialization

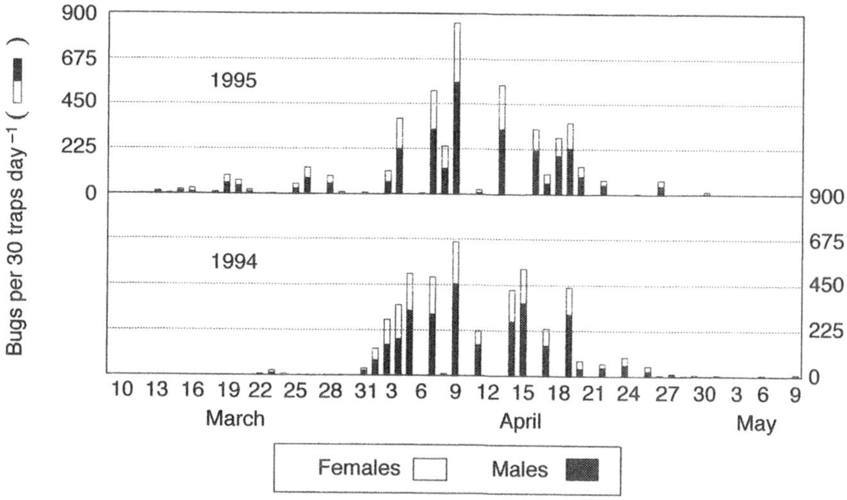

Fig. 14.3. Spined soldier bug adults, *Podisus maculiventris*, caught alive in pheromone-baited traps, March–May, 1994 and 1995, Agricultural Experiment Station, Beltsville, Maryland, USA.

(P.D. Greany, Gainesville, 1997, personal communication). Torres *et al.* (1997) showed that treating artificial diets with pheromone gland extracts accelerated finding and feeding on artificial diets by *Podisus* and *Supputius* nymphs. For another asopine, *Eocanthecona furcellata* (Wolff), (*E*)-phytol originating from chlorophyll in the prey diet stimulates the proboscis-extension behaviour leading to feeding (Yasuda, 1997, 1998). Since asopines evolved from phytophagous ancestors, the response to (*E*)-phytol may be a relic behaviour. Nevertheless, incorporation of (*E*)-phytol and/or synthetic pheromones with artificial diets for *P. maculiventris* and related predators may improve the acceptance of the diets.

At the Beltsville Agricultural Research Center, we have been pursuing another approach to producing and manipulating predators for augmentative biological control (Aldrich, 1998). Our goal has been to develop hardware and protocols enabling individual growers and insectary operators to mass-produce predators from wild adults caught in pheromone-baited traps. For example, in 1994 enough wild *P. maculiventris* adults were captured in 30 pheromone-baited traps during 2 weeks in early spring to produce over 400,000 immature predators (Fig. 14.3). We have also shown that spined soldier bug nymphs will walk toward commercial Spined Soldier Bug Attractors™ (Sterling International Inc., Liberty Lake, Washington, USA) (Fig. 14.4), indicating that the synthetic pheromone can be used to disperse nymphs from points of augmentation (Sant'Ana *et al.*, 1997). In the final year of a 3-year pilot test we combined pheromone-mediated harvest of adult predators with pheromone manipulation of their offspring. Cages were made using a mesh size that allowed young predators to escape while retaining the adults, and these cages were placed in the centre of treatment plots to serve as predator nurseries. Wild adult predators caught in pheromone-baited traps were transferred

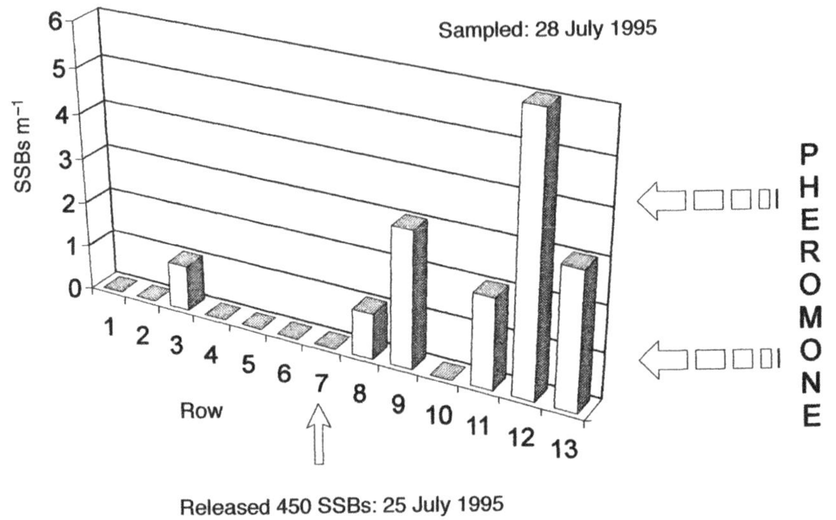

Fig. 14.4. Position of spined soldier bug nymphs (SSBs), *Podisus maculiventris*, 3 days after being released on row number 7 of snap bean plots with pheromone dispensers placed outside row number 13 (one experimental replicate from Sant'Ana *et al.*, 1997).

to nursery cages (in the future, encapsulated artificial diet will be supplied), and synthetic pheromone dispensers were positioned at the perimeter of plots to attract immature predators from the nursery cages into the surrounding rows of potato plants. Colorado potato beetle infestations were significantly suppressed using this approach (J.R. Aldrich and W.W. Cantelo, 1998, unpublished results).

Conclusions

Semiochemicals offer a means to manage the vast arsenal of generalist predators, and to monitor their populations in agroecosystems. Furthermore, the realization that imported biocontrol agents can adversely affect non-target endemic species (Simberloff and Stiling, 1996) makes managing native beneficials more important than ever. Indirect interactions and intraguild predation can negate the intended results of augmentation programmes; for example, application of artificial honeydew ultimately led to a decline in the aphid parasitoid population (Evans and England, 1996), and experimental addition of certain predaceous Heteroptera generated sufficient lacewing larval mortality to release aphid populations from regulation (Rosenheim *et al.*, 1993). Therefore, augmentation programmes must be carefully monitored, and the trophic webs in agroecosystems should be well understood (Rosenheim *et al.*, 1995).

More general predators have been commercialized than specialized predators (Weeden *et al.*, 1997), and this trend is likely to continue because there is a broader-based market for generalists, and they can be mass-produced more economically than specialists (e.g. Hough-Goldstein and McPherson, 1996). Inclusion of pheromones or prey-associated kairomones in artificial diets and oviposition sites seems likely to improve the efficacy of predator mass-propagation. Combining pheromone-mediated harvest of wild predators with semiochemically bolstered mass-rearing will multiply the supply of predators, and help to avoid inbreeding. Augmentation of immatures should minimize the problem of emigration of predators to non-target sites.

Insecticidal disruption of general predators can lead to catastrophic crop failures (e.g. Summy *et al.*, 1996). Luring predators away from infestations prior to pesticide treatments with semiochemical attractants may be the most immediately practical way to conserve predators (e.g. Evans and Richards, 1997). The direct approach to augment predator populations by semiochemically drawing wild predators to agroecosystems may only be effective when pest populations are low, and this tactic is likely to take longer and require more convincing experimentation to gain grower acceptance. As our comprehension of predator and parasitoid semiochemistry evolves, new opportunities should emerge to apply this knowledge for practical benefit. For example, the day may not be far off when it is feasible to accelerate the natural process of host shifts by artificial selection, based on appropriate semiochemical information, to 'teach' endemic beneficials to recognize alien hosts (Aldrich, 1995b). The looming challenge, then, for biocontrol researchers in the next millennium is to continue the momentum of semiochemical discovery, and to implement semiochemical husbandry of predators and parasitoids.

Acknowledgements

I especially thank the editors for inviting me to contribute to this volume. Photographs of insects were kindly supplied by Dr Scott Salom (Virginia Polytechnic Institute and State University; *T. dubius*), Dr Mary Louise Flint (University of California Statewide IPM Project: Jack K. Clark, photographer; *C. vittatus*), and Dr Michael P. Hoffmann (Department of Entomology, Cornell University; several insect photographs from http://www.nysaes.cornell.edu/ent/ biocontrol/). I am also grateful to Carolyn Smith, Jen-Yi Lee, Belle Le, and especially Maria Bernadete Vieira for help in preparing this manuscript. I acknowledge Drs Erick Riddick (ICEL), Patrick Greany, Jean-Claude Grégoire, and one vicious anonymous reviewer for criticisms on how to improve the manuscript. Mention of commercial products does not constitute an endorsement by the US Department of Agriculture.

References

Abassi, S. al, Birkett, M.A., Pettersson, J., Pickett, J.A. and Woodcock, C.M. (1998) Ladybird beetle odour identified and found to be responsible for attraction between adults. *Cellular and Molecular Life Sciences* 54, 876–879.

Akre, R.D., Greene, A., MacDonald, J.F., Landolt, P.J. and Davis, H.G. (1981) *The Yellowjackets of America North of Mexico*. US Department of Agriculture Handbook No. 552, Washington, DC, 102 pp.

Aldrich, J.R. (1995a) Chemical communication in true bugs and exploitation by parasitoids and commensals. In: Cardé, R.T. and Bell, W.J. (eds) *Chemical Ecology of Insects II*. Chapman & Hall, New York, pp. 318–363.

Aldrich, J.R. (1995b) Testing the 'new associations' biological control concept with a predator (Heteroptera: Pentatomidae: *Podisus maculiventris*) and a parasitoid (Diptera: Tachinidae: *Euclytia flava*). *Journal of Chemical Ecology* 21, 1031–1042.

Aldrich, J.R. (1998) Status of semiochemical research on predatory Heteroptera. In: Coll, M. and Ruberson, J.R. (eds) *Predatory Heteroptera: Their Ecology and Use in Biological Control*. Thomas Say Foundation, Entomological Society of America, Lanham, Maryland, pp. 33–48.

Aldrich, J.R. and Barros, T.M. (1995) Chemical attraction of male crab spiders (Araneae, Thomisidae) and kleptoparasitic flies (Diptera: Milichiidae and Chloropidae). *Journal of Arachnology* 23, 212–214.

Aldrich, J.R. and Puapoomchareon, P. (1996) Management of predaceous hemipterans with semiochemicals: practice and potential. *Proceedings of the 20th International Congress of Entomology*, Florence, p. 625.

Aldrich, J.R., Kochansky, J.P. and Abrams, C.B. (1984) Attractant for a beneficial insect and its parasitoids: pheromone of the predatory spined soldier bug, *Podisus maculiventris* (Hemiptera: Pentatomidae). *Environmental Entomology* 13, 1031–1036.

Aldrich, J.R., Kochansky, J.P. and Sexton, J.D. (1985) Chemical attraction of the eastern yellowjacket, *Vespula maculifrons* (Hymenoptera: Vespidae). *Experientia* 41, 420–422.

Aldrich, J.R., Lusby, W.R. and Kochansky, J.P. (1986a) Identification of a new predaceous stink bug pheromone and its attractiveness to the eastern yellowjacket. *Experientia* 42, 583–585.

Aldrich, J.R., Oliver, J.E., Lusby, W.R. and Kochansky, J.P. (1986b) Identification of male specific exocrine secretions from predatory stink bugs (Hemiptera, Pentatomidae). *Archives of Insect Biochemistry and Physiology* 3, 1–12.

Aldrich, J.R., Oliver, J.E., Waite, G.K., Moore, C.J. and Waters, R.M. (1996) Identification of the presumed pheromone blend from the Australasian predaceous bug, *Oechalia schellenbergii* (Heteroptera: Pentatomidae). *Journal of Chemical Ecology* 22, 729–738.

Aldrich, J.R., Leal, W.S., Nishida, R., Khrimian, A.P., Lee, C.-J. and Sakuratani, Y. (1997a) Semiochemistry of aposematic seed bugs. *Entomologia Experimentalis et Applicata* 84, 127–135.

Aldrich, J.R., Zanuncio, J.C., Vilela, E.F., Torres, J.B. and Cave, R.D. (1997b) Field tests of predaceous pentatomid pheromones and semiochemistry of *Podisus* and *Supputius* species (Hemiptera: Pentatomidae: Asopinae). *Anais da Sociedade Entomologica do Brasil* 26, 1–14.

Allan, R.A., Elgar, M.A. and Capon, R.J. (1996) Exploitation of an ant chemical alarm signal by the zodariid spider *Habronestes bradleyi* Walckenaer. *Proceedings of the Royal Society of London Series B* 263, 69–73.

Banfield, M.G. (1994) Indirect aphid control with low concentration of EBF. US Government Patent No. 5, 308, 613.

Bitsch, J. (1984) Anatomy of adult Chrysopidae. In: Canard, M., Séméria, Y. and New, T.R. (eds) *Biology of Chrysopidae*. Dr W. Junk Publishers, The Hague, pp. 29–36.

Budenberg, W.J. and Powell, W. (1992) The role of honeydew as an ovipositional stimulant for two species of syrphids. *Entomologia Experimentalis et Applicata* 64, 57–61.

Carter, M.C. and Dixon, A.F.G. (1984) Honeydew: an arrestant stimulus for coccinellids. *Ecological Entomology* 9, 383–387.

Coll, M. and Ruberson, J.R. (1998) Predatory Heteroptera: an important yet neglected group of natural enemies. In: Coll, M. and Ruberson, J.R. (eds) *Predatory Heteroptera: Their Ecology and Use in Biological Control*. Thomas Say Foundation, Entomological Society of America, Lanham, Maryland, pp. 1–6.

Daane, K.M., Yokota, G.Y., Zheng, Y. and Hagen, K.S. (1996) Inundative release of common green lacewings (Neuroptera: Chrysopidae) to suppress *Erythroneura variabilis* and *E. elegantula* (Homoptera: Cicadellidae) in vineyards. *Environmental Entomology* 25, 1224–1234.

Denlinger, D.L. (1994) The beetle tree. *American Entomologist* 40, 168–171.

Dicke, M. (1998) Evolution of induced indirect defense of plants. In: Harvell, C.D. and Tollrian, R. (eds) *Evolution of Inducible Defenses*. Princeton University Press, Princeton, New Jersey, pp. 62–88.

Dixon, A.F.G., Hemptinne, J.-L. and Kindlmann, P. (1997) Effectiveness of ladybirds as biological control agents: patterns and processes. *Entomophaga* 42, 71–83.

Doumbia, M., Hemptinne, J.-L. and Dixon, A.F.G. (1998) Assessment of patch quality by ladybirds: role of larval tracks. *Oecologia* 113, 197–202.

Dunkelblum, E., Mendel, Z., Gries, G., Gries, R., Zegelman, L., Hassner, A. and Mori, K. (1996) Antennal response and field attraction of the predator *Elatophilus hebraicus* (Hemiptera: Anthocoridae) to sex pheromones and analogues of three *Matsucoccus* spp. (Homoptera: Matsucoccidae). *Bioorganic and Medicinal Chemistry* 4, 489–494.

Emden, H.F. van and Hagen, K.S. (1976) Olfactory reactions of the green lacewing, *Chrysopa carnea*, to tryptophan and certain breakdown products. *Environmental Entomology* 5, 469–473.

Evans, E.W. and England, S. (1996) Indirect interactions in biological control of insects: pests and natural enemies in alfalfa. *Ecological Applications* 6, 920–930.

Evans, E.W. and Richards, D.R. (1997) Managing the dispersal of ladybird beetles (Col.: Coccinellidae): use of artificial honeydew to manipulate spatial distributions. *Entomophaga* 42, 93–102.

Flint, H.M., Salter, S.S. and Walters, S. (1979) Caryophyllene: an attractant for the green lacewing. *Environmental Entomology* 8, 1123–1125.

Flint, H.M., Merkle, J.R. and Sledge, M. (1981) Attraction of male *Collops vittatus* in the field by caryophyllene alcohol. *Environmental Entomology* 10, 301–304.

Glenister, C.S. (1998) Predatory heteropterans in augmentative biological control: an industry perspective. In: Coll, M. and Ruberson, J.R. (eds) *Predatory Heteroptera: Their Ecology and Use in Biological Control.* Thomas Say Foundation, Entomological Society of America, Lanham, Maryland, pp. 199–208.

Greany, P.D. and Hagen, K.S. (1981) Prey selection. In: Nordlund, D.A., Jones, R.L. and Lewis, W.J. (eds) *Semiochemicals: Their Role in Pest Control.* John Wiley & Sons, New York, pp. 121–135.

Grégoire, J.-C. (1988) The greater European spruce beetle. In: Berryman, A.A. (ed.) *Dynamics of Forest Insect Populations*. Plenum Publishing, New York, pp. 455–478.

Grégoire, J.-C., Couillien, D., Drumont, A., Meyer, H. and Francke, W. (1992a) Semiochemicals and the management of *Rhizophagus grandis* Gyll. (Col., Rhizophagidae) for the biocontrol of *Dendroctonus micans* Kug. (Col., Scolytidae). *Journal of Applied Entomology* 114, 110–112.

Grégoire, J.-C., Couillien, D., Krebber, R., König, W.A., Meyer, H. and Francke, W. (1992b) Orientation of *Rhizophagus grandis* (Coleoptera: Rhizophagidae) to oxygenated monoterpenes in a species-specific predator–prey relationship. *Chemoecology* 3, 14–18.

Hagen, K.S., Greany, P., Sawall, E.F. Jr and Tassan, R.L. (1976) Tryptophan in artificial honeydews as a source of an attractant for adult *Chrysopa carnea. Environmental Entomology* 5, 458–468.

Hagler, J.R. and Naranjo, S.E. (1996) Using gut content immunoassays to evaluate predaceous biological control agents: a case study. In: Symondson, W.O.C. and Liddell, J.E. (eds) *The Ecology of Agricultural Pests: Biochemical Approaches*. Chapman & Hall, London, pp. 383–399.

Hayes, J.L., Strom, B.L., Roton, L.M. and Ingram, L.L. Jr (1994) Repellent properties of the host compound 4-allylanisole to the southern pine beetle. *Journal of Chemical Ecology* 20, 1595–1615.

Hedin, P.A., Phillips, V.A. and Dysart, R.J. (1988) Identification of some volatile constituents of the lady beetle *Coleomegilla maculata lengi* [Timberlake]: search for pheromones. *Journal of the Mississippi Academy of Sciences* 33, 59–66.

Hemptinne, J.-L., Dixon, A.F.G. and Coffin, J. (1992) Attack strategy of ladybird beetles (Coccinellidae): factors shaping their numerical response. *Oecologia* 90, 238–245.

Hemptinne, J.-L., Dixon, A.F.G. and Lognay, G. (1996) Searching behaviour and mate recognition by males of the two-spot ladybird beetle, *Adalia bipunctata*. *Ecological Entomology* 21, 165–170.

Hendrichs, J., Katsoyannos, B.I., Wornoaypron, V. and Hendrichs, M.A. (1994) Odour-mediated foraging by yellowjacket wasps (Hymenoptera: Vespidae): predation on leks of pheromone-calling Mediterranean fruit fly males (Diptera: Tephritidae). *Oecologia* 99, 88–94.

Hendrichs, M.A. and Hendrichs, J. (1998) Perfumed to be killed: interception of Mediterranean fruit fly (Diptera: Tephritidae) sexual signaling by predatory foraging wasps (Hymenoptera: Vespidae). *Annals of the Entomological Society of America* 91, 228–234.

Hislop, R.G. and Prokopy, R.J. (1981) Mite predator responses to prey and predator-emitted stimuli. *Journal of Chemical Ecology* 7, 895–904.

Hokkanen, H.M.T. and Pimentel, D. (1989) New associations in biological control: theory and practice. *Canadian Entomologist* 121, 829–840.

Hough-Goldstein, J. (1998) Use of predatory pentatomids in integrated management of the Colorado potato beetle (Coleoptera: Chrysomelidae). In: Coll, M. and Ruberson, J.R. (eds) *Predatory Heteroptera: Their Ecology and Use in Biological Control.* Thomas Say Foundation, Entomological Society of America, Lanham, Maryland, pp. 209–223.

Hough-Goldstein, J. and Whalen, J. (1993) Inundative release of predatory stink bugs for control of Colorado potato beetle. *Biological Control* 3, 343–347.

Hough-Goldstein, J.A. and McPherson, D. (1996) Comparison of *Perillus bioculatus* and *Podisus maculiventris* (Hemiptera: Pentatomidae) as potential control agents of the Colorado potato beetle (Coleoptera: Chrysomelidae). *Journal of Economic Entomology* 89, 116–123.

Hoy, M.A. and Smilanick, J.M. (1979) A sex pheromone produced by immature and adult females of the predatory mite, *Metaseiulus occidentalis*, Acarina: Phytoseiidae. *Entomologia Experimentalis et Applicata* 26, 291–300.

Huang, H.T. and Yang, P. (1987) The ancient cultured citrus ant. *BioScience* 39, 665–671.

James, D.G., Moore, M.J. and Aldrich, J.R. (1994) Identification, synthesis and bioactivity of a male-produced aggregation pheromone in the assassin bug, *Pristhesancus plagipennis* (Hemiptera: Reduviidae). *Journal of Chemical Ecology* 20, 3281–3295.

Janssen, A., Bruin, J., Jacobs, G., Schraag, R. and Sabelis, M.W. (1997) Predators use volatiles to avoid prey patches with conspecifics. *Journal of Animal Ecology* 66, 223–232.

Kan, E. (1989) Assessment of aphid colonies by hoverflies. III. Pea aphids and *Episyrphus balteatus* (de Geer) (Diptera: Syrphidae). *Journal of Ethology* 7, 1–6.

Kielty, J.P., Allen-Williams, L.J., Underwood, N. and Eastwood, E.A. (1996) Behavioral responses of three species of ground beetle (Coleoptera: Carabidae) to olfactory cues associated with prey and habitat. *Journal of Insect Behavior* 9, 237–250.

Kindlmann, P. and Dixon, A.F.G. (1993) Optimal foraging in ladybird beetles (Coleoptera: Coccinellidae) and its consequences for their use in biological control. *European Journal of Entomology* 90, 443–450.

King, G.K. and Meinwald, J. (1996) Review of the defensive chemistry of Coccinellids. *Chemical Reviews* 96, 1105–1122.

Kochansky, J.P., Aldrich, J.R. and Lusby, W.R. (1989) Synthesis and pheromonal activity of 6,10,13-trimethyl-1-tetradecanol for predatory stink bug, *Stiretrus anchorago* (Heteroptera: Pentatomidae). *Journal of Chemical Ecology* 15, 1717–1728.

Köpf, A., Rank, N.E., Roininen, H. and Tahvanainen, J. (1997) Defensive larval secretions of leaf beetles attract a specialist predator *Parasyrphus nigritarsis*. *Ecological Entomology* 22, 176–183.

Landolt, P.J. (1998) Chemical attractants for trapping the yellowjackets *Vespula germanica* (Fab.) and *Vespula pensylvanica* (Saussure) (Hymenoptera: Vespidae). *Environmental Entomology* 27, 1229–1234.

Lövei, G.L. and Sunderland, K.D. (1996) Ecology and behavior of ground beetles (Coleoptera: Carabidae). *Annual Review of Entomology* 41, 231–256.

Mahr, S.E.R. (1997) Spined soldier bug. *Midwest Biological Control News Online*. Department of Entomology, University of Wisconsin, Madison, USA.

Majerus, M.E.N. (1994) *Ladybirds*. Harper Collins Publishers, London, 367 pp.

Mayer, M.S. and McLaughlin, J.R. (1991) *Handbook of Insect Pheromones and Sex Attractants*. CRC Press, Boca Raton, Florida, 1083 pp.

Mazor, M., Gothilf, S. and Galun, R. (1987) The role of ammonia in the attraction of females of the Mediterranean fruit fly to protein hydrolysate baits. *Entomologia Experimentalis et Applicata* 43, 25–29.

McEwen, P.K., Jervis, M.A. and Kidd, N.A.C. (1994) Use of a sprayed L-tryptophan solution to concentrate numbers of the green lacewing *Chrysoperla carnea* in olive tree canopy. *Entomologia Experimentalis et Applicata* 70, 97–99.

Meiracker, R.A.F. van den, Hammond, W.N.O. and van Alphen, J.M. (1990) The role of kairomones in prey finding by *Diomus* sp. and *Exochomus* sp., two coccinellid predators of the cassava mealybug, *Phenacoccus manihoti*. *Entomologia Experimentalis et Applicata* 56, 209–217.

Mendel, Z., Zegelman, L., Hassner, A., Assael, F., Harel, M., Tam, S. and Dunkelblum, E. (1995) Outdoor attractancy of males of *Matsucoccus josephi* (Homoptera: Matsucoccidae) and *Elatophilus hebraicus* (Hemiptera: Anthocoridae) to synthetic female sex pheromone of *Matsucoccus josephi*. *Journal of Chemical Ecology* 21, 331–341.

Mendel, Z., Adar, K., Nestel, D. and Dunkelblum, E. (1997) Sex pheromone traps as a tool for the study of population trends of the predator of a scale insect and for the identification of potential predators for biological control. *IOBC/WPRS Bulletin* 20, 231–240.

Merlin, J., Lemaitre, O. and Grégoire, J.-C. (1996) Chemical cues produced by conspecific larvae deter oviposition by the coccidophagous ladybird beetle, *Cryptolaemus montrouzieri*. *Entomologia Experimentalis et Applicata* 79, 147–151.

Miller, D.R., Gibson, K.E., Raffa, K.F., Seybold, S.J., Teale, S.A. and Wood, D.L. (1997) Geographic variation in response of pine engraver, *Ips pini*, and associated species to pheromone, lanierone. *Journal of Chemical Ecology* 23, 2013–2031.

Moore, B.P., Brown, W.V. and Rothschild, M. (1990) Methylalkylpyrazines in aposematic insects, their hostplants and mimics. *Chemoecology* 1, 43–51.

Moore, C.J., Aldrich, J.R. and James, D.G. (1993) Some aspects of the scent gland chemistry of the assassin bug *Pristhesancus plagipennis* Walker (Hemiptera: Reduviidae). In: Corey, S.A., Dall, K.J. and Milne, W.M. (eds) *Pest Control and Sustainable Agriculture*. CSIRO Publications, Canberra, Australia, pp. 304–306.

Mori, K. and Wu, J. (1991) Synthesis of the four stereoisomers of 6,10,13-trimethyl-1-tetradecanol, the aggregation pheromone of predatory stink bug, *Stiretrus anchorago*. *Liebig's Annals of Chemistry* 1991, 783–788.

Nakamuta, K. (1991) Aphid alarm pheromone component, (*E*)-β-farnesene, and local search by a predatory lady beetle, *Coccinella septempunctata bruckii* Mulsant (Coleoptera: Coccinellidae). *Applied Entomology and Zoology* 26, 1–7.

Nalepa, C.A., Kidd, K.A. and Ahlstrom, K.R. (1996) Biology of *Harmonia axyridis* (Coleoptera: Coccinellidae) in winter aggregations. *Annals of the Entomological Society of America* 89, 681–685.

Nalepa, C.A., Kidd, K.A. and Hopkins, D.I. (1999) The multicolored Asian lady beetle: orientation to aggregation sites. *Journal of Entomological Science* (in press).

Obata, S. (1986) Mechanisms of prey finding in the aphidophagous ladybird beetle, *Harmonia axyridis* (Coleoptera: Coccinellidae). *Entomophaga* 31, 303–311.

Oliveira, P.S. (1997) The ecological function of extrafloral nectaries: herbivore deterrence by visiting ants and reproductive output in *Caryocar brasiliense* (Caryocaraceae). *Functional Ecology* 11, 323–330.

O'Neil, R.J., Nagarajan, K., Wiedenmann, R.N. and Legaspi, J.C. (1996) A simulation model of *Podisus maculiventris* (Say) (Heteroptera: Pentatomidae) and Mexican bean beetle, *Epilachna varivestis* (Mulsant) (Coleoptera: Coccinellidae), population dynamics in soybean, *Glycine max* (L.). *Biological Control* 6, 330–339.

Overmyer, S.L. and Jeanne, R.L. (1998) Recruitment to food by the German yellowjacket, *Vespula germanica*. *Behavioral and Ecological Sociobiology* 42, 17–21.

Payne, T.L., Dickens, J.C. and Richerson, J.V. (1984) Insect predator–prey coevolution via enantiomeric specificity in a kairomone-pheromone system. *Journal of Chemical Ecology* 10, 487–492.

Phillips, T.W., Parajulee, M.N. and Weaver, D.K. (1995) Toxicity of terpenes secreted by the predator *Xylocoris flavipes* (Reuter) to *Tribolium castaneum* (Herbst) and *Oryzaephilus surinamensis* (L.). *Journal of Stored Products Research* 31, 131–138.

Pickett, J.A., Wadhams, L.J., Woodcock, C.M. and Hardie, J. (1992) The chemical ecology of aphids. *Annual Review of Entomology* 37, 67–90.

Principi, M.M. and Canard, M. (1984) Feeding habits. In: Canard, M., Séméria, Y. and New, T.R. (eds) *Biology of Chrysopidae*. Dr W. Junk Publishers, The Hague, pp. 76–92.

Raffa, K.F. and Dahlsten, D.L. (1995) Differential responses among natural enemies and prey to bark beetle pheromones. *Oecologia* 102, 17–23.

Raffa, K.F. and Klepzig, K.D. (1989) Chiral escape of bark beetles from predators responding to a bark beetle pheromone. *Oecologia* 80, 566–569.

Riddick, E.W., Dively, G. and Barbosa, P. (1998) Effect of a seed-mix deployment of Cry3A-transgenic and nontransgenic potato on the abundance of *Lebia grandis* (Coleoptera: Carabidae) and *Coleomegilla maculata* (Coleoptera: Coccinellidae). *Annals of the Entomological Society of America* 91, 647–653.

Riechert, S.E. and Lawrence, K. (1997) Test for predation effects of single versus multiple species of generalist predators: spiders and their insect prey. *Entomologia Experimentalis et Applicata* 84, 147–155.

Roach, S.H. and Thomas, W.M. (1991) Overwintering and spring emergence of three coccinellid species in the coastal plain of South Carolina. *Environmental Entomology* 20, 540–544.

Rock, G.C., Monroe, R.J. and Yeargan, D.R. (1976) Demonstration of a sex pheromone in the predaceous mite *Neoseiulus fallacis*. *Environmental Entomology* 5, 264–266.

Rosenheim, J.A., Wilhoit, L.R. and Armer, C.A. (1993) Influence of intraguild predation among generalist insect predators on the suppression of a herbivore population. *Oecologia* 96, 439–449.

Rosenheim, J.A., Kaya, H.K., Ehler, L.E., Marois, J.J. and Jaffee, B.A. (1995) Intraguild predation among biological-control agents: theory and evidence. *Biological Control* 5, 303–335.

Ross, K.W., Gibson, K.E., Thier, R.W. and Munson, A.S. (1996) Optimal dose of an antiaggregation pheromone (3-methylcyclohex-2-en-1-one) for protecting live Douglas-fir from attack by *Dendroctonus pseudotsugae* (Coleoptera: Scolytidae). *Journal of Economic Entomology* 89, 1204–1207.

Ruzicka, Z. (1994) Oviposition-deterring pheromone in *Chrysopa oculata* (Neuroptera: Chrysopidae). *European Journal of Entomology* 91, 361–370.
Ruzicka, Z. (1996) Oviposition-deterring pheromone in Chrysopidae (Neuroptera): intra- and interspecific effects. *European Journal of Entomology* 93, 161–166.
Ruzicka, Z. (1997a) Persistence of the oviposition-deterring pheromone in *Chrysopa oculata* (Neur.: Chrysopidae). *Entomophaga* 42, 109–114.
Ruzicka, Z. (1997b) Recognition of oviposition-deterring allomones by aphidophagous predators (Neuroptera: Chrysopidae, Coleoptera: Coccinellidae). *European Journal of Entomology* 94, 431–434.
Salom, S.M., Billings, R.F., Upton, W.W., Dalusky, M.J., Grosman, D.M., Payne, T.L., Berisford, C.W. and Shaver, T.N. (1992) Effect of verbenone enantiomers and racemic *endo*-brevicomin on response of *Dendroctonus frontalis* (Coleoptera: Scolytidae) to attractant-baited traps. *Canadian Journal of Forestry Research* 22, 925–931.
Sant'Ana, J., Bruni, R., Abdul-Baki, A.A. and Aldrich, J.R. (1997) Pheromone-induced movement of nymphs of the predator *Podisus maculiventris* (Heteroptera: Pentatomidae). *Biological Control* 10, 123–128.
Sasaki, T., Aoki, T., Hayashi, H. and Ishikawa H. (1990) Amino acid composition of the honeydew of symbiotic and aposymbiotic pea aphids *Acyrthosiphon pisum*. *Journal of Insect Physiology* 36, 35–40.
Schuh, R.T. and Slater, J.A. (1995) *True Bugs of the World (Hemiptera: Heteroptera): a Classification and Natural History*. Cornell University Press, Ithaca, New York, 336 pp.
Schulz, S. (1997) The chemistry of spider toxins and spider silk. *Angewandte Chemie International Edition in English* 36, 314–326.
Schulz, S. and Toft, S. (1993) Identification of a sex pheromone from a spider. *Science* 260, 1635–1637.
Scutareanu, P., Drukker, B., Bruin, J., Posthumus, M. and Sabelis, M.W. (1997) Volatiles from *Psylla*-infested pear trees and their possible involvement in attraction of anthocorid predators. *Journal of Chemical Ecology* 23, 2241–2260.
Sengonca, C. and Liu, B. (1994) Responses of the different instar predator, *Coccinella septempunctata* L. (Coleoptera: Coccinellidae), to the kairomones produced by the prey and non-prey insects as well as the predator itself. *Journal of Plant Diseases and Protection* 101, 173–177.
Shaltiel, L. and Ayal, Y. (1998) The use of kairomones for foraging decisions by an aphid parasitoid in small host aggregations. *Ecological Entomology* 23, 319–329.
Shimoda, T., Takabayashi, J., Ashihara, W. and Takafuji, A. (1997) Response of predatory insect *Scolothrips takahashii* toward herbivore-induced plant volatiles under laboratory and field conditions. *Journal of Chemical Ecology* 23, 2033–2048.
Shonouda, M.L. (1996) Crude aqueous-extract (kairomone) from *Aphis fabae* Scop. (Hom., Aphididae) and its effect on the behaviour of the predator *Metasyrphus corollae* Fabr. (Dipt., Syrphidae) female. *Journal of Applied Entomology* 120, 489–492.
Shonouda, M.L., Bombosch, S., Shalaby, A.M. and Osman, S.I. (1998a) Biological and chemical characterization of a kairomone excreted by the bean aphid, *Aphis fabae* Scop. (Hom., Aphididae) and its effect on the predator *Metasyrphus corollae* Fabr. I. Isolation, identification and bioassay of aphid-kairomone. *Journal of Applied Entomology* 122, 15–23.
Shonouda, M.L., Bombosch, S., Shalaby, A.M. and Osman, S.I. (1998b) Biological and chemical characterization of a kairomone excreted by the bean aphid, *Aphis fabae* Scop. (Hom., Aphididae) and its effect on the predator *Metasyrphus corollae* Fabr. II. Behavioural response of the predator M. *corollae* to the aphid kairomone. *Journal of Applied Entomology* 122, 25–28.
Simberloff, D. and Stiling, P. (1996) How risky is biological control? *Ecology* 77, 1965–1974.

Stowe, M.D., Tumlinson, J.H. and Heath, R.R. (1987) Chemical mimicry: bolas spiders emit components of moth prey species sex pheromones. *Science* 236, 964–967.

Stubbs, A.E. and Falk, S.J. (1993) *British Hoverflies: an Illustrated Identification Guide*. British Entomological and Natural History Society, London, 253 pp.

Summy, D.R., Raulston, J.R., Spurgeon, D. and Vargas, J. (1996) An analysis of the beet armyworm outbreak on cotton in the lower Rio Grande valley of Texas during the 1995 production season. *Proceedings of the Beltwide Cotton Production Conference* 1996, 837–842.

Takabayashi, J. and Dicke, M. (1996) Plant–carnivore mutualism through herbivore-induced carnivore attractants. *Trends in Plant Science* 1, 109–113.

Taneja, J. and Guerin, P.M. (1997) Ammonia attracts the haematophagous bug *Triatoma infestans*: behavioural and neurophysiological data on nymphs. *Journal of Comparative Physiology* A 181, 21–34.

Teerling, C.R. (1995) Chemical ecology of western flower thrips. In: Parker, B.L., Skinner, M. and Lewis, T. (eds) *Thrips Biology and Management*. Plenum Press, New York, pp. 439–447.

Teerling, C.R, Gillespie, D.R. and Borden, J.H. (1993) Utilization of western flower thrips alarm pheromone as a prey-finding kairomone by predators. *Canadian Entomologist* 125, 431–437.

Thomas, D.B. (1992) *Taxonomic Synopsis of the Asopine Pentatomidae (Heteroptera) of the Western Hemisphere*. The Thomas Say Foundation, Entomological Society of America, Lanham, Maryland, 156 pp.

Torres, J.B., Zanuncio, J.C., Saavedra, J.L.D. and Aldrich, J.R. (1997) Extrato de glandulas de feromônio na atração e estimulação alimentar de ninfas de *Podisus nigrispinus* (Dallas) e *Supputius cincticeps* (Stål). *Anais da Sociedade Entomologica do Brasil* 26, 463–469.

Veilleux, M., Coderre, D., Poliquin, L. and Boileau, S. (1993) Effect of visual and olfactory stimuli on the aggregation behavior of the spotted lady beetle *Coleomegilla maculata lengi* Timberlake (Coccinellidae). *Aphidophaga* 5, 33.

Wainhouse, D., Beech-Garwood, P.A., Howell, R.S., Kelly, D. and Orozco, M.P. (1992) Field response of predator *Rhizophagus grandis* to prey frass and synthetic attractants. *Journal of Chemical Ecology* 18, 1693–1705.

Weeden, C.R., Shelton, A.M. and Hoffmann, M.P. (1997) Biological control: a guide to natural enemies in North America. New York State Agricultural Experiment Station, Cornell University: http://www.nysaes.cornell.edu/ent/biocontrol.

Weslien, J. (1992) The arthropod complex associated with *Ips typographus* (L.) (Coleoptera, Scolytidae): species composition, phenology, and impact on bark beetle productivity. *Entomologica Fennica* 3, 205–213.

Weston, R.J., Woolhouse, A.D., Spurr, E.B., Harris, R.J. and Suckling, D.M. (1997) Spiroacetals and other venom constituents as potential wasp attractants. *Journal of Chemical Ecology* 23, 553–568.

Whitcomb, W.H. and Godfrey, K.E. (1991) The use of predators in insect control. In: Pimentel, D. and Hanson, A.A. (eds) *CRC Handbook of Pest Management in Agriculture*, 2nd edn. CRC Press, Boca Raton, Florida, pp. 215–241.

White, S.M. (1985) *Collops hirtellus* (Coleoptera: Melyridae) males and *Formica* spp. (Hymenoptera: Formicidae) attracted by a wood preservative. *Entomological News* 96, 49–52.

Wigger, H. (1996) Population-dynamics and predator–prey-interaction between the predator *Nemosoma elongatum* and the bark beetle species *Pityogenes chalcographus* (Coleoptera: Ostomidae, Scolytidae). *Entomologia Generalis* 21, 55–67.

Yasuda, T. (1997) Chemical cues from *Spodoptera litura* larvae elicit prey-locating behavior by the predatory stink bug, *Eocanthecona furcellata*. *Entomologia Experimentalis et Applicata* 82, 349–354.

Yasuda, T. (1998) Role of chlorophyll content of prey diets in prey-locating behavior of a generalist predatory stink bug, *Eocanthecona furcellata*. *Entomologia Experimentalis et Applicata* 86, 119–124.

Yasui, Y. (1992) The existence of sex pheromone eliciting precopulatory mate guarding behavior in a male predatory mite, *Macrocheles muscaedomesticae* (Scopoli). *Journal of Ethology* 10, 15–19.

Yeargan, K.V. (1994) Biology of bolas spiders. *Annual Review of Entomology* 39, 81–99.

Yeargan, K.V. (1998) Overview and the prevalence of heteropteran predators in agroecosystems. In: Coll, M. and Ruberson, J.R. (eds) *Predatory Heteroptera: Their Ecology and Use in Biological Control.* Thomas Say Foundation, Entomological Society of America, Lanham, Maryland, pp. 7–19.

Yeargan, K.V. and Quate, L.W. (1996) Juvenile bolas spiders attract psychodid flies. *Oecologia* 106, 266–271.

15 Parasitoids

YOOICHI KAINOH

Institute of Agriculture and Forestry, University of Tsukuba, Tsukuba, Ibaraki 305–8572, Japan

Introduction

Chemical messages emitted and/or perceived by parasitoids are critically important during their lives. These messages may be divided into two categories: pheromones that mediate intraspecific communication; and allelochemicals that induce chemical interactions between parasitoids and other species, mainly between a parasitoid and its host (kairomones) or a parasitoid and its host plant (synomones) in a tritrophic context (Vet and Dicke, 1992). The pheromones produced by parasitoids are the subject of this chapter. Those that have been intensively studied are the sex pheromones and marking pheromones.

Although ethological aspects (Matthews, 1975; van den Assem, 1986, 1996) and mating systems (Godfray, 1994) of parasitoids have been studied extensively, parasitoid sex pheromones have received little attention. Eller *et al.* (1984) reviewed articles in which the existence of a sex pheromone in parasitoid species was demonstrated and listed examples from seven families: Aphelinidae, Chalcididae, Cynipidae, Pteromalidae, Scelionidae, Braconidae and Ichneumonidae. Quicke (1997) reviewed parasitoid sex pheromones and listed several species in which sex pheromone components have been identified.

Parasitoids are known to have the ability to distinguish unparasitized from parasitized hosts (host discrimination) (van Lenteren, 1981) and marking pheromones play an important role in facilitating this distinction. Vinson (1985) reviewed marking behaviour and marking pheromones, hence the properties of marking pheromones are only briefly summarized here.

Other pheromones (i.e. alarm, trail and epideictic pheromones) have not been found in parasitoids, although an aggregation pheromone has been demonstrated in the pupal parasitoid, *Brachymeria* spp. (Simser and Coppel, 1980b), and this is briefly described in the final section of this chapter. To my knowledge, there have been no studies on dipteran parasitoid pheromones.

Sex Pheromones

It has been suggested that sex pheromones should be less common in haplodiploid species since the production of a long-range sex pheromone presumably incurs metabolic costs, so sex pheromones are unlikely to have evolved in species where females gain little advantage from mating (Godfray, 1994). We cannot compare the cost of sex pheromone production between haplodiploid and diploid species in the same situation, since the active range of the sex pheromones varies from long-range volatiles to compounds detected by contact (Table 15.1). In most of the species listed in Table 15.1, female parasitoids attract males, but in a few species male sex pheromones that attract females occur. In *Aphytis* spp., females attract males (Rao and DeBach, 1969) and, at the same time, males attract females over a short range (*c.* 20 mm). This attractant, produced by the male, acts as an aphrodisiac in quietening females and rendering them receptive for copulation (Khasimuddin and DeBach, 1975). Also, males have been found to attract females in the blind and flightless *Melittobia* spp. (Matthews *et al.*, 1985).

Most of the sex pheromones isolated from insect parasitoids are volatile and olfactometers have been used to demonstrate their effects in the laboratory. Sex pheromones have also been shown to succeed in attracting parasitoids to sticky traps in the field.

In this chapter, parasitoid sex pheromones are divided into three categories based on their active range: volatile, contact and aggregation pheromones. Some of the cases where intensive research has been performed will be described.

Volatile sex pheromones

Attraction of males into field traps containing females indicates that a sex pheromone is a volatile chemical (e.g. Lewis *et al.*, 1971). Wind tunnel and olfactometer experiments which demonstrate the attraction of males to females (e.g. Cole, 1970; Powell and Zhang, 1983; McAuslane *et al.*, 1990) also show that the chemicals are volatile.

Macrocentrus grandii (Goidanich), a braconid larval parasitoid of the European corn borer, *Ostrinia nubilalis* (Hübner), was shown to have a multicomponent pheromone system by using a wind tunnel and field bioassays (Swedenborg and Jones, 1992a). The combination of a Florisil hexane fraction and the male- or female-derived 50%-ether-in-hexane fraction was the most attractive source, but a 5%-ether-in-hexane fraction also showed activity. Active components were later identified as (Z)-4-tridecenal (Fig. 15.1) from females (Swedenborg and Jones, 1992b) and (3*R*,5*S*,6*R*)-3,5-dimethyl-6-(methylethyl)-3,4,5,6-tetrahydropyran-2-one (Fig. 15.2) from both males and females, which worked as a synergist (Swedenborg *et al.*, 1993). Both enantiomers, (3*R*,5*S*,6*R*) and (3*S*,5*R*,6*S*), were subsequently tested and synthetic *SRS* was found to be more active than synthetic *RSR*. Consequently, a combination of the *SRS* enantiomer of the lactone and (Z)-4-tridecenal was most active in field tests (Swedenborg *et al.*, 1994). As in *M. grandii*, a multicomponent sex pheromone system has been suggested in several other parasitoid species (Robacker and Hendry, 1977; Weseloh, 1977; Hrdy and Sedivy, 1979; Kainoh *et al.*, 1991; Shu and Jones, 1993; Syvertsen *et al.*, 1995).

Table 15.1. Parasitoid sex pheromones.

Family	Species	Host	Chemicals	Reference
Alloxystidae	*Alloxysta victrix* (Westwood) (hyperparasitoid)	*Aphidius uzbekistanicus* (Luzhetski) (Hymenoptera: Aphidiidae)	Volatile 6-methyl-5-heptene-2-one (MHO) (Y-tube olfactometer)	Micha *et al.* (1993) Höller *et al.* (1994)
Aphelinidae	*Aphenlinus asychis* Walker	*Diuraphis noxia* (Mordvilko) (Homoptera: Aphididae)	Non-volatile, trail sex pheromone	Fauvergue *et al.* (1995)
Aphelinidae	*Aphytis lingnanensis* Compere	*Aonidiella aurantii* (Maskell) (Homoptera: Diaspididae)	Volatile Source: thorax, base of the wings	Rao and DeBach (1969)
Aphelinidae	*A. maculicornis* (Masi) *A. mytilaspidis* Le Baron	*Aonidiella aurantii* (Maskell) (Homoptera: Diaspididae)	Volatile male sex pheromone (effective range: *c.* 20 mm)	Khasimuddin and DeBach (1975)
Braconidae	*Apanteles (=Cotesia) glomeratus* L.	*Pieris rapae crucivora* Boisduval (Lepidoptera: Pieridae)	Volatile (visual stimuli elicit mounting and copulation)	Obara and Kitano (1974) Kitano (1975)
Braconidae	*A. medicaginis* Muesebeck	*Colias philodice eurytheme* (Lepidoptera: Pieridae)	Volatile (attraction over 90 cm in a wind tunnel)	Cole (1970)
Braconidae	*A. melanoscelus* (Ratzeburg)	*Lymantria dispar* L. (Lepidoptera: Lymantriidae)	Relatively non-volatile	Weseloh (1977)
Braconidae	*Aphidius colemani* Viereck	*Diuraphis noxia* (Mordvilko) (Homoptera: Aphididae)	Volatile (4-choice olfactometer, attraction in field traps)	Reed *et al.* (1994)
Braconidae	*A. nigripes* (Ashmead)	*Macrosiphum euphorbiae* (Thomas) (Homoptera: Aphididae)	Volatile (attraction in field traps)	McNeil and Brodeur (1995)
Braconidae	*A. rhopalosiphi* De Stefani-Perez	Cereal aphids (Homoptera: Aphididae)	Volatile (attraction in field traps)	Höller (1991) Decker *et al.* (1993)
Braconidae	*A. uzbekistanicus* Luzhetski, *A. ervi* Haliday	Cereal aphids (Homoptera: Aphididae)	Volatile (attraction in a wind tunnel)	Powell and Zhang (1983)
Braconidae	*Ascogaster reticulatus* Watanabe	*Adoxophyes* sp. (Lepidoptera: Tortricidae)	Contact chemical (*Z*)-9-hexadecenal	Kainoh *et al.* (1991)
Braconidae	*Cardiochiles nigriceps* Viereck	*Heliothis virescens* (Fabricius) (Lepidoptera: Noctuidae)	Volatile (attraction in field traps) Source: Dufour's gland (*Z,Z*)-7,13-heptacosadiene	Lewis *et al.* (1971) Vinson (1978) Syvertsen *et al.* (1995)
Braconidae	*Cotesia rubecula* Marshall	*Pieris rapae* (L.) (Lepidoptera: Pieridae))	Volatile (attraction in a wind tunnel)	Field and Keller (1994)
Braconidae	*Diaeretiella rapae* (McIntosh)	*Brevicoryne brassicae* (L.) (Homoptera: Aphididae)	Volatile (attractive in a Y-tube olfactometer)	Read *et al.* (1970) Askari and Alishah (1979)

Continued

Table 15.1 (*Continued*)

Family	Species	Host	Chemicals	Reference
Braconidae	*Habrobracon juglandis* Ashmead (= *Bracon hebetor* Say)	*Ephestia* spp. (Lepidoptera: Pyralidae)	Volatile (attraction in a tube, at least 20 cm) Source: abdomen	Grosch (1948)
Braconidae	*Macrocentrus grandii* (Goidanich)	*Ostrinia nubilalis* (Hübner) (Lepidoptera: Pyralidae)	(*Z*)-4-tridecenal and (3*S*,5*R*,6*S*)-3,5-dimethyl-6-(methylethyl) – 3,4,5,6-tetrahydropyran-2-one	Swedenborg and Jones (1992a,b) Swedenborg *et al.* (1993) Swedenborg *et al.* (1994)
Braconidae	*Meteorus pulchricornis* (Wesmael)	*Lymantria dispar* L. (Lepidoptera: Lymantriidae)	Volatile (ether extract of virgin females attracted males)	Askari and Coppel (1978)
Braconidae	*Microplitis croceipes* (Cresson)	*Heliothis* spp. (Lepidoptera: Noctuidae)	Volatile (attraction in field traps) Hexane soluble (wind tunnel)	Powell and King (1984) Jones (1989)
Braconidae	*Opius alloeus* Muesebeck	Apple maggot *Rhagoletis pomonella* (Walsh)	Males responded to filter paper treated with a crushed female	Boush and Baerwald (1967)
Braconidae	*Pholetesor ornigis* (Weed)	*Phyllonorycter blancardella* (F.) (Lepidoptera: Gracillariidae)	Volatile (attraction in field traps baited with live virgin females)	Ridgway and Mahr (1986)
Braconidae	*Praon volucre* Haliday	(Homoptera: Aphididae)	Volatile (attraction in field traps) (Olfactometer, EAG)	Decker *et al.* (1993) Nazzi *et al.* (1996)
Chalcididae	*Brachymeria intermedia* (Nees)	*Lymantria dispar* L.	Volatile (+ male courtship song)	Leonard and Ringo (1978)
Chalcididae	*B. intermedia* (Nees), *B. lasus* (Walker)	*Lymantria dispar* L.	Volatile (attraction in field traps) (9 cm Petri dish assay)	Mohamed and Coppel (1987a)
Chalcididae	*B. lasus, B. intermedia*	*Lymantria dispar* L.	Volatile (active space: 3 cm) Source: Dufour's gland	Simser and Coppel (1980a)
Cynipidae	*Charips brassicae* (hyperparasitoid)	*Diaeretiella rapae* (McIntosh) (Hymenoptera: Braconidae)	Volatile (Y-tube olfactometer)	Read *et al.* (1970)
Eulophidae	*Diglyphus isaea* Walker	*Liriomyza trifolii* Burgess (Diptera: Agromyzidae)	Contact sex pheromone	Finidori-Logli *et al.* (1996)
Eulophidae	*Melittobia* spp.	*Trypoxylon politum* Say (Hymenoptera: Sphecidae)	Volatile (males attract females)	Matthews *et al.* (1985)
Eulophidae	*Tetrastichus hagenowii* Ratzeburg	*Periplaneta* spp. (Orthoptera: Blattidae)	Volatile (ether extract elicits zigzag approach and wing vibration)	Takahashi and Sugai (1982)
Ichneumonidae	*Campoletis sonorensis* (Cameron)	*Heliothis virescens* (Fabricius) (Lepidoptera: Noctuidae)	Volatile (attraction in an olfactometer) (host plant influence in a wind tunnel)	Vinson (1972a) McAuslane *et al.* (1990)

Ichneumonidae	*Diadromus pulchellus* WSM *D. collaris* Gr.	*Acrolepiopsis assectella* (Lepidoptera: Acrolepiidae)	EAG response to female odour	Lecomte and Pouzat (1985)
Ichneumonidae	*Diplazon pectoratorius* (Thunberg)	Aphidophagous Syrphidae	Volatile (attraction in a T-tube olfactometer)	Rotheray (1981)
Ichneumonidae	*Eriborus terebrans* (Gravenhorst)	*Ostrinia nubilalis* (Hübner) (Lepidoptera: Pyralidae)	Polar component(s) and hydrocarbon(s) as synergist	Shu and Jones (1993)
Ichneumonidae	*Exetastes cinctipes* Retzius	*Mamestra brassicae* L. (Lepidoptera: Noctuidae)	*(Z)*-8-dodecenyl acetate, and *(Z)*-11-tetradecenyl acetate (possible sex pheromone)	Hrdy and Sedivy (1979)
Ichneumonidae	*Itoplectis conquisitor* (Say)	Lepidopteran pupae	Neral and geranial as components of sex pheromone system	Robacker *et al.* (1976) Robacker and Hendry (1977)
Ichneumonidae	*Phaeogenes invisor* Thunberg	*Tortrix viridana* (L.) pupae (Lepidoptera: Tortricidae)	Volatile (attraction in a wind tunnel over 2 m)	Cole (1970)
Ichneumonidae	*Syndipnus rubiginosus* Walley	*Pikonema alaskensis* (Rohwer) (Hymenoptera: Tenthredinidae)	Volatile (attractive in the field) ethyl *(Z)*-9-hexadecenoate	Eller *et al.* (1984)
Pteromalidae	*Nasonia vitripennis* (Walker)	Cyclorraphous fly pupae	Volatile (short distance) (Males release another pheromone with mouth parts which provokes female receptivity)	King *et al.* (1969) van den Assem *et al.* (1980, 1981)
Pteromalidae	*Anisopteromalus calandrae* (Howard)	*Callosobruchus chinensis* (L.) (Coleoptera: Bruchidae)	Volatile (active space: 2.5 cm)	Yoshida (1978)
Trichogrammatidae	*Trichogramma brassicae* Bezdenko	Lepidopteran eggs	Substrate-borne (hexane soluble)	Pompanon *et al.* (1997)

Fig. 15.1. The chemical structure of (*Z*)-4-tridecenal (Swedenborg and Jones, 1992b), the sex pheromone from female *Macrocentrus grandii*.

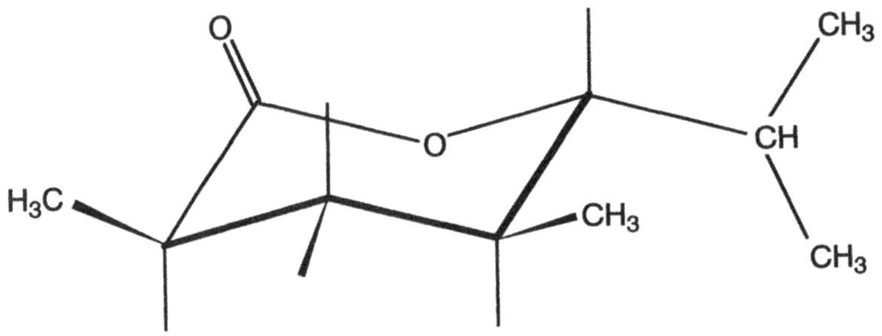

Fig. 15.2. The chemical structure of (3*R*,5*S*,6*R*)-3,5-dimethyl-6-(methylethyl)-3,4,5,6-tetrahydropyran-2-one (Swedenborg *et al.*, 1993), the sex pheromone from both sexes of *M. grandii*.

Campoletis sonorensis (Cameron) males exhibited four main elements of courtship behaviour (orientation, wing vibration, antennal touching and attempted copulation) when exposed to females or odour-laden air from females in an olfactometer (Vinson, 1972a). There was evidence that the sequence of these behaviours was elicited by a volatile sex pheromone from females and an ethyl acetate extract of females was active in eliciting these elements of the courtship behaviour.

The sex pheromone of *Syndipnus rubiginosus* Walley was identified as ethyl (Z)-9-hexadecenoate (ethyl palmitoleate) (Fig. 15.3) and a synthetic pheromone was subsequently tested in the field (Eller *et al.*, 1984). Catches in traps containing the synthetic pheromone were comparable with those in traps baited with virgin females. Cross-attraction studies with a related species, *S. gaspesianus* (Provancher), showed that females of each species attracted only conspecific males, indicating that the two species have different pheromone components or component ratios. Traps baited with the sex pheromone of *S. rubiginosus* were placed in white spruce plantations, and the relationship between catches of the parasitoid and its host, the yellowheaded spruce sawfly, *Pikonema alaskensis* (Rohwer), as well as subsequent defoliation caused by the sawfly, was studied (Morse and Kulman, 1985). There was

Fig. 15.3. The chemical structure of ethyl (*Z*)-9-hexadecenoate (Eller *et al.*, 1984), the sex pheromone of *Syndipnus rubiginosus*.

a significant relationship between spruce damage and sawfly population levels only when parasitoid numbers were also included.

There are a few species in which both males and females are attractive to males. Extracts of newly emerged virgin females and of newly emerged virgin males of the ichneumonid parasitoid *Itoplectis conquisitor* (Say) were highly active in eliciting responses in older males, although newly emerged males did not respond to the extract (Robacker *et al.*, 1976). Newly emerged males may be unresponsive for a period so that they have time to disperse before becoming sexually mature. Neral and geranial were identified as constituents of this multicomponent pheromone system (Fig. 15.4) (Robacker and Hendry, 1977).

The sex pheromone of the hyperparasitoid *Alloxysta victrix* (Westwood), which attacks primary parasitoids of aphids, was also released from both males and females. Whilst it attracted males, the pheromone repelled females, thereby acting as a spacing pheromone for the latter (Micha *et al.*, 1993). This compound was identified as 6-methyl-5-heptene-2-one (MHO) (Fig. 15.5) (Micha *et al.*, 1993; Höller *et al.*, 1994). It is notable that MHO appears to elicit dispersal of the host (primary) parasitoid, *Aphidius uzbekistanicus* Luzhetski, which leaves the area, thereby increasing its reproductive success by reducing the chances that its offspring will be attacked (Höller *et al.*, 1994).

Courtship in *Cardiochiles nigriceps* Viereck consists of male attraction to females, followed by male antennation of females and then mounting and copulation. Wing fanning by the male plays an important role in orientation towards a female located 3–5 cm away (Vinson, 1978). The females produce a sex pheromone just prior to emergence from the cocoon, the activity of which persists throughout adult life after mating. In *Cotesia glomeratus* (L.), the female sex pheromone is already being secreted at the black-eye stage during pharate adult development (Tagawa, 1977). The release of sex pheromone before emergence may serve to attract males so that they are available for mating immediately after female emergence, thereby increasing the chances of a female being mated before she begins oviposition. The release of pheromone after mating in C. *nigriceps* may play a role in the aggregation of both sexes (Lewis *et al.*, 1971).

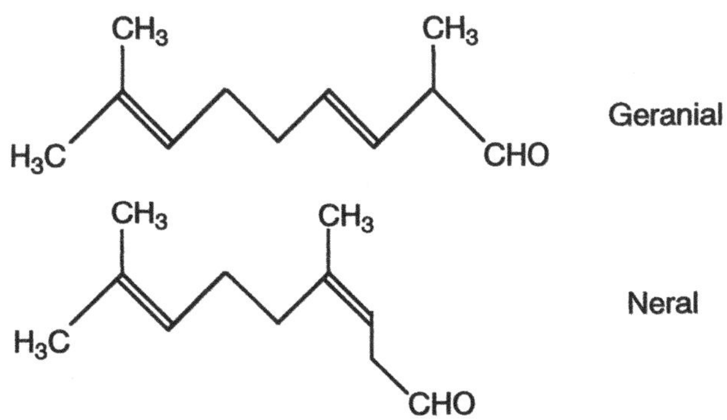

Fig. 15.4. The chemical structures of geranial and neral (Robacker and Hendry, 1977), components of the sex pheromone of *Itoplectis conquisitor*.

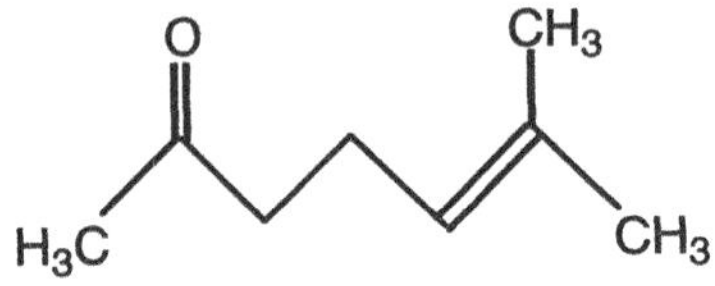

Fig. 15.5. The chemical structure of 6-methyl-5-heptene-2-one (MHO) (Micha *et al.*, 1993; Höller *et al.*, 1994), the sex pheromone of *Alloxysta victrix*.

Contact sex pheromones

In addition to the attraction of *C. nigriceps* males to females, the hexane extract of females elicited antennation by males (Vinson, 1978). Antennation and mounting by the males were also mediated by the hydrocarbon fraction of the female's Dufour's gland and cuticle. Males would follow a line to which the hydrocarbon fraction was applied (S.B. Vinson, College Station, 1998, personal communication). Syvertsen *et al.* (1995) identified the cuticular hydrocarbons of both sexes, revealing a mixture of alkanes and alkenes, in addition to a series of alkadienes specific to females. Bioassays with three available alkadienes indicated that contact behaviours (antennation and mounting) are in part mediated by (Z,Z)-7,13-heptacosadiene (Fig. 15.6) and at least one other alkadiene in combination with other hydrocarbons found in both sexes.

Diglyphus isaea Walker males recognized females (Fig. 15.7) at close range, i.e. within a few millimetres. The signal triggering male courtship behaviour is located on the female cuticle (Finidori-Logli *et al.*, 1996). The principal compounds inducing male behaviour are contained in the non-hydrocarbon fraction of the female cuticular extract and the authors suggested that the encounter between the sexes could depend on a long-range perception of volatile compounds from the host plant. *D. isaea* uses a very close-range sex pheromone, because both males and females are attracted to the same location through their responses to host plant volatiles, and mating then occurs on the host plants.

Sex pheromones as trail cues

Ascogaster reticulatus Watanabe is an egg-larval parasitoid of the smaller tea tortrix, *Adoxophyes* sp. Mating activity peaks during the latter half of the light period (Kainoh, 1986) and a behavioural sequence has been identified (Kamano *et al.*, 1989; summarized in Fig. 15.8). The hexane extract of adult females was applied in

Fig. 15.6. The chemical structure of (*Z,Z*)-7,13-heptacosadiene (Syvertsen *et al.*, 1995), the sex pheromone of *Cardiochiles nigriceps*.

Fig. 15.7. Female *Diglyphus isaea.* (Courtesy of Koppert Biological Systems.)

a line across the bottom of a Petri dish and a male was introduced into the dish. When the wasp contacted the extract with its antennae, it followed the line across the dish by antennating it. In another experiment, males actively searched on a glass surface over which females had previously walked (Kamano *et al.*, 1989). These results indicated that the male responded to a sex pheromone left by females as a trail when they walked across a surface. The sex pheromone deposited by the females activated male searching behaviour and increased the chances of mating. This compound was analysed and identified as (Z)-9-hexadecenal (Z9:HDAL; Fig. 15.9) (Kainoh *et al.*, 1991). This is a sex pheromone component common to several noctuid moths (*Heliothis* spp., *Gortyna* spp., *Orthosia* spp., *Scotogramma* spp.) and pyralid moths (*Chilo* spp., *Diatraea* spp., *Scirpophaga* spp.) (Mayer and McLaughlin, 1991). In these moth species, sex pheromones are released from calling females and act as volatile cues for mate location by males. In contrast, the Argentine ant, *Iridomyrmex humilis* (Mayr), uses Z9:HDAL as one of its trail pheromones (Cavill *et al.*, 1979, 1980), where this compound acts as a volatile cue. *A. reticulatus* females, however, deposit Z9:HDAL on the substrate whilst walking around and it is probable that males detect Z9:HDAL as a contact chemical, since no positive responses were recorded in olfactometer experiments (Y. Kainoh, 1998, unpublished results). An electroantennogram (EAG) response was obtained with authentic Z9:HDAL by treating male antennae with puffs of pheromone-impregnated air; the largest response was obtained from the ventral side of the antennae (Hidoh *et al.*, 1992). These results indicated that the pheromone stimulus is detected by the antennae as an olfactory response, but a mechanical contact of antennae to the substrate is required for the behavioural response of male *A. reticulatus*. The results were also consistent with the behaviour, i.e. palpating the pheromone-treated part of the substrate with the ventral side of the antennae,

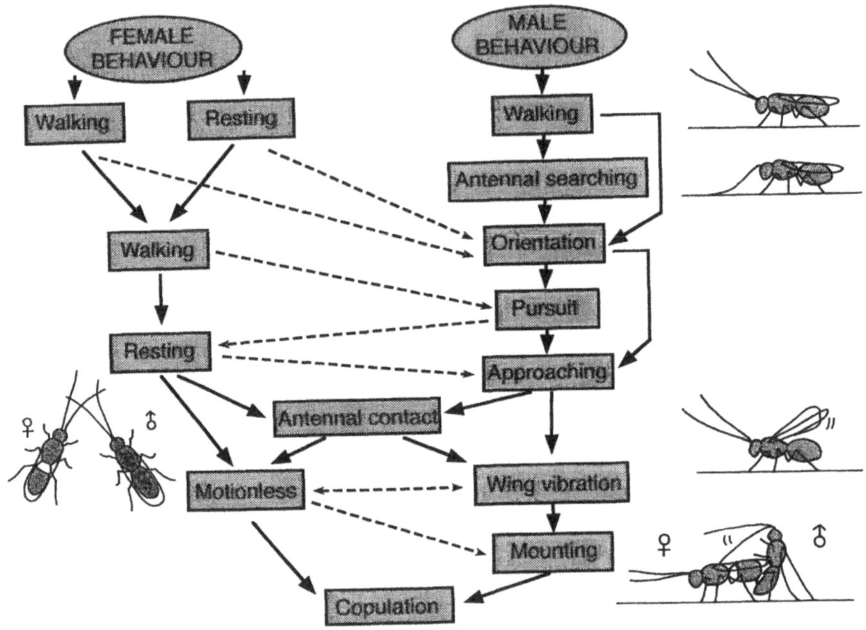

Fig. 15.8. Behavioural sequence in mating of *Ascogaster reticulatus*. Solid lines show behavioural transition, and dotted lines show possible influence of factors on eliciting behaviour in the opposite sex. (Redrawn from Kamano *et al.*, 1989.)

which therefore appears to be the location of the sensilla for this pheromone. The geometric isomer (E9:HDAL) was not active at all in a behavioural bioassay while the activity of Z9:HDAL was maximal at 5–500 pg per line (Fig. 15.10) (Kainoh *et al.*, 1991). E9:HDAL also elicited lower EAG responses than the (Z)-isomer (Hidoh *et al.*, 1992).

The same type of sex pheromone has been found in an aphelinid, *Aphelinus asychis* Walker, in which virgin females deposited sex pheromone on the substrate, to which males responded by intensively searching on and near the marked area (Fauvergue *et al.*, 1995). The males did not respond to plant leaves exposed to mated females or other males, and the pheromone activity was lost in less than 24 h. Also, in this case, males did not respond to odours of virgin females in the olfactometer, which indicated that this sex pheromone acts as either a contact chemical or a very close-range (< 1 mm) volatile.

Fig. 15.9. The chemical structure of (*Z*)-9-hexadecenal (Kainoh *et al.*, 1991), the sex pheromone of *A. reticulatus*.

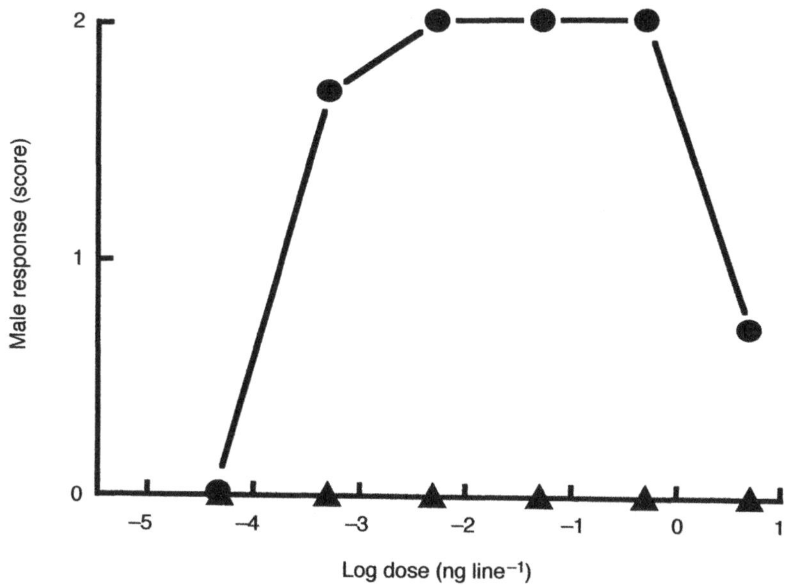

Fig. 15.10. Dose–response of (*Z*)-9-hexadecenal (circles) as a sex pheromone in *A. reticulatus*. The geometric isomer (*E*)-9-hexadecenal (triangles) was not active. Scoring was based on the distance the wasps walked with antennal searching along a treated line in a 9 cm Petri dish within 3 min. Score 1: 1–9 cm; Score 2: more than 9 cm.

Another substrate-borne sex pheromone was demonstrated in *Trichogramma brassicae* Bezdenko, in which males concentrated their walking in an area where virgin females had previously walked (Pompanon *et al.*, 1997). Again, male response decreased over time. It was suggested that the substrate-borne pheromones could increase mating opportunities for virgin females searching for oviposition sites, and may also increase mating chances at emergence sites by reducing male dispersion.

Sex pheromone glands

The sex pheromone gland in parasitic Hymenoptera is poorly known, though the Dufour's gland seems to be commonly implicated in some species (Vinson, 1978; Simser and Coppel, 1980a). In other species, a range of different locations, such as the base of the second valvifer (Tagawa, 1977, 1983), the base of a genital capsule (Field and Keller, 1994) and the hind tibia (Kainoh and Oishi, 1993), have been suggested.

Although no specific body parts of *C. nigriceps* females or specific chemical extracts of females were effective in eliciting full sexual display in males, the Dufour's gland and a hexane extract from it both elicited antennation by males (Vinson, 1978). Thereafter, Syvertsen *et al.* (1995) demonstrated that the sexual recognition of females by males of this species is mediated by the hydrocarbon fraction of the Dufour's gland and that the sex-specific alkadienes play a role, in combination with some of the alkenes or alkanes present.

With regard to the sex pheromone source of *Cotesia* (= *Apanteles*) spp., there have been conflicting opinions based on studies of closely related species. Weseloh (1976) reported that the Dufour's gland of female *Cotesia melanoscela* (Ratzeburg) elicited male sexual behaviour, whilst in *C. glomerata* Tagawa (1977) found that a group of secretory cells located at the base of the second valvifer was a source of the sex pheromone. Weseloh (1980) subsequently revised his opinion and suggested that the sex pheromone originated from the paired epidermal glands on the eighth (or ninth) abdominal tergum, and the columnar cells of these glands contained lipid deposits, microvilli and numerous mitochondria, which are typical of pheromone gland cells in other insects. Tagawa (1983) examined seven other *Cotesia* species and found that, in all of them, females had secretory cells at the base of the second valvifer. He also found epidermal glands, which Weseloh (1980) had indicated as pheromone glands, not only in females but also in males, and he concluded that *Cotesia* species have the female-specific second valvifer glands in common as the location of sex pheromone glands. In further research with *Cotesia rubecula* (Marshall), Field and Keller (1994) presented males with a choice of female body parts in a flight tunnel and concluded that the genital capsule is the source of female sex pheromones. The definitive location of the female sex pheromone gland in this genus has yet to be elucidated.

Use of parasitoid sex pheromones

Lewis *et al.* (1971) devised a trap to estimate population levels of *Cardiochiles nigriceps* and found that traps containing females were attractive to both males and females. They suggested that parasitoid traps have the potential for evaluating parasitoid populations, and provide a method of studying parasitoid–host relationships on various host plants. There are some other practical examples of traps with virgin females being used to monitor field populations of parasitoids (Brodeur and McNeil, 1994). The sex pheromone of the ichneumonid *Syndipnus rubiginosus* Walley, identified as ethyl palmitoleate (Eller *et al.*, 1984), was used for monitoring populations of the parasitoid, as well as its sawfly host, in an ecological study (Morse and Kulman, 1985). Another possible use would be to detect the establishment and subsequent spread of introduced parasitoid species in classical biological control programmes.

McNeil and Brodeur (1995) postulated other uses for parasitoid sex pheromones in addition to monitoring parasitoid populations, such as to enhance the impact of parasitoids on aphid populations. For example, the sex pheromones of aphids are very attractive to their female parasitoids and can be used to attract those females into crop fields (Powell *et al.*, 1993, 1998; Hardie *et al.*, 1994; see also Chapter 16, this volume). However, this may result in a very strong female-biased sex ratio, which would then produce a male-biased population in the next generation. Use of parasitoid sex pheromones simultaneously with those of aphids may ensure that both sexes are attracted into the habitat, thereby increasing parasitoid efficiency.

Marking Pheromones

To avoid superparasitism (deposition of more eggs than can be supported by the host), parasitoid females are known to deposit a marking pheromone in the area of oviposition (patch or area marking), on the surface of the host (external marking), or inside the host (internal marking). Marking is performed by parasitoids before, during or after oviposition in order to prevent the wastage of parasitoid eggs or hosts, to save time and to initiate dispersal after investigatory probes (van Lenteren, 1981). Salt (1934) first demonstrated that *Trichogramma evanescens* Westwood could discriminate between parasitized and unparasitized host eggs. He then proved that this marking was mediated by both external and internal marking pheromones (Salt, 1937).

Table 15.2 shows the list of parasitoid species in which the presence of marking pheromones has been demonstrated. In several species, both external marking and internal marking have been reported (Salt, 1937; Greany and Oatman, 1972; Guillot and Vinson, 1972; Vinson, 1972b; van Lenteren and Bakker, 1975, 1978).

Area or patch marking pheromones

In *T. evanescens*, a water extract collected from the substrate on which females had been allowed to walk deterred other females from parasitizing hosts in areas treated with this extract during choice tests (Salt, 1937). This indicated that the female parasitoids deposited a marking pheromone, not only on the host but also on the substrate on which they walked.

When a parasitoid searches for its host, it appears to mark the area of searching to avoid subsequent wasteful visits. Price (1970) found that females of the ichneumonid parasitoid, *Pleolophus basizonus* (Gravenhorst), deposited a mark as they searched the leaf litter and other females recognized and avoided these areas. This marking was recognized not only by conspecifics, but also by other ichneumonid species attacking the same host sawfly pupae.

External marking pheromones

Clean hosts and hosts that had previously been walked over by another parasitoid were exposed to *T. evanescens* females in choice tests. More clean eggs were accepted and more treated eggs rejected than expected by chance, which indicated that the parasitoids avoided attacking hosts that had been walked on, just as they avoid attacking parasitized hosts (Salt, 1937). This marking effect was mediated by water-soluble substance(s).

In two braconids, *C. nigriceps* and *Microplitis croceipes* (Cresson), the activity of marking pheromones was demonstrated by topically applying solutions of the accessory female reproductive glands, i.e. poison gland, calyx fluid and alkaline (Dufour's) gland, on to host larvae. Only a solution from the alkaline gland was effective in deterring females from attacking host larvae (Vinson and Guillot, 1972).

Table 15.2. Parasitoid marking pheromones.

Family	Species	Host	Stage	Chemicals	Reference
Bethylidae	*Cephalonomia stephanoderis* Betrem	*Hypothenemus hampei* (Ferrari) (Coleoptera: Scolytidae)	Pupa	Internal or external marking	Barrera *et al.* (1994)
Braconidae	*Dapsilarthra rufiventris* (Nees)	*Phytomyza ranunculi* Schrank (Diptera: Agromyzidae)	Larva	Area marking on the leaf washed off by 50% EtOH	Sugimoto *et al.* (1986)
Braconidae	*Orgilus lepidus* Muesebeck	*Phthorimaea operculella* (Zeller) (Lepidoptera: Gelechiidae)	Larva	Area marking: extractable with MeOH:$CHCl_3$(2:1); internal one	Greany and Oatman (1972)
Braconidae	*Cardiochiles nigriceps* Viereck	*Heliothis virescens* (Fabricius) (Lepidoptera: Noctuidae)	Larva	External marking: hydrocarbons from Dufour's gland	Vinson (1972b) Guillot *et al.* (1974)
Braconidae	*Ascogaster reticulatus* Watanabe	*Adoxophyes* sp. (Lepidoptera: Tortricidae)	Egg	Internal marking: venom	Yamaguchi (1987)
Braconidae	*Microplitis croceipes* (Cresson)	*Heliothis virescens* (Fabricius) (Lepidoptera: Noctuidae)	Larva	External marking: Dufour's gland	Vinson and Guillot (1972)
Cynipidae	*Pseudeucoila bochei* Weld	*Drosophila melanogaster* (Diptera: Drosophilidae)	Larva	Internal marking; area marking	van Lenteren and Bakker (1975, 1978)
Ichneumonidae	*Pleolophus basizonus* (Gravenhorst)	Sawfly (Hymenoptera: Tenthredinidae)	Pupa	Trail odour (patch marking)	Price (1970)
Ichneumonidae	*Phaeogenes cynarae* Bragg	*Platyptilia carduidactyla* (Riley) (Lepidoptera: Pterophoridae)	Pupa	Internal marking: venom	Bragg (1974)
Ichneumonidae	*Nemeritis canescens* (Grav.)	*Ephestia kuehniella* (Zeller) (Lepidoptera: Pyralidae)	Larva	Area marking: heneicosane from Dufour's gland	Mudd *et al.* (1982) Harrison *et al.* (1985)
Ichneumonidae	*Campoletis perdistinctus* (Viereck)	*Heliothis virescens* (Fabricius) (Lepidoptera: Noctuidae)	Larva	External marking: Dufour's gland Internal marking: lateral oviduct	Guillot and Vinson (1972)
Megaspilidae	*Dendrocerus carpenteri* (Curtis) (hyperparasitoid)	Aphid parasitoids (Hymenoptera: Aphidiidae, Aphelinidae)	Larva Pupa	External marking: two-component system	Höller *et al.* (1991)
Mymaridae	*Caraphractus cinctus* Walker	*Agabus, Ilybius, Dytiscus* (Coleoptera: Dytiscidae)	Egg	Internal marking: poison apparatus (?)	Jackson (1966, 1969)
Pteromalidae	*Muscidifurax zaraptor* K. and L.	*Musca domestica* L. (Diptera: Muscidae)	Pupa	Internal marking	Wylie (1971)
Pteromalidae	*Nasonia vitripennis* (Walker).	*M. domestica* L.	Pupa	Internal marking	Wylie (1965)
Scelionidae	*Telenomus fariai* Costa Lima	*Triatoma phillidipennis* Stål (Hemiptera: Reduviidae)	Egg	External marking: water-soluble	Bosque and Rabinovich (1979)
Trichogrammatidae	*Trichogramma evanescens* Westwood	(Lepidoptera)	Egg	Internal and external marking Area marking	Salt (1937)

Dendrocerus carpenteri (Curtis), a solitary ectohyperparasitoid of aphid primary parasitoids, avoids superparasitism by means of two externally perceivable chemical markers that are deposited on the aphid mummy shell (Höller *et al.*, 1991). The two-component external marking pheromone system (i.e. one of which is highly active and short-lived, while the other is moderately active and persistent) is thought to be advantageous for female parasitoids because the ratio can provide information on the age of the mark and on the producer of the mark (the same female or a conspecific individual).

Internal marking pheromones

When water-washed parasitized hosts and water-washed healthy hosts were exposed to *T. evanescens* females, they probed both types of hosts equally, but primarily laid eggs in unparasitized hosts (Salt, 1937), which showed that they detected an internal mark after inserting the ovipositor.

In addition to external markers, Guillot and Vinson (1972) identified the source of material which enables the ichneumonid *Campoletis perdistinctus* (Viereck) to discriminate internally between a parasitized and unparasitized host.

Using Parafilm as an artificial egg surface, it was shown that female *A. reticulatus* parasitoids inject a venom at the time of egg-laying, which deters further oviposition by other female parasitoids (Yamaguchi, 1987).

Marking pheromone gland

Dufour's gland seems to be the source of external marking pheromones in some larval parasitoids, such as *C. nigriceps* (Guillot *et al.*, 1974), *M. croceipes* (Vinson and Guillot, 1972), *C. perdistinctus* (Guillot and Vinson, 1972) and *Nemeritis canescens* (Gravenhorst) (Mudd *et al.*, 1982). In contrast, the venom gland seems to be the source of internal marking pheromones in *Caraphractus cinctus* Walker (Jackson, 1966, 1969), *Phaeogenes cynarae* Bragg (Bragg, 1974) and *A. reticulatus* (Yamaguchi, 1987). In *C. perdistinctus*, among the candidate organs (i.e. poison gland, alkaline gland, lateral oviduct) which were injected into the host larva by means of glass micropipettes, only the solution of the lateral oviduct deterred oviposition – the ovipositor was inserted but no eggs were released (Guillot and Vinson, 1972).

Aggregation Pheromones

Pheromone traps baited with female *C. nigriceps* captured males and females, and males attracted other males but did not attract females (Lewis *et al.*, 1971). The females which were collected in the field seemed to be mated and to attract both sexes. It was suggested that the aggregation response may increase the females' efficiency in finding the most suitable host areas.

An aggregation pheromone of parasitoids was reported also from *Brachymeria lasus* (Walker), which is thought to aid the aggregation of overwintering adults and the avoidance of predators (Simser and Coppel, 1980b). Hollow styrofoam cubes were conditioned by exposing them to adult wasps in a container for 2 weeks. In a laboratory bioassay, adult wasps preferentially selected cubes conditioned by virgin females over unconditioned cubes when choosing overnight aggregation sites. In further studies, a vial lined with filter paper was placed in a rearing cage for conditioning: both *B. lasus* and *B. intermedia* aggregated on the filter papers conditioned by conspecific adults and by *B. intermedia*; but *B. intermedia* adults did not aggregate on *B. lasus*-conditioned filter paper (Mohamed and Coppel, 1987b). The aggregation pheromone of *B. intermedia* was identified as 3-hexanone (Fig. 15.11), but *B. lasus* did not respond to this chemical. Other unidentified chemical(s) may be involved in the attraction of *B. lasus* to *B. intermedia*-conditioned paper (Mohamed and Coppel, 1987b). Volatile aggregation pheromones in these *Brachymeria* spp. may play a role in assembling both sexes at particular sites, thereby aiding mate location. In addition, the aggregation pheromone may serve to aggregate females at sites with high host densities, thereby increasing parasitization rates.

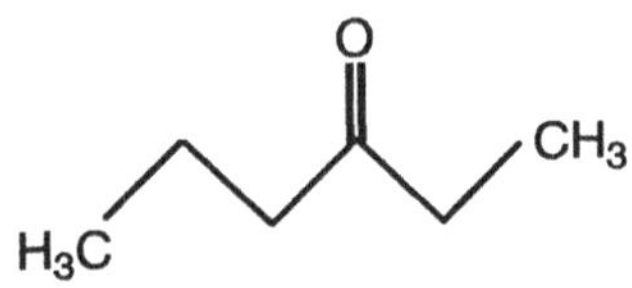

Fig. 15.11. The chemical structure of 3-hexanone aggregation pheromone of *Brachymeria intermedia*. (Mohamed and Coppel, 1987b.)

Conclusions

Sex pheromones, marking pheromones and aggregation pheromones have all been identified from parasitoids. Sex pheromones found in female parasitoids are mostly volatile, attract males from a distance and have been demonstrated by using olfactometers and/or pheromone-releasing field traps. In other cases, contact sex pheromones from females mediate male behaviour (antennal contact, mounting) at close range. Trail sex pheromones are deposited by some females and males respond to the affected substrate by intensive searching behaviour, which may increase their chances of encountering nearby females.

Marking pheromones play an important role by allowing females to discriminate between parasitized and unparasitized hosts. These pheromones are deposited on the substrate during walking (area marking), on the host (external marking), or inside the host during oviposition (internal marking). Aggregation pheromones were found in only a few chalcidid species, and these may play a role in assembling the sexes for mating.

Acknowledgements

I am indebted to Drs S.B. Vinson, Wilf Powell and Jun Tagawa for reviewing an earlier draft of the manuscript.

References

Askari, A. and Alishah, A. (1979) Courtship behavior and evidence for a sex pheromone in *Diaeretiella rapae* (Hymenoptera: Braconidae), the cabbage aphid primary parasitoid. *Annals of the Entomological Society of America* 72, 749–750.

Askari, A. and Coppel, H.C. (1978) Observation on courtship and mating behavior of *Meteorus pulchricornis*, a gypsy moth parasitoid. *Annals of the Entomological Society of America* 71, 364–366.

Assem, J. van den (1986) Mating behaviour in parasitic wasps. In: Waage, J.K. and Greathead, D.J. (eds) *Insect Parasitoids*. Academic Press, London, pp.137–167.

Assem, J. van den (1996) Mating behaviour. In: Jervis, M. and Kidd, N. (eds) *Insect Natural Enemies*. Chapman & Hall, London, pp.163–221.

Assem, J. van den, Jachmann, F. and Simbolotti, P. (1980) Courtship behaviour of *Nasonia vitripennis* (Hym., Pteromalidae): some qualitative, experimental evidence for the role of pheromones. *Behaviour* 75, 301–307.

Assem, J. van den, Jachmann, F. and de Jong, K.A. (1981) Courtship behaviour of *Nasonia vitripennis*: head nodding, mouth-part extrusion and pheromone discharge by abdomectomized males. *Entomologia Experimentalis et Applicata* 30, 215–218.

Barrera, J.F., Gomez, J. and Alauzet, C. (1994) Evidence for a marking pheromone in host discrimination by *Cephalonomia stephanoderis* (Hym., Bethylidae). *Entomophaga* 39, 363–366.

Bosque, C. and Rabinovich, J.E. (1979) Population dynamics of *Telenomus fariai* (Hymenoptera: Scelionidae), a parasite of Chagas' disease vectors. VII. Oviposition behavior and host discrimination. *Canadian Entomologist* 111, 171–180.

Boush, G.M. and Baerwald, R.J. (1967) Courtship behavior and evidence for a sex pheromone in the apple maggot parasite, *Opius alloeus* (Hymenoptera: Braconidae). *Annals of the Entomological Society of America* 60, 865–866.

Bragg, D.E. (1974) Ecological and behavioral studies of *Phaeogenes cynarae*: ecology, host specificity, searching and oviposition, and avoidance of superparasitism. *Annals of the Entomological Society of America* 67, 931–936.

Brodeur, J. and McNeil, J.N. (1994) Seasonal ecology of *Aphidius nigripes* (Hymenoptera: Aphidiidae), a parasitoid of *Macrosiphum euphorbiae* (Homoptera: Aphididae). *Environmental Entomology* 23, 292–298.

Cavill, G.W.K., Robertson, P.L. and Davies, N.W. (1979) An Argentine ant aggregation factor. *Experientia* 35, 989–990.

Cavill, G.W.K., Davies, N.W. and McDonald, F.J. (1980) Characterization of aggregation factors and associated compounds from the Argentine ant, *Iridomyrmex humilis*. *Journal of Chemical Ecology* 6, 371–383.

Cole, L.R. (1970) Observations on the finding of mates by male *Phaeogenes invisor* and *Apanteles medicaginis* (Hym: Ichneumonidae). *Animal Behavior* 18, 184–189.

Decker, U.M., Powell, W. and Clark, S.J. (1993) Sex pheromones in the cereal aphid parasitoids *Praon volucre* and *Aphidius rhopalosiphi*. *Entomologia Experimentalis et Applicata* 69, 33–39.

Eller, F.J., Bartelt, R.J., Jones, R.L. and Kulman, H.M. (1984) Ethyl (Z)-9-hexadecenoate: a sex pheromone of *Syndipnus rubiginosus*, a sawfly parasitoid. *Journal of Chemical Ecology* 10, 291–300.

Fauvergue, X., Hopper, K.R. and Antolin, M.F. (1995) Mate finding via a trail sex pheromone by a parasitoid wasp. *Proceedings of the National Academy of Sciences USA* 92, 900–904.

Field, S.A. and Keller, M.A. (1994) Localization of the female sex pheromone gland in *Cotesia rubecula* Marshall (Hymenoptera: Braconidae). *Journal of Hymenopteran Research* 3, 151–156.

Finidori-Logli, V., Bagnères, A.G., Erdmann, D., Francke, W. and Clement, J.L. (1996) Sex recognition in *Diglyphus isaea* Walker (Hymenoptera, Eulophidae) – role of an uncommon family of behaviorally active compounds. *Journal of Chemical Ecology* 22, 2063–2079.

Godfray, H.C.J. (1994) *Parasitoids – Behavioral and Evolutionary Ecology*. Monographs in Behavior and Ecology, Princeton University Press, Princeton, New Jersey, 473 pp.

Greany, P.D. and Oatman, E.R. (1972) Analysis of host discrimination in the parasite *Orgilus lepidus* (Hymenoptera: Braconidae). *Annals of the Entomological Society of America* 65, 377–383.

Grosch, D.S. (1948) Experimental studies on the mating reaction of male *Habrobracon*. *Journal of Comparative Physiology and Psychology* 41, 188–195.

Guillot, F.S. and Vinson, S.B. (1972) Sources of substances which elicit a behavioural response from the insect parasitoid, *Campoletis perdistinctus*. *Nature* 235, 169–170.

Guillot, F.S., Joiner, R.L. and Vinson, S.B. (1974) Host discrimination: isolation of hydrocarbons from Dufour's gland of a braconid parasitoid. *Annals of the Entomological Society of America* 67, 720–721.

Hardie, J., Hick, A.J., Höller, C., Mann, J., Merritt, L., Nottingham, S.F., Powell, W., Wadhams, L.J., Witthinrich, J. and Wright, A.F. (1994) The responses of *Praon* spp. parasitoids to aphid sex pheromone components in the field. *Entomologia Experimentalis et Applicata* 71, 95–99.

Harrison, E.G., Fisher, R.C. and Ross, K.M. (1985) The temporal effects of Dufour's gland secretion in host discrimination by *Nemeritis canescens*. *Entomologia Experimentalis et Applicata* 38, 215–220.

Hidoh, O., Kawashima, T., Fukami, J. and Kainoh, Y. (1992) EAG responses of parasitoids, *Ascogaster reticulatus* Watanabe (Hymenoptera: Braconidae), to the female sex pheromone. *Applied Entomology and Zoology* 27, 587–589.

Höller, C. (1991) Evidence for the existence of a species closely related to the cereal aphid parasitoid *Aphidius rhopalosiphi* De Stefani-Perez based on host ranges, morphological characters, isoelectric focusing banding patterns, cross-breeding experiments and sex pheromone specificities (Hymenoptera, Braconidae, Aphidiinae). *Systematic Entomology* 16, 15–28.

Höller, C., Williams, H.J. and Vinson, S.B. (1991) Evidence for a two-component external marking pheromone system in an aphid hyperparasitoid. *Journal of Chemical Ecology* 17, 1021–1035.

Höller, C., Micha, S.G., Francke, W. and Pickett, J.A. (1994) Enemy-induced dispersal in a parasitic wasp. *Experientia* 50, 182–185.

Hrdy, I. and Sedivy, J. (1979) Males of *Exetastes cinctipes* (Hymenoptera, Ichneumonidae) attracted to 8-dodecenyl and 11-tetradecenyl acetates. *Acta Entomologica Bohemoslovaca* 76, 59–61.

Jackson, D.J. (1966) Observations on the biology of *Caraphractus cinctus* Walker (Hymenoptera: Mymaridae), a parasitoid of the eggs of Dytiscidae (Coleoptera) III. The adult life and sex ratio. *Transactions of the Royal Entomological Society of London* 118, 23–49.

Jackson, D.J. (1969) Observations on the female reproductive organs and the poison apparatus of *Caraphractus cinctus* Walker (Hymenoptera: Mymaridae). *Zoological Journal of the Linnean Society* 48, 59–81.

Jones, R.L. (1989) Semiochemicals mediating *Microplitis croceipes* – habitat, host, and mate finding behavior. *Southwestern Entomologist* 12, 53–57.

Kainoh, Y. (1986) Mating behavior of *Ascogaster reticulatus* Watanabe (Hymenoptera: Braconidae), an egg-larval parasitoid of the smaller tea tortrix moth, *Adoxophyes* sp. (Lepidoptera: Tortricidae). I. Diel patterns of emergence and mating, and some conditions for mating. *Applied Entomology and Zoology* 21, 1–7.

Kainoh, Y. and Oishi, Y. (1993) Source of sex pheromone of the egg-larval parasitoid, *Ascogaster reticulatus* Watanabe (Hymenoptera: Braconidae). *Journal of Chemical Ecology* 19, 963–969.

Kainoh, Y., Nemoto, T., Shimizu, K., Tatsuki, S., Kusano, T. and Kuwahara, Y. (1991) Mating behavior of *Ascogaster reticulatus* Watanabe (Hymenoptera: Braconidae), an egg-larval parasitoid of the smaller tea tortrix, *Adoxophyes* sp. (Lepidoptera: Tortricidae). III. Identification of the sex pheromone. *Applied Entomology and Zoology* 26, 543–549.

Kamano, Y., Shimizu, K., Kainoh, Y. and Tatsuki, S. (1989) Mating behavior of *Ascogaster reticulatus* Watanabe (Hymenoptera: Braconidae), an egg-larval parasitoid of the smaller tea tortrix, *Adoxophyes* sp. (Lepidoptera: Tortricidae). II. Behavioral sequence and a role of sex pheromone. *Applied Entomology and Zoology* 24, 363–369.

Khasimuddin, S. and DeBach, P. (1975) Mating behavior and evidence of a male sex pheromone in species of the genus *Aphytis*. *Annals of the Entomological Society of America* 68, 893–896.

King, P.E., Askew, R.R. and Sanger, C. (1969) The detection of parasitized hosts by males of *Nasonia vitripennis* (Walker) (Hymenoptera: Pteromalidae) and some possible implications. *Proceedings of the Royal Entomological Society of London (A)* 44, 85–90.

Kitano, H. (1975) Studies on the courtship behavior of *Apanteles glomeratus* L. 2. Role of the male wings during courtship and the releaser of mounting and copulatory behavior in the males. *Kontyu* 43, 513–521. (In Japanese with English summary.)

Lecomte, C. and Pouzat, J. (1985) EAG responses of two ichneumonid parasitoids, *Diadromus pulchellus* and *D. collaris*, to odours emitted by plants, the phytophagous-host *Acrolepiopsis assectella* and the sexual partner. *Entomologia Experimentalis et Applicata* 39, 295–306.

Lenteren, J.C. van (1981) Host discrimination by parasitoids. In: Nordlund, D.A., Jones, R.L. and Lewis, W.J. (eds) *Semiochemicals – Their Role in Pest Control*. John Wiley & Sons, New York, pp. 153–179.

Lenteren, J.C. van and Bakker, K. (1975) Discrimination between parasitized and unparasitized hosts in the parasitic wasp *Pseudeucoila bochei*: a matter of learning. *Nature* 254, 417–419.

Lenteren, J.C. van and Bakker, K. (1978) Behavioural aspects of the functional responses of a parasite (*Pseudeucoila bochei* Weld) to its host (*Drosophila melanogaster*). *Netherlands Journal of Zoology* 28, 213–233.

Leonard, S.H. and Ringo, J.M. (1978) Analysis of male courtship patterns and mating behavior of *Brachymeria intermedia*. *Annals of the Entomological Society of America* 71, 817–826.

Lewis, W.J., Snow, J.W. and Jones, R.L. (1971) A pheromone trap for studying populations of *Cardiochiles nigriceps*, a parasite of *Heliothis virescens*. *Journal of Economic Entomology* 64, 1417–1421.

Matthews, R.W. (1975) Courtship in parasitic wasps. In: Price, P.W. (ed.) *Evolutionary Strategies of Parasitic Insects and Mites*. Plenum Press, New York, pp. 66–86.

Matthews, R.W., Yukawa, J. and Gonzalez, J.M. (1985) Sex pheromones in male *Melittobia* parasitic wasps: female response to conspecific and congeneric males of 3 species. *Journal of Ethology* 3, 59–62.

Mayer, M.S. and McLaughlin, J.R. (1991) *Handbook of Insect Pheromones and Sex Attractants*. CRC Press, Boca Raton, Florida, 1083 pp.

McAuslane, H.J., Vinson, S.B. and Williams, H.J. (1990) Influence of host plant on mate location by the parasitoid *Campoletis sonorensis* (Hymenoptera: Ichneumonidae). *Environmental Entomology* 19, 26–31.

McNeil, J.N. and Brodeur, J. (1995) Pheromone-mediated mating in the aphid parasitoid, *Aphidius nigripes* (Hymenoptera, Aphidiidae). *Journal of Chemical Ecology* 21, 959–972.

Micha, S.G., Stammel, J. and Höller, C. (1993) 6-Methyl-5-hepten-2-one, a putative sex and spacing pheromone of the aphid hyperparasitoid, *Alloxysta victrix* (Hymenoptera: Alloxystidae). *European Journal of Entomology* 90, 439–442.

Mohamed, M.A. and Coppel, H.C. (1987a) Pheromonal basis of courtship behavior in two gypsy moth parasitoids: *Brachymeria intermedia* (Nees) and *Brachymeria lasus* (Walker) (Hymenoptera: Chalcididae). *Journal of Chemical Ecology* 13, 1099–1113.

Mohamed, M.A. and Coppel, H.C. (1987b) Pheromonal basis for aggregation behavior of parasitoids of the gypsy moth: *Brachymeria intermedia* (Nees) and *Brachymeria lasus* (Walker) (Hymenoptera: Chalcididae). *Journal of Chemical Ecology* 13, 1385–1393.

Morse, B.W. and Kulman, H.M. (1985) Monitoring damage by yellowheaded spruce sawflies with sawfly and parasitoid pheromones. *Environmental Entomology* 14, 131–133.

Mudd, A., Fisher, R.C. and Smith, M.C. (1982) Volatile hydrocarbons in the Dufour's gland of the parasite *Nemeritis canescens* (Grav.) (Hymenoptera; Ichneumonidae). *Journal of Chemical Ecology* 8, 1035–1042.

Nazzi, F., Powell, W., Wadhams, L.J. and Woodcock, C.M. (1996) Sex pheromone of aphid parasitoid *Praon volucre* (Hymenoptera, Braconidae). *Journal of Chemical Ecology* 22, 1169–1175.

Obara, M. and Kitano, H. (1974) Studies on the courtship behavior of *Apanteles glomeratus* L. I. Experimental studies on releaser of wing-vibrating behavior in the male. *Kontyu* 42, 208–214.

Pompanon, F., Deschepper, B., Mourer, Y., Fouillet, P. and Bouletreau, M. (1997) Evidence for a substrate-borne sex pheromone in the parasitoid wasp *Trichogramma brassicae*. *Journal of Chemical Ecology* 23, 1349–1360.

Powell, J.E. and King, E.G. (1984) Behavior of adult *Microplitis croceipes* (Hymenoptera: Braconidae) and parasitism of *Heliothis* spp. (Lepidoptera: Noctuidae) host larvae in cotton. *Environmental Entomology* 13, 272–277.

Powell, W. and Zhang, Z.L. (1983) The reactions of two cereal aphid parasitoids, *Aphidius uzbekistanicus* and *A. ervi* to host aphids and their food-plants. *Physiological Entomology* 8, 439–443.

Powell, W., Hardie, J., Hick, A.J., Höller, C., Mann, J., Merritt, L., Nottingham, S.F., Wadhams, L.J., Witthinrich, J. and Wright, A.F. (1993) Responses of the parasitoid *Praon volucre* (Hymenoptera: Braconidae) to aphid sex pheromone lures in cereal fields in autumn: implication for parasitoid manipulation. *European Journal of Entomology* 90, 435–438.

Powell, W., Pennacchio, F., Poppy, G.M. and Tremblay, E. (1998) Strategies involved in the location of hosts by the parasitoid *Aphidius ervi* Haliday (Hymenoptera: Braconidae: Aphidiinae). *Biological Control* 11, 104–112.

Price, P.W. (1970) Trail odours: recognition by insects parasitic on cocoons. *Science* 170, 546–547.

Quicke, D.L.J. (1997) *Parasitic Wasps*. Chapman & Hall, London, 470 pp.

Rao, S.V. and DeBach, P. (1969) Experimental studies on hybridization and sexual isolation between some *Aphytis* species (Hymenoptera: Aphelinidae) I. Experimental hybridization and interpretation of evolutionary relationships among the species. *Hilgardia* 39, 515–553.

Read, D.P., Feeny, P.P. and Root, R.B. (1970) Habitat selection by the aphid parasite *Diaeretiella rapae* (Hym.: Braconidae) and hyperparasite *Charips brassicae* (Hym.: Cynipidae). *Canadian Entomologist* 102, 1567–1578.

Reed, H.C., Tan, S., Reed, D.K., Elliot, N.C., Burd, J.D. and Walker, T. (1994) Evidence for a sex attractant in *Aphidius colemani* Viereck with potential use in field studies. *Southwestern Entomology* 19, 273–278.

Ridgway, N.M. and Mahr, D.L. (1986) Monitoring adult flight of *Pholetesor ornigis* (Hymenoptera: Braconidae), a parasitoid of the spotted tentiform leafminer (Lepidoptera: Gracillariidae). *Environmental Entomology* 15, 331–334.

Robacker, D.C. and Hendry, L.B. (1977) Neral and geranial: components of the sex pheromone of the parasitic wasp, *Itoplectis conquisitor*. *Journal of Chemical Ecology* 3, 563–577.

Robacker, D.C., Weaver, K.M. and Hendry, L.B. (1976) Sexual communication and associative learning in the parasitic wasp *Itoplectis conquisitor* (Say). *Journal of Chemical Ecology* 2, 39–48.

Rotheray, G.E. (1981) Courtship, male swarms and a sex pheromone of *Diplazon pectoratorius* (Thunberg) (Hymenoptera: Ichneumonidae). *Entomological Gazette* 32, 193–196.

Salt, G. (1934) Experimental studies in insect parasitism. II. Superparasitism. *Proceedings of the Royal Entomological Society of London Ser. B* 114, 455–476.

Salt, G. (1937) The sense used by *Trichogramma* to distinguish between parasitized and unparasitized hosts. *Proceedings of the Royal Entomological Society of London Ser. B.* 122, 57–75.

Shu, S. and Jones, R.L. (1993) Evidence for a multicomponent sex pheromone in *Eriborus terebrans* (Gravenhorst) (Hym.: Ichneumonidae), a larval parasitoid of the European corn borer. *Journal of Chemical Ecology* 19, 2563–2576.

Simser, D.H. and Coppel, H.C. (1980a) Female-produced sex pheromone in *Brachymeria lasus* and *B.intermedia* (Hym.: Chalcididae). *Entomophaga* 25, 373–380.

Simser, D.H. and Coppel, H.C. (1980b) Aggregation behavior of *Brachymeria lasus* (Walker) in the laboratory. *Environmental Entomology* 9, 486–488.

Sugimoto, T., Uenishi, M. and Machida, F. (1986) Foraging for patchily-distributed leaf-miners by the parasitoid, *Dapsilarthra rufiventris* (Hymenoptera: Braconidae). I. Discrimination of previously searched leaflets. *Applied Entomology and Zoology* 21, 500–508.

Swedenborg, P.D. and Jones, R.L. (1992a) Multicomponent sex pheromone in *Macrocentrus grandii* Goidanich (Hymenoptera: Braconidae). *Journal of Chemical Ecology* 18, 1901–1912.

Swedenborg, P.D. and Jones, R.L. (1992b) (Z)-4-Tridecenal, a pheromonally active air oxidation product from a series of (Z,Z)-9,13 dienes in *Macrocentrus grandii* Goidanich (Hymenoptera: Braconidae). *Journal of Chemical Ecology* 18, 1913–1931.

Swedenborg, P.D., Jones, R.L., Liu, H.-W. and Krick, T.P. (1993) (3*R*,5*S*,6*R*)-3,5-Dimethyl-6-(methylethyl)-3,4,5,6-tetrahydropyran-2-one, a third sex pheromone component for *Macrocentrus grandii* (Goidanich) (Hymenoptera: Braconidae) and evidence for its utility at eclosion. *Journal of Chemical Ecology* 19, 485–502.

Swedenborg, P.D., Jones, R.L., Zhou, H.-Q., Shin, I. and Liu, H.-W. (1994) Biological activity of (3*R*,5*S*,6*R*)- and (3*S*,5*R*,6*S*)-3,5-dimethyl-6-(methylethyl)-3,4,5,6-tetra-hydropyran-2-one, a pheromone of *Macrocentrus grandii* (Goidanich) (Hymenoptera: Braconidae). *Journal of Chemical Ecology* 20, 3373–3380.

Syvertsen, T.C., Jackson, L.L., Blomquist, G.J. and Vinson, S.B. (1995) Alkadienes mediating courtship in the parasitoid *Cardiochiles nigriceps* (Hymenoptera: Braconidae). *Journal of Chemical Ecology* 21, 1971–1989.

Tagawa, J. (1977) Localization and histology of the female sex pheromone-producing gland in the parasitic wasp, *Apanteles glomeratus*. *Journal of Insect Physiology* 23, 49–56.

Tagawa, J. (1983) Female sex pheromone glands in the parasitic wasps, genus *Apanteles*. *Applied Entomology and Zoology* 18, 416–427.

Takahashi, S. and Sugai, T. (1982) Mating behavior of the parasitoid wasp *Tetrastichus hagenowii* (Hymenoptera: Eulophidae). *Entomologia Generalis* 7, 287–293.

Vet, L.E.M. and Dicke, M. (1992) Ecology of infochemical use by natural enemies in a tritrophic context. *Annual Review of Entomology* 37, 141–172.

Vinson, S.B. (1972a) Courtship behavior and evidence for a sex pheromone in the parasitoid *Campoletis sonorensis* (Hymenoptera: Ichneumonidae). *Environmental Entomology* 1, 409–414.

Vinson, S.B. (1972b) Competition and host discrimination between two species of tobacco budworm parasitoid. *Annals of the Entomological Society of America* 65, 229–236.

Vinson, S.B. (1978) Courtship behavior and source of a sexual pheromone from *Cardiochiles nigriceps*. *Annals of the Entomological Society of America* 71, 832–837.

Vinson, S.B. (1985) The behaviour of parasitoids. In: Kerkut, G.A. and Gilbert, L.I. (eds) *Comprehensive Insect Physiology, Biochemistry and Pharmacology*, Vol. 9. *Behaviour*. Pergamon Press, Oxford, pp. 417–469.

Vinson, S.B. and Guillot, F.S. (1972) Host marking: source of a substance that results in host discrimination in insect parasitoids. *Entomophaga* 17, 241–245.

Weseloh, R.M. (1976) Dufour's gland: source of sex pheromone in a hymenopterous parasitoid. *Science* 193, 695–697.

Weseloh, R.M. (1977) Mating behavior of the gypsy moth parasite, *Apanteles melanoscelus*. *Annals of the Entomological Society of America* 70, 549–554.

Weseloh, R.M. (1980) Sex pheromone gland of the gypsy moth parasitoid, *Apanteles melanoscelus*: revaluation and ultrastructural survey. *Annals of the Entomological Society of America* 73, 576–580.

Wylie, H.G. (1965) Discrimination between parasitized and unparasitized house fly pupae by females of *Nasonia vitripennis* (Walk.). *Canadian Entomologist* 97, 279–286.

Wylie, H.G. (1971) Oviposition restraint of *Muscidifurax zaraptor* (Hymenoptera: Pteromalidae) on parasitized *Nasonia vitripennis* housefly pupae. *Canadian Entomologist* 103, 1537–1544.

Yamaguchi, H. (1987) The role of venom in host-discrimination of *Ascogaster reticulatus* Watanabe. *Japanese Journal of Applied Entomology and Zoology* 31, 80–82. (In Japanese with English summary.)

Yoshida, S. (1978) Behavior of males in relation to the female sex pheromone in the parasitoid wasp, *Anisopteromalus calandrae* (Hym., Pteromalidae). *Entomologia Experimentalis et Applicata* 23, 152–162.

16 Parasitoid Hosts

WILF POWELL

Department of Entomology and Nematology, IACR-Rothamsted, Harpenden, Herts AL5 2JQ, UK

Introduction

To locate their hosts, parasitoids rely upon a range of cues, many of which are volatile chemicals originating from the host itself, host products or the substrate in or on which the host occurs (Weseloh, 1981; Vinson, 1984). Therefore, most arthropod hosts of insect parasitoids have evolved to remain as inconspicuous as possible to avoid detection by their enemies. However, intraspecific communication by insects usually necessitates the release of volatile pheromones which reveal the producers' presence, and many parasitoids have developed the ability to use these pheromones as host-location kairomones (Weseloh, 1981; Vet and Dicke, 1992; Stowe *et al.*, 1995). This has been described as 'chemical eavesdropping' (Stowe *et al.*, 1995). Because pheromones have evolved as chemical signals between individuals of the same species, they are usually highly species-specific, which makes them extremely reliable semiochemical cues for parasitoids seeking particular hosts.

Pheromones mediate a number of different behaviours amongst insects, which are extensively described in the other chapters of this book, and most of these are potentially valuable as kairomonal cues for parasitoids to exploit during host location. There is now evidence, involving a variety of insects, of kairomonal activity for several types of pheromones, including sex pheromones, aggregation pheromones, epideictic pheromones and alarm pheromones. Both hymenopteran and dipteran parasitoids have been shown to respond to host pheromones, which may function as long-range or short-range volatile cues or as contact cues. Evidence for the kairomonal function of host pheromones is reviewed in this chapter, arranged according to pheromone type and then according to host type.

Sex Pheromones as Parasitoid Kairomones

Amongst insects, pheromones are widely used as a means of ensuring that males and females locate each other for the purpose of sexual reproduction. These sex pheromones are produced by members of one sex specifically to attract members of the other sex. Many are highly volatile chemicals, having evolved as long-range communication signals, and so would appear to be ideal candidates as kairomones for foraging parasitoids. However, the majority of parasitoids attack immature stages of their hosts, rather than reproductive adults, in which case sex pheromones are unlikely to be reliable host location cues. Nevertheless, there is considerable evidence that sex pheromones do act as kairomones, especially for parasitoids attacking two distinct host types: insect eggs and homopteran plant pests such as aphids and scale insects (Fig. 16.1).

Parasitoids of insect eggs

Egg parasitoids are able to use chemical cues associated with adult female insects as a means of locating host eggs. Laing (1937) observed that *Trichogramma evanescens* Westwood females were attracted to chemical traces left by female grain moths, *Sitotroga cerealella* (Olivier). It was then discovered that lepidopteran scales deposited during oviposition provide a source of contact kairomones that stimulate host-searching behaviour by female *Trichogramma* (Lewis *et al.*, 1972; Noldus and van Lenteren, 1985b; Zaborski *et al.*, 1987; Noldus, 1989; see below). In addition, there is now evidence that lepidopteran sex pheromones can also have kairomonal effects on foraging egg parasitoids (Lewis *et al.*, 1982; Nordlund *et al.*, 1983; Zaki,

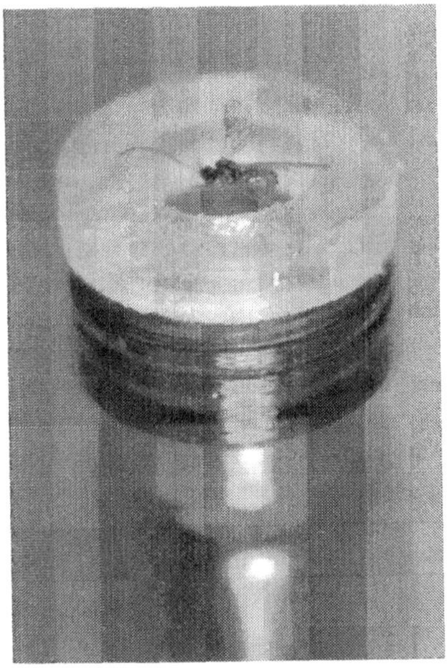

Fig. 16.1. Female *Praon volucre* (Haliday), a generalist parasitoid of aphids, attracted to a vial releasing (+)-(4a*S*,7*S*,7a*R*)-nepetalactone, a major component of many aphid sex pheromones. (Courtesy of IACR-Rothamsted Experimental Station, UK.)

1985; Noldus, 1988; Noldus *et al.*, 1990, 1991a; Frenoy *et al.*, 1992; Klemm and Schmutterer, 1993; Arakaki *et al.*, 1996, 1997) (Table 16.1).

The abdominal tips of virgin female *Heliothis zea* (Boddie), squashed on filter paper, and a synthetic blend ((Z)-7-hexadecenal, (Z)-9-hexadecenal and (Z)-11-hexadecenal) of *H. zea* sex pheromone both significantly increased percentage parasitism by *Trichogramma pretiosum* Riley of *H. zea* eggs attached to cowpea plants in glasshouse trials (Lewis *et al.*, 1982). Similarly, both abdominal tips of virgin females and components ((Z)-9-tetradecenyl acetate and (Z)-9-dodecenyl acetate) of the sex pheromone of *Spodoptera frugiperda* (J.E. Smith) increased parasitization rates of *S. frugiperda* eggs by the scelionid *Telenomus remus* Nixon in laboratory and glasshouse trials (Nordlund *et al.*, 1983). Increased rates of parasitism by *Trichogramma* spp. of *H. zea* eggs placed on cotton plants were recorded in the field when a synthetic blend of the host pheromone was released in the crop (Lewis *et al.*, 1982). This effect was greater on the second day of the field trial than on the first and the authors concluded that this was due to attraction and consequent redistribution of the parasitoids rather than just stimulation of those already present.

However, Noldus (1988) tested the responses of experienced *T. pretiosum* females to calling virgin female *H. zea* in an olfactometer and concluded that the parasitoids were showing an arrestment response to the pheromone. Subsequent wind tunnel studies also indicated that the sex pheromones of *H. zea* and *Mamestra brassicae* L. acted as arrestment cues for *T. pretiosum* and *T. evanescens*, respectively, suggesting that the increased rates of parasitism recorded in the field by Lewis *et al.* (1982) were the result of interception of parasitoids moving through plots together with extended retention of those already present, rather than attraction from a distance (Noldus *et al.*, 1991a). Arrestment behaviour consists of increased residence times, walking times and path lengths as well as reduced walking speeds – behaviours that would increase the probability of encountering any host eggs present in the area (Noldus *et al.*, 1990, 1991a).

What is the adaptive benefit to an egg parasitoid of these kairomonal responses to host sex pheromones? It has been suggested that the sex pheromones act as host-location kairomones because they are associated with areas where eggs are likely to be found, and that arrestment of the parasitoid and stimulation of intensive searching behaviour in these areas increase the probability of encounters with host eggs, or with other cues more closely associated with the precise location of eggs (Noldus *et al.*, 1991a). In many cases, a habitat patch in which virgin female hosts are calling is indeed likely to contain eggs laid by females that have already mated (Noldus and van Lenteren, 1985a) but some moths, such as the pine processionary moth *Thaumetopoea pityocampa* Denis and Schiffermüller, may oviposit in areas far removed from their mating site (Demolin, 1969; Battisti, 1989). Therefore, it is not surprising that the eulophid egg parasitoid *Tetrastichus servadeii* Domenichini, which is a specialist on *Th. pityocampa*, was never caught in field traps releasing pityolure, the synthetic sex pheromone of its host (Battisti, 1989). The polyphagous egg parasitoid *Ooencyrtus pityocampae* Mercet (Encyrtidae), however, was attracted to the pityolure traps. This species is phoretic on host females, a habit exhibited by a number of egg parasitoids (Vinson, 1985), and this could explain its response to the *Th. pityocampa* sex pheromone, since phoresy would ensure

Table 16.1. Responses of egg parasitoids to sex pheromones from lepidopteran hosts.

Parasitoid	Host	Stimulus	Response	Reference
Trichogrammatidae				
Trichogramma brassicae Bezdenko	*Ostrinia nubilalis* Hübner	Virgin female hosts	Increased foraging activity	Frenoy *et al.* (1991) Frenoy *et al.* (1992)
T. brassicae	*O. nubilalis*	Main sex pheromone component	No response	Renou *et al.* (1992)
T. brassicae	*O. nubilalis*	Synthetic sex pheromone blend	No response	Kaiser *et al.* (1989)
T. brassicae	*O. nubilalis*	Pheromone plus host egg and food plant odours	Increased foraging activity	Kaiser *et al.* (1989)
Trichogramma evanescens Westwood	*Mamestra brassicae* L.	Virgin female hosts	Increased foraging activity (arrestment)	Noldus *et al.* (1991a,b) Noldus and van Lenteren (1985a)
T. evanescens	*M. brassicae*	Main sex pheromone component	No response	Noldus and van Lenteren (1985a)
T. evanescens	*M. brassicae*	Pheromone gland extract	No response	Noldus and van Lenteren (1985a)
T. evanescens	*Pieris brassicae* L.	Virgin female hosts	Attraction/arrestment	Noldus and van Lenteren (1983, 1985a)
T. evanescens	*Pectinophora gossypiella* Saunders *Spodoptera littoralis* (Boisduval) *Earias insulana* Boisduval	Synthetic sex pheromone components	Attraction	Zaki (1985)
Trichogramma pretiosum Riley	*Heliothis zea* (Boddie)	Virgin female abdominal tips Synthetic sex pheromone	Increased rate of parasitization	Lewis *et al* (1982)
T. pretiosum	*H. zea*	Virgin female hosts	Increased foraging activity (arrestment)	Noldus (1988) Noldus *et al.* (1991a)
T. pretiosum	*Spodoptera frugiperda* (J.E. Smith)	Synthetic sex pheromone	No response	Noldus *et al.* (1990)
T. pretiosum	*Plutella xylostella* (L.)	Synthetic sex pheromone	Increased rate of parasitization	Klemm and Schmutterer (1993)
Scelionidae				
Telenomus euproctidis Wilcox	*Euproctis taiwana* (Shiraki) *Euproctis pseudoconspersa* (Strand)	Virgin female hosts Synthetic sex pheromone	Attraction in the field	Arakaki *et al.* (1996, 1997)
Telenomus remus Nixon	*S. frugiperda*	Virgin female abdominal tips Synthetic sex pheromone	Increased rate of parasitization	Nordlund *et al.* (1983)
Encyrtidae				
Ooencyrtus pityocampae Mercet	*Thaumetopoea pityocampa* Denis & Schiffermüller	Synthetic sex pheromone (pityolure)	Attraction in the field	Battisti (1989)

encounters with host eggs once virgin or recently mated females were located (Battisti, 1989). The scelionid egg parasitoid *Telenomus euproctidis* Wilcox, which is phoretic on female *Euproctis taiwana* (Shiraki) moths in Japan, occurred more frequently on virgin than on mated female moths when these were exposed concurrently in the field, and large numbers were also caught in traps releasing a synthetic component ((Z)-16-methyl-9-heptadecenyl isobutyrate) of the *E. taiwana* sex pheromone (Arakaki *et al.*, 1996).

The ability of parasitoids to distinguish between the sex pheromones of closely related host species was demonstrated by Arakaki *et al.* (1997). *Telenomus euproctidis* parasitizes the eggs of two allopatric moths, *E. taiwana* and *Euproctis pseudoconspersa* (Strand). In regions of Japan where *E. taiwana* predominates, many more female parasitoids occurred on virgin female *E. taiwana* than on the alternative host. Also, traps emitting an *E. taiwana* pheromone component ((Z)-16-methyl-9-heptadecenyl isobutyrate) attracted more parasitoids than did traps emitting an *E. pseudoconspersa* pheromone component (10,14-dimethylpentadecyl isobutyrate) when both were exposed simultaneously. However, *E. pseudoconspersa* was significantly more attractive in regions where it was the more abundant host, indicating that the parasitoid could distinguish between the sex pheromones of the two moths and showed a strong preference for that of the species most abundant in the area (Arakaki *et al.*, 1997). Parasitoids from each of the two regions readily cross-mated and produced female offspring, so it is unlikely that two different parasitoid species were involved.

In olfactometer tests, *T. evanescens* showed variable responses to synthetic sex pheromones (or individual species-specific components of the sex pheromones) of *Pectinophora gossypiella* Saunders, *Spodoptera littoralis* (Boisduval) and *Earias insulana* Boisduval, responses being greatest for *P. gossypiella* pheromone components and weakest for those of *E. insulana* (Zaki, 1985). When *T. evanescens* females were presented with eggs of either *S. littoralis* or *Helicoverpa* (= *Heliothis*) *armigera* (Hübner) that had been sprayed with *P. gossypiella* sex pheromones, the rate of parasitization was significantly greater than that on untreated eggs, but the effect was much weaker for eggs of *Ephestia* (= *Anagasta*) *kuehniella* Zeller or those of *P. gossypiella* itself when these were similarly treated (Zaki, 1985). However, synthetic pheromones of *S. littoralis* and *E. insulana* failed to increase rates of parasitization when sprayed on the eggs of other species.

Most of the responses to host sex pheromones so far recorded have involved species attacking the eggs of moths, which generally release pheromones at night, whereas the parasitoids are diurnal foragers. Noldus and van Lenteren (1985a) hypothesized that pheromone molecules are adsorbed on to surfaces of the vegetation immediately surrounding calling female moths, from which they slowly evaporate, allowing detection by the diurnal parasitoids. This hypothesis was later tested by exposing Brussels sprout leaves to air that had passed over a calling female *M. brassicae* and observing the foraging behaviour of *T. evanescens* on treated and untreated leaves (Noldus *et al.*, 1991b). Parasitoids foraged for longer periods and spent a greater proportion of time walking on pheromone-treated leaves than on untreated leaves and these effects were still evident on leaves treated 4 h previously; residence times were also significantly longer on leaves treated 24 h earlier, compared with controls (Noldus *et al.*, 1991b).

Attempts have been made to disrupt the mating behaviour of pests by treating large areas with synthetic sex pheromones, and the potential effects of this strategy on natural enemies, including egg parasitoids, have been investigated in some cases. Parasitoids were reared from egg masses of the gypsy moth, *Lymantria dispar* (L.), collected from forest plots treated with the synthetic sex pheromone disparlure (Brown and Cameron, 1979), but no effect on percentage parasitism by the encyrtid *Ooencyrtus kuwanai* (Howard) was detected. Similarly, treatment of coniferous forest plots with a synthetic sex pheromone ((Z)-6-heneicosen-11-one) of the Douglas-fir tussock moth, *Orgyia pseudotsugata* (McDunnough), did not affect numbers of *Telenomus californicus* Ashmead and *Tetrastichus* sp. reared from moth egg masses (Sower and Torgersen, 1979).

Not all egg parasitoid responses to virgin female hosts have been shown to be responses to sex pheromones and not all host sex pheromones elicit responses from parasitoids. In olfactometer studies, *T. evanescens* responded to virgin female *Pieris brassicae* L. but not to mated females or males, although no sex pheromone of *P. brassicae* has been identified (Noldus and van Lenteren, 1983, 1985a). Conversely, *T. pretiosum* did not respond to a synthetic blend ((Z)-7-dodecenyl acetate, 11-dodecenyl acetate, dodecanyl acetate, (Z)-9-tetradecenyl acetate and (Z)-11-hexadecenyl acetate) of the sex pheromone of *S. frugiperda* (Noldus *et al.*, 1990). *T. evanescens* responded to calling virgin female M. *brassicae* but did not respond to the main component ((Z)-11-hexadecenyl acetate) of the M. *brassicae* sex pheromone or to a crude hexane extract of its pheromone gland (Noldus and van Lenteren, 1985a). *Trichogramma brassicae* Bezdenko showed no response in a linear olfactometer to the main component ((Z)-11-tetradecenyl acetate) of the sex pheromone of its host *Ostrinia nubilalis* Hübner, regardless of concentration, or to a synthetic blend of the pheromone (Frenoy *et al.*, 1992; Renou *et al.*, 1992), even though it responded to calling virgin female moths (Frenoy *et al.*, 1992). In this case, the kairomone appears to be a compound associated with pheromone production but not the pheromone itself and oleic acid has been identified as the possible kairomonal compound involved (Frenoy *et al.*, 1992; Renou *et al.*, 1992). Female *Trichogramma maidis* Pint. & Voeg. did not respond to a synthetic blend ((Z)-11-tetradecenyl acetate and (E)-11-tetradecenyl acetate) of the sex pheromone of O. *nubilalis*, when this was presented alone in a four-arm olfactometer, but they did respond when the pheromone was presented along with volatiles from host eggs and maize plants, suggesting possible synergy between different olfactory cues (Kaiser *et al.*, 1989).

Larval/pupal parasitoids

Although the kairomonal activity of host sex pheromones may have apparent adaptive advantages for certain egg parasitoids, it is difficult to perceive advantages for parasitoids attacking larval or pupal stages. Therefore, it is not surprising that there is very little evidence of larval/pupal parasitoid responses to virgin adult female hosts or their sex pheromones.

Disparlure, the synthetic sex pheromone of the gypsy moth, *L. dispar*, was developed for use as a mating disruption control strategy in forest plantations, and laboratory studies were done to measure its effects on oviposition by the pupal parasitoid *Brachymeria intermedia* (Nees) (Cameron and Rhoads, 1973). Although parasitoids produced more female offspring in the presence of the pheromone, high concentrations significantly reduced the number of progeny. The authors concluded that, at the concentrations used in the field, disparlure was unlikely to have any significant effect on the parasitoid.

A similar conclusion was reached by Tingle and Mitchell (1982) concerning the effects on larval parasitoids of a synthetic blend ((*Z*)-11-hexadecenal, (*Z*)-9-tetradecenal, (*Z*)-9-hexadecenal and (*Z*)-11-hexadecenol) of the sex pheromone of the tobacco budworm, *Heliothis virescens* (Fabricius), which was being used for mating disruption in tobacco fields. Percentage parasitism of larvae collected from treated and untreated fields was not significantly different and no individual effects were detected for any of the six parasitoids involved – *Cardiochiles nigriceps* (Viereck), *Cotesia* (= *Apanteles*) *marginiventris* (Cresson), *Pristomerus spinator* (Fabricius), *Netelia sayi* (Cushman), *Eucelatoria rubentis* (Coq.) and *Archytas marmoratus* Townsend. Synthetic sex pheromones ((*Z*)-9-tetradecenyl acetate and (*Z*)-9-dodecenyl acetate) of *S. frugiperda* had no significant effects on levels of parasitism by the larval parasitoids *Chelonis insularis* Cresson and *Temelucha difficilis* Dasch. when applied as individual components in the field (Mitchell *et al.*, 1984), or on mating and oviposition rates of *C. marginiventris* in the laboratory (Tingle *et al.*, 1989). Also, no significant direct effects on parasitism of the European pine shoot moth, *Rhyacionia zozana* (Kearfott), by the larval parasitoid *Glypta zozanae* Walley & Barron and the pupal parasitoid *Mastrus aciculatus* (Provancher), or on parasitoid sex ratios, were detected during mating disruption trials in pine plantations using synthetic blends ((*E*)-9-dodecenyl acetate and (*E*)-9-dodecenol) of the *R. zozana* sex pheromone (Niwa and Daterman, 1989). Very few parasitoids were caught on sticky traps releasing the pheromone, indicating the apparent absence of any kairomonal activity.

Nevertheless, a few instances of larval parasitoids responding to host sex pheromones have been reported. In olfactometer tests, synthetic sex pheromones of the moths *Agrotis ipsilon* (Hufnagel) ((*Z*)-7-dodecenyl acetate, (*Z*)-9-tetradecenyl acetate and (*Z*)-11-hexadecenyl acetate) and *S. littoralis* ((*Z*)-9, (*E*)-11, tetradecadienyl acetate) attracted females of the braconids *Apanteles ruficrus* (Haliday) and *Microplitis rufiventris* Kok., respectively (Zaki, 1996). Also in olfactometer studies, the odour of sexually active female bruchid beetles, *Bruchidius atrolineatus* (Pic.), increased the locomotor activity of the larval/pupal parasitoid *Eupelmus vuilleti* (CRW), although there was no evidence of attraction (Cortesero *et al.*, 1993). The sex pheromone of the female beetle was thought to be the kairomone involved, stimulating intense parasitoid foraging behaviour in a habitat (cowpea plants) where adult and larval hosts often coexist (Cortesero *et al.*, 1993).

Parasitoids of scale insects and aphids

One of the earliest records of parasitoids using the sex pheromone of their host as a kairomone involved parasitoids of the California red scale, *Aonidiella aurantii* (Mask.) (Chapter 10). Sticky traps baited with virgin female scales and placed in citrus groves to trap male scales attracted large numbers of the aphelinids *Aphytis melinus* DeBach and *Aphytis coheni* DeBach (Sternlicht, 1973). Subsequent olfactometer studies confirmed that the parasitoids were highly responsive to fruit infested with female scales (Sternlicht, 1973), and pheromone traps have since been developed to monitor both *A. aurantii* and its *Aphytis* parasitoids in citrus groves (Samways, 1988). Traps baited with synthetic sex pheromones (7-methyl-3-methylene-7-octenyl propanoate and (*E*)-3,7-dimethyl-2,7-octadienyl propanoate) of the San José scale, *Quadraspidiotus perniciosus* (Comstock) (see also Chapter 10, this volume), caught the aphelinid *Encarsia* (= *Prospaltella*) *perniciosa* (Tower) in fruit orchards and so use of pheromone traps for control of the scale had to be carefully timed to avoid disruption of the parasitoid (Rice and Jones, 1982).

Not all scale insect parasitoids use host sex pheromones as kairomones. Sticky traps of different colours, with or without the sex pheromone of *Q. perniciosus*, caught ten aphelinid parasitoid species in apple orchards, at least six of which were known to be associated with *Q. perniciosus*, but only *E. perniciosa* showed a significant response to the pheromone (McLain *et al.*, 1990). Two isomers of the *Q. perniciosus* sex pheromone were used in traps in Switzerland; one isomer ((*E*)-3,7-dimethyl-2,7-octadienyl propanoate) caught more male scales while the other ((*Z*)-3,7,dimethyl-2,7-octadienyl propanoate) caught more *E. perniciosa* (Hippe and Mani, 1995).

Sexual females of holocyclic aphid species produce sex pheromones for mate attraction (see also Chapter 9, this volume) and the main chemical components of the pheromones of several species have been isolated and identified (Dawson *et al.*, 1987, 1990; Pickett *et al.*, 1992). Water traps releasing synthetic aphid sex pheromone component (+)-(4aS,7S,7a*R*)-nepetalactone caught large numbers of female *Praon* spp. (Braconidae) when placed in natural woodland or in cereal fields (Hardie *et al.*, 1991, 1994; Powell *et al.*, 1993). Subsequent olfactometer studies confirmed attraction of female *Praon volucre* (Haliday), a polyphagous species, to synthetic aphid sex pheromone components (Lilley *et al.*, 1994), whilst wind tunnel studies have demonstrated responses by female aphid parasitoids from a range of both generalist and specialist species (Table 16.2) (Glinwood, 1998). Different species varied in their responses to different pheromone components or blends. *Aphidius rhopalosiphi* (de Stefani-Perez), a specialist parasitoid of cereal aphids, responded most strongly to (+)-(4aS,7S,7a*R*)-nepetalactone, which is the only compound detected in the sex pheromone of its main host, the grain aphid *Sitobion avenae* (Fabricius), whereas *Aphidius ervi* Haliday and *Aphidius eadyi* Stary, Gonzales & Hall also responded very strongly to a 1:1 mixture of (+)-(4aS,7S,7a*R*)-nepetalactone and (−)-(1*R*,4aS,7S,7a*R*)-nepetalactol, which corresponds to the pheromone of their main host, *Acyrthosiphon pisum* (Harris).

In field trials using potted trap plants, synthetic aphid sex pheromones significantly increased rates of parasitization by *P. volucre* and *A. rhopalosiphi* on *S. avenae* (Lilley *et al.*, 1994; Glinwood, 1998; Glinwood *et al.*, 1998), and by *A. ervi* and *A. eadyi* on the pea aphid, *Ac. pisum* (Glinwood, 1998; Glinwood *et al.*, 1998). In recent field trials, using trap plants with and without pheromone, more than 80% of the parasitized aphids recovered from the plants occurred on those with pheromone (Glinwood, 1998; Glinwood *et al.*, 1998). Interestingly, the *Aphidius* species have not been caught in significant numbers in water traps releasing

Table 16.2. Responses of parasitoids to sex pheromones from homopteran hosts.

Parasitoid	Host	Stimulus	Response	Reference
Aphelinidae				
Aphytis coheni DeBach	*Aonidiella aurantii* (Mask.)	Virgin female hosts	Attraction in the field	Sternlicht (1973)
Aphytis melinus DeBach				
Aphytis africanus Quednau	*A. aurantii*	Synthetic sex pheromone	Attraction in the field	Samways (1988)
A. melinus				
Encarsia perniciosa (Tower)	*Quadraspidiotus perniciosus* (Comstock)	Synthetic sex pheromone	Attraction in the field	Rice and Jones (1982)
				McClain *et al.* (1990)
				Hippe and Mani (1995)
Braconidae (Aphidiinae)				
Diaeretiella rapae (McIntosh)	*Brevicoryne brassicae* (L.)	Synthetic sex pheromone	Attraction in the field	Hardie *et al.* (1991, 1994)
Praon abjectum (Haliday)	Various aphid species			Powell *et al.* (1993)
Praon dorsale (Haliday)	Various aphid species			Gabrys *et al.* (1997)
Praon volucre (Haliday)	Various aphid species			Glinwood (1998)
P. volucre	*Sitobion avenae* (Fabricius)	Synthetic sex pheromone	Attraction in olfactometer	Lilley *et al.* (1994)
			Attraction in wind tunnel	Glinwood (1998)
			Increased rates of parasitization	
Aphidius eadyi Stary, Gonzalez & Hall	*Acyrthosiphon pisum* (Harris)	Synthetic sex pheromone	Attraction in wind tunnel	Glinwood (1998)
Aphidius ervi Haliday	*A. pisum*			
Aphidius rhopalosiphi De Stefani Perez	*S. avenae*			
Ephedrus plagiator (Nees)	Various aphid species			
Praon myzophagum Mackauer	Various aphid species			
A. eadyi	*A. pisum*	Synthetic sex pheromone	Increased rates of parasitization	Glinwood *et al.* (1998)
A. ervi	*A. pisum*			
A. rhopalosiphi	*S. avenae*			

pheromone, unlike *Praon* species (Hardie *et al.*, 1991, 1994; Powell *et al.*, 1993), and recent field studies indicated that the former appear to be arrested in the vicinity of the pheromone which stimulates intensive searching behaviour before they reach the point source of the kairomone (Glinwood, 1998). *Diaeretiella rapae* (McIntosh), a brassica aphid specialist, was also caught in water traps releasing aphid sex pheromone components, placed in brassica crops in England and Poland (Gabrys *et al.*, 1997; Glinwood, 1998). When *D. rapae* was very abundant in the crops, pheromone traps also caught more males than did control traps, but the males were probably being attracted by the sex pheromones of *D. rapae* females as these aggregated around the pheromone traps.

Strategies for using synthetic aphid sex pheromones to enhance the impact of parasitoids on pest aphids in agricultural field crops are currently being investigated (Powell *et al.*, 1998). In annual crops such as cereals, early-season synchrony between colonizing aphids and parasitoids can play a key role in limiting aphid population growth. Aphid sex pheromones are being used in an attempt to induce this synchrony by aggregating parasitoids into field margin overwintering sites in the autumn, when parasitoids are dispersing from harvested fields, and by encouraging them back into nearby crops the following spring (Powell *et al.*, 1998).

Responses of parasitoids to scale insect and aphid sex pheromones are summarized in Table 16.2.

Aggregation Pheromones as Parasitoid Kairomones

Some insect pheromones, which may be produced by members of either or both sexes, attract both sexes of the same species to a particular area, and these have been termed aggregation pheromones (Shorey, 1973; Borden, 1985). They generally function as long-range communication signals, and so are potentially useful as host location cues for parasitoids. Aggregation pheromones are often released by pre-reproductive adults, of either sex, that have located suitable resources for feeding and/or reproduction, although some insects aggregate at sites that offer protection from adverse climatic conditions (Borden, 1985). When the consequent aggregation of both sexes at the resource site leads to mating and oviposition, the aggregation pheromone involved may also function as a type of sex pheromone (Tumlinson *et al.*, 1992). In some groups, such as the scolytid bark beetles, aggregation behaviour is initiated by the combined release of insect pheromones and plant volatiles induced by insect damage (Chapter 6). Evidence for the use of aggregation pheromones as host-location kairomones exists for parasitoids of bark beetles, heteropteran bugs and, more recently, fruit flies.

Parasitoids of bark beetles

Whilst studying the biology of *Tomicobia tibialis* Ashmead, a pteromalid parasitoid of adult scolytid bark beetles, Bedard (1965) noted that both sexes were attracted to pine logs freshly attacked by *Ips paraconfusus* (LeConte), and he suggested that the attractant was a pheromone released by male beetles (see also Chapter 6, this volume). Subsequent studies, using traps with small logs containing either adult male or female beetles of one of several species, showed that *T. tibialis* was attracted exclusively to male beetles of the genus *Ips*, again suggesting that the kairomone involved was the aggregation pheromone released by unmated males (Rice, 1968, 1969). The parasitoids mate on the logs and then oviposit in adult beetles (Bedard, 1965), and Rice (1968) pointed out that suitable hosts are available to the parasitoids only during the period of pheromone production, making it an effective host-location kairomone. The aggregation pheromone produced by male *Ips pini* (Say) varies geographically within North America and *T. tibialis* was able to distinguish between logs infested with males from different regions (Lanier *et al.*, 1972). However, traps releasing ipsdienol (2-methyl-6-methylene-2,7-octadien-4-ol), a synthetic aggregation pheromone of *I. pini*, caught few parasitoids, even though many were attracted to logs infested by male beetles in the same trial (Seybold *et al.*, 1992).

In contrast, traps baited with multilure, the synthetic aggregation pheromone of *Scolytus multistriatus* (Marsham), attracted a variety of parasitoid species in field trials, including *Cheiropachus colon* (L.), *Cerocephala eccoptogastri* Masi, *Entodon leucogramma* (Ratzeburg), *Spathius benefactor* Matthews and *Dendrosoter protruberans* (Nees), which include egg parasitoids, larval parasitoids and secondary parasitoids (Kennedy, 1979, 1984). Several combinations of the individual components of multilure were also tested; those containing multistriatin were particularly attractive to *C. colon* (Kennedy, 1984). The egg parasitoid *E. leucogramma* appeared to be attracted to a combination of an elm-produced volatile and the beetle pheromone; the tree-produced chemical attracted it to the tree and the pheromone led it to the oviposition site (Kennedy, 1984). Examples of parasitoid responses to bark beetle aggregation pheromones are given in Table 16.3.

Parasitoids of heteropteran bugs

The first evidence for kairomonal activity of heteropteran bug aggregation pheromones involved the southern green stink bug, *Nezara viridula* (L.), and its tachinid parasitoid *Trichopoda pennipes* (Fabricius). In field tests, large numbers of female *T. pennipes* were attracted to cages containing male *N. viridula* but not to cages containing females (Mitchell and Mau, 1971; Harris and Todd, 1980). Because male, female and fifth instar *N. viridula* were also attracted to caged males, it was concluded that males release an aggregation pheromone which also acts as a kairomone for the parasitoid (Harris and Todd, 1980). Differential parasitization of geographically isolated populations of *N. viridula* by *T. pennipes* suggested the existence of pheromone strains of the stink bug, and this was later chemically confirmed (Aldrich *et al.*, 1987, 1989; Todd, 1989; Aldrich, 1995b; see Chapter 14). This parasitoid is native to North America but it readily transferred from its native hosts, *Acrosternum* spp., to the introduced *N. viridula* (Aldrich *et al.*, 1989). The typical aggregation pheromone blend ((*Z*)-α-bisabolene, *E*- and *Z*-1,2-epoxides of (*Z*)-α-bisabolene, (*E*)-nerolidol and *n*-nonadecane) of *N. viridula* is chemically very similar to that of *Acrosternum* spp., comprising distinctive ratios of the same sesquiterpenes, and it has been suggested that this stimulated the rapid adoption of the introduced bug as a host (Aldrich *et al.*, 1989). The same authors also

Table 16.3. Responses of parasitoids to bark beetle aggregation pheromones.

Parasitoid	Host	Stimulus	Response	Reference
Pteromalidae				
Tomicobia tibialis Ashmead	*Ips* spp.	Pine logs containing male beetles	Attraction in the field	Bedard (1965) Rice (1968, 1969) Lanier *et al.* (1972) Seybold *et al.* (1992)
T. tibialis	*Ips pini*	Synthetic pheromone (ipsdienol)	No response	Seybold *et al.* (1992)
Cerocephala eccoptogastri Masi *Cheiropachus colon* (L.)	*Scolytus multistriatus* (Marsham)	Synthetic pheromone (multilure)	Attraction in the field	Kennedy (1979, 1984)
Eulophidae				
Entodon leucogramma (Ratzeburg)	*S. multistriatus*	Synthetic pheromone (multilure)	Attraction in the field	Kennedy (1979, 1984)
Braconidae				
Dendrosoter protruberans (Nees) *Spathius benefactor* Matthews	*S. multistriatus*	Synthetic pheromone (multilure)	Attraction in the field	Kennedy (1979, 1984)

speculated that parasitoid pressure has led to a change in the relative amounts of *E*- and *Z*-isomers of one of the *N. viridula* pheromone components (1,2-epoxy-(*Z*)-α-bisabolene); *T. pennipes* was preferentially attracted to the *Z*-isomer (Aldrich *et al.*, 1989). Selection by females for large males of *N. viridula*, promoted by their greater reproductive success, appears to be counteracted by their increased attractiveness to parasitoids, presumably because they release more kairomone (McLain and Marsh, 1990).

Other parasitoids of heteropteran bugs that appear to use male-produced aggregation pheromones of their hosts as kairomones include the tachinids *Euclytia flava* (Townsend) and *Hemyda aurata* Robineau-Desvoidy on the spined soldier bugs, *Podisus maculiventris* (Say) and *Podisus neglectus* (Westwood) (Aldrich *et al.*, 1984; Aldrich, 1995a,b), the tachinids *E. flava*, *Gymnosoma* spp. and *Euthera* spp. on *Euschistus* spp. stink bugs (Aldrich *et al.*, 1991), the scelionid egg parasitoid *Telenomus calvus* Johnson on *P. maculiventris* and *P. neglectus* (Aldrich *et al.*, 1984; Aldrich, 1995b), the scelionid *Gryon pennsylvanicum* (Ashmead) on the coreid plant bug, *Leptoglossus australis* (Fabricius) (Yasuda and Tsurumachi, 1995) and the encyrtid egg parasitoid *Ooencyrtus nezarae* Ishii on the alydid plant bug, *Riptortus clavatus* (Thunberg) (Leal *et al.*, 1995). A review of parasitoid exploitation of heteropteran bugs is given by Aldrich (1995b) and some examples of parasitoid responses to bug aggregation pheromones are listed in Table 16.4.

Since mating and oviposition often occur in the area where the aggregation pheromones are emitted, they are obviously good host-location kairomones for egg parasitoids. The presence of caged males near *L. australis* egg masses placed in the field significantly increased parasitization rates by *G. pennsylvanicum* (Yasuda and Tsurumachi, 1995). Leal *et al.* (1995) speculated that the habit of female *R. clavatus* of scattering some of their eggs on non-food plants is a mechanism for avoiding parasitism by *O. nezarae*. The aggregation pheromone then appears to act as a food-plant location pheromone for larvae hatching from these eggs. *Telenomus calvus* is phoretic on female *P. maculiventris*, a habit that allows it access to newly oviposited eggs. This is vital, since the parasitoid can only successfully develop in eggs no older than 12 h (Orr *et al.*, 1986), and location of virgin or newly mated female hosts is facilitated by kairomonal attraction to the male-produced aggregation pheromone (see also Chapter 11, this volume).

Parasitoids of fruit flies

Male *Drosophila* spp. fruit flies produce an aggregation pheromone that is passed to the female during mating and then transferred to the substrate during oviposition (Bartelt *et al.*, 1985; Schaner *et al.*, 1987; Wiskerke *et al.*, 1993; Hedlund *et al.*, 1996). The same major chemical component ((*Z*)-11-octadecenyl acetate) has been identified from the pheromones of *Drosophila melanogaster* (Meigen) and *Drosophila simulans* (Sturtevant) and is probably common to several other species (Wiskerke *et al.*, 1993; Hedlund *et al.*, 1996). Wind tunnel studies revealed that the fly aggregation pheromone acts as a host-location kairomone for naïve females of the eucoilid parasitoids *Leptopilina heterotoma* (Thomson) and *Leptopilina boulardi* (Barbotin *et al.*), but the braconid parasitoid *Asobara tabida* (Nees) was only

Table 16.4. Responses of parasitoids to heteropteran bug aggregation pheromones.

Parasitoid	Host	Stimulus	Response	Reference
Tachinidae				
Trichopoda pennipes (F.)	*Nezara viridula* (L.)	Unmated male hosts	Attraction in the field	Mitchell and Mau (1971)
				Harris and Todd (1980)
T. pennipes	*N. viridula*	Pheromone extract	Attraction in the field	Aldrich *et al.* (1987)
Euclytia flava (Townsend)	*Podisus maculiventris* (Say)	Synthetic pheromone	Attraction in the field	Aldrich *et al.* (1984)
Hemyda aurata Robineau-Desvoidy	*Podisus neglectus* (Westwood)			Aldrich (1995a,b)
E. flava	*Euschistus* spp.	Synthetic pheromone	Attraction in the field	Aldrich *et al.* (1991)
Euthera spp.				
Gymnosoma spp.				
Scelionidae				
Telenomus calvus Johnson	*P. maculiventris*	Synthetic pheromone	Attraction in the field	Aldrich *et al.* (1984), Aldrich (1995b)
	P. neglectus			
Gryon pennsylvanicum (Ashmead)	*Leptoglossus australis* (F.)	Male hosts	Attraction in the field	Yasuda and Tsurumaci (1995)
			Increased rate of parasitization	
Encyrtidae				
Ooencyrtus nezarae Ishii	*Riptortus clavatus* (Thunberg)	Synthetic pheromone	Attraction in the field	Leal *et al.* (1995)

attracted after females had been given an oviposition experience in the presence of the pheromone (Wiskerke *et al.*, 1993; Hedlund *et al.*, 1996). There was a positive correlation between the amount of pheromone present on a patch of substrate and the level of response by *L. heterotoma*, suggesting that the parasitoid may be able to assess patch profitability by means of the kairomone (Wiskerke *et al.*, 1993). It has also been proposed that the kairomone allows parasitoids to locate hosts in novel or less preferred substrates (Wiskerke *et al.*, 1993).

Epideictic Pheromones as Parasitoid Kairomones

Some insects employ pheromones (e.g. oviposition-deterring pheromones or host-marking pheromones: Chapters 1 and 8) that influence their spatial distribution by inducing the dispersal of conspecifics, thereby optimizing the exploitation of resources. These epideictic pheromones may also function as host-location kairomones for parasitoids (Prokopy, 1981). Those associated with oviposition, in particular, have potential value for foraging egg parasitoids or those that can attack newly hatched hosts.

The oviposition-deterring pheromone deposited on fruit by the tephritid fly, *Rhagoletis pomonella* (Walsh), elicited intensive searching behaviour by the braconid parasitoid *Opius lectus* Gahan, which attacks the eggs and young larvae of the fly (Prokopy and Webster, 1978). The parasitoid only responded to the pheromone in the presence of fruit but a closely related species, *Opius alloeus* Muesebeck, showed no response at all. *Opius alloeus* attacks third instar larvae of *R. pomonella* and so forages on the fruit after the oviposition-deterring pheromone has dissipated (Prokopy and Webster, 1978). The butterfly *Pieris brassicae* deposits an oviposition-deterring pheromone on its eggs, and this may act as a kairomone for the egg parasitoid *Trichogramma evanescens*, which was attracted to extracts made by washing host eggs in water or methanol (Noldus and van Lenteren, 1985b). The scales deposited by many Lepidoptera during oviposition have a kairomonal effect on egg parasitoids, particularly *Trichogramma* spp., and there is evidence that volatiles from these scales function as epideictic pheromones (Prokopy, 1981).

An epideictic pheromonal function has also been attributed to secretions from the mandibular glands of the pyralid moths, *Ephestia kuehniella* and *Plodia interpunctella* (Hübner) (Corbet, 1971; Phillips and Strand, 1994). Components from the mandibular secretions of *E. kuehniella* had a kairomonal effect on the ichneumonid parasitoid *Venturia* (= *Nemeritis*) *canescens* (Grav.), eliciting oviposition behaviour. The chemicals showing this kairomonal activity (4-hydroxy-2-oleoylcyclohexane-1,3-dione and 2-oleoylcyclohexane-1,3-dione) were also identified from the mandibular glands of several other pyralids infesting stored products, including *P. interpunctella* (Corbet, 1971; Mudd and Corbet, 1973, 1982).

There are other insect secretions that have kairomonal effects on parasitoids and for which an epideictic pheromonal role has been proposed but not proved. These include a chemical trail left on the substrate by wandering fourth instar larvae of the cucujid beetle pest of stored products, *Cryptolestes ferrugineus* (Stephens), which acts as a host-location kairomone for its bethylid parasitoid *Cephalonomia waterstoni* Gahan (Howard and Flinn, 1990), and volatile secretions from the dorsal abdominal glands of several pentatomid stink bugs which act as kairomones for tachinid and scelionid parasitoids (Aldrich *et al.*, 1995).

Alarm Pheromones as Parasitoid Kairomones

Insects that aggregate at some stage in their life often release pheromones that initiate defensive behaviour in other members of the same species within the aggregation. These have been termed alarm pheromones and have been detected most commonly in eusocial insects, such as bees, wasps, ants and termites, and in phytophagous insects that feed in dense colonies, such as some aphids and heteropteran bugs (Blum, 1985). The defensive behaviour triggered by alarm pheromones may take the form of aggression, typical of eusocial insects, or evasion. To be effective, alarm pheromones need to induce rapid responses by individuals close to the emitter and so tend to be short-range signals. They are, therefore, potentially less useful than the long-range sex and aggregation pheromones as host-location kairomones for parasitoids. Indeed, they may inhibit parasitoid efficiency by improving the host's ability to escape.

Aphids release a pheromone in the cornicle secretion which they exude in response to disturbance or attack by predators or parasitoids (Nault and Montgomery, 1979; Pickett and Griffiths, 1980). This pheromone causes other aphids feeding nearby, usually on the same leaf or shoot, to move away or even drop from the plant, although the strength of this response is much greater in some aphid species than in others (see also Chapter 9, this volume). By inducing this defensive behaviour in the host, alarm pheromones disrupt parasitoid foraging behaviour, but there is also a little evidence that parasitoids themselves can detect and respond to alarm pheromones. A contact kairomone in aphid cornicle secretions stimulated oviposition attack responses by the braconid parasitoids *Lysiphlebus testaceipes* (Cresson) and *Aphidius ervi* Haliday (Grasswitz and Paine, 1992; Battaglia *et al.*, 1993). This response was not elicited by the main component of the alarm pheromone, (*E*)-β-farnesene, when this compound was presented by itself, but the kairomonal activity of cornicle secretion may depend upon an interaction between the alarm pheromone and other chemical components (Grasswitz and Paine, 1992). Nevertheless, the cereal aphid parasitoid, *Aphidius uzbekistanicus* Luzhetski, appeared to be attracted to (*E*)-β-farnesene in an olfactometer (Micha, 1995).

The only other evidence for kairomonal activity by an alarm pheromone involves eusocial ants. Both males and females of the phorid parasitoid *Apocephalus paraponerae* Borgmeier were attracted to two chemical products (4-methyl-3-heptanone and 4-methyl-3-heptanol) from the mandibular glands of workers of the tropical ponerine ant *Paraponera clavata* (Fabricius) (Feener *et al.*, 1996). Mandibular gland secretions function as alarm pheromones in many ants, including species closely related to *P. clavata* (Hölldobler and Wilson, 1990), but have a reduced function in *P. clavata* itself (Hermann *et al.*, 1984). Feener *et al.* (1996) proposed that the reduced response of *P. clavata* workers to these mandibular gland products has evolved to minimize levels of parasitism by *A. paraponerae*. The parasitoid rapidly locates dying and freshly killed hosts, where mating then takes place, and it has been suggested that the alarm pheromone components released from the mandibular glands are an excellent indicator to male parasitoids of the eventual presence of receptive females (Brown and Feener, 1991; Feener *et al.*, 1996). Although *P. clavata* is not an agricultural pest or beneficial, there are other ant species that are, and further investigation of the effects of their defensive pheromone systems on predators and parasitoids could be worthwhile.

Conclusions

During the evolutionary development of insect host/parasitoid relationships, some parasitoids have acquired the ability to detect the pheromones used by their hosts for intraspecific communication and exploit them as host location cues. Kairomonal activity has been detected most often for sex and aggregation pheromones, many of which operate as long-range communication signals, although there is also some evidence of parasitoid responses to host epideictic and alarm pheromones. The host development stage that is attacked (egg, larva, pupa or adult) will influence the relative importance of host pheromones as foraging cues, compared with those from other sources such as the substrate associated with the host (Vet *et al.*, 1995). Because there are often temporal and spatial associations between sex pheromone production and oviposition, egg parasitoids are more likely to benefit by responding to host sex pheromones than are larval or pupal parasitoids, although the latter may also benefit when all stages of the host occupy the same restricted habitat (e.g. some stored-products pests). Other chemical cues, such as plant volatiles induced by herbivore damage, which may be important to parasitoids of actively feeding hosts, are usually of little value to egg parasitoids, which therefore have to rely more on host-derived kairomones (Vet *et al.*, 1995). Interestingly, there is evidence that aphid parasitoids respond to both aphid-induced plant volatiles and to aphid sex pheromones (Powell *et al.*, 1998), although little is known about potential interactions between the two cues if both are present at the same time.

Understanding the kairomonal function of insect pheromones, as attractants, arrestants or searching stimulants for predators and parasitoids, offers valuable opportunities for developing their use in biological control and integrated pest management. However, it is essential to link this behavioural knowledge with ecological knowledge so that the use of kairomones, or any other behaviour-modifying chemicals, can be targeted at the specific elements of a particular host/parasitoid relationship that are perceived as constraints on effective pest control.

References

Aldrich, J.R. (1995a) Testing the 'new associations' biological control concept with a tachinid parasitoid (*Euclytia flava*). *Journal of Chemical Ecology* 21, 1032–1042.

Aldrich, J.R. (1995b) Chemical communication in the true bugs and parasitoid exploitation. In: Cardé, R.T. and Bell, W.J. (eds) *Chemical Ecology of Insects 2*. Chapman & Hall, New York, pp. 318–363.

Aldrich, J.R., Kochansky, J.P. and Abrams, C.B. (1984) Attractant for a beneficial insect and its parasitoids: pheromone of the predatory spined soldier bug, *Podisus maculiventris* (Hemiptera: Pentatomidae). *Environmental Entomology* 13, 1031–1036.

Aldrich, J.R., Oliver, J.E., Lusby, W.R., Kochansky, J.P. and Lockwood, J.A. (1987) Pheromone strains of the cosmopolitan pest, *Nezara viridula* (Heteroptera: Pentatomidae). *Journal of Experimental Zoology* 244, 171–175.

Aldrich, J.R., Lusby, W.R., Marron, B.E., Nicolaou, K.C., Hoffmann, M.P. and Wilson, L.T. (1989) Pheromone blends of green stink bugs and possible parasitoid selection. *Naturwissenschaften* 76, 173–175.

Aldrich, J.R., Hoffmann, M.P., Kochansky, J.P., Lusby, W.R., Eger, J.E. and Payne, J.A. (1991) Identification and attractiveness of a major pheromone component for nearctic *Euschistus* spp. stink bugs (Heteroptera: Pentatomidae). *Environmental Entomology* 20, 477–483.

Aldrich, J.R., Rosi, M.C. and Bin, F. (1995) Behavioral correlates for minor volatile compounds from stink bugs (Heteroptera: Pentatomidae). *Journal of Chemical Ecology* 21, 1907–1920.

Arakaki, N., Wakamura, S. and Yasuda, T. (1996) Phoretic egg parasitoid, *Telenomus euproctidis* (Hymenoptera: Scelionidae), uses sex pheromone of Tussock moth *Euproctis taiwana* (Lepidoptera: Lymantriidae) as a kairomone. *Journal of Chemical Ecology* 22, 1079–1085.

Arakaki, N., Wakamura, S., Yasuda, T. and Yamagishi, K. (1997) Two regional strains of a phoretic egg parasitoid, *Telenomus euproctidis* (Hymenoptera: Scelionidae), that use different sex pheromones of two allopatric Tussock moth species as kairomones. *Journal of Chemical Ecology* 23, 153–161.

Bartelt, R.J., Schaner, A.M. and Jackson, L.L. (1985) *cis*-Vaccenyl acetate as an aggregation pheromone in *Drosophila melanogaster*. *Journal of Chemical Ecology* 11, 1747–1756.

Battaglia D., Pennacchio, F., Marincola, G. and Tranfaglia, A. (1993) Cornicle secretion of *Acyrthosiphon pisum* (Homoptera: Aphididae) as a contact kairomone for the parasitoid *Aphidius ervi* (Hymenoptera: Braconidae). *European Journal of Entomology* 90, 423–428.

Battisti, A. (1989) Field studies on the behaviour of two egg parasitoids of the pine processionary moth *Thaumetopoea pityocampa*. *Entomophaga* 34, 29–38.

Bedard, W.D. (1965) The biology of *Tomicobia tibialis* (Hymenoptera: Pteromalidae) parasitizing *Ips confusus* (Coleoptera: Scolytidae) in California. *Contributions from Boyce Thompson Institute* 23, 77–81.

Blum, M.S. (1985) Alarm pheromones. In: Kerkut, G.A. and Gilbert, L.I. (eds) *Comprehensive Insect Physiology, Biochemistry and Pharmacology*, Vol. 9, *Behaviour*. Pergamon Press, Oxford, pp. 193–224.

Borden, J.H. (1985) Aggregation pheromones. In: Kerkut, G.A. and Gilbert, L.I. (eds) *Comprehensive Insect Physiology, Biochemistry and Pharmacology*, Vol. 9, *Behaviour*. Pergamon Press, Oxford, pp. 257–285.

Brown, B.V. and Feener, D.H. (1991) Behavior and host location cues of *Apocephalus paraponerae* (Diptera: Phoridae), a parasitoid of the giant tropical ant, *Paraponera clavata*. *Biotropica* 23, 182–187.

Brown, M.W. and Cameron, E.A. (1979) Effects of disparlure and egg mass size on parasitism by the gypsy moth egg parasite, *Ooencyrtus kuwanai*. *Environmental Entomology* 8, 77–80.

Cameron, E.A. and Rhoads, L.D. (1973) Oviposition by *Brachymeria intermedia* in the presence of disparlure. *Environmental Entomology* 2, 485–486.

Corbet, S.A. (1971) Mandibular gland secretion of larvae of the flour moth, *Anagasta kuehniella* contains an epideictic pheromone and elicits oviposition movements in a hymenopteran parasite. *Nature* 232, 481–484.

Cortesero, A.M., Monge, J.P. and Huignard, J. (1993) Response of the parasitoid *Eupelmus vuilleti* to the odours of the phytophagous host and its host plant in an olfactometer. *Entomologia Experimentalis et Applicata* 69, 109–116.

Dawson, G.W., Griffiths, D.C., Janes, N.F., Mudd, A., Pickett, J.A., Wadhams, L.J. and Woodcock, C.M. (1987) *Nature* 325, 614–616.

Dawson, G.W., Griffiths, D.C., Merritt, L.A., Mudd, A., Pickett, J.A., Wadhams, L.J. and Woodcock, C.M. (1990) Aphid semiochemicals – a review, and recent advances on the sex pheromone. *Journal of Chemical Ecology* 16, 3019–3030.

Demolin, G. (1969) Comportement des adultes de *Thaumetopoea pityocampa* Schiff. Dispersion spatiale, importance écologique. *Annales des Sciences Forestières* 26, 81–102.

Feener, D.H., Jacobs, L.F. and Schmidt, J.O. (1996) Specialized parasitoid attracted to a pheromone of ants. *Animal Behavior* 51, 61–66.

Frenoy, C., Farine, J.P., Hawlitzky, N. and Durier, C. (1991) Role of kairomones in the relations between *Ostrinia nubilalis* Hübner (Lep., Pyralidae) and *Trichogramma brassicae* Bezdenko (Hym., Trichogrammatidae). *Redia* 74 (Appendice), 143–151.

Frenoy, C., Durier, C. and Hawlitzky, N. (1992) Effect of kairomones from egg and female adult stages of *Ostrinia nubilalis* (Hübner) (Lepidoptera, Pyralidae) on *Trichogramma brassicae* Bezdenko (Hymenoptera, Trichogrammatidae) female kinesis. *Journal of Chemical Ecology* 18, 761–773.

Gabrys, B.J., Gadomski, H.J., Klukowski, Z., Pickett, J.A., Sobota, G.T., Wadhams, L.J. and Woodcock, C.M. (1997) Sex pheromone of cabbage aphid *Brevicoryne brassicae*: identification and field trapping of male aphids and parasitoids. *Journal of Chemical Ecology* 23, 1881–1890.

Glinwood, R.T. (1998) Laboratory and field responses of parasitoids to aphid sex pheromones. PhD thesis, University of Nottingham, UK.

Glinwood, R.T., Powell, W. and Tripathi, C.P.M. (1998) Increased parasitisation of aphids on trap plants alongside vials releasing synthetic aphid sex pheromone, and effective range of the pheromone. *Biocontrol, Science and Technology* 8, 607–614.

Grasswitz, T.R. and Paine, T.D. (1992) Kairomonal effect of an aphid cornicle secretion on *Lysiphlebus testaceipes* (Cresson) (Hymenoptera: Aphidiidae). *Journal of Insect Behavior* 5, 447–457.

Hardie, J., Nottingham, S.F., Powell, W. and Wadhams, L.J. (1991) Synthetic aphid sex pheromone lures female parasitoids. *Entomologia Experimentalis et Applicata* 61, 97–99.

Hardie, J., Hick, A.J., Höller, C., Mann, J., Merritt, L., Nottingham, S.F., Powell, W., Wadhams, L.J., Witthinrich, J. and Wright, A.F. (1994) The responses of *Praon* spp. parasitoids to aphid sex pheromone components in the field. *Entomologia Experimentalis et Applicata* 71, 95–99.

Harris, V.E. and Todd, J.W. (1980) Male-mediated aggregation of male, female and 5th-instar southern green stink bugs and concomitant attraction of a tachinid parasite, *Trichopoda pennipes*. *Entomologia Experimentalis et Applicata* 27, 117–126.

Hedlund, K., Vet, L.E.M. and Dicke, M. (1996) Generalist and specialist parasitoid strategies of using odours of adult drosophilid flies when searching for larval hosts. *Oikos* 77, 390–398.

Hermann, H.R., Blum, M.S., Wheeler, J.W., Overal, W.L., Schmidt, J.O. and Chao, J.-T. (1984) Comparative anatomy and chemistry of the venom apparatus and mandibular glands in *Dinoponera grandis* (Guérin) and *Paraponera clavata* (F.) (Hymenoptera: Formicidae: Ponerinae). *Annals of the Entomological Society of America* 77, 272–279.

Hippe, C. and Mani, E. (1995) Flight monitoring of the San José scale, *Quadraspidiotus perniciosus* (Comstock) (Diaspididae) and of its parasitoid *Encarsia perniciosi* (Tower) (Aphelinidae) in North-west Switzerland. *Israel Journal of Entomology* 29, 184.

Hölldobler, B. and Wilson, E.O. (1990) *The Ants*. Harvard University Press, Cambridge, Massachusetts, 732 pp.

Howard, R.W. and Flinn, P.W. (1990) Larval trails of *Cryptolestes ferrugineus* (Coleoptera: Cucujidae) as kairomonal host-finding cues for the parasitoid *Cephalonomia waterstoni* (Hymenoptera: Bethylidae). *Annals of the Entomological Society of America* 83, 239–245.

Kaiser, L., Pham-Delègue, M.H., Bakchine, E. and Masson, C. (1989) Olfactory responses of *Trichogramma maidis* Pint. et Voeg.: effects of chemical cues and behavioral plasticity. *Journal of Insect Behavior* 2, 701–712.

Kennedy, B.H. (1979) The effect of multilure on parasites of the European elm bark beetle, *Scolytus multistriatus*. *Bulletin of the Entomological Society of America* 25, 116–118.

Kennedy, B.H. (1984) Effect of multilure and its components on parasites of *Scolytus multistriatus* (Coleoptera: Scolytidae). *Journal of Chemical Ecology* 10, 373–385.

Klemm, U. and Schmutterer, H. (1993) Wirkungen von Niempräparaten auf die Kohlmotte *Plutella xylostella* L. und ihre natürlichen Feinde aus der Gattung *Trichogramma*. *Zeitschrift für Pflanzenkrankheiten und Pflanzenschutz* 100, 113–128.

Laing, J. (1937) Host-finding by insect parasites. 1. Observations on the finding of hosts by *Alysia manducator*, *Mormoniella vitripennis* and *Trichogramma evanescens*. *Journal of Animal Ecology* 6, 298–317.

Lanier, G.N., Birch, M.C., Schmitz, R.F. and Furniss, M.M. (1972) Pheromones of *Ips pini* (Coleoptera: Scolytidae): variation in response among three populations. *Canadian Entomologist* 104, 1917–1923.

Leal, W.S., Higuchi, H., Mizutani, N., Nakamori, H., Kadosawa, T. and Ono, M. (1995) Multifunctional communication in *Riptortus clavatus* (Heteroptera: Alydidae): conspecific nymphs and egg parasitoid *Ooencyrtus nezarae* use the same adult attractant pheromone as chemical cue. *Journal of Chemical Ecology* 21, 973–985.

Lewis, W.J., Jones, R.L. and Sparks, A.N. (1972) A host-seeking stimulant for the egg parasite, *Trichogramma evanescens*: its source and a demonstration of its laboratory and field activity. *Annals of the Entomological Society of America* 65, 1087–1089.

Lewis, W.J., Nordlund, D.A., Gueldner, R.C., Teal, P.E.A. and Tumlinson, J.H. (1982) Kairomones and their use for management of entomophagous insects. XIII. Kairomonal activity for *Trichogramma* spp. of abdominal tips, excretion, and a synthetic sex pheromone blend of *Heliothis zea* (Boddie) moths. *Journal of Chemical Ecology* 8, 1323–1331.

Lilley, R., Hardie, J. and Wadhams, L.J. (1994) Field manipulation of *Praon* populations using semiochemicals. *Norwegian Journal of Agricultural Sciences* Suppl. 16, 221–226.

McClain, D.C., Rock, G.C. and Woolley, J.B. (1990) Influence of trap colour and San José scale (Homoptera: Diaspididae) pheromone on sticky trap catches of 10 aphelinid parasitoids (Hymenoptera). *Environmental Entomology* 19, 926–931.

McLain, D.K. and Marsh, N.B. (1990) Male copulatory success: heritability and relationship to mate fecundity in the southern green stinkbug, *Nezara viridula* (Hemiptera: Pentatomidae). *Heredity* 64, 161–167.

Micha, S.G. (1995) Die Bedeutung von Semiochemikalien im tritrophischen System aus Hafer (*Avena sativa*), Große Getreideblattlaus (*Sitobion avenae*) und Blattlausparasitoid (*Aphidius uzbekistanicus*). PhD thesis, Christian-Albrechts-Universität, Kiel, Germany.

Mitchell, E.R., Waddill, V.H. and Ashley, T.R. (1984) Population dynamics of the fall armyworm (Lepidoptera: Noctuidae) and its larval parasites on whorl stage corn in pheromone-permeated field environments. *Environmental Entomology* 13, 1618–1623.

Mitchell, W.C. and Mau, R.F.L. (1971) Response of the female southern green stink bug and its parasite *Trichopoda pennipes*, to male stink bug pheromones. *Journal of Economic Entomology* 64, 856–859.

Mudd, A. and Corbet, S.A. (1973) Mandibular gland secretion of larvae of stored products pests *Anagasta kuehniella*, *Ephestia cautella*, *Plodia interpunctella* and *Ephestia elutella*. *Entomologia Experimentalis et Applicata* 16, 291–293.

Mudd, A. and Corbet, S.A. (1982) Response of the ichneumonid parasite *Nemeritis canescens* to kairomones from the flour moth, *Ephestia kuehniella*. *Journal of Chemical Ecology* 8, 843–850.

Nault, L.R. and Montgomery, M.E. (1979) Aphid alarm pheromones. *Miscellaneous Publications of the Entomological Society of America* 11, 23–31.

Niwa, C.G. and Daterman, G.E. (1989) Pheromone mating disruption of *Rhyacionia zozana* (Lepidoptera: Tortricidae): influence on the associated parasite complex. *Environmental Entomology* 18, 570–574.

Noldus, L.P.J.J. (1988) Response of the egg parasitoid *Trichogramma pretiosum* to the sex pheromone of its host *Heliothis zea*. *Entomologia Experimentalis et Applicata* 48, 293–300.

Noldus, L.P.J.J. (1989) Semiochemicals, foraging behaviour and quality of entomophagous insects for biological control. *Journal of Applied Entomology* 108, 425–451.

Noldus, L.P.J.J. and Lenteren, J.C. van (1983) Kairomonal effects on searching for eggs of *Pieris brassicae*, *Pieris rapae* and *Mamestra brassicae* of the parasite *Trichogramma evanescens* Westwood. *Mededelingen Faculteit Landbouwwetenschappen Rijksuniversiteit Gent* 48, 183–194.

Noldus, L.P.J.J. and Lenteren, J.C. van (1985a) Kairomones for the egg parasite *Trichogramma evanescens* Westwood. I. Effect of volatile substances released by two of its hosts, *Pieris brassicae* L. and *Mamestra brassicae* L. *Journal of Chemical Ecology* 11, 781–791.

Noldus, L.P.J.J. and Lenteren, J.C. van (1985b) Kairomones for the egg parasite *Trichogramma evanescens* Westwood. II. Effect of contact chemicals produced by two of its hosts, *Pieris brassicae* L. and *Pieris rapae* L. *Journal of Chemical Ecology* 11, 793–800.

Noldus, L.P.J.J., Lewis, W.J. and Tumlinson, J.H. (1990) Beneficial arthropod behavior mediated by airborne semiochemicals. IX. Differential response of *Trichogramma pretiosum*, an egg parasitoid of *Heliothis zea*, to various olfactory cues. *Journal of Chemical Ecology* 16, 3531–3544.

Noldus, L.P.J.J., Lenteren, J.C. van and Lewis, W.J. (1991a) How *Trichogramma* parasitoids use moth sex pheromones as kairomones: orientation behaviour in a wind tunnel. *Physiological Entomology* 16, 313–327.

Noldus, L.P.J.J., Potting, R.P.J. and Barendregt, H.E. (1991b) Moth sex pheromone adsorption to leaf surface: bridge in time for chemical spies. *Physiological Entomology* 16, 329–344.

Nordlund, D.A., Lewis, W.J. and Gueldner, R.C. (1983) Kairomones and their use for management of entomophagous insects. XIV. Response of *Telenomus remus* to abdominal tips of *Spodoptera frugiperda*, (Z)-9-tetradecene-1-ol acetate and (Z)-9-dodecene-1-ol acetate. *Journal of Chemical Ecology* 9, 695–701.

Orr, D.B., Russin, J.S. and Boethel, D.J. (1986) Reproductive biology and behavior of *Telenomus calvus* (Hymenoptera: Scelionidae), a phoretic egg parasitoid of *Podisus maculiventris* (Hemiptera: Pentatomidae). *Canadian Entomologist* 118, 1063–1072.

Phillips, T.W. and Strand, M.R. (1994) Larval secretions and food odors affect orientation in female *Plodia interpunctella*. *Entomologia Experimentalis et Applicata* 71, 185–192.

Pickett, J.A. and Griffiths, D.C. (1980) Composition of aphid alarm pheromones. *Journal of Chemical Ecology* 6, 349–360.

Pickett, J.A., Wadhams, L.J., Woodcock, C.M. and Hardie, J. (1992) The chemical ecology of aphids. *Annual Review of Entomology* 37, 67–90.

Powell, W., Hardie, J., Hick, A.J., Höller, C., Mann, J., Merritt, L., Nottingham, S.F., Wadhams, L.J., Witthinrich, J. and Wright, A.F. (1993) Responses of the parasitoid *Praon volucre* (Hymenoptera: Braconidae) to aphid sex pheromone lures in cereal fields in autumn: implications for parasitoid manipulation. *European Journal of Entomology* 90, 435–438.

Powell, W., Pennacchio, F., Poppy, G.M. and Tremblay, E. (1998) Strategies involved in the location of hosts by the parasitoid *Aphidius ervi* Haliday (Hymenoptera: Braconidae: Aphidiinae). *Biological Control* 11, 104–112.

Prokopy, R.J. (1981) Epideictic pheromones that influence spacing patterns of phytophagous insects. In: Nordlund, D.A., Jones, R.L. and Lewis, W.L. (eds) *Semiochemicals*. John Wiley & Sons, New York, pp.181–214.

Prokopy, R.J. and Webster, R.P. (1978) Oviposition-deterring pheromone of *Rhagoletis pomonella*, a kairomone for its parasitoid *Opius lectus*. *Journal of Chemical Ecology* 4, 481–494.

Renou, M., Nagnan, P., Berthier, A. and Durier, C. (1992) Identification of compounds from the eggs of *Ostrinia nubilalis* and *Mamestra brassicae* having kairomonal activity on *Trichogramma brassicae*. *Entomologia Experimentalis et Applicata* 63, 291–303.

Rice, R.E. (1968) Observations on host selection by *Tomicobia tibialis* Ashmead (Hymenoptera: Pteromalidae). *Contributions from Boyce Thompson Institute* 24, 53–56.

Rice, R.E. (1969) Response of some predators and parasites of *Ips confusus* (Lec.) (Coleoptera: Scolytidae) to olfactory attractants. *Contributions from Boyce Thompson Institute* 24, 189–194.

Rice, R.E. and Jones, R.A. (1982) Collections of *Prospaltella perniciosi* Tower (Hymenoptera: Aphelinidae) on San José scale (Homoptera: Diaspididae) pheromone traps. *Environmental Entomology* 11, 876–880.

Samways, M.J. (1988) Comparative monitoring of red scale *Aonidiella aurantii* (Mask.) (Hom., Diaspididae) and its *Aphytis* spp. (Hym., Aphelinidae) parasitoids. *Journal of Applied Entomology* 105, 483–489.

Schaner, A.M., Bartelt, R.J. and Jackson, L.L. (1987) (Z)-11-Octadecenyl acetate, an aggregation pheromone in *Drosophila simulans*. *Journal of Chemical Ecology* 13, 1777–1786.

Seybold, S.J., Teale, S.A., Wood, D.L., Zhang, A., Webster, F.X., Lindahl, K.Q. and Kubo, I. (1992) The role of lanierone in the chemical ecology of *Ips pini* (Coleoptera: Scolytidae) in California. *Journal of Chemical Ecology* 18, 2305–2329.

Shorey, H.H. (1973) Behavioral responses to insect pheromones. *Annual Review of Entomology* 18, 349–380.

Sower, L.L. and Torgersen, T.R. (1979) Field application of synthetic Douglas-fir tussock moth sex pheromone did not reduce egg parasitism by two Hymenoptera. *Canadian Entomologist* 111, 751–752.

Sternlicht, M. (1973) Parasitic wasps attracted by the sex pheromone of their coccid host. *Entomophaga* 18, 339–342.

Stowe, M.K., Turlings, T.C.J., Loughrin, J.H., Lewis, W.J. and Tumlinson, J.H. (1995) The chemistry of eavesdropping, alarm, and deceit. *Proceedings of the National Academy of Sciences USA* 92, 23–28.

Tingle, F.C. and Mitchell, E.R. (1982) Effect of synthetic pheromone on parasitization of *Heliothis virescens* (F.) (Lepidoptera: Noctuidae) in tobacco. *Environmental Entomology* 11, 913–916.

Tingle, F.C., Mitchell, E.R. and Heath, R.R. (1989) Mating and oviposition by *Cotesia (= Apanteles) marginiventris* (Hymenoptera: Braconidae) in presence of synthetic pheromone of *Spodoptera frugiperda* (Lepidoptera: Noctuidae). *Journal of Chemical Ecology* 15, 2045–2050.

Todd, J.W. (1989) Ecology and behavior of *Nezara viridula*. *Annual Review of Entomology* 34, 273–292.

Tumlinson, J.H., Turlings, T.C.J. and Lewis, W.J. (1992) The semiochemical complexes that mediate insect parasitoid foraging. *Agricultural Zoology Reviews* 5, 221–252.

Vet, L.E.M. and Dicke, M. (1992) Ecology of infochemical use by natural enemies in a tritrophic context. *Annual Review of Entomology* 37, 141–172.

Vet, L.E.M., Lewis, J.W. and Cardé, R.T. (1995) Parasitoid foraging and learning. In: Cardé, R.T. and Bell, W.J. (eds) *Chemical Ecology of Insects 2*. Chapman & Hall, New York, pp. 65–101.

Vinson, S.B. (1984) Parasitoid–host relationships. In: Bell, W.J. and Cardé, R.T. (eds) *Chemical Ecology of Insects*. Chapman & Hall, London, pp. 205–233.

Vinson, S.B. (1985) The behaviour of parasitoids. In: Kerkut, G.A. and Gilbert, L.I. (eds) *Comprehensive Insect Physiology, Biochemistry and Pharmacology*, Vol. 9, *Behaviour*. Pergamon Press, Oxford, pp. 417–470.

Weseloh, R.M. (1981) Host location by parasitoids. In: Nordlund, D.A., Jones, R.L. and Lewis, W.J. (eds) *Semiochemicals, Their Role in Pest Control*. John Wiley & Sons, New York, pp. 79–95.

Wiskerke, J.S.C., Dicke, M. and Vet, L.E.M. (1993) Larval parasitoid uses aggregation pheromone of adult hosts in foraging behaviour: a solution to the reliability–detectability problem. *Oecologia* 93, 145–148.

Yasuda, K. and Tsurumachi, M. (1995) Influence of male adults of the leaf-footed plant bug, *Leptoglossus australis* (Fabricius) (Heteroptera: Coreidae), on host-searching of the egg parasitoid, *Gryon pennsylvanicum* (Ashmead) (Hymenoptera: Scelionidae). *Applied Entomology and Zoology* 30, 139–144.

Zaborski, E., Teal, P.E.A. and Laing, J.E. (1987) Kairomone-mediated host finding by the spruce budworm egg parasite, *Trichogramma minutum*. *Journal of Chemical Ecology* 13, 113–122.

Zaki, F.N. (1985) Reactions of the egg parasitoid *Trichogramma evanescens* Westw. to certain insect sex pheromones. *Zeitschrift für Angewandte Entomologie* 99, 448–453.

Zaki, F.N. (1996) Effect of some kairomones and pheromones on two hymenopterous parasitoids *Apanteles ruficrus* and *Microplitis demolitor* (Hym., Braconidae). *Journal of Applied Entomology* 120, 555–557.

17 Bees

Jeff Pettis,[1] Tanya Pankiw[2] and Erika Plettner[3]

[1] *USDA–ARS Honey Bee Research Laboratory, Building 476, Beltsville Agricultural Research Center East, Beltsville, MD 20705, USA;* [2] *Department of Entomology, University of California, Davis, CA 95616, USA;* [3] *Department of Medicinal Chemistry, University of Utah, 30 S. 2000 E., Skaggs Hall, Salt Lake City, UT 84112–5820, USA*

Introduction

Modern agriculture depends heavily upon several managed bee species for pollination. Large monoculture plantings necessitate that pollinators be present in sufficient number, at specific times during the year, for adequate pollination. Prior to the mechanized era of farming, local native pollinators were present in sufficient numbers to meet the needs of small farms (Torchio, 1991; Batra, 1995). Native bees are represented by a wide array of taxa and any one specific taxon may be present for only a brief period when a particular group of plants is in bloom. Habitat destruction and pesticide use are cited as primary reasons for native bee decline (Buchmann and Nabhan, 1996) but efforts are underway to conserve native pollinators (Matheson *et al.*, 1996). Humans have robbed honey bees and stingless bees of honey for thousands of years (Crane, 1983) but only recently have we managed them specifically for pollination. Today, with increased acreage and the planting of non-indigenous crops, it is vital that managed bees of numerous taxa be available to meet pollination demands. Additionally, with the year-round production of food in greenhouses, pollinators must be available at times when they might normally be dormant. Pheromones have just begun to play a role in bee management and pollination. As the understanding and identification of bee pheromones increases, the further utilization of pheromones in the management of bees will be possible.

Bee pheromones are somewhat unique in that, as bees have evolved into larger social colonies, the pheromones used within their nests have become more complex. These pheromones can be classified into two groups, based on how they act on the recipient (see also Chapter 13, this volume). Releaser pheromones are more common and act directly on the nervous system of the recipient to elicit an immediate change in behaviour; while primer pheromones act much more slowly upon the recipient and cause a physiological change which may then alter the behaviour of the individual (Free, 1987). Sex and alarm pheromones are common examples of releaser pheromones, while queen-produced pheromones that inhibit reproduction by workers or other queens would be examples of primer pheromones. The actions of numerous releaser and primer pheromones have been documented but few have been fully identified. Both releaser and primer pheromones hold promise for enhancing colony productivity and pollination as the understanding and identification of them continue to increase.

The majority of bees currently used in commercial pollination include two solitary genera – the leaf cutter bee (*Megachile* spp. Latreille) and the blue orchard bee (*Osmia* spp. Panzer; Fig. 17.1) – and two social colony-nesting genera: bumblebees (*Bombus* spp. Latreille) and honey bees (*Apis* spp. L.). While these four groups comprise the majority of pollinators used worldwide, several other groups are currently being used (e.g. *Nomia* spp. Latreille: Torchio, 1991) and others will probably be used in the near future as understanding of their biology and management increases (Batra, 1995; Buchmann and Nabhan, 1996; Matheson *et al.*, 1996). Roubik (1989, Table 2.4 therein) provides an excellent overview to the wide array of chemicals produced by various bee species that may play a role in foraging behaviour. Numerous species produce chemical signals for aggregation and mating and these signals could be used to manipulate these species for pollination (Batra, 1978). This chapter will focus on four genera and include our present understanding of their pheromone identification and biological activity. Additionally, areas will be identified where these pheromones have been, and are likely to be, used in enhancing bee management and pollination.

Bee Pheromones: Identification and Biological Activity

*Leafcutter bees (*Megachile *spp.)*

Leafcutter bees (*Megachile* spp.) are widely used for the pollination of lucerne (alfalfa). These solitary bees have been the focus of numerous studies on improving management techniques for overwintering, reducing parasitism and increasing brood survival, and the number of females produced (reviewed by Richards, 1987; Free, 1993). Chemical communication in solitary species usually involves mate recognition, aggregation and flower-marking pheromones but to date little is known of the chemical ecology of leafcutter bees. The important role of leafcutter bees in lucerne pollination worldwide warrants further research to increase their survival and productivity.

*Blue orchard bees (*Osmia *spp.)*

The blue orchard bee (*Osmia* spp.) (Fig. 17.1) is gaining worldwide attention for its efficient pollination of orchard crops and has been used in Japan for many years (Maeta and Kitamura, 1981; Batra, 1982). *Osmia* spp. are being examined as potential pollinators on a variety of crops in addition to pome fruits, including almonds (Torchio, 1981) and greenhouse crops (Kristjanson and Rasmussen, 1991). As in the case of leafcutter bees, there has been little research into the

Fig. 17.1. Solitary nesting bees like this blue orchard bee (*Osmia* sp.) use pheromones primarily to attract mates and for nest marking, while their more social cousins, the bumblebees and the honey bees, communicate using a wide variety of pheromones. The use of bee pheromones in agriculture is still in its infancy but offers great promise in enhancing pollination and colony management as identification and understanding of these compounds continue to grow. (USDA photograph no. 682X598–16.)

nature of chemical communication that the solitary blue orchard bee might employ. Recent advances in artificial nesting material and the understanding of nesting preferences have increased the ability to manage *Osmia*. Increased attention and research should yield additional advances in our ability to utilize this pollinator.

*Bumblebees (*Bombus *spp.)*

A wide variety of compounds have been identified as pheromones of *Bombus* spp. and many have demonstrated biological activity. *Bombus* males are known to mark territories (reviewed by Duffield *et al.*, 1984) and the Dufour's gland has been shown to produce species-specific compounds that are likely to be active in nestmate recognition and brood cell lining (Oldham *et al.*, 1994). Additionally, Foster (1992) proposed that nestmate recognition is an avoidance mechanism against inbreeding in bumblebees. Foraging bumblebees may use flower-marking pheromones and one species has been shown to mark a foraging trail on the ground (Cameron and Whitfield, 1996). Our use of pheromones to manipulate bumblebees would be likely to concern chemical cues to speed colony build-up or prolong colony life span for continued use in greenhouses. Currently methods have been developed to rear colonies at any time of the year (Roseler, 1985; Eijnde *et al.*,

1991). Advances may be made in the near future in mass rearing and timing of colony build-up along with prolonging colony life span by preventing the production of reproductives (van Heemert *et al.*, 1990). The role of pheromones in bumblebee commercialization will increase only if our understanding of the biological activity of these substances is further studied.

Honey bees (Apis spp.)

The honey bee is the most widely studied of all bees, due in part to our ability to rear these bees in large numbers and to study honey bees under laboratory conditions. Additionally, humans have a long association with honey bees as a source of honey and beeswax (Crane, 1983). The pheromone studies of *Apis* spp. have involved the identification and biological activity of a variety of compounds. These pheromones include releaser pheromones used for alarm and orientation, along with primer pheromones that suppress colony reproduction. The majority of research has been focused on the European honey bee (*Apis mellifera* L.) and the Asian honey bee (*Apis cerana* Fabricius): both species build multiple combs and are cavity-nesting, as is *Apis koshevnikovi* Buttel-Reepen. The other *Apis* species, including *Apis florea* Fabricius and *Apis andreniformis* Smith and the species complex of *Apis dorsata* Fabricius, all build a single comb (Engel and Schultz, 1997). The diversity within *Apis* has allowed studies between species to compare the evolution of pheromone complexity and biological activity. The majority of pheromone identifications have been carried out with *A. mellifera* first, and then efforts were made to explore the other species. Unless otherwise noted in the text, the term honey bee will refer to the European bee, *A. mellifera*.

Honey bee worker pheromones

ALARM AND RECRUITMENT PHEROMONES. Several worker-produced signal chemicals elicit alarm and colony defence responses in fellow workers. Alarmed bees are agitated, assume aggressive postures, and charge toward and gather round the source of the alarm signal. The behaviour is distinguished from defensive behaviour by a lack of stinging. Alarm mandibular gland secretions are widespread among arthropods and are believed to be a primitive function (Duffield *et al.*, 1984). The mandibular glands of older guard and forager honey bees produce an alarm pheromone, 2-heptanone (Fig. 17.2, **1**) (Kerr *et al.*, 1974), while the major defensive compounds include (*Z*)-11-eicosenol (Fig. 17.2, **2**), isopentyl acetate (Fig. 17.2, **3**) and a blend of low molecular weight alcohols and their corresponding acetates, all produced in the membranes associated with the stinger. Over 80 compounds from the sting apparatus of honey bees have been identified, of which approximately 15 elicit some level of aggressive response in workers (Collins and Blum, 1982, 1983). Several sting compounds are common to various honey bee species. *Apis mellifera*, *A. cerana*, *A. dorsata* and *A. florea* all produce isopentyl acetate, *n*-octyl acetate (Fig. 17.2, **4**) and, with the exception of *A. cerana*, 2(*E*)-decenyl acetate (Fig. 17.2, **5**) (Morse *et al.*, 1967; Koeniger *et al.*, 1979).

The first defensive substance isolated and identified was isopentyl acetate (Boch *et al.*, 1962) which, when presented alone, elicits more stinging activity than

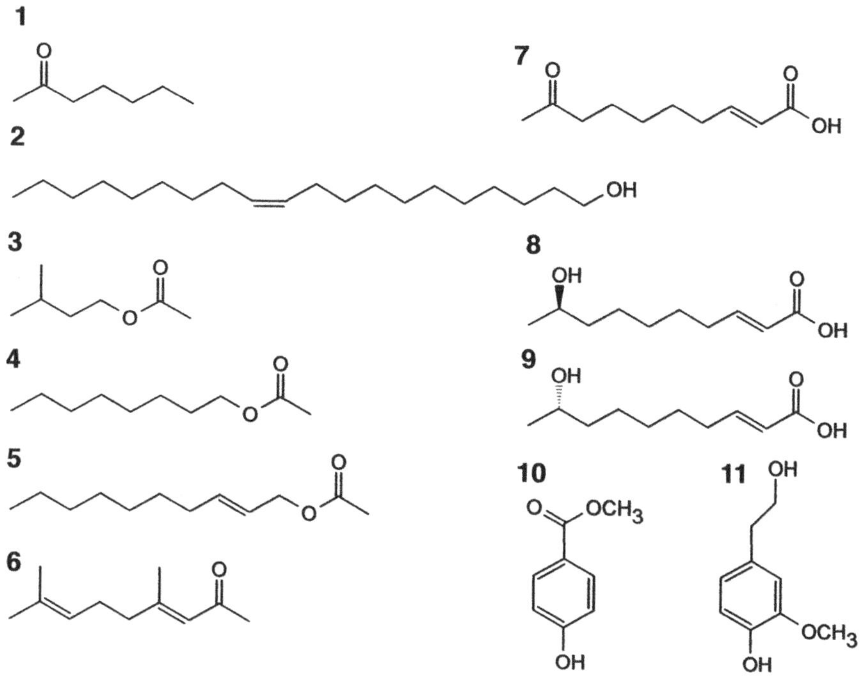

Fig. 17.2. Structures of some bee pheromones: (**1**) 2-heptanone; (**2**) (*Z*)-11-eicosenol; (**3**) isopentyl acetate; (**4**) *n*-octyl acetate; (**5**) 2(*E*)-decenyl acetate; (**6**) (*E*)-citral; (**7**) 9-keto-2(*E*)-decenoic acid; (**8**) (*R*)9-hydroxy-2(*E*)-decenoic acid; (**9**) (*S*)9-hydroxy-2(*E*)-decenoic acid; (**10**) methyl *p*-hydroxybenzoate (HOB); (**11**) 4-hydroxy-methoxyphenyl ethanol (HVA).

any of the other defence chemicals. Isopentyl acetate also acts as a target-marking pheromone, guiding other defenders to the sting site. The most abundant compound in sting extracts, (Z)-11-eicosenol, also induces stinging, but not to the same degree as isopentyl acetate (Pickett *et al.*, 1982). These two chemicals are synergistic in their natural ratios and just as likely to induce stinging as an equivalent dose of sting extract. Furthermore, the blend is active for a longer time than isopentyl acetate alone, possibly because (Z)-11-eicosenol slows down evaporation of isopentyl acetate (Pickett *et al.*, 1982). The two defence chemicals elicit a loud buzzing response from honey bees (Lefebvre and Beattie, 1991), possibly more to scare vertebrate predators than for intraspecies communication. The application of smoke to honey bees, implemented by honey-robbing humans for centuries, generally reduces honey bee olfaction and sensitivity to the defence chemicals (Visscher *et al.*, 1995).

Trace components in sting extracts include 2-nonanol, benzyl alcohol, 2-nonyl acetate, benzyl acetate, 1-acetoxy-2-nonene, 1-acetoxy-2-octene and other six- to eight-carbon aliphatic alcohols and their acetates (Allan *et al.*, 1987; Collins and Blum 1982, 1983). These compounds all elicit some behaviours associated with alarm but require higher than natural doses when presented individually. Wing fanning, most strongly induced by benzyl acetate and 2-nonanol, facilitates rapid dispersal and amplification of the alarm signal (Collins and Blum, 1982). The minor sting components may influence the ability of a colony to amplify an alarm signal and thus regulate the overall level of aggression.

Defensive behaviour in honey bees has a genetic basis (Guzmán-Novoa and Page, 1993) positively correlated with the production of pheromones (Boch and Rothenbuhler, 1974; Collins *et al.*, 1984, 1987, 1989; Kerr *et al.*, 1974). The *africanized* honey bee subspecies, derived from European (*Apis mellifera*) and African (*Apis mellifera scutellata* Lepeletier) subspecies, display significantly greater aggressive responses to alarm and defence chemicals and produce significantly greater quantities of nine sting chemicals than do European honey bees (Boch and Rothenbuhler, 1974; Collins *et al.* 1984). However, all nine components are involved in the augmentation and amplification of the defensive response.

The orientation odours produced by the Nasanov gland are composed of seven volatile chemicals (Boch and Shearer, 1964; Pickett *et al.*, 1980) secreted into a groove in the seventh dorsal abdominal segment (Snodgrass, 1956). Exposure of the gland, along with wing-fanning, releases and disperses the volatile chemicals, providing an important orientation signal. The minor component (*E*)-citral (Fig. 17.2, **6**) is most biologically active (Williams *et al.*, 1981, 1982) because it is most volatile. Volatiles from synthetic blends of Nasanov change with time as (*E*)-citral is exhausted, while volatiles collected from excised glands do not change in composition because an enzyme on the gland surface oxidizes geraniol to (*E*)-citral (Pickett *et al.*, 1980). This ensures that the composition of the blend remains constant.

Nasanoving workers signal colony and nest-entrance location to returning foragers, new flyers and disoriented workers. Foragers may cover areas of at least 112 km^2 (Seeley, 1985), then return to a nest site and can locate a small entrance typically of only 10–20 cm^2, or an open nest occupying a volume of merely 20–30 l (Winston, 1987). Nasanov odours released at the colony entrance, or on the surface of the nest in the case of exposed nests, assist returning workers in establishing the location of the nest. Nasanov chemicals also serve important orientation functions during colony-swarming and nest-seeking. After issue from the parent colony, a swarm usually regroups on a tree limb or some other substrate to maintain the integrity of the group while a new nest cavity is being sought by scouts. Orienting workers release Nasanov pheromone to facilitate swarm cluster formation (Avitable *et al.*, 1975), and later nest-seeking scout bees signal entrance locations of new nest sites (Free *et al.*, 1982).

Finally, Nasanov pheromone is used by foragers to assist in orienting to relatively odourless resources such as water and artificial feeding stations (Free and Williams, 1970, 1972). Foragers are rarely observed nasanoving at flowers, and do not use this pheromone as a recruitment odour (Wells *et al.*, 1993), but under experimental conditions workers will release Nasanov odours at odourless feeding stations containing extraordinarily high sugar concentrations (Wilhelm and Pflumm, 1983). Footprint pheromone, left by walking bees, augments forager attraction to Nasanov pheromone at feeding stations (Williams *et al.*, 1981) and nest entrances (Butler *et al.*, 1969). However, the source and identity of footprint pheromone are unknown. A compound produced by the sting gland, (Z)-11-

eicosenol, is also attractive to foraging workers but, unlike footprint pheromone, it does not augment the attractiveness of Nasanov pheromone (Free *et al.*, 1982). Ferguson and Free (1979) described odours they called forage-marking pheromone. These chemicals appear to emanate from the dorsal surface of the abdomen and act to increase forager attraction to food sources. Bees have been shown to scent-mark flowers with chemicals of uncertain origin to avoid revisiting flowers, and an individual is more strongly repelled by her own mark than the mark of another (Giurfa, 1993). It is possible that the footprint, sting-produced and forage-marking odours are the same pheromones, but more thorough investigations are necessary to clarify the activity, sources and identity of these odours.

WORKER PRIMER PHEROMONES. All worker pheromones discussed above elicit an immediate behavioural response: alarm, defence, or attraction. Workers also produce primer pheromones that induce physiological, endocrine and behavioural changes. Evidence of a primer pheromone from empty comb is increasing and has been shown to stimulate nectar hoarding by workers in small colonies. Indirect contact with comb at temperatures above 5°C stimulates nectar foraging, while honey in the comb reduces stimulation, suggesting that the stimulant is volatile (Rinderer, 1981). Decanal, decanol and nonanol have been suggested as possible candidates, all of which are found in newly constructed comb (Blum and Fales, 1988).

Honey bee brood pheromones

Honey bee development passes through four basic stages: egg, larva, pupa and adult; with eggs, larvae and pupae collectively referred to as brood. All three castes – workers, drones and queens – develop in the wax combs of the colony, each in a different type of cell. Worker cells cannot be distinguished from food storage cells. Drone cells are larger in diameter and the wax-covered pupae appear as dome-like protrusions from the surface of the wax comb. Queen cells are elongate and conical in shape, oriented downward along the comb face or periphery. Larvae are fed primarily with food produced by the hypopharyngeal and mandibular glands of nurse bees deposited in a pool at the bottom of the cell and a minor amount of pollen may be fed directly to fourth and fifth instar larvae. Diploid larvae are female and developmentally bipotent, depending on nutrition: those individuals fed an enriched diet of primarily mandibular gland secretions prior to the fourth moult develop into queens, while those fed primarily on hypopharyngeal secretions become workers capable of producing haploid eggs. Haploid eggs develop into males called drones. Drone larvae are fed more food than workers because they are larger and receive more carbohydrates, vitamins and trace minerals, as well as more diverse proteins. Toward the end of the larval stage, at about the time the larva has oriented its head to the top of the cell, adult workers seal the cell with a thin layer of wax, a behaviour commonly referred to as capping the cell. This task ends the tending of brood by adults and marks the beginning of the pupal stage. The principal activities during the pupal stage are muscle and organ system formation since most of the external features are established by this time.

Nurse bees recognize the caste and development stage of brood and tend accordingly. Brood stimulates pollen foraging and brood care, and inhibits ovary development (see below) in adult workers (Winston, 1987). The incubation of brood is initiated by the recognition of a triglyceride from pupae which causes worker bees to cluster over the brood (Koeniger and Veith, 1983). Several fatty acid esters extracted from larval cuticle induce workers to cap cells and build queen cells; and the esters act as a kairomone that attracts the varroa mite, an external parasite of honey bee larvae (Le Conte *et al.*, 1989, 1990; Trouiller *et al.*, 1991, 1992, 1994; Le Conte 1994). Workers are induced to cap cells when model larvae are placed in the bottom of a cell with a pheromonal blend containing the most abundant methyl esters (Le Conte, 1994); in contrast, model larvae impregnated with a blend of 64% ethyl esters induce workers to build queen cells (Le Conte, 1994). Placing of the model larva near the top of the cell is more likely to trigger cell capping or shaping behaviours (Le Conte, 1994), as is the position of a real larva (Goetz and Koeniger, 1992). The methyl esters extracted from the cuticle have a low volatility; therefore, the closer the larvae are to contacting adults, the more likely it is that adults will be able to detect the signal. Thus there is an interaction between pheromone blend and larva orientation which signals brood stage to adults.

The Le Conte group has undertaken fundamental work on elucidating the brood pheromone components, but additional comparative studies using synthetic blends and whole extracts with live controls are needed to uncouple behaviours induced by pheromones alone and/or other signals. Most of the responses so far induced by the synthetic brood pheromone are releaser-type responses: capping cells and making queen cells. However, brood pheromone has primer influences as well, such as inhibiting worker ovary development (Jay, 1970; Jay and Jay, 1976; Kropáková and Haslbachová, 1970, 1971), and stimulating pollen foraging (Al-Tikrity *et al.*, 1972). There has been some demonstration of the primer influences of the synthetic brood pheromone on ovary development (Arnold *et al.*, 1994) and hypopharyngeal glands (Mohammedi *et al.*, 1996). Recently, pollen foraging was shown to be directly regulated by the amount of brood (or brood extract) present in the hive (Pankiw *et al.*, 1998b) demonstrating that stimulation of pollen foraging is a real possibility.

Honey bee queens are distinct from workers physically as well as physiologically, and are characterized by the presence of large, well developed ovaries, and a spermatheca for storing semen from several matings. Workers possess functional ovaries and the ability to lay haploid male eggs, but the absence of mating behaviour and specific queen genitalia prevent the production of diploid female offspring. Normally, worker ovaries are suppressed from producing eggs. However, if a colony becomes queenless and all replacement queens reared from the last remaining eggs are destroyed, the ovaries of the colony's workers will gradually enlarge. Fifteen days after the loss of the original queen about 5% of the workers possess ovaries with mature eggs, and by 30 days this proportion increases to about 50% (Jay, 1968). Compounds from the mandibular glands of queens (particularly 9-keto-2(*E*)-decenoic acid: Fig. 17.2, **7**) have long been thought to be the source of worker ovary development suppression, but early studies produced ambiguous results. Ovary development is at best only partially inhibited by the heads of queens or compounds from queen mandibular glands (de Groot and Voogd, 1954; Butler, 1959; Verheijen-Voogd, 1959; Butler *et al.*, 1962; Butler and Fairey, 1963; Velthuis and van Es, 1964; Velthuis, 1970a). Worker ovary development can be somewhat suppressed by queens with excised mandibular glands (Velthuis, 1970b; Velthuis and van Es, 1964), and with extracts of entire queen bodies (de Groot and Voogd, 1954; Voogd, 1955; Butler, 1957; Verheijen-Voogd, 1959; van Erp, 1960). However, many of these early experiments were flawed by anecdotal reporting or unclear statistical tests. Other contemporary studies concluded that queen mandibular gland substances were not involved in worker ovary development inhibition (Pain, 1961; Velthuis and van Es, 1964; Velthuis, 1970b). Willis *et al.* (1990) directly compared worker ovary development in queenless colonies receiving daily doses of queen mandibular gland pheromone to queenright colonies and found that the pheromone does not inhibit ovary development. The presence of brood provides a stronger inhibition of worker ovary development (Jay, 1970; Kropáková and Haslbachová, 1970, 1971; Jay and Jay, 1976). In this way, worker reproduction is controlled by worker larvae in the nest and thus workers control their own reproduction. The presence of young brood inhibits the production of queen cells and thus the queen circumvents direct reproductive control over workers using brood signals to indicate primarily her fecundity, and secondarily her presence (Pettis *et al.*,1997).

Honey bee queen pheromones

A queen primarily signals her presence (reviewed by Winston and Slessor, 1992) and ontogenetic and reproductive stages by her mandibular gland pheromone (QMP) (Plettner *et al.*, 1993; Pankiw *et al.*, 1996). A five-component blend of QMP has been identified and synthesized (Slessor *et al.*, 1988) and is composed of three acids: ODA = 9-keto-2(*E*)-decenoic acid (Fig. 17.2, **7**) and *R*- and *S*-isomers of 9-HDA = 9-hydroxy-2(*E*)-decanoic acid (Fig. 17.2, **8** and **9**); and two aromatic compounds: HOB = methyl *p*-hydroxybenzoate (Fig. 17.2, **10**) and HVA = 4-hydroxy-3-methoxyphenyl ethanol (Fig. 17.2, **11**). The within-nest attraction of workers to the queen is called 'retinue behaviour'. A retinue is a dynamic composition of workers which almost always surrounds the queen. A specific blend and quantity of the five mandibular gland components elicits a retinue response. Workers engaged in this behaviour respond equally well to gland extracts as to the five-component blend of queen mandibular pheromone (Slessor *et al.*, 1988; Pankiw *et al.*, 1994). Not only do egg-laying workers develop in queenless colonies, but also a very small number of these workers produce elevated quantities of mandibular gland components (Plettner *et al.*, 1993) and elicit weak queen-like retinue behaviours from workers. These bees are called 'false queens'. Indeed, false queens, laying workers and intercaste individuals (resulting from the abnormal development of an individual reared as a queen from the fourth instar) all have mandibular gland compositions that fall along a continuum of queen- to worker-like states with respect to the biosynthetic capability of the gland (Plettner *et al.*, 1993). However, there is no correlation between gland production and ovarian development among intercastes, false queens and laying workers (Plettner *et al.*, 1993).

There is considerable variation among individuals in queen pheromone production of total and individual components (Slessor *et al.*, 1990; Pankiw *et al.*, 1996). Worker retinue responses to the pheromone are also highly variable and although worker responses and queen production are not correlated (Pankiw *et al.*, 1996), there is a genetic component to pheromone production and response (Pankiw *et al.*, 1994). Strains of workers that are highly responsive to queen pheromone with respect to retinue behaviour are also more likely to spend proportionately more time building and provisioning queen cells than less responsive worker strains in queenless colonies (Pankiw, 1997).

A queen indicates species, race, caste, ontogeny and reproductive state by her mandibular gland composition. Queens of *Apis mellifera*, *A. dorsata*, *A. florea* and *A. andreniformis* all produce ODA and 9-HDA. The open-nesting species (*A. dorsata*, *A. florea* and *A. andreniformis*) have a mandibular pheromone blend that contains only aliphatic components, while the cavity-nesting species, *A. mellifera* and *A. cerana*, produce aromatic compounds (Plettner *et al.*, 1997). Additionally, the cavity-nesting species have greater queen/worker differences than open-nesting ones, suggesting that corresponding differential functions of the mandibular glands in the colony have been a significant factor in the evolution within the *Apis* genus. The composition of the five-component mandibular gland pheromone of mated *africanized* and that of the European queen pheromone are similar except *africanized* queens produce greater quantities of 9-HDA and a reduced total quantity of pheromone (Pankiw *et al.*, 1996). In fact, queens of either race will head colonies composed of workers from both races (Winston, 1992) and the European pheromone blend is equally effective at preventing swarming in both races (Pettis *et al.*, 1994). The lower total quantity of pheromone found in *africanized* queens may contribute to the tendency of this race to swarm with smaller colony sizes (reviewed by Winston, 1987). The higher proportion of 9-HDA in *africanized* queen glands may reflect an adaptation to greater swarming activity in this race, because 9-HDA is an important component for swarm clustering behaviour (Winston *et al.*, 1982). Queen pheromone ontogeny shifts from a high proportion of ODA to 9-HDA in young unmated queens to roughly equal proportions of ODA and 9-HDA in mated queens whose glands contain significantly greater quantities of pheromone. The glands of older unmated, drone-laying queens contain quantities of the five-component pheromone that are intermediate between young unmated and mated queens, reflecting their intermediate reproductive state and age. A biosynthetic shift occurs after mating that results in a greater proportion of 9-HDA, HOB and HVA production with a decreased proportion of ODA (Pankiw *et al.*, 1996). Many of the differences between workers and queens can be attributed to the caste-selective biosynthetic pathways by which queens produce mandibular gland pheromone (Plettner *et al.*, 1996).

Queen pheromone is distributed throughout the nest by a series of transfers from queen to workers, from the queen to the wax comb, and from worker to worker (Naumann *et al.*, 1991, 1992). Approximately 40–400 μg (0.2–2 Qeq = queen equivalents) of queen pheromone is distributed daily (Naumann *et al.*, 1991) to workers. The queen's body surface carries on average, at one time, about 1/1000 of the quantity of pheromone found in her glands (10^{-3} Qeq) (Slessor *et al.*, 1988; Naumann *et al.*, 1991). Approximately 10% of the retinue workers remove substantial amounts of queen pheromone. A worker that licks the queen for up to 60 s may remove around 11% of the pheromone found on her body (Naumann *et al.*, 1991). Some of these workers act as messengers (Seeley, 1979; Juška *et al.*,

1981), transmitting the queen pheromone to other workers while contacting each other's antennae and mouthparts (Naumann *et al.*, 1991). Transmission in the nest approximates a first-order rate equation, meaning that short-interval transfers are proportional to the quantity present at the source during that interval (Naumann *et al.*, 1991), and all of the five components of the blend are transmitted as a unit (Naumann *et al.*, 1992).

Thirty minutes after the queen or pheromone source has been removed from the colony, the amount of pheromone circulating in the nest decreases below a detectable level (Naumann *et al.*, 1991, 1992), and this approximates to the same time that workers become agitated after queen removal (Naumann *et al.*, 1990). Two hours after queen removal the amount of pheromone remaining on comb falls below the workers' detection level (10^{-7} Qeq) (Naumann *et al.*, 1991), coinciding with the time when queen rearing is initiated in queenless colonies. Colonies rear new queens to: (i) replace dead queens or those incapacitated due to disease or injury; (ii) reproduce at the colony level by swarming; and (iii) replace a queen showing reduced fecundity.

The first reason for queen rearing is called 'emergency queen rearing'. Within hours of queen loss colonies will initiate new queen cells and have approximately 6 days to rear a new queen after a sudden queen loss. After 6 days, the last eggs deposited by the 'old' queen are 3-day-old larvae, approaching an age at which feeding the larvae an enriched diet will not result in queen development. Queenless colonies receiving daily doses of 1 Qeq of synthetic mandibular pheromone rear few or no queens for 4 days following queen removal but 6 days later there is no significant difference in the number of queen cells in pheromone-treated or untreated queenless colonies (Winston *et al.*, 1989, 1990). Queen pheromone added to queenless colonies within 24 h of queen loss inhibits queen rearing but not when added 4 days after queen cells have been initiated (Melathopolous *et al.*, 1996). Thus, queen pheromone inhibits the initiation of queen rearing but not the maintenance and nurturing of established queen larvae.

Workers are stimulated to produce new queens for colony-level reproduction or swarming when: (i) colony size reaches approximately 20,000 individuals; (ii) the brood nest is congested, restricting egg laying; (iii) the worker age distribution is skewed to a higher proportion of younger workers; and (iv) there is a reduced perception of the queen's presence. The role of queen pheromone is especially important in the reduced perception of queen presence. Queen rearing in preparation for swarming is enigmatic because new queens must be reared in the presence of the old queen. However, it is known that the presence of a queen inhibits the production of new queens (Melathopolous *et al.*, 1996). One hypothesis postulates that the transmission of queen pheromone is reduced as colonies grow and become congested with increasing worker numbers. This hypothesis makes two predictions: that colonies given supplemental amounts of queen pheromone while growing to swarming size will show delayed queen rearing and swarming; and that well distributed pheromone will enhance swarming inhibition. Testing the first prediction, colonies growing to swarming size daily receiving 10 Qeq of queen pheromone swarm an average of 25 days later than control colonies receiving no pheromone (Winston *et al.*, 1991). Testing the second prediction, 1 Qeq of pheromone dispensed in a spray delays swarming for 16 additional days compared with 25 days for the 10 Qeq delivered on a stationary glass slide (Winston *et al.*, 1991). A significantly smaller proportion of workers receive pheromone in very congested colonies, especially those workers situated at the periphery of the nest (Naumann *et al.*, 1993).

Queen pheromone is an important factor in the regulation of colony reproduction, but this should not be assumed to be a form of queen control over worker activities. The queen serves as a point-source of pheromone, apparently not actively able to regulate or alter the quality and quantity of pheromone produced (Slessor *et al.*, 1990; Pankiw *et al.*, 1996). Workers may be more effective in regulating the influence of the pheromone than the queen by their transmission activities, attenuation of pheromone titre by their population size, and their genotypic-based response level (Pankiw *et al.*, 1994).

The role of queen pheromone in queen supersedure is poorly understood. It has been proposed that a failing queen may produce less QMP than a normal queen, thus initiating queen cells and the supersedure process by worker bees (Winston, 1987). However, 2-year-old queens had similar amounts of QMP to young 6-month-old queens, giving no indication of reduced pheromone output with increased age (Pankiw *et al.*, 1996). Evidence from Pettis *et al.* (1997) suggested that a fecundity signal is present in combs containing young brood and that this as yet unidentified signal can act to inhibit the initiation of queen rearing. The addition of young brood to queenless colonies resulted in reduced queen cell construction over colonies receiving QMP alone (Pettis *et al.*, 1997). The authors postulate that it is the interaction of the signal from young brood comb and QMP that inhibits queen rearing.

The above descriptions of queen pheromone primer effects concern influences on the reproductive system of the honey bee colony. However, queen pheromone as a primer pheromone also acts on juvenile hormone (JH) biosynthesis in adult workers. Generally, honey bee workers perform age-related tasks. Younger workers engage in within-nest tasks such as cell cleaning, brood tending and comb building. Older workers perform outside-nest tasks such as guarding and foraging. Workers undergo physiological changes as they age such that their response threshold to various environmental stimuli changes (Robinson, 1992). These behavioural changes are accompanied by shifts in endocrine gland activity, specifically JH production (Robinson, 1992). JH titres increase as honey bee workers age, and treatment with JH mimic, JH, or JH analogue will induce precocious foraging (reviewed by Robinson, 1992). In contrast, same-age workers with lower JH titres show delayed foraging. Extracts of queen mandibular pheromone and ODA inhibit the rate of JH biosynthesis *in vitro* (Kaatz *et al.*, 1992). Workers in queenright colonies daily supplemented with 1 Qeq of queen pheromone showed delayed foraging ontogeny and associated lower JH titres (Pankiw *et al.*, 1998a). Queen pheromone, therefore, can delay the ontogeny of foraging apparently by inhibiting JH production, thereby acting as a regulator of division of labour in honey bees.

Bee Pheromones: Uses in Pollination and Colony Management

The use of bee pheromones in pollination and colony management is still in its infancy, but is likely to involve their use to attract bees to a crop or to stimulate foraging. The use of foraging stimulants has been proposed for some time (Free, 1978) and is likely to be a reality in the near future. Some progress in using pheromone sprays on crops has been made (Winston and Slessor, 1993), but at present these products are short lived in the field and can vary widely in their effectiveness with different crops and environmental conditions. Additionally, using pheromones as crop repellents has been proposed (Johansen and Mayer, 1990) but such a strategy may only be employed if pressure is brought to bear on producers to find new ways to limit pollinator losses due to pesticide spraying. The remainder of this chapter will deal with current uses of bee pheromones in agriculture as well as discussion as to their use in the future.

Crop attractants

All bees are attracted to flowers by colour, odour or the nectar and pollen rewards they offer (Seeley, 1985). Bees can discriminate between those plants that produce higher rewards and those that are less rewarding (Winston, 1987). This discrimination can cause pollination problems when a crop is grown in an area where a more rewarding wildflower may be in bloom. Even using excess colonies of honey bees may not provide an answer, as the non-target wildflowers may still receive the bulk of pollinator visits to the detriment of the target crop. Where crops are unattractive or less attractive to bees, pheromones have been tried as crop attractants. Examples of this type of crop attractant are 'Bee-Here'™, 'BeeLine'™ and 'BeeScent'™, most of which contain a synthetic blend of the Nasonov pheromone or other attractants of the honey bee A. *mellifera*. The use of these products or their constituents has been limited; although they can attract additional foragers (Waller, 1970; McGourty, 1992), numerous trials have failed to demonstrate significant yield increases (Mayer and Johansen, 1982; Rajotte and Fell, 1982; Mayer *et al.*, 1989). The most recent pollination enhancement product is 'Fruit Boost'™, which contains a synthetic blend of the mandibular gland pheromone of queen honey bees. Fruit Boost has been demonstrated to work by increasing recruitment of bees to the crop and by bees spending more time in the crop when compared with untreated areas (Higo *et al.*, 1995). In field trials, Fruit Boost has proved effective in increasing yields in pears, cranberries and highbush blueberries (Currie *et al.*, 1992a,b). Research is continuing to develop means to prolong the field life of Fruit Boost and to define optimal application conditions. The use of these and other similar products is likely to be most effective on high value crops, such as almonds and seed crops, and those that are inherently unattractive to bees, such as kiwi (Winston and Slessor, 1993).

Foraging stimulants

The use of pheromones to stimulate increased foraging has been speculated upon for years (Free, 1978). When extra brood is added to honey bee colonies they will increase pollen foraging to feed the additional brood (Free, 1987). The addition of brood pheromones has been demonstrated to increase food deposition by nurse bees that feed the developing brood (Le Conte *et al.*, 1995). Brood care requires pollen; thus, the addition of synthetic brood pheromones could be used to induce colonies to increase pollen collection. Pankiw *et al.* (1998b) were able to demonstrate that brood extracts increased pollen foraging; the next step is to make brood pheromone commercially viable as a pollination aid. Honey bee pollen foragers are often better pollinators than nectar foragers (Free, 1987). Therefore, the use of brood pheromones in honey bee colonies to enhance pollination is a real possibility. Honey bee queen mandibular gland pheromone stimulated foraging when added to small newly established colonies (Higo *et al.*, 1992) but subsequent studies with larger colonies showed no stimulatory effects. The use of foraging stimulants in honey bee colonies should occur within 10 years. Additionally, as bumblebee rearing techniques improve, the use of brood pheromones to stimulate bumblebee foraging may be feasible.

Foraging repellents

The need for foraging repellents arises because many crops are treated with insecticide during bloom, when bees are actively visiting the crop. The evidence for natural foraging repellents exists (Free *et al.*, 1985; Giurfa, 1993) and several synthetic compounds have demonstrated repellency to honey bees (Melksham *et al.*, 1988). Pesticides themselves may be repellent or substances can be added to the pesticide mixture to repel bees from the crops during spray application (Melksham *et al.*, 1988; Solomon and Hooker, 1989). The use of bee repellents has been limited and the beekeepers are normally responsible for keeping abreast of pesticide use near their apiaries.

Disposable pollination units

The use of a disposable unit of bulk honey bees without a queen or brood to pollinate a crop and then be left to dwindle and die has been tested (Thorp *et al.*, 1973, 1974; Erickson *et al.*, 1977). A disposable pollination unit (DPU) would make pollination in remote areas possible and would eliminate the need for the beekeeper to return to the area (Cantwell *et al.*, 1972). One drawback of disposable pollination units has been stimulating these units to forage without the presence of a queen or brood (Thorp *et al.*, 1973). The addition of queen mandibular gland pheromone to DPUs did increase foraging life but improvements are still needed (Jaycox, 1970; Currie *et al.*, 1994). As further brood foraging stimulants are identified and the release of queen and brood pheromones is refined, it may indeed be possible to produce a unit of bees that will actively pollinate over a 3- to 4-week period and be of use in the pollination industry.

Colony management

The use of pheromones within hives to augment or control colony growth is easily within the scope of current knowledge. A synthetic blend of queen pheromone

(QMP) is currently being produced and sold as 'Bee Boost'™ (Phero Tech Inc., Abbottsford, British Columbia, Canada) for use within colonies and may have a variety of applications. Bee Boost can successfully act as a queen substitute in the shipment of bulk bees to keep bees calm during shipment (Naumann *et al.*, 1990). Also, Bee Boost has been applied in the management of small colonies used to mate queen bees (Pettis *et al.*, 1993). An important management problem facing beekeepers is the need to eliminate swarming, which results in decreased honey yields. Queen pheromone could be used to suppress swarming (Winston *et al.*, 1991) and may work if an efficient delivery system can be devised. Uses for other pheromones in colony management include the potential of brood pheromones in combination with QMP to suppress queen rearing and perhaps swarming, and of brood pheromones to increase pollen foraging (Winston and Slessor, 1993; Pankiw *et al.*, 1998b). The use of pheromones in colony management is only limited by our understanding of how and at what level these pheromones function within the colony.

Bee pheromones have increased in complexity and number as bees have evolved from a solitary to a colony-nesting lifestyle (Free, 1987). The leafcutter and blue orchard bees are both solitary (Batra, 1984) and their pheromone use is less than that of the social, colony-nesting bumblebees and honey bees. The identification of solitary bee pheromones and understanding of their basic biology has been limited. However, *Osmia* spp. and the Megachilidae both fill important pollination niches and thus research to improve their efficiency as pollinators should continue. The use of pheromones in bumblebees could involve the application of attractants in the field but is perhaps more likely to involve pheromones within the hive. In the past 10 years the commercial rearing of bumblebees has seen great improvements (van der Eijnde *et al.*, 1991). To find the switch that causes a bumblebee colony to produce reproductives and stop the production of workers would prolong the life of an individual colony, perhaps leading to large perennial colonies. It is reasonable to assume that, as the identification and understanding of bumblebee pheromones increases, ways of improving pollination will be found. With honey bees, there has been some success in using pheromones to manipulate the colony and as foraging repellents and attractants. The next two decades are likely to see the routine use of pheromones in the management of bees to increase pollination.

References

Allan, S.A., Slessor, K.N., Winston, M.L. and King, G.G.S. (1987) The influence of age and task specialization on the production and perception of honey bee pheromones. *Journal of Insect Physiology* 33, 917–922.

Al-Tikrity, W.S., Benton, A.W., Hillman, R.C. and Clarke, W.W. (1972) The relationship between the amount of unsealed brood in honey bee colonies and their pollen collection. *Journal of Apicultural Research* 11, 9–12.

Arnold, G., Le Conte, Y., Trouiller, J., Hervet, H., Chappe, B. and Masson, C. (1994) Inhibition of worker honeybee ovaries development by a mixture of fatty acid esters from larvae. *Comptes Rendus de l'Académie des Sciences, Paris, Sciences de la Vie* 317, 511–515.

Avitable, A., Morse, R.A. and Boch, R. (1975) Swarming honey bees guided by pheromones. *Annals of the Entomological Society of America* 68, 1079–1082.

Batra, S.W.T. (1978) Aggression, territoriality, mating and nest aggregation of some solitary bees (Hymenoptera: Halictidae, Megachilidae, Colletidae, Anthophoridae). *Journal of the Kansas Entomological Society* 51, 547–559.

Batra, S.W.T. (1982) The hornfaced bee for efficient pollination of small farm orchards. In: Kerr, W.H. Jr and Knutson, L.V. (eds) *Research for Small Farms*. USDA Miscellaneous Publication 1422, pp. 117–120.

Batra, S.W.T. (1984) Solitary bees. *American Scientist* 250, 120–127.

Batra, S.W.T. (1995) Bees and pollination in our changing environment. *Apidologie* 26, 361–370.

Blum, M.S. and Fales, H.M. (1988) Eclectic chemisociality of the honey bee: a wealth of behaviors, pheromones, and exocrine glands. *Journal of Chemical Ecology* 14, 2099–2108.

Boch, R. and Rothenbuhler, W.C. (1974) Defensive behaviour and production of alarm pheromone in honey bees. *Journal of Apicultural Research* 13, 217–221.

Boch, R. and Shearer, D.A. (1964) Identification of neolic and geranic acids in the Nasanov pheromone of the honey bee. *Nature* 202, 320–321.

Boch, R., Shearer, D.A. and Stone, B.C. (1962) Identification of iso-amyl acetate as an active component in the sting pheromone of the honey bee. *Nature* 195, 1018–1020.

Buchmann, S.L. and Nabhan, G.P. (1996) *The Forgotten Pollinators*. Island Press, Washington, DC, 256 pp.

Butler, C.G. (1957) The control of ovary development in worker honey bees (*Apis mellifera* L.). *Experientia* 13, 256–257.

Butler, C.G. (1959) The source of the substance produced by a queen honey bee (*Apis mellifera* L.) which inhibits development of ovaries of the workers of her colony. *Proceedings of the Royal Entomological Society of London* (A) 34, 137–138.

Butler, C.G. and Fairey, E. M. (1963) The role of the queen in preventing oogenesis in worker honey bees. *Journal of Apicultural Research* 2, 14–18.

Butler, C.G., Callow, R K. and Johnston, N.C. (1962) The isolation and synthesis of queen substance, 9-oxodec-*trans*-2-enoic acid, a honey bee pheromone. *Proceedings of the Royal Entomological Society of London* (B) 155, 417–432.

Butler, C.G., Fletcher, J.C. and Watler, D. (1969) Nest-entrance marking with pheromones by the honey bee *Apis mellifera* L. and a wasp, *Vespula vulgaris* L. *Animal Behaviour* 17, 142–147.

Cameron, S.A. and Whitfield, J.B. (1996) Use of walking trails by bees. *Nature* 379, 125.

Cantwell, G.E., Shimanuki, H. and Retzer, H.J. (1972) Airdropping disposable packages of honey bees into cranberry bogs. *American Bee Journal* 112, 328–329.

Collins, A.M. and Blum, M.S. (1982) Bioassay of compounds derived from the honey bee sting. *Journal of Chemical Ecology* 8, 463–470.

Collins, A.M. and Blum, M.S. (1983) Alarm responses caused by newly identified compounds derived from the honey bee sting. *Journal of Chemical Ecology* 9, 57–65.

Collins, A.M., Rinderer, T.E., Harbo, J.R. and Brown, M.A. (1984) Heritabilities and correlations for several characteristics in the honey bee. *The Journal of Heredity* 75, 135–140.

Collins, A.M., Brown, M.A., Rinderer, T.E., Harbo, J.R. and Tucker, K.W. (1987) Heritabilities of honey-bee alarm pheromone production. *The Journal of Heredity* 78, 29–31.

Collins, A.M., Rinderer, T.E., Daly, H.V., Harbo, J.R. and Pesante, D. (1989) Alarm pheromone production by two honey bee (*Apis mellifera*) types. *Journal of Chemical Ecology* 15, 1747–1756.

Crane, E. (1983) *The Archaeology of Beekeeping*. Cornell University Press, Ithaca, New York, 352 pp.
Currie, R.W., Winston, M.L., Slessor, K.N. and Mayer, D.F. (1992a) Impact of synthetic queen mandibular pheromone sprays on honey bee (*Apis mellifera* L.) pollination of fruit crops. *Journal of Economic Entomology* 85, 1293–1299.
Currie, R.W., Winston, M.L. and Slessor, K.N. (1992b) Impact of synthetic queen mandibular pheromone sprays on honey bee (*Apis mellifera* L.) pollination of berry crops. *Journal of Economic Entomology* 85, 1300–1306.
Currie, R.W., Winston, M.L. and Slessor, K.N. (1994) The effect of honey bee (*Apis mellifera* L.) synthetic queen mandibular compound on queenless 'disposable' pollination units. *American Bee Journal* 134, 200–202.
Duffield, R.M., Wheeler, J.W. and Eickwort, G.C. (1984) Sociochemicals of bees. In: Bell, W.J. and Cardé, R.T. (eds) *Chemical Ecology of Insects*. Sinauer Associates, Sunderland, Massachusetts, pp. 387–428.
Eijnde, J. van den, Ruijter, A. de and Steen, J. van der (1991) Method for rearing *Bombus terrestris* continuously and the production of bumblebee colonies for pollination purposes. *Acta-Horticulturae* 288, 154–158.
Engel, M.S. and Schultz, T.R. (1997) Phylogeny and behaviour in honey bees (Hymenoptera: Apidae). *Annals of Entomological Society of America* 90, 43–53.
Erickson, E.H., Thorp, R.W. and Briggs, D.L. (1977) The use of disposable pollination units in almond orchards. *Journal of Apicultural Research* 16, 107–111.
Erp, A. van (1960) Mode of action of the inhibitory substance of the honey bee queen. *Insectes Sociaux* 7, 207–211.
Ferguson, A.W. and Free, J.B. (1979) Production of forage-marking pheromone by the honey bee. *Journal of Apicultural Research* 18, 128–135.
Foster, R.L. (1992) Nestmate recognition as an inbreeding avoidance mechanism in bumble bees (Hymenoptera: Apidae). *Journal of the Kansas Entomological Society* 65, 238–243.
Free, J.B. (1978) Progress toward the use of pheromones to stimulate pollination by honey bees. *Proceedings of the IV International Symposium on Pollination*, 1978, 7–22.
Free, J.B. (1987) *Pheromones of Social Bees*. Cornell University Press, Ithaca, New York, 207 pp.
Free, J.B. (1993) *Insect Pollination of Crops*, 2nd edn. Academic Press, London, 636 pp.
Free, J.B. and Williams, I.H. (1970) Exposure of the Nasonov gland by honey bees (*Apis mellifera*) collecting water. *Behaviour* 37, 286–290.
Free, J.B. and Williams, I.H. (1972) The role of Nasanov pheromone in crop communication by honey bees (*Apis mellifera* L.). *Behaviour* 41, 314–318.
Free, J.B., Williams, I.H., Pickett, J.A., Ferguson, A.W. and Martin, A.P. (1982) Attractiveness of (Z)-11-eicosen-1-ol to foraging honey bees. *Journal of Apicultural Research* 21, 151–156.
Free, J.B., Pickett, J.A., Ferguson, A.W. Simpkins, J.R. and Smith, M.C. (1985) Repelling foraging honey bees with alarm pheromones. *Journal of Agricultural Science, Cambridge* 105, 255–260.
Giurfa, M. (1993) The repellent scent-mark of the honey bee *Apis mellifera ligustica* and its role as communication cue during foraging. *Insectes Sociaux* 40, 59–67.
Goetz, B. and Koeniger, N. (1992) Structural features trigger capping of brood cells in honey bees. *Apidologie* 23, 211–216.
Groot, A.P. de and Voogd, S. (1954) On the ovary development in queenless worker bees (*Apis mellifera* L.). *Experientia* 10, 384–385.
Guzmán-Novoa, E. and Page, R.E. (1993) Backcrossing *Africanized* honey bee queens to European drones reduces colony defensive behaviour. *Annals of the Entomological Society of America* 86, 352–355.

Heemert, C. van, de Ruijter, A., Eijnde, J. van den and Steen, J. van der (1990) Year-round production of bumble bee colonies for crop pollination. *Bee World* 71, 54–56.

Higo, H.A., Colley, S.J., Winston, M.L. and Slessor, K.N. (1992) Effects of honey bee (*Apis mellifera* L.) queen mandibular gland pheromone on foraging and brood rearing. *Canadian Entomologist* 124, 409–418.

Higo, H.A.,Winston, M.L. and Slessor, K.N. (1995) Mechanisms by which (Hymenoptera: Apidae) queen pheromone sprays enhance pollination. *Annals of the Entomological Society of America* 88, 366–373.

Jay, S.C. (1968) Factors influencing ovary development of worker honey bees under natural conditions. *Canadian Journal of Zoology* 46, 345–347.

Jay, S.C. (1970) The effect of various combinations of immature queens and worker bees on the ovary development of worker honey bees in colonies with and without queens. *Canadian Journal of Zoology* 48, 169–173.

Jay, S.C. and Jay, D.H. (1976) The effect of various types of brood comb on the ovary development of worker honey bees. *Canadian Journal of Zoology* 54, 1724–1726.

Jaycox, E.R. (1970) Honey bee queen pheromones and worker foraging behavior. *Annals of the Entomological Society of America* 63, 222–228.

Johansen, C.A. and Mayer, D.F. (1990) *Pollinator Protection*. Wicwas Press, Cheshire, 199 pp.

Juška, A., Seeley, T.D. and Velthuis, H.H.W. (1981) How honey bee queen attendants become ordinary workers. *Journal of Insect Physiology* 27, 515–519.

Kaatz, H.-H., Hildebrandt, H. and Engels, W. (1992) Primer effect of queen pheromone on juvenile hormone biosynthesis in adult worker honey bees. *Journal of Comparative Physiology B* 162, 588–592.

Kerr, W.E., Blum. M.S., Pisani, J. and Stort, A.C. (1974) Correlation between amounts of 2-heptanone and iso-amyl acetate in honey bees and their aggressive behaviour. *Journal of Apicultural Research* 13, 173–176.

Koeniger, N. and Veith, H.J. (1983) Identification of a triglyceride as a brood pheromone of the honey bee (*Apis mellifera* L.). *Apidologie* 14, 59.

Koeniger, N., Weiss, J. and Maschewitz, U. (1979) Alarm pheromones of the sting in the genus *Apis*. *Journal of Insect Physiology* 25, 467–476.

Kristjansson, K. and Rasmussen, K. (1991) Pollination of sweet pepper (*Capsicum annuum* L.) with the solitary bee *Osmia cornifrons* (Radoszkowski). *Acta Horticulturae* 288, 173–179.

Kropácová, S. and Haslbachová, H. (1970) The development of ovaries in worker honey bees in queenright colonies examined before and after swarming. *Journal of Apicultural Research* 9, 65–70.

Kropácová, S. and Haslbachová, H. (1971) The influence of queenlessness and of unsealed brood on the development of ovaries in worker honey bees. *Journal of Apicultural Research* 10, 57–61.

Le Conte, Y. (1994) The recognition of larvae by worker honey bees. *Naturwissenschaften* 81, 462–465.

Le Conte, Y., Arnold, G., Trouiller, J., Masson, C., Chappe, B. and Ourisson, G. (1989) Attraction of the parasitic mite *Varroa* to the drone larvae of honey bees by simple aliphatic esters. *Science* 245, 638–639.

Le Conte, Y., Arnold, G., Trouiller, J. and Masson, C. (1990) Identification of a brood pheromone in honey bees. *Naturwissenschaften* 77, 334–336.

Le Conte, Y., Sreng L. and Poitout, S. H. (1995) Brood pheromone can modulate the feeding behaviour of *Apis mellifera* workers (Hymenoptera: Apidae). *Journal of Economic Entomology* 88, 798–804.

Lefebvre, M.G. and Beattie, A.J. (1991) Sound responses of honey bees to six chemical stimuli. *Journal of Apicultural Research* 30, 156–161.

Maeta, Y. and Kitamura, T. (1981) Pollination efficiency by *Osmia cornifrons* (Radoszkowski) in relation to required number of nesting bees for economic fruit production. *Honeybee Science* 2, 65–72.

Matheson, A., Buchmann, S.L., O'Toole, C., Westrich, P. and Williams, I.H. (1996) *The Conservation of Bees*. Linnean Society Symposium Series No. 18, Academic Press, London, 245 pp.

Mayer, D.F. and Johansen, C.A. (1982) Field evaluation of chemical pollinator attractants on tree fruits. *American Bee Journal* 122, 287–289.

Mayer, D.F., Britt, R.L. and Lunden, J.D. (1989) Evaluation of BeeScent as a honey bee attractant. *American Bee Journal* 129, 41–42.

McGourty, M.A. (1992) A field evaluation of 'Bee-Here'™, a pheromone-based forage attractant for honey bees, *Apis mellifera* L., applied to red clover *Trifolium pratense* L. Master of Agriculture research paper, Department of Entomology, Oregon State University.

Melathopolous, A.P., Winston, M.L., Pettis, J.S. and Pankiw, T. (1996) Effect of queen mandibular pheromone on initiation and maintenance of queen cells in the honey bee (*Apis mellifera* L.). *Canadian Entomologist* 128, 263–272.

Melksham, K.J., Jacobsen, N. and Rhodes, J. (1988) Compounds which affect the behaviour of the honey bee, *Apis mellifera* L: a review. *Bee World* 69, 104–124.

Mohammedi, A., Crauser, D., Paris, A. and Le Conte, Y. (1996) Effect of a brood pheromone on honeybee hypopharyngeal glands. *Comptes Rendus de l'Académie des Sciences Paris, Sciences de la Vie* 319, 769–772.

Morse, R.A., Shearer, D.A., Boch, R. and Benton, A.W. (1967) Observations on alarm substances in the genus *Apis*. *Journal of Apicultural Research* 6, 113–118.

Naumann, K., Winston, M.L., Wyborn, M.H. and Slessor, K.N. (1990) Effects of synthetic honey bee (Hymenoptera: Apidae) queen mandibular-gland pheromone on workers in packages. *Journal of Economic Entomology* 83, 1271–1275.

Naumann, K., Winston, M.L., Slessor, K.N., Prestwich, G.D. and Webster, F.X. (1991) Production and transmission of honey bee queen (*Apis mellifera* L.) mandibular gland pheromone. *Behavioural Ecology and Sociobiology* 29, 321–332.

Naumann, K., Winston, M.L., Slessor, K.N., Prestwich, G.D. and Latli, B. (1992) Intra-nest transmission of aromatic honey bee mandibular gland pheromone components: movement as a unit. *Canadian Entomologist* 124, 917–934.

Naumann, K., Winston, M.L. and Slessor, K.N. (1993) Movement of honey bee (*Apis mellifera* L.) queen mandibular gland pheromone in populous and unpopulous colonies. *Journal of Insect Behavior* 6, 211–223.

Oldham, N.J., Billen, J. and Morgan, D.E. (1994) On the similarity of the Dufour secretion and the cuticular hydrocarbons of some bumblebees. *Physiological Entomology* 19, 115–123.

Pain, J. (1961) Sur la phéromone des reines d'abeilles et ses effets physiologiques. *Annal Abeille* 24, 12.

Pankiw, T. (1997) Queen rearing by high and low queen mandibular pheromone responding worker honey bees (*Apis mellifera* L.). *Canadian Entomologist* 129, 679–690.

Pankiw, T., Winston, M.L. and Slessor, K.N. (1994) Variation in worker response to honey bee (*Apis mellifera* L.) queen mandibular pheromone (Hymenoptera: Apidae). *Journal of Insect Behaviour* 7, 1–15.

Pankiw, T., Winston, M.L., Plettner, E., Slessor, K.N., Pettis, J.S. and Taylor, O.R. (1996) Mandibular gland components of European and *Africanized* honey bee queens (*Apis mellifera* L.). *Journal of Chemical Ecology* 22, 605–615.

Pankiw, T., Huang, Z.-Y., Winston, M.L. and Robinson, G.E. (1998a) Queen mandibular gland pheromone influences worker honey bee (*Apis mellifera* L.) foraging ontogeny and juvenile hormone titers. *Journal of Insect Physiology* 44, 685–692.

Pankiw, T., Page, R.E. Jr and Fondrk, M.K. (1998b) Brood pheromone stimulates pollen foraging in honey bees (*Apis mellifera*). *Behavioral Ecology and Sociobiology* 44, 193–198.
Pettis, J.S., Winston, M.L., Malyon, M. and Slessor, K.N. (1993) The use of honey bee queen mandibular gland pheromone in mating nuclei management. *American Bee Journal* 133, 725–727.
Pettis, J.S., Winston, M.L. and Collins, A.M. (1994) Suppression of queen rearing in European and *Africanized* honey bee *Apis mellifera* L. by synthetic queen mandibular gland pheromone. *Insectes Sociaux* 42, 1–9.
Pettis, J.S., Higo, H.A., Pankiw, T. and Winston, M.L. (1997) Queen rearing suppression in the honey bee – evidence for a fecundity signal. *Insectes Sociaux* 44, 311–322.
Pickett, J.A., Williams, I.H., Martin, A.P. and Smith, M.C. (1980) Nasanov pheromone of the honey bee *Apis mellifera* L. (Hymenoptera, Apidae). I. Chemical characterization. *Journal of Chemical Ecology* 6, 425–434.
Pickett, J.A., Williams, I.H., Smith, M.C. and Martin, A.P. (1982) (Z)-11-Eicosen-1-ol, an important new pheromonal component from the sting of the honey bee *Apis mellifera* L. (Hymenoptera: Apidae). *Journal of Chemical Ecology* 8, 163–176.
Plettner, E., Slessor, K.N., Winston, M.L., Robinson, G.E. and Page, R.E. (1993) Mandibular gland components and ovarian development as measures of caste differentiation in the honey bee (*Apis mellifera* L.). *Journal of Insect Physiology* 39, 235–240.
Plettner, E., Slessor, K.N., Winston, M.L. and Oliver, J.E. (1996) Caste-selective pheromone biosynthesis in honey bees. *Science* 271, 1851–1853.
Plettner, E., Otis, G.W., Wimalarantne, P.D.C., Winston, M.L., Slessor, K.N., Pankiw, T. and Punchihewa, P.W.K. (1997) Species- and caste-determined mandibular gland signals in honey bees (*Apis*). *Journal of Chemical Ecology* 23, 363–377.
Rajotte, E.G. and Fell, R.D. (1982) A commercial bee attractant ineffective in enhancing apple pollination. *HortScience* 17, 230–231.
Richards, K.W. (1987) Alfalfa leafcutter bee management in Canada. *Bee World* 68, 168–178.
Rinderer, T.E. (1981) Volatiles from empty comb increase hoarding by the honey bee. *Animal Behaviour* 29, 1275–1276.
Robinson, G.E. (1992) Regulation of division of labour in insect societies. *Annual Review of Entomology* 37, 637–665.
Roseler, P.F. (1985) A technique for year-round rearing of *Bombus terrestris* (Apidae, Bombini) colonies in captivity. *Apidologie* 16, 165–170.
Roubik, D.W. (1989) *Ecology and Natural History of Tropical Bees*. Cambridge University Press, Cambridge, 485 pp.
Seeley, T.D. (1979) Queen substance dispersal by messenger workers in honey bee colonies. *Behavioral Ecology and Sociobiology* 5, 391–415.
Seeley, T.D. (1985) *Honey Bee Ecology*. Princeton University Press, Princeton, New Jersey, 394 pp.
Slessor, K.N., Kaminski, L.-A., King, G.G.S., Borden, J.H. and Winston, M.L. (1988) Semiochemical basis of the retinue response to queen honey bees. *Nature* 332, 345–356.
Slessor, K.N., Kaminski, L.-A., King, G.G.S. and Winston, M.L. (1990) Semiochemicals of the honey bee queen mandibular glands. *Journal of Chemical Ecology* 16, 851–860.
Snodgrass, R.E. (1956) *Anatomy of the Honey Bee*. Cornell University Press, Ithaca, New York, 328 pp.
Solomon, M.G. and Hooker, K.J.M. (1989) Chemical repellents for reducing pesticide hazard to honey bees in apple orchards. *Journal of Apicultural Research* 28, 223–227.
Thorp, R.W., Erickson, E.H., Moeller, F.E., Levin, M.D., Stanger, W. and Briggs, D.L. (1973) Flight activity and uniformity comparisons between honey bees in disposable pollination units (DPU's) and overwintered colonies. *Environmental Entomology* 2, 525–529.

Thorp, R.W., Erickson, E.H., Moeller, F.E., Lenin, M.D., Stanger, W. and Briggs, D.L. (1974) Disposable pollination units tested for almond pollination in California. *American Bee Journal* 114, 58–60.

Torchio, P.F. (1981) Field experiments with *Osmia lignaria propinqua* Cresson as a pollinator in almond orchards. I. 1975 studies (Hymenoptera: Megachilidae). *Journal of the Kansas Entomological Society* 54, 815–823.

Torchio, P.F. (1991) Bees as crop pollinators and the role of solitary species in changing environments. *Acta Horticulturae* 288, 49–61.

Trouiller, J., Arnold, G., Chappe, B., Le Conte, Y. and Masson, C. (1991) Temporal pheromonal and kairomonal secretions in the brood of honey bees. *Naturwissenschaften* 78, 368–370.

Trouiller, J., Arnold, G., Chappe, B., Le Conte, Y. and Masson, C. (1992) Semiochemical basis of infestation of honey bee brood by *Varroa jacobsoni*. *Journal of Chemical Ecology* 18, 2041–2053.

Trouiller, J., Arnold, G., Chappe, B., Le Conte, Y., Billion, A. and Masson, C. (1994) The kairomonal esters attractive to the *Varroa jacobsoni* mite in the queen brood. *Apidologie* 25, 314–321.

Velthuis, H.H.W. (1970a) Ovarian development in *Apis mellifera* worker bees. *Entomologia Experimentalis et Applicata* 13, 377–394.

Velthuis, H.H.W. (1970b) Queen substance from the abdomen of the honey bee queen. *Zeitschrift für Vergleichende Physiologie* 70, 210–222.

Velthuis, H.H.W. and Es, J. van (1964) Some functional aspects of the mandibular glands of the queen honey bee. *Journal of Apicultural Research* 3, 11–16.

Verheijen-Voogd, C. (1959) How worker bees perceive the presence of their queen. *Zeitschrift für Vergleichende Physiologie* 41, 527–582.

Visscher, P.K., Vetter, R.S. and Robinson, G.E. (1995) Alarm pheromone perception in honey bees is decreased by smoke. *Journal of Insect Behavior* 8, 11–18.

Voogd, S. (1955) Inhibition of ovary development in worker honey bees by extraction fluid of the queen. *Experientia* 11, 181–182.

Waller, G.D. (1970) Attracting honey bees to alfalfa with citral, geraniol, and anise. *Journal of Apicultural Research* 9, 9–12.

Wells, P., Wells, H., Vu, V., Vadehra, N., Lee, C., Han, R., Han, K. and Chang, L. (1993) Does honey bee Nasanov pheromone attract foragers? *Bulletin of the Southern California Academy of Sciences* 92, 70–77.

Wilhelm, K.T. and Pflumm, W.W. (1983) Über den Einfluss blumenhafter Düfte auf des Duftmarkieren der Sammelbiene. *Apidologie* 14, 183–190.

Williams, I.H., Pickett, J.A. and Martin, A.P. (1981) The Nasanov pheromone of the honey bee, *Apis mellifera* L. (Hymenoptera: Apidae). II. Bioassay of the components using foragers. *Journal of Chemical Ecology* 7, 225–237.

Williams, I.H., Pickett, J.A. and Martin, A.P. (1982) Nasanov pheromone of the honey bee *Apis mellifera* L. (Hymenoptera, Apidae). IV. Comparative electroantennogram responses. *Journal of Chemical Ecology* 8, 567–574.

Willis, L.G., Winston, M.L. and Slessor, K.N. (1990) Queen honey bee mandibular pheromone does not affect worker ovary development. *Canadian Entomologist* 122, 1093–1099.

Winston, M.L. (1987) *The Biology of the Honey Bee*. Harvard University Press, Cambridge, Massachusetts, 267 pp.

Winston, M.L. (1992) The *Africanized* honey bee. *Annual Review of Entomology* 37, 173–193.

Winston, M.L. and Slessor, K.N. (1992) The essence of royalty: honey bee queen pheromone. *American Scientist* 80, 374–385.

Winston, M.L. and Slessor, K.N. (1993) Applications of queen honey bee pheromone for beekeeping and crop pollination. *Bee World* 74, 111–128.

Winston, M.L., Slessor, K.N., Smirle, M.J. and Kandil, A.A. (1982) The influence of queen-produced substance, 9-HDA, on swarm clustering behaviour in the honey bee *Apis mellifera* L. *Journal of Chemical Ecology* 8, 1283–1288.

Winston, M.L., Slessor, K.N., Willis, L.G., Naumann, K., Higo, H.A., Wyborn, M.H. and Kaminski, L.-A. (1989) The influence of queen mandibular pheromones on worker attraction to swarm clusters and inhibition of queen rearing in the honey bee (*Apis mellifera* L.). *Insectes Sociaux* 36, 15–27.

Winston, M.L., Higo, H.A. and Slessor, K.N. (1990) Effect of various dosages of queen mandibular gland pheromone on the inhibition of queen rearing in the honey bee (Hymenoptera: Apidae). *Annals of the Entomological Society of America* 83, 234–238.

Winston, M.L., Higo, H.A., Colley, S.J., Pankiw, T. and Slessor, K.N. (1991) The role of queen mandibular pheromone and colony congestion in honey bee (*Apis mellifera* L.) reproductive swarming. *Journal of Insect Behavior* 4, 649–659.

Species Index

INSECTS

BACTERIA

FUNGI

NEMATODES

OTHER ARACHNIDS

PLANTS

PROTOZOA

Subject Index

Printed and bound by CPI Group (UK) Ltd, Croydon, CR0 4YY
16/09/2026
14963433-0004